Advances in Experimental Medicine and Biology

Volume 856

More information about this series at http://www.springer.com/series/5584

Chantra Eskes • Maurice Whelan
Editors

Validation of Alternative Methods for Toxicity Testing

Springer

Editors
Chantra Eskes
SeCAM Services & Consultation
on Alternative Methods
Magliaso, Switzerland

Maurice Whelan
European Commission
Joint Research Centre (JRC)
Ispra, Italy

ISSN 0065-2598 ISSN 2214-8019 (electronic)
Advances in Experimental Medicine and Biology
ISBN 978-3-319-33824-8 ISBN 978-3-319-33826-2 (eBook)
DOI 10.1007/978-3-319-33826-2

Library of Congress Control Number: 2016943083

Printed on acid-free paper

This Springer imprint is published by Springer Nature
The registered company is Springer International Publishing AG Switzerland

Foreword

Why do we need to validate alternative test methods?

The validation of alternative methods ultimately serves the decision-making process towards the safe use of chemicals. Whether they are based on *in vitro* tests, computer models or combinations of both, validated methods can be used to determine the properties of chemicals used in all sorts of products and processes, including pharmaceuticals, cosmetics, household products, food and industrial manufacturing.

Hazard property information influences risk management decisions at numerous stages of the life cycle of a chemical. For example, during the research and development stage of a new chemical, industry uses non-test methods such as (quantitative) structure–activity relationships to predict its hazards and estimate the risks involved with its use to decide whether the chemical should move towards production. Industry and authorities use results from laboratory tests and non-test methods to classify and label chemicals, which in turn, can trigger specific risk management measures, such as the use of personal protective equipment by workers handling those chemicals or even marketing restrictions to protect consumers and the environment.

These kinds of risk management decisions have to be taken for all the many thousands of chemicals on the market in so many different sectors, even if only one result is available for each relevant hazard endpoint. It is therefore important that authorities, industry and the public at large, have the assurance that the results of the methods used are reliable and relevant. Furthermore, only on these grounds can the data generated be exchanged and accepted across countries for regulatory purposes. This is why demonstration of relevance and reliability are the requirements for the validation and regulatory use of OECD Test Guidelines. Also, both the Test Guidelines (developed following validation studies) and their accompanying guidance documents, generally provide sufficient details to allow all studies to be replicated in any state-of-the-art laboratory.

Research laboratories are continuously developing new methods that better characterise the hazardous properties of chemicals (e.g., for new effects such as endocrine disruption) or alternative methods that do not use laboratory animals (e.g., *in vitro* methods or toxicogenomics). But decision-makers often do not feel confident to use the results from these methods for risk-reduction decisions before they have been demonstrated to be scientifically valid. Furthermore, many non-animal testing-based methods do not sufficiently establish the link with the predicted adverse outcome in humans or wildlife.

But regulatory toxicology is changing. Toxicologists are now seeking to understand the mode of action of chemicals or the adverse outcome pathway that they trigger, i.e., how they interact at a molecular level resulting in effects at the organ or organism level. With increasing knowledge about the modes of action or the adverse outcome pathways that chemicals can trigger, decision-makers are more comfortable using results from alternative methods if it can be shown that the results are linked to key events along the chain of events that constitute the adverse outcome pathway.

This also means that, ultimately, individual animal test methods will be replaced by a number of *in chemico*, *in vitro* and/or *in silico* methods that collectively allow the gathering of information needed to characterise the hazardous property of a chemical. In parallel, as alternative methods become more sophisticated, they will better predict adverse effects in a specific species of interest—e.g., humans.

While this new approach to safety testing will challenge the current approach taken to standardise and validate test methods for regulatory purposes, the objectives of validation will remain the same. The novel test methods used to identify the modes of action will need to be validated in the sense that their reliability and relevance will need to be demonstrated when used to make regulatory decisions. Validation of alternative test methods will therefore remain one of the cornerstones of a successful toxicological (r)evolution.

Environment, Health and Safety Division Bob Diderich
OECD,
Paris Cedex 16, France

Preface

This book provides a comprehensive overview of the best practices and new perspectives regarding the validation of alternative methods for animal procedures used in toxicity testing. Alternative methods cover a wide range of non-animal techniques and technologies, including: *in vitro* assays based on various biological tests and measurement systems; chemoinformatics approaches; computational modelling; and different ways of weighting and integrating information to make predictions of a toxicological effect or endpoint. Validation of an alternative method or approach aims not only to establish the reproducibility and robustness of an alternative method but also to determine its capacity to correctly predict effects of concern in a species of interest. This latter aspect is one of the most critical considerations when striving to replace or reduce animal testing and promoting new approaches in toxicology that are more relevant for human hazard assessment. This book covers the validation of experimental and computational methods and integrated approaches to testing and assessment. Furthermore, validation strategies are discussed for methods employing the latest technologies such as tissue-on-a-chip systems, induced human pluripotent stem cells, bioreactors, transcriptomics and methods derived from pathway-based concepts in toxicology.

Validation of Alternative Methods for Toxicity Testing provides practical insights into state-of-the-art approaches that have resulted in successfully validated and accepted alternative methods. In addition, it explores the evolution of validation principles and practices that will ensure that validation continues to be fit for purpose and has the greatest international impact and reach. Indeed, validation needs to keep pace with the considerable scientific advancements being made in biology and toxicology, the availability of increasingly sophisticated tools and techniques, and the growing societal and regulatory demands for better protection of human health and the environment.

This book is a unique resource for scientists and practitioners working in the field of applied toxicology and safety assessment who are interested in the development and application of new relevant and reliable non-animal approaches for toxicity testing and in understanding the principles and practicalities of validation as critical steps in promoting their regulatory acceptance and use.

Magliaso, Switzerland Chantra Eskes
Ispra, Italy Maurice Whelan

Acknowledgments

The quest for the development and implementation of alternative methods to animal testing really took hold in the 1980s, driven by both heightened ethical concerns surrounding animal testing and the scientific advances being made in the *in vitro* field. Since then, additional motivation has emerged including an increasing emphasis on the need for more human-based and scientifically relevant models for use in basic biomedical research and safety assessment. However, only through the development and implementation of validation principles, establishing the relevance and reliability of new methods for specific applications, have the regulatory acceptance and use of alternative methods been possible. The editors of this book would like to acknowledge the huge contribution and sustained commitment of so many pioneers, too numerous to mention here, who have progressed the field to the point where we can now truly believe in better safety assessment without the use of animals.

Contents

Contributors

Yeyejide Adeleye Unilever Safety and Environmental Assurance Centre, Bedfordshire, UK

Il Young Ahn Toxicological Evaluation and Research Department, Korean Center for the Validation of Alternative Methods (KoCVAM), National Institute of Food and Drug Safety Evaluation, Cheongju-si, South Korea

Paula M. Alves iBET, Instituto de Biologia Experimental e Tecnológica, Oeiras, Portugal

Instituto de Tecnologia Química e Biológica, Universidade Nova de Lisboa, Oeiras, Portugal

Melvin E. Andersen ScitoVation, Research Triangle Park, NC, USA

Jiri Aubrecht Pfizer Global Research and Development, Groton, CT, USA

Nicholas Ball Toxicology and Environmental Research and Consulting (TERC), Environment, Health & Safety (EH&S), The Dow Chemical Company, Horgen, Switzerland

João Barroso European Commission, Joint Research Centre (JRC), Ispra, Italy

Sonja Beken Division Evaluators, DG PRE Authorisation, Federal Agency for Medicines and Health Products (FAMHP), Brussels, Belgium

Camilla Bernasconi European Commission, Joint Research Centre (JRC), Ispra, Italy

Bertrand Desprez European Commission, Joint Research Centre (JRC), Ispra, Italy

Ann-Charlotte Bostroem European Commission, Joint Research Centre (JRC), Ispra, Italy

Gerard Bowe European Commission, Joint Research Centre (JRC), Ispra, Italy

Catarina Brito iBET, Instituto de Biologia Experimental e Tecnológica, Oeiras, Portugal

Instituto de Tecnologia Química e Biológica, Universidade Nova de Lisboa, Oeiras, Portugal

Julien Burton European Commission, Joint Research Centre (JRC), Ispra, Italy

Cristiane Caldeira Brazilian Center for Validation of Alternative Methods (BraCVAM), and National Institute of Quality Control in Health (INCQS), Rio de Janeiro, Brazil

Paul L. Carmichael Unilever Safety and Environmental Assurance Centre, Bedfordshire, UK

Warren Casey Division of the National Toxicology Program, National Institute of Environmental Health Sciences, Research Triangle Park, DC, USA

Interagency Coordinating Committee on the Validation of Alternative Methods (ICCVAM), Washington, DC, USA

Rebecca A. Clewell ScitoVation, Research Triangle Park, NC, USA

Sandra Coecke European Commission, Joint Research Centre (JRC), Ispra, Italy

Thomas Cole European Commission, Joint Research Centre (JRC), Ispra, Italy

Raffaella Corvi European Commission, Joint Research Centre (JRC), Ispra, Italy

Rodger D. Curren Institute for *In Vitro* Sciences, Inc., Gaithersburg, MD, USA

Eva-Maria Dehne Department of Medical Biotechnology, Technische Universität Berlin, Institute of Biotechnology, Berlin, Germany

Nathalie Delrue Environment, Health and Safety Division, Organisation for Economic Cooperation and Development, Paris, France

Chantra Eskes SeCAM Services and Consultation on Alternative Methods, Magliaso, Switzerland

Salvador Fortaner European Commission, Joint Research Centre (JRC), Ispra, Italy

Varvara Gouliarmou European Commission, Joint Research Centre (JRC), Ispra, Italy

Anne Gourmelon Environment, Health and Safety Division, Organisation for Economic Cooperation and Development, Paris, France

Andrew Gray UK GLP Monitoring Authority, MHRA, London, UK

Claudius Griesinger European Commission, Joint Research Centre (JRC), Ispra, Italy

Jiabin Guo Evaluation and Research Centre for Toxicology, Institute of Disease Control and Prevention, Academy of Military Medical Sciences, Beijing, China

Emilio Mendoza-de Gyves European Commission, Joint Research Centre (JRC), Ispra, Italy

Thomas Hartung Johns Hopkins Bloomberg School of Public Health, Center for Alternatives to Animal Testing (CAAT), Baltimore, MD, USA
University of Konstanz, CAAT-Europe, Konstanz, Germany

Simone Haupt Life and Brain, Bonn, Germany

SEURAT-1 Stem Cell Group, Paris, France

Lyn Healy Haematopoietic Stem Cell Laboratory, The Francis Crick Institute, London, UK

SEURAT-1 Stem Cell Group, Paris, France

Erin Hill Institute for *In Vitro* Sciences, Inc., Gaithersburg, MD, USA

Sebastian Hoffmann seh consulting + services, Paderborn, Germany

Reyk Horland Department of Medical Biotechnology, Technische Universität Berlin, Institute of Biotechnology, Berlin, Germany

Magnus Ingelman-Sundberg Karolinska Institutet, Solna, Sweden

SEURAT-1 Stem Cell Group, Paris, France

Paul Jennings Division of Physiology, Medical University of Innsbruck, Innsbruck, Austria

SEURAT-1 Stem Cell Group, Paris, France

Elisabeth Joossens European Commission, Joint Research Centre (JRC), Ispra, Italy

Peter Kasper Federal Institute for Drugs and Medical Devices (BfArM), Bonn, Germany

Darren Kidd Covance Laboratories Limited, North Yorkshire, UK

SEURAT-1 Stem Cell Group, Paris, France

Hajime Kojima Japanasese Center for the Validation of Alternative Methods (JaCVAM), National Institute of Health Sciences, Tokyo, Japan

Jan-Willem van der Laan Pharmacology, Toxicology and Biotechnology Department, Medicines Evaluation Board (MEB), Utrecht, The Netherlands

Ingrid Langezaal European Commission, Joint Research Centre (JRC), Ispra, Italy

Gilles Lemaitre I-Stem, INSERM/UEVE U861, Evry, France

SEURAT-1 Stem Cell Group, Paris, France

Jin Li Unilever Safety and Environmental Assurance Centre, Bedfordshire, UK

Chae Hyung Lim Toxicological Evaluation and Research Department, Korean Center for the Validation of Alternative Methods (KoCVAM), National Institute of Food and Drug Safety Evaluation, Cheongju-si, South Korea

Roman Liska European Commission, Joint Research Centre (JRC), Ispra, Italy

Susanna Louhimies Directorate General for Environment, European Commission, Brussels, Belgium

Jochem Louisse Wageningen University and Research Centre, Wageningen, The Netherlands

SEURAT-1 Stem Cell Group, Paris, France

Uwe Marx Department of Medical Biotechnology, Technische Universität Berlin, Institute of Biotechnology, Berlin, Germany

Patrick D. McMullen ScitoVation, Research Triangle Park, NC, USA

Anne Milcamps European Commission, Joint Research Centre (JRC), Ispra, Italy

Siegfried Morath European Commission, Joint Research Centre (JRC), Ispra, Italy

Wlamir Moura Brazilian Center for Validation of Alternative Methods (BraCVAM) and National Institute of Quality Control in Health (INCQS), Rio de Janeiro, Brazil

Glenn Myatt Leadscope, Columbus, OH, USA

SEURAT-1 Stem Cell Group, Paris, France

Akiyoshi Nishikawa Japanasese Center for the Validation of Alternative Methods (JaCVAM), National Institute of Health Sciences, Tokyo, Japan

Nicholaos Parissis European Commission, Joint Research Centre (JRC), Ispra, Italy

HyeKyung Park Toxicological Evaluation and Research Department, Korean Center for the Validation of Alternative Methods (KoCVAM), National Institute of Food and Drug Safety Evaluation, Cheongju-si, South Korea

Grace Patlewicz Dupont Haskell Global Centers for Health and Environmental Sciences, Newark, DE, USA

National Center for Computational Toxicology (NCCT), US Environmental Protection Agency (EPA), Research Triangle Park, NC, USA

Shuangqing Peng Evaluation and Research Centre for Toxicology, Institute of Disease Control and Prevention, Academy of Military Medical Sciences, Beijing, China

Aldert Piersma Center for Health Protection, National Institute for Public Health and the Environment RIVM, Bilthoven, The Netherlands

Institute for Risk Assessment Sciences, Utrecht University, Utrecht, The Netherlands

Christian Pinset I-Stem, INSERM/UEVE U861, Evry, France

SEURAT-1 Stem Cell Group, Paris, France

Albrecht Poth Eurofins BioPharma Product Testing, Munich, Germany

Octavio Presgrave Brazilian Center for Validation of Alternative Methods (BraCVAM) and National Institute of Quality Control in Health (INCQS), Rio de Janeiro, Brazil

Anna-Bal Price European Commission, Joint Research Centre (JRC), Ispra, Italy

SEURAT-1 Stem Cell Group, Paris, France

Maurits-Jan Prinz Directorate General for Internal Market, Industry, Entrepreneurship and SMEs, European Commission, Brussels, Belgium

Hans A. Raabe Institute for *In Vitro* Sciences, Inc., Gaithersburg, MD, USA

Sofia P. Rebelo iBET, Instituto de Biologia Experimental e Tecnológica, Oeiras, Portugal

Instituto de Tecnologia Química e Biológica, Universidade Nova de Lisboa, Oeiras, Portugal

Vittorio Reina European Commission, Joint Research Centre (JRC), Ispra, Italy

Andrea Robitski University of Leipzig, Leipzig, Germany

SEURAT-1 Stem Cell Group, Paris, France

Annett Janusch Roi European Commission, Joint Research Centre (JRC), Ispra, Italy

Tim Singer Environmental Health Science and Research Bureau, Health Canada, Ottawa, ON, Canada

Soo Jung Sohn Toxicological Evaluation and Research Department, Korean Center for the Validation of Alternative Methods (KoCVAM), National Institute of Food and Drug Safety Evaluation, Cheongju-si, South Korea

Glyn N. Stacey UK Stem Cell Bank, Advanced Therapies Division, NIBSC-MHRA, London, UK

SEURAT-1 Stem Cell Group, Paris, France

Martin Stephens Johns Hopkins Bloomberg School of Public Health, Center for Alternatives to Animal Testing (CAAT), Baltimore, MD, USA

Mireia Vilardell European Commission, Joint Research Centre (JRC), Ispra, Italy

Carl Westmoreland Unilever Safety and Environmental Assurance Centre, Bedfordshire, UK

Maurice Whelan European Commission, Joint Research Centre (JRC), Ispra, Italy

Iwona Wilk-Zasadna European Commission, Joint Research Centre (JRC), Ispra, Italy

Anja Wilmes Division of Physiology, Medical University of Innsbruck, Innsbruck, Austria

SEURAT-1 Stem Cell Group, Paris, France

Andrew P. Worth European Commission, Joint Research Centre (JRC), Ispra, Italy

Xingfen Yang Guangdong Province Centre for Disease Control and Prevention, Guangzhou, China

Ying Yang Guangdong Province Centre for Disease Control and Prevention, Guangzhou, China

Chiara Zorzoli European Commission, Joint Research Centre (JRC), Ispra, Italy

Valérie Zuang European Commission, Joint Research Centre (JRC), Ispra, VA, Italy

About the Editors

Chantra Eskes, Ph.D., Eng. is an *in vitro* toxicologist with over 20 years of experience in the development, optimization, validation, peer review and regulatory acceptance of alternative methods to animal toxicity testing. She currently acts as a Nominated Expert at the Organisation for Economic Co-operation and Development (OECD), the President of the European Society of *In Vitro* Toxicology (ESTIV) and the Executive Secretary of the Animal Cell Technology Industrial Platform on the production of biopharmaceuticals (ACTIP). She is also founder and manager of a company offering independent consultation services regarding alternative methods for scientific, regulatory and industrial tailored requirements. Her areas of activity include food sciences, neurotoxicity, topical toxicity, chemicals, cosmetics, detergent and cleaning products and biopharmaceuticals.

Maurice Whelan is head of the Chemicals Safety and Alternative Methods Unit of the Directorate for Health, Consumers and Reference Materials of the European Commission Joint Research Centre (JRC), Ispra, Italy. He also heads the European Union Reference Laboratory for Alternatives to Animal Testing (EURL ECVAM) of the JRC, established under EU Directive 2010/63 on the protection of animals used for scientific purposes to build on the 20 years of activities of ECVAM, the European Centre for the Validation of Alternative Methods. Priorities of his work include the development, validation and promotion of alternative approaches to animal testing both for regulatory safety assessment of chemicals (including nanomaterials) and for applications in biomedical research. Whelan is the EU co-chair of the OECD Advisory Group on Molecular Screening and Toxicogenomics that is responsible for the OECD programme on Adverse Outcome Pathways, and he is a member of the Steering Committee of the European Partnership for Alternative Approaches to Animal Testing (EPAA). He was awarded his Ph.D. in 1993 in Mechanical Engineering (design of orthopaedic knee prostheses) by the University of Limerick (Ireland) and holds an external appointment of visiting Professor of Bioengineering at the University of Liverpool (UK).

Chapter 1
Introduction

Chantra Eskes and Maurice Whelan

Abstract Alternative approaches to animal testing are gaining momentum with an increasing number of test methods obtaining international acceptance, thanks in large part to the validation efforts conducted on these assays. The principles and process of validation were first established in the 1990s in Europe and USA, and further gained international recognition ensuring the broader acceptance of alternative test methods at a regulatory level. If these principles were successful in pioneering the regulatory acceptance of alternative methods for less complex endpoints, an evolution of concepts is needed to embrace emerging technologies and the increased complexity of endpoints. Innovative concepts and approaches of scientific validation can help to ensure the continued regulatory and international acceptance of novel alternative methods and technologies for toxicity testing such as human-based *in vitro* models derived from induced pluripotent stem cells and significant advances in bioengineering. This chapter provides a historical overview of the establishment and evolution of the principles of the scientific validation of alternative methods for toxicity testing as well as the challenges and opportunities for adapting those principles to keep pace with scientific progress whilst ensuring human safety and best serve the needs of society.

1 The Need for Validation

Alternative methods refer to procedures that can replace the need for animal experiments, reduce the number of animals required, or diminish the amount of distress or pain experienced by animals (Smyth 1978). This definition embodies the "Three Rs" concept proposed by Russell and Burch in The Principles of Humane Experimental Technique (Russell and Burch 1959), which was considered by many

C. Eskes (✉)
SeCAM Services and Consultation on Alternative Methods (SeCAM), Magliaso, Switzerland
e-mail: chantra.eskes@secam-ce.eu

M. Whelan
European Commission, Joint Research Centre (JRC), Ispra, Italy

C. Eskes, M. Whelan (eds.), *Validation of Alternative Methods for Toxicity Testing*, Advances in Experimental Medicine and Biology 856,
DOI 10.1007/978-3-319-33826-2_1

countries in defining regulatory requirements concerning the protection of animals used for scientific purposes (Council Directive 86/609/EEC 1986; Directive 2010/63/EU 2010; Brazil 2008).

During the last quarter of the twentieth century, public concern over ethical aspects regarding the use of animals for scientific purposes has steadily increased, especially in the USA and in Europe. Humane societies have questioned in particular the need for animals in product-safety testing, medical research and science education (Wilhelmus 2001). For example, eye irritation testing procedures on rabbits has often been used as a symbol for cruelty by animal welfare activists, since at times such procedures can be very painful and result in visible suffering, trauma and reactions in the rabbit eyes. In April 1980, a group of animal welfare activists specifically targeted the rabbit eye test by publishing a full-page advertisement in the New York Times stating "*How many rabbits does Revlon blind for beauty's sake?*", followed by a second advertisement published in October 1980. Such campaigns resulted in grant investments to support the development of alternatives to the rabbit eye test (Wilhelmus 2001).

In order to ensure the acceptance of the developed alternatives to animal testing, regulatory action was also taken. In Europe for example, the original Directive on the protection of laboratory animals for experimental and other scientific purposes stated that "*An* (animal) *experiment shall not be performed if another scientifically satisfactory method of obtaining the result sought, not entailing the use of an animal, is reasonably and practicably available*" (Directive 86/609/EEC).

The final acceptance of an alternative test method may depend on various factors such as national regulatory requirements, the test method purposes, uses and applicability. However, demonstrating the scientific validity of an *in vitro* method is usually required for its use within the regulatory framework especially for detecting both hazardous and non-hazardous effects as a replacement, reduction or refinement of animal testing (OECD Guidance Document No. 34 2005; Regulation (EC) No 1907/2006). As such, for an alternative method to gain regulatory acceptance, it is current practice to demonstrate that the method is scientifically satisfactory, i.e., valid, for the purpose sought. This is generally carried out through a validation process through which the scientific validity of a test method can be demonstrated.

2 Historical Developments

The criteria and processes for the validation of a test method were first developed in the 1990s. In Europe, the European Centre for the Validation of Alternative Methods (ECVAM) was created in 1991 as part of the European Commission's Joint Research Centre (JRC), to respond to the requirement from the original EU Directive on the protection of animals for scientific purposes, namely that "*The Commission and Member States should encourage research into the development and validation of alternative techniques (…) and shall take such other steps as they consider appropriate to encourage research in this field*" (Directive 86/609/EEC). This was followed in the United States by the creation in 1997 of the Interagency Coordinating

Committee on the Validation of Alternative Methods (ICCVAM), and subsequently in Japan in 2005 with the establishment of the Japanese Center for the Validation of Alternative Methods (JaCVAM). Reflecting the growing awareness of the importance of validation worldwide, internationally agreed principles of validation were adopted by the Organization for Economic Co-operation and Development (OECD) in 2005 (OECD Guidance Document No. 34 2005). More recently, the implementation of the EU Directive 2010/63 on the protection of animals used for scientific purposes (Directive 2010/63/EU 2010), which came into full force in 2013, has reinforced Europe's commitment to place the 3Rs at the heart of EU policy and to strengthen legislative provision to minimize the reliance on animal procedures in different contexts whenever possible. Moreover, outreaching countries have since also established national centers for the validation of alternative methods such as the South Korean Center for the Validation of Alternative Methods (KoCVAM) established in 2010 and the Brazilian Centre for the Validation of Alternative Methods (BraCVAM) established in 2011 (see Chap. 14).

Based upon the experiences gained during earlier multi-laboratory evaluation studies on e.g. eye irritation, and in consultation with various international experts, ECVAM published under the enriching leadership of Michael Balls, recommendations on the principles, practical and logistical aspects of validating alternative test methods (Balls et al. 1990, 1995; Curren et al. 1995). These documents represent the first basic principles for the validation of alternative methods including the management and design of a validation study that were later integrated at an international level (OECD Guidance Document No. 34 2005).

An alternative method for the replacement (or partial replacement) of an animal test is defined as the combination of a "test system", which provides a means of generating physicochemical or *in vitro* data for the chemicals of interest, and a "prediction model (PM)" or "data interpretation procedure" (Archer et al. 1997). The prediction model or data interpretation procedure plays an important role in the acceptance process, as it allows converting the obtained data (e.g., *in vitro* or physicochemical) into predictions of toxicological endpoints in the species of interest e.g., animals or humans (OECD Guidance Document No. 34 2005).

Test method validation is defined as the process whereby the relevance and reliability of the method are characterized for a particular purpose (OECD Guidance Document No. 34 2005; Balls et al. 1990). In the context of a replacement test method, relevance refers to the scientific basis of the test system and to the predictive capacity of the test method as compared to a reference method. Reliability refers to the reproducibility of test results, both within and between laboratories, and over time. The "purpose" of an alternative method refers to its intended application, such as the regulatory testing of chemicals for a specific toxicological endpoint (e.g., eye irritation). Adequate validation (i.e., to establish scientific validity) of an alternative test requires demonstration that, for its stated purpose:

- the test system has a sound scientific basis;
- the predictions made are sufficiently accurate; and
- the results generated by the test system are sufficiently reproducible within and between laboratories, and over time.

Furthermore, some of the key principles of the validation process encompass (Balls et al. 1990):

- An alternative method can only be judged valid if the method is reliable and relevant;
- The prediction model should be defined in advance by the test developer;
- The aspired performance criteria should be set in advance by the management team (for a prospective validation study);
- Performance is assessed by using coded chemicals;
- There should be independence in:
  - the management of the study,
  - the selection, coding and distribution of test chemicals,
  - the data collection and statistical analysis;
- Laboratory procedures should comply with GLP criteria.

In addition, a prevalidation scheme has been recommended to ensure that a method included in a formal validation study adequately fulfils the criteria defined for inclusion in such a study, so that financial and human resources are used most efficiently with a greater likelihood that the expectations will be met. The prevalidation process includes three main phases: protocol refinement, protocol transfer and protocol performance (Curren et al. 1995).

In 2004, a "Modular Approach to the ECVAM Principles on Test Validity" was proposed with the objective to make the validation process more flexible by breaking down the various steps of validation into seven independent modules, and defining for each module the information needed for assessing the scientific validity of a test method (Hartung et al. 2004). One of the main advantages of the Modular Approach to Validation is the possibility to complete the different modules in any sequence, allowing the use of data both gathered retrospectively and generated prospectively as required. This approach has the potential to increase the evidence gathered on a specific test method whilst decreasing the time necessary if only prospective data were to be considered. The seven modules are:

1. Test definition;
2. Within-laboratory reproducibility;
3. Transferability;
4. Between-laboratory reproducibility;
5. Predictive capacity;
6. Applicability domain; and
7. Definition of performance standards.

A consequence of the replacement in 2010 of Directive 86/609/EEC with Directive 2010/63/EU was the formalization and broadening of the role of ECVAM, reflected in its name being changed by the JRC to the European Union Reference Laboratory for Alternatives to Animal Testing (EURL ECVAM, see also http://ihcp.jrc.ec.europa.eu/our_labs/eurl-ecvam). EURL ECVAM's duties and tasks (Article 48/ Annex VII of Directive 2010/63) now encompass the coordination and promotion of the development, validation and use of alternative methods; acting as a focal point for the exchange of information; setting up, maintaining and managing public databases and information systems on alternative methods; and promoting dialogue between legislators, regulators, and all relevant stakeholders with a view to the development, validation, regulatory acceptance, international recognition, and application of alternative approaches.

Regarding the USA, the NIH Revitalization Act of 1993 (Public Law 103-43) required the National Institute of Environmental Health Sciences (NIEHS) to establish criteria for the validation and regulatory acceptance of alternative toxicological testing methods, and that NIEHS recommend a process to achieve the regulatory acceptance of scientifically valid alternative test methods. To respond to requirements of this Act, NIHS created ICCVAM initially as an *ad hoc* committee in 1994, and subsequently as a standing committee in 1997 (see also http://www.iccvam.niehs.nih.gov) with the aim to (i) implement a process by which new test methods of interest could be evaluated and (ii) coordinate interactions among US agencies related to the development, validation, acceptance, and national and international harmonization of toxicological test methods. ICCVAM was then formally established as a permanent interagency committee of the NIEHS under the National Toxicology program (NTP) Interagency Center for the Evaluation of Alternative Toxicological Methods (NICEATM) in 2000 by the ICCVAM Authorization Act Public Law 106-545.

Criteria for validation and regulatory acceptance of alternative test methods were published in 1997 by ICCVAM-NIEHS (Validation and Regulatory Acceptance of Toxicological Test Methods 1997). The definition and principles of scientific validity are similar to those adopted in the European Union, although a specific format of data compilation is required including for example: test method protocol components, intra- and inter- laboratory reproducibility, test method accuracy, protocol transferability, information on the selection of reference substances, information on the reference species, supporting data and quality, animal welfare considerations and practical considerations.

The Japanese Center for the Validation of Alternative Methods (JaCVAM, see also http://jacvam.jp/en) was established in 2005 as part of the Biological Safety Research Center (BSRC) of the National Institute of Health Sciences (NIHS). Its key objectives are to ensure that new or revised test methods are validated, peer reviewed, and officially accepted by regulatory agencies (Kojima 2007). For this purpose, JaCVAM assesses the utility, limitations, and suitability for use of alternative test methods in regulatory studies for determining the safety of chemicals and other materials. JaCVAM also performs validation studies when necessary. Furthermore, JaCVAM establishes guidelines for new alternative experimental methods through international collaboration.

As validation is an important step within the regulatory acceptance of alternative methods, international efforts have been undertaken to favor the harmonization of its processes and principles with the ultimate goal of promoting harmonization of international acceptance and recognition of alternative methods. In particular, through a process of consultation with validation bodies and key stakeholders, the OECD adopted internationally agreed validation principles and criteria for the regulatory acceptance of alternative test methods. Such internationally agreed principles are described in the OECD Guidance Document No. 34 on *"The Validation and International Acceptance of New or Updated Test Methods for Hazard Assessment"* (OECD Guidance Document No. 34 2005). The OECD GD 34 details internationally agreed principles and criteria on how validation studies of new or updated test methods should be performed. It represents a document of key importance for promoting harmonized approaches and procedures for the validation and regulatory acceptance of alternative methods at the international level (see also Chap. 2).

3 Current Challenges and Opportunities

If the validation principles and processes established in the 1990s were successful in achieving international acceptance of a number of alternative test methods, the scientific advances made in the recent years in the area of *in vitro* toxicology call for an evolution of the traditional validation principles. Indeed, considerable progress was dictated by new technologies and discoveries, as well as by the increasing complexity of the endpoints assessed. For instance, the 2012 Nobel Prize Shinya Yamanaka opened the door for the reprogramming of mature cells to become pluripotent, the so-called induced pluripotent stem cells, which allow the use of human-based cells reprogrammed in any organ-type cell for the evaluation of toxicity. Furthermore, a number of scientific groups have developed new complex bioengineering technologies such as the human-on-a-chip models which allow combining various organ-specific cell types and obtaining a more holistic response to toxicants whilst providing a more complex model mimicking the *in vivo* toxicity. In the US, the use of high-throughput *in vitro* screening assays, systems biology and predictive *in silico* approaches have also been recently used within the twenty-first century NTP program to improve the hazard evaluation of environmental chemicals. Furthermore, the evaluation of more complex endpoints require not only complex models but also their integration into e.g., integrated approaches for testing and assessment as well as consideration of the mechanistic adverse-outcome pathways of toxicity, that call for new considerations regarding the approaches for the scientific validation of alternatives to toxicity testing. Finally, collaboration of the validation centers in the various geographical regions is critical to ensure the harmonized international acceptance of alternative methods, the removal of barriers and the promotion of harmonized human safety assessment across the globe.

This book provides two distinct yet complementary perspectives on the approaches used for the scientific validation of alternative methods. The first is more retrospective and describes the state-of-the-art in validation including the underlying principles and practical approaches that have been successful over the years in gaining international regulatory acceptance of alternative methods. The second, more forward-looking perspective addresses the need to foster innovation and ensure progressive evolution of validation concepts and practices that are fit for the purpose of aiding the translation of emerging technologies and sophisticated methodologies in the field of alternative methods into internationally accepted solutions for regulatory toxicity testing.

References

Archer G, Balls M, Bruner LH, Curren RD, Fentem JH, Holzhütter H-G, Liebsch M, Lovell DP, Southee JA (1997) The validation of toxicological prediction models. ATLA 25:505

Balls M, Blaauboer B, Brusick D, Frazier J, Lamb D, Pemberton M, Reinhardt C, Roberfroid M, Rosenkranz H, Schmid B, Spielmann H, Stammati A-L, Walum E (1990) Report and recommendations of the CAAT/ERGATT workshop on the validation of toxicity test procedures. ATLA 18:313

Balls M, Blaauboer BJ, Fentem JH, Bruner L, Combes RD, Ekwall B, Fielder RJ, Guillouzo A, Lewis RW, Lovell DP, Reinhardt CA, Repetto G, Sladowski D, Spielmann H, Zucco F (1995) Practical aspects of the validation of toxicity test procedures. The report and recommendations of ECVAM workshop 5. ATLA 23:129

Brazil (2008) Law no. 11.794 on the scientific use of animals, November 08

Council Directive 86/609/EEC of 24 November 1986 on the approximation of laws, regulations and administrative provisions of the Member States regarding the protection of animals used for experimental and other scientific purposes (1986) Official J L358:1

Curren RD, Southee JA, Spielmann H, Liebsch M, Fentem JH, Balls M (1995) The role of prevalidation in the development, validation and acceptance of alternative methods. ATLA 23:211

Directive 2010/63/EU of the European Parliament and of the council of 22 September 2010 on the protection of animals used for scientific purposes (2010) Official J Eur Union L276:33

Hartung T, Bremer S, Casati S, Coecke S, Corvi R, Fortaner S, Gribaldo L, Halder M, Hoffmann S, Roi AJ, Prieto P, Sabbioni E, Scott L, Worth A, Zuang V (2004) A modular approach to the ECVAM principles on test validity. ATLA 32:467

Kojima H (2007) JaCVAM: an organization supporting the validation and peer review of new alternatives to animal testing. AATEX 14(special issue):483–485

OECD Guidance Document No. 34 on "*the Validation and International Acceptance of New or Updated Test Methods for Hazard Assessment*" (2005) OECD Series on Testing and Assessment. Organization for Economic Cooperation and Development, Paris, France

Regulation (EC) No 1907/2006 of the European Parliament and of the Council of 18 December 2006 concerning the Registration, Evaluation, Authorisation and Restriction of Chemicals (REACH), establishing a European Chemicals Agency, amending Directive 1999/45/EC and repealing Council Regulation (EEC) No 793/93 and Commission Regulation (EC) No 1488/94 as well as Council Directive 76/769/EEC and Commission Directives 91/155/EEC, 93/67/EEC, 93/105/EC and 2000/21/EC (2006) Official J Eur Union L396:1

Russell WMS, Burch RL (1959) The principles of humane experimental technique. Methuen, London

Smyth DH (1978) Alternatives to animal experiments. Scolar Press-Royal Defence Society, London

Validation and Regulatory Acceptance of Toxicological Test Methods: A Report of the *Ad Hoc* Interagency Coordinating Committee on the Validation of Alternative Methods (1997) NIH publication n. 97-3981. National Institute of Environmental Health Sciences, Research Triangle Park, NC, USA

Wilhelmus KR (2001) The Draize eye test. Surv Ophthalmol 45:493–515

Chapter 2
Validation in Support of Internationally Harmonised OECD Test Guidelines for Assessing the Safety of Chemicals

Anne Gourmelon and Nathalie Delrue

Abstract Ten years elapsed since the OECD published the Guidance document on the validation and international regulatory acceptance of test methods for hazard assessment. Much experience has been gained since then in validation centres, in countries and at the OECD on a variety of test methods that were subjected to validation studies. This chapter reviews validation principles and highlights common features that appear to be important for further regulatory acceptance across studies. Existing OECD-agreed validation principles will most likely generally remain relevant and applicable to address challenges associated with the validation of future test methods. Some adaptations may be needed to take into account the level of technique introduced in test systems, but demonstration of relevance and reliability will continue to play a central role as pre-requisite for the regulatory acceptance. Demonstration of relevance will become more challenging for test methods that form part of a set of predictive tools and methods, and that do not stand alone. OECD is keen on ensuring that while these concepts evolve, countries can continue to rely on valid methods and harmonised approaches for an efficient testing and assessment of chemicals.

Keywords OECD validation principles • Test Guidelines • Integrated approaches • Mutual acceptance

---

A. Gourmelon (✉) • N. Delrue
Environment, Health and Safety Division,
Organisation for Economic Cooperation and Development,
2, rue André-Pascal, Paris, 75775, France
e-mail: anne.gourmelon@oecd.org

C. Eskes, M. Whelan (eds.), *Validation of Alternative Methods for Toxicity Testing*, Advances in Experimental Medicine and Biology 856,
DOI 10.1007/978-3-319-33826-2_2

1 Introduction to the OECD Test Guidelines Programme

1.1 Context and Goal

Since 1981, OECD countries have tasked the Environment, Health and Safety Programme to develop harmonized methods for the testing of chemicals. The methods are intended to generate valid and high quality data to support chemical safety regulations in member countries. The OECD Guidelines for the testing of chemicals are a collection of the most relevant internationally agreed testing methods used by governments, industry and independent laboratories to assess the safety of chemical products. OECD Test Guidelines are covered by the OECD Council Decision on the Mutual Acceptance of Data (MAD) stating that test data generated in any member country—or partner country adhering to MAD—in accordance with OECD Test Guidelines and Principles of Good Laboratory Practice (GLP) shall be accepted in other member countries and adhering partner countries for assessment purposes and other uses relating to the protection of human health and the environment (OECD 1981). This Decision minimises the costs associated with testing chemicals by avoiding duplicative testing, and utilises more effectively scarce test facilities and specialist manpower in countries. Having harmonised Test Guidelines also avoids non-tariff barriers to international trade of chemicals through a level playing of environmental protection across countries.

Started in 1981, the collection of OECD Test Guidelines is augmented every year with new and updated Test Guidelines that have undergone a number of stages to demonstrate their validity in order to be accepted by regulatory authorities. The motivations for continuously improving testing standards at OECD level are keeping the pace with progress in science, responding to countries' regulatory needs, addressing animal welfare and improving cost-effectiveness of test methods. At various stages of Test Guidelines development, OECD-wide networks of scientists in government, academia, and industry provide input. The OECD Test Guidelines Programme is also fed by the work of validation centres established in certain countries or regions which establish and/or review the scientific validity of test methods proposed for the development of Test Guidelines. It is indeed essential that test methods undergo a critical appraisal of their relevance and reliability through experimental demonstration in laboratories who are potential future users, so that the utility of the method for a specific purpose, as well as its limitations, can be defined and understood by users and regulators. The use of Test Guidelines that are based on validated test methods promotes the generation of dependable data for human and animal health and environmental safety. In 2005, the OECD published a Guidance Document for test method validation outlining general principles, important considerations, illustrative examples, potential challenges and the results of experience gained (OECD 2005).

1.2 Participation (WNT, Nominated Experts, Industry Experts, Animal Welfare Organisations)

The development of OECD Test Guidelines is overseen by the Working Group of the National Coordinators of the Test Guidelines Programme (WNT). National Coordinators represent regulatory authorities in OECD member countries and countries adhering to MAD. Representatives from identified interest groups (industry and animal welfare non-governmental organisations, green NGOs) and from some additional countries having an economically important chemical industry also attend annual meetings of the WNT as invited experts, and can participate in technical expert groups. National Coordinators take decisions on Test Guidelines for approval (including updates of existing Test Guidelines) and decide on project proposals to include on the work plan. Experts in technical groups are nominated by their National Coordinators, Business and Industry Advisory Council to OECD (BIAC), the International Council on Animal Protection in OECD programmes (ICAPO) and the European Environmental Bureau (EEB). Expert groups are specialised by area of hazard assessment (e.g. reproductive toxicity, genotoxicity, toxicity to the aquatic environment, environmental fate), and thus can work on several projects of the work plan that fall under the same area.

Experts participating in technical groups are nominated to provide their technical expertise in the area. Many experts participate over many years in the technical groups. This ensures consistency in the work done over time; however new expertise is always sought to ensure the best available science is taken into account and used in test method development. It is important that Test Guidelines development and regulatory science benefit from progress made in scientific research through networks and consortia of academic and industry laboratories. Gathering expertise and input from academia, industry, environmental and animal welfare organisations is essential for the OECD work on chemical safety to remain relevant for countries. Although industry and environmental organisations have been involved from the start in TG development, the participation of animal welfare NGOs is more recent, starting in the early 2000, and was encouraged by countries' uptake of ethical considerations in the use of laboratory animals for safety testing of chemicals. Occasionally, for specific areas of hazard assessment (e.g. endocrine disrupters), other interest groups are also involved. Furthermore, the European Commission, although not a member "country", participates in all the activities; indeed a large number of research activities in Europe relevant to the work of the Test Guidelines Programme are undertaken and coordinated by the European Union Reference Laboratory—European Centre for the Validation of Alternative Methods (EURL-ECVAM). Finally, countries like the People's Republic of China and the Russian Federation are invited to contribute to the work of the Test Guidelines Programme.

1.3 Workflow and Decision-Making Processes

National Coordinators can propose new projects. Such proposals have to be motivated by a regulatory need in more than one country or region (to benefit from international harmonisation), by a progress in science, by animal welfare considerations (e.g. making it possible to use fewer animals or to reduce duration of a test for example), or by an improvement in the cost-effectiveness of a test method. Proposals are reviewed and commented on by all members of the WNT a few months before the annual WNT meeting. At the meeting itself, the National Coordinators take a consensus decision on whether or not to include the project on the work plan following discussions. Project proposals can be submitted at different stages of test method development. In cases where the test method has already been validated, information and documents supporting the validation and the development of a Test Guideline are brought to the attention of the WNT upon submission of the project proposal. The WNT takes its decision to include the proposal in the work plan based on all available information.

If the project is accepted and the test method has already been validated, the lead country will take the first steps to make the first draft Test Guideline, while the Secretariat asks the WNT to nominate experts to a group, unless an existing group is competent and can take the new project on board. When the draft Test Guideline is sufficiently ready, it is circulated for a commenting round. The National Coordinators, industry, environmental organisations and ICAPO usually consult their expert networks when providing comments. In case of diverging views expressed by national experts, National Coordinators can take a national position. The Secretariat collects and compiles comments received and works with the lead country to address issues raised and revise the draft Test Guideline. Typically, following two rounds of WNT comments, the draft documents are mature enough for submission and eventually approval by the WNT, but there may be exceptions. The OECD Guidance Document 1 on the Development of Guidelines for the Testing of Chemicals, updated in 2009 (OECD 2009a), describes in more details the process and procedures for the development of OECD Test Guidelines and related documents (see Fig. 2.1). When Test Guidelines are approved by the WNT, they are subsequently endorsed by higher policy-level bodies of the Organisation until final adoption by the OECD Council and publication. Guidance documents approved by the WNT do not go to OECD Council for adoption (because they are not covered by the OECD Council Decision on the Mutual Acceptance of Data) and they are published under the responsibility of the policy body overseeing the work on chemical safety at OECD.

Projects may be included in the work plan at various stages of test method development, and the validity of the test method may not necessarily be fully established. In such cases, the project starts with experimental validation across laboratories, organised by the lead country(ies), with the assistance of the expert group or a Validation Management Group (VMG), with support from the OECD Secretariat as appropriate. When a project starts with a proposal for a test method that has not yet been validated, the whole process until approval of a Test Guideline takes more time, as the experimental validation is the most resource-intensive stage of the project.

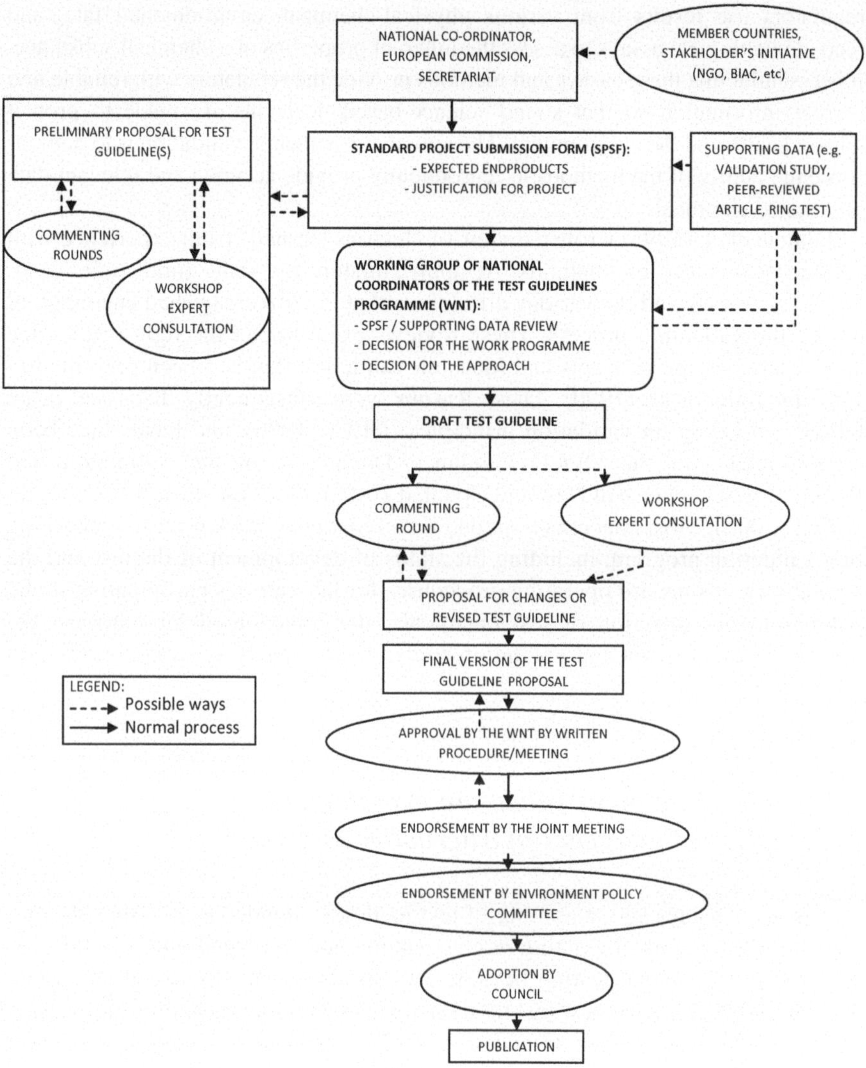

Fig. 2.1 OECD Test Guidelines development flow diagram (from Guidance Document 1 (OECD 2006))

2 Importance of Validation in the Development Process of Test Guidelines

Regulatory authorities are charged by law with protecting human health and the environment. The purpose of validation is to ensure that regulators obtain reliable and useful information for their decision making, and that data generated can be exchanged and mutually accepted across countries. In the case of test chemical, regulators use results from various physical-chemical, environmental fate, and (eco-)toxicological assays to assess the inherent properties of a chemical substance. It is essential that these assays and methods provide the regulators with reliable and correct information so that sound science-based decisions are made to protect human health and the environment. The aim of experimental validation is to demonstrate the ability of the methods to reproducibly provide accurate and relevant data on a tested chemical.

Fentem et al. (1995) wrote a review of "lessons learned" from experience with validating *in vitro* test methods. At approximately the same time, Balls et al. (1995) also reviewed the various difficulties that *in vitro* assays had encountered during the validation process. These reviews examined in the light of practice and experience the concepts and ideas on validation that had been presented in 1990 by Balls et al. (1990). These lessons were subsequently discussed at an OECD workshop on validation principles (OECD 1996), and have since been incorporated into the OECD Guidance Document on the Validation and Regulatory Acceptance of New and Updated Test Methods for Hazard Assessment (OECD 2005). Most concerns regarded the preparatory work prior to embarking in a validation program, including the status of development of the test and the availability of standard operating procedures for laboratories participating in the validation, the selection of test chemicals, the selection of laboratories, the design of the experimental validation study, and the analysis and interpretation of results.

2.1 Formalisation of Validation Programmes with the Emergence of Alternative Methods

About two decades ago, a number of test methods intended as possible alternative or replacements of existing *in vivo* test methods emerged, initially for hazards for which animal testing became less and less ethically acceptable (e.g., topical toxicity). These new test models measured endpoints and/or biomarkers *in vivo*, *ex vivo* or *in vitro*, intended to predict response to a chemical stressor on a hazard endpoint. These new assays were often designed and intended as surrogates of traditional endpoints or models. Relevance, transferability and reliability, including reproducibility over time, needed to be established through empirical demonstration or validation by means of inter-laboratory studies. The validation and the determination of the predictive capacity of these new models for *in vivo* effects was a pre-requisite to their acceptance and use in a regulatory context. For alternative test methods to be up taken by chemical regulations, consensus was needed around clear principles and criteria, transparent practice in reporting and review of results that establish the scientific validity of a method.

2.2 Readiness of a Test Method for an Inter-laboratory Validation Programme

Although the perception of the level of readiness of a test method to enter a validation program may vary among experts/developers, the development and standardisation of the candidate test method and the availability of detailed procedure descriptions are critical to the success of a validation programme; in the absence of these, participating laboratories may have insufficient guidance for proper conduct of the test, may not keep records of important parameters, possibly leading to unexplained variations in the results. While controlled deviations in the conduct of the test are possible and useful to understand how robust the test is to small variations from the recommended procedure, monitoring of parameters and recording of effects are essential to characterize the dynamic range of the test.

Also important is the selection of the chemicals to test in the various phases of the validation, i.e. intra-laboratory, or multi- laboratories. Data generated in one laboratory are generally collected, and discussions take place on the set of chemicals to choose when evaluating the transferability of the test method, when assessing the between-laboratories reproducibility. Practical considerations are relevant for the selection of chemicals: easy access and availability, cost, known composition, analytical method available if needed, especially in the case of aquatic toxicity testing. The test chemicals should be as representative as possible of the intended applicability domain: range of physical-chemical properties, mode(s) of action, potency of chemicals to detect/identify or characterise the expected response from the test (i.e. not only potent or strong chemicals should be used).

Laboratories participating in the validation programme should be characterised by their experience in using the test or similar test procedures. It is acceptable and interesting to include naive laboratories in validation studies in order to know the level of proficiency that may be required for the successful conduct of the test, but it is important to know in advance who has experience and who has not, and how much training and guidance may be needed to transfer the know-how. In addition, an optimal design of the validation study will ensure an efficient use of resources: not all participating laboratories have to test all chemicals, it is usually considered sufficient to have three or four laboratories testing the same chemical in order to be able to assess inter-laboratory reproducibility.

Finally, the analysis and interpretation of results deserves specific attention at the stage of test method development; it is important to have predefined, clear and understandable data interpretation rules and procedures for the statistical analysis of data, rather than *a posteriori* adjusting data to an expected outcome of the test.

At the OECD level, guidance on technical aspects in the conduct of validation studies was formalised in a guidance document developed and agreed by the relevant players involved in validation (OECD 2005), to set the expected standard on good practice for validation studies, and to ensure future success and regulatory acceptance across countries of resulting Test Guidelines. This was particularly critical for *in vitro* methods intended to replace, partly or fully, existing *in vivo* test methods.

2.3 Experience at OECD with the Validation of Various Types of Test Methods

2.3.1 Test Methods for Ecotoxicity Testing

Assays measuring effects *in vivo* (in mammalian or aquatic species) have been credited for a long time for their assumed relevance and predictivity of effects to human health or wildlife species. Biological and toxicological relevance of such animal models were relatively well accepted *de facto* for hazard identification, with some exceptions. Similarly for the environment, fish, daphnia and alga have for decades represented the biodiversity of aquatic environments and formed the basis for testing chemicals to protect the aquatic environment. Demonstration of the capacity of these assays to generate valid data has very much focused on their capacity to be repeatable in laboratories implementing them. Ring-test have been organised when assays were gradually becoming more complex to implement or interpret, in particular with the introduction of e.g. more quantitative measurements, or scoring systems having inherent potential subjectivity. Countries organised some of these ring-tests at the OECD level, ensuring that the same standard operating procedures were used across participating laboratories and that data were collected and analysed in the same way (OECD 1997, 2010a). This practice of ring-testing rapidly became routine in the area of ecotoxicity testing; good practice and sound scientific principles were applied, and importantly, study results were reported transparently to regulators in support of the proposed new or updated test methods. Most of these assays were however not intended as replacement methods, and their relevance for a given protection goal was implicit.

2.3.2 Test Methods Containing Refined Procedures to Animal Testing

In the area of alternative methods, the diversity of so-called alternatives has given rise to a variety of approaches to validation. For acute toxicity for instance, a number of refinement methods based on the use of fewer animals (up-and-down procedure, acute toxic class method, fixed dose procedure) have demonstrated through statistical analysis, as the main piece of information supporting the validation status, the robustness and sensitivity of data generated using the alternative procedure (e.g. OECD 2009b). Relevance of the test procedure was not challenged in this type of alternative methods as they remained refinement of existing animal experiments.

2.3.3 Test Methods for the Detection of Endocrine Active Substances

The development of Test Guidelines for the detection of endocrine active substances emerged at the time OECD was developing a comprehensive set of validation principles and guidance for the validation and regulatory acceptance of new and updated test methods for hazard assessment (OECD 2005). This was a challenge for those involved in validation studies: while validation studies for *in vivo* and *in vitro* assays were being designed, countries were building consensus around important principles of validation in parallel, and setting good practice for how to conduct validation. The resulting guidance was generalised across new and updated *in vitro* and *in vivo* test methods. The area of endocrine disruption testing and assessment has succeeded in bringing together toxicologists and ecotoxicolgists to organise validation studies following the same principles, and testing the same chemicals. Three validation management groups (VMGs) were established approximately at the same time at OECD under the Test Guidelines Programme: the VMG-mammalian, the VMG-eco (for ecotoxicity testing) and the VMG-non animal (for *in vitro* assays). Practical challenges arose in some areas; for instance, it was not a common practice in aquatic toxicity testing to use coded chemicals. Also, some disciplines of toxicology have been required to provide clear guidance and formalise best practice through consensus OECD guidance document in areas such as histopathology for various types of organs and taxa.

Differences between types of studies (e.g. oral administration of a dose to a rat or mouse versus waterborne exposure system for fish) made it difficult for aquatic toxicity studies to show as low coefficients of variation as rodent studies. The chemical delivery to the test system in aquatic toxicity studies and the ability of the laboratory to maintain the exposure level over an extended period of time are major challenge for the success of validation studies assessing the inter-laboratory reproducibility. As a result, the inter-laboratory variability is typically higher in aquatic toxicity studies.

Furthermore, experience in laboratories and level of standardisation of test procedures varied substantially between assays that had a history of 50-years of use in the pharmaceutical industry when they entered validation studies at OECD (e.g. uterotrophic bioassay), and assays in fish measuring vitellogenin as a biomarker for estrogenicity of chemicals, which had been performed for a maximum of five years in the most advanced laboratories.

Finally, to conclude on differences between ecotoxicity and toxicology, the diversity of environmental species used in regulatory testing in OECD countries is intended to represent the biological diversity of ecosystems. This diversity makes it challenging to develop a harmonised Test Guideline that can accommodate all species using the same test procedure, but is essential for the regulatory acceptance of the Test Guideline when the goal is to protect indigenous fauna. This requirement to use countries preferred species in OECD validation studies created additional constraint on the design of the validation. Nowadays *a posteriori*, other approaches would be pursued, e.g. Performance-Based Test Guidelines, which tend to simplify the emergence of additional similar and alternative methods by setting essential components of the test method, clear goals and expected performance of the given method.

2.3.4 Test Methods Describing *In Vitro* Alternatives to Animal Testing

There is now more experience in the validation and regulatory acceptance of *in vitro* procedures, and certainly the OECD GD 34 (2005) has been beneficial in that respect, as well as all the experience gained by validation centres such as ICCVAM in the United States, ZEBET in Germany, ECVAM in the European Union and JaCVAM in Japan. Several OECD Test Guidelines have been published in the last 10 years that witness progress made in the conduct of validation studies, leading to their regulatory acceptance. Challenges are often different from *in vivo* studies, for one part because purposes are different. By providing clear mechanistic information, *in vitro* methods may pave the way to Integrated Approaches to Testing and Assessment (IATA), where data from various *in vitro* tests combined with other source of information, may lead to a reduction in the use of animals and ultimate replacement of animal testing.

Performance standards (PS) have been developed for some Test Guidelines (e.g. TG 435, TG 439, TG 455) to address two issues relating to *in vitro* test methods: (1) *in vitro* test methods often use proprietary components such as cell lines, and abuse of monopoly situations should be avoided, and (2) the emergence of similar test methods is expected to be frequent due to innovation in this area. The concept of PS was already elaborated in the OECD Guidance Document 34 on validation (OECD 2005).

Indeed, several existing *in vitro* Test Guidelines contain elements that are covered by patents and/or licensing agreements that cannot be reproduced or re-engineered, and for which fees have to be paid by the user. In the validation study, this is not an issue as such, as everyone can be requested to use the same cell line or commercial kit in order to minimise sources of variability in the results. However, the OECD policy is to enable a broad and unrestricted use of the test method at reasonable expenses for the purpose of protecting human health and the environment; situations of abuse of a monopoly for a given test method, where a single commercial provider could take a disproportionate financial advantage, are therefore avoided. For that purpose, performance standards are developed facilitating the validation of other similar test methods.

Additionally, PS can also be developed for proposed test methods that are mechanistically and functionally similar to each other. The PS include the following three elements:

- Essential test method components,
- A minimum list of reference chemicals, and
- The level of accuracy and reliability that a similar test method should demonstrate.

They are developed for the validation of future alternative or "me-too" test methods that will have to be adopted by OECD in order to be covered by the Mutual Acceptance of Data. The performance standards are based on one or several validated test methods. Any other similar "me-too" test method, whether it contains intellectual property elements or not, should meet the minimum criteria set in these PS in order to be considered for inclusion in an existing OECD Test Guideline.

The concept of Performance-Based Test Guideline (or PBTG) was developed as an elaboration of PS, in view of the variety of methods that could address the same endpoint through the same mode of action (e.g. binding to the estrogen receptor). However, test systems are not necessarily strictly similar (e.g. systems using radio-labeled elements versus non-radiolabeled systems). A PBTG (e.g. TG 455) is a TG that only provides a generic description of how the test method operates and is based on at least two validated and accepted test methods (designated the Validated Reference Method (VRM), or just "reference test method"). The test methods themselves are described in further details in annexes.

The PBTG concept has also been promoted to prevent the duplication of similar Test Guidelines covering similar test methods; it should allow faster validation of test methods addressing the same endpoint. There is still limited experience at OECD on the implementation of these new approaches that offer greater flexibility vis-à-vis innovative methods, provided they are well described, characterised, communicated, and used appropriately.

As a new test method is used, the usefulness of the test method may be expanded. It is appropriate from time to time to review and reassess the performance characteristics of established test methods. Data generated could be subjected to the same validation principles as described for a new test method if the proposed changes are significant, but it may also be appropriate to undertake a more limited assessment or review of reliability and accuracy using the established PS. The extent of the validation study or type of review that would be appropriate should be commensurate to the extent of changes proposed. In recent updates in 2013 and 2014 of OECD TG 431 on *in vitro* skin corrosion using reconstituted human epidermis, amendments have been proposed to enable the use of the test methods included in the TG for the sub-categorisation of corrosive chemicals. A statistical performance analysis (OECD 2013) has been carried out to define the predictive capacity of the methods for this purpose, without impacting the rest of the TG.

3 OECD Guidance Document on the Validation Principles and Regulatory Acceptance of New and Updated Test Methods

The development of the OECD Guidance Document 34 started in 1998 as a follow-up to the 1996 Solna Workshop on "Harmonisation of Validation and Acceptance Criteria for Alternative Toxicological Test Methods". Whereas the principles and criteria for validation and regulatory acceptance of new and revised test methods, agreed in Solna (OECD 1996) were generally accepted, the principles needed to be expanded and additional guidance provided.

The principles of the OECD Guidance Document 34 apply generally to new and updated *in vivo* or *in vitro* test methods, for effects on human health or the environment; however, some principles are more sound in the context of *in vitro* test methods that are intended as alternatives or replacement of an existing *in vivo* test. The

OECD Guidance Document 34 principles include the following points as described below: (1) the availability of a rationale for the test method; (2) description of the relationship between the test method's endpoint(s) and the biological phenomenon of interest; (3) the availability of a detailed protocol for the test method; (4) demonstration of the intra-, and inter-laboratory reproducibility of the test method; (5) demonstration of the test method's performance based on the testing of reference chemicals representative of the types of substances for which the test method will be used; 6) evaluation of the performance of the test method in relation to relevant information from the species of concern, and existing relevant toxicity data; (7) the data supporting the validity of a test method should be obtained in accordance with the principles of GLP; and (8) all data supporting the assessment of the validity of the test method should be available for expert review.

3.1 Rationale for the Test Method

A rationale for the test method should be available, and should include a clear statement on the regulatory needs in one or more countries, and the scientific justification supporting the method. The rationale can be: (1) the absence of an existing test method to address the hazard endpoint of interest, (2) the possibility to have an alternative test method that can be safer or provide better, more reliable information, or use fewer or no animals or be more cost-effective for the same level of human health or environmental protection. Here, considerations of the 3Rs (replacement, reduction, refinement) principles should be addressed.

3.2 Relationship Between the Test method's Endpoint(s) and the Biological Phenomenon of Interest

The relationship between the test method's endpoint(s) and the biological phenomenon of interest should be described. This second principle of validation is especially relevant for *in vitro* test methods intended to replace or predict an effect *in vivo*. For *in vivo* methods, the relationship is usually more direct, although in the case of biomarker endpoints, a justification based on mechanistic considerations leading to an adverse outcome is expected. It is not always possible, nor essential, for further regulatory acceptance of the test method being validated to have a deep understanding of all possible chemical interactions to their targets at various levels of biological organisation; however, existing knowledge of the relationship linking the test system being validated and response measured to the *in vivo* adverse effect should be described (e.g. similarity between the *in vitro* test system and the target tissue *in vivo*, associative or correlative relationship between the endpoint measured in the system being validated and the biological effect it intends to predict). Integrative test systems being validated (e.g. organ-level test systems such as *ex vivo* eye test) typically require less justification about their biological relevance to the biological effect of interest measured *in vivo*, while more simple *in chemico* or *in vitro* systems will require greater justification of their relationship to the target biological effect of interest. For simpler test systems, based on e.g. a cell line, a very clear understanding of their applicability and limitations (e.g. absence of metabolism) is necessary to reach regulatory acceptance.

For *in vivo* test methods, the relationship of the endpoint measured in the test system being validated (e.g. egg numbers in a fish test to predict reproductive fitness, hepatocyte enlargement via histopathology evaluation to predict liver toxicity) is often more implicit and intuitive for the determination of the toxicity *in vivo*.

As science and techniques progress, regulators may be faced with test systems that are quite sophisticated (e.g. reconstituted 3D tissue or organ) and resemble or mimic biological processes in the target organ, including its metabolic capacity. In that case, the biological relationship will be relatively straightforward to demonstrate. In other cases, as progress is made in the understanding of mechanisms of action, future test systems may be simplified to such extent that a demonstration of the biological relevance will be as critical as the demonstration of the reproducibility of test results obtained using that particular test system. This issue is easily conceivable in the case of *in chemico* test systems for which a well-calibrated experiment will be reliable over time and between laboratories due to limited number of sources of biological variability, but the demonstration of the relationship of the response measured to an effect *in vivo* will be the main challenge of the validation. In these cases, the test system will not likely be a stand-alone method, and the context of use, the applicability, limitations and possible combinations with other test systems in a more complex framework, will require careful consideration.

3.3 Detailed Test Method Protocol

A detailed protocol for the test method should be available. This principle calls for transparency in the test procedure proposed, as a pre-requisite to the success of the validation. In order for laboratories to participate in the validation, a detailed protocol including a description of the material needed, a description of what is measured and how it is measured, a description of how data need to be recorded and analysed, a description of the criteria for the acceptance of results, a template to record data are essential to enable the user to adhere to the protocol and to have means to control deviations from the prescribed procedures and report them. For the validation studies, it is important for participating laboratories to have the agreed standard operating procedures in hand prior to starting the study in order to minimise the sources of variation in the conduct of the study. Changes to the protocol that occur in the middle of the experiments will systematically lead to failure of the validation. If certain aspects of the protocol are flexible, these needs to be indicated as such in the protocol ahead of the validation studies. Problems encountered in validation studies sometimes resulted from a lack of standardisation of the protocol, leaving choice to various interpretations for those applying the test method. Obviously, there is a trade-off between having a very detailed protocol that participants have to adhere to in order to generate homogenous results across laboratories (method may then be seen as not robust in case slight deviations from protocol have a major impact on reproducibility), and having a less prescriptive protocol with some degrees of freedom in the implementation of a specific procedure that will have limited consequences on the reproducibility of the test method. From experience at OECD with the validation of a variety of *in vitro* and *in vivo* test methods, there is an increasing degree of freedom authorised in the implementation of the protocol as one goes from short *in vitro* test method validation to long *in vivo* test validation. Deviations from the test procedures will not always result in failed experiments, and the learning from deviations can inform about the robustness of the test method. The resulting OECD Test Guideline should be sufficiently robust and contain the essential elements of the test method that allow minor deviations from the validated protocol to produce reliable results. A protocol that is not sufficiently robust has limited chances of being accepted for safety testing by regulators in OECD member and partner countries. Alternatively, it will only be used in a limited number of very experienced and proficient laboratories around the world, thus limiting its broad access and opportunities for testing facilities in countries.

A clear way to analyse the response measured by the test system and a clear decision criteria are important parts of the protocol and need to be validated. The validation of these aspects of the protocol demonstrates how stable over time and between laboratories the defined decision criteria are; decision criteria should be unambiguous, reliable and sufficiently protective of human health or the environment in case of small variations are observed in the results.

The requirement for having a detailed protocol publicly available has been adapted to accommodate methods containing elements of intellectual property, for which complete disclosure is not possible in order to protect innovation. Obviously, clarity and transparency regarding essential components of the test method are still needed, for the method to be applied in a reproducible way.

3.4 Intra- and Inter-laboratory Reproducibility of the Test Method

The intra- and inter-laboratory reproducibility of the test method should be demonstrated. This is an aspect of the validation studies that has received much attention. Issues related to the minimum number of participating laboratories needed have been the subject of discussions. For example for test methods that are already well standardised, 3–4 laboratories applying the same test procedures, using the same chemicals (3–4 independent repetitions of the test) may be sufficient. If the between-laboratory results from the testing facilities are much scattered and do not overlap, an analysis is needed. Possible explanations may be: (1) the protocol is not ready for validation and there may be a need to review whether the test procedures have been sufficiently standardised to enable the assay to be reproduced across laboratories; further inter-calibration of equipment and test material might help; (2) the number of laboratories could be increased for a better characterisation of the spread of possible responses generated by the test. Conversely, performance can also show that between-laboratory results from e.g., four testing facilities do overlap to a great extent and could have been demonstrated with fewer numbers of participating laboratories. Importantly, one needs to know a priori what is the expected range of response values, what is the natural variability of the response measured, and how does this range relates to the magnitude of the response for a range of test chemicals (from weakly active to potent test chemicals). Ideally, the natural range of variability of the response can be indicated in the Test Guideline and each testing facility can build its own historical control database accordingly.

Repeatability of the results over time within the same testing facility is also part of the reproducibility assessment of the test method. Generally, for complex test methods, a number of proficiency chemicals are defined post-validation on the basis of the applicability domain and the dynamic range (i.e. spread of responses in the dataset) of the test method. Proficiency chemicals are then recommended in the OECD Test Guideline, serving as a benchmark of responsiveness of the test system when establishing the method for routine use. The proficiency chemicals are also recommended when a testing facility goes through changes in e.g. change of equipment.

The coefficient of variation (CV) or the Standard Deviation (SD) of the measured endpoint can be used for example to assess how reproducible a method is. It is not possible to give the absolute value of what an acceptable CV or SD is, because it will depend on the nature of the endpoint measured. For example for body or organ weight measurement data, intra-laboratory CVs below 20–30 % are considered achievable and acceptable (Fentem et al. 1998). However, for e.g. hormone measurement, variability is typically much higher from one test organism to another, resulting in larger SD or CV for a given group. The inter-laboratory variability will usually be higher than the intra-laboratory variability and should be considered together with other information on the performance of the test. For that reason, building an internal historical control database over time is important as an internal benchmark of the stability of the test system.

3.5 Test Chemicals

Demonstration of the test method's performance should be based on the testing of reference chemicals representative of the types of substances for which the test method will be used. The number of so-called reference chemicals and their representativity (in terms of modes of action, range of physical-chemical properties, range of possible application/use, etc.) has been and continues to be the subject of much discussion. Usually the validation management group is tasked to make a proposal on the number and identity of test chemicals, that should be representative of what toxicity(ies) the assay is expected to correctly detect (e.g. one or several modes of action, strong and weekly active chemicals, liquids and solids). In the area of traditional ecotoxicity testing, an apical endpoint is measured (e.g. survival, growth or reproduction) and no unique mode of action is involved, just baseline toxicity. In these cases a few moderately toxic chemicals are tested in a large number of testing facilities; this is usually considered adequate and sufficient to demonstrate the reproducibility of the method. However, for *in vitro* assays intended to be used as alternatives and replacement of existing *in vivo* assays that measure a range of activities (positive, negative, strongly active, weakly active, acting via several mechanisms or modes of action), it will be very important to have a good understanding of the predictive capacity of the test method being validated using a broad selection of chemicals. Ratios of sensitivity and specificity are calculated, and this cannot be done in a meaningful way if the number of chemicals is not e.g. statistically justified and/or too low (e.g., lower than about 20 for each type of expected outcome). Additionally, these chemicals need to represent a range of activities that characterise the capacity of the assay. For instance, if the assay is intended to discriminate positive versus negative chemicals for the targeted biological effect, then a range of strong and borderline positives and negatives need to be tested in the validation; if the assay is intended to discriminate among strong, moderate, weak and negative chemicals, then chemicals representative of the dynamic range of the assay need to be tested.

Coding or blind testing of chemicals is a good practice to eliminate bias where it can influence the outcome of the assay. However, the entire assay does not necessarily have to be performed blindly. Quite often, measurements that are obtained by using electronic-type equipment (i.e. where there is no possibility of subjective reading or assessment) do not necessitate coding or blind testing as an absolute requirement. Nevertheless, it is always possible to code part of the study without jeopardising the safety in the laboratory, nor involving heavy and costly management of the study. Blind evaluation of histopathology has shown to be challenging for experts involved who typically need to compare slides and have an understanding of what is the normal aspect and the lesions or findings that can be expected. Guidance documents have been developed at the OECD to share and communicate best practice in the review and peer-review of histopathology slides; these endeavours enabled to catalogue pathological findings and associate such findings with a scoring system to facilitate a semi-quantitative analysis of endpoints, with the view to decrease subjectivity of the evaluation (OECD 2009c, 2010b, c).

3.6 Performance of the Test Method

The performance of test method should be evaluated in relation to relevant information from the species of concern, and existing relevant toxicity data. This principle is particularly pertinent for alternative test methods that are intended to substitute an existing test, and for which the predictive capacity needs to be as high as possible, typically ranges of 85–90 % predictivity are achievable. For the protection of human health or the environment, the rate of false negatives should be as low as possible, to facilitate the regulatory acceptance of the validated test method. Also the selection of good *in vivo* data for the biological effect of interest is essential for an undisputable characterisation of predictivity. The *in vivo* data are typically obtained from animal tests, which can occasionally present an issue if the animal model used (e.g. rabbit) has anatomical difference compared to human (e.g. eye sac present in rabbit and not in humans), which may result in more severe effects resulting from the test than those that could be expected from human exposure to the test chemical (e.g. eye irritation test). Consequently, the database against which an alternative method is validated is only as good as the model's relevance can be, given the differences between animals and humans.

One concern with surrogate models used to make predictions over qualitative and quantitative toxicological properties of a chemical substance is the potential loss in the dynamic range of possible responses or effects compared to the target organism's response (i.e. human being, environmental species). By simplifying the test system to a tissue or a cell, sensitivity and specificity to the chemical stressor inevitably decrease. For regulators who have to ensure a sufficient level of protection, the rate of false negatives is critical for future acceptance of a test method; the rate of false positives, indicating the specificity of the method, is also important and should remain as low as possible, but less critical for the purpose of protecting human health or the environment. A structured and formal validation programme, where many chemicals having good quality *in vivo* data are carefully selected, helps generating sensitivity and specificity measures. These measures can then help determine how the validated test method can be used with confidence in a regulatory context.

3.7 Accordance with the Principles of GLP

Ideally, all data supporting the validity of a test method should be obtained in accordance with the principles of GLP. At the OECD, member countries and some non-member countries have decided to adhere to the system of Mutual Acceptance of Data by applying Good Laboratory Practice and using OECD Test Guidelines, because they see benefits for them. GLP includes quality assurance of studies performed. For regulators who have not been involved in the conduct of a study, but who bear a responsibility in how regulatory decisions are made, a system that ensures to third party(ies) that the study(ies) supporting a hazard conclusion were conducted and documented following agreed standards, is important to enable data exchange and acceptance globally. Nevertheless, GLP certification of a laboratory participating in a validation study is not a requirement.

3.8 Expert Review

All data supporting the assessment of the validity of the test method should be available for expert review. The validation report usually provides access to data summarised in a way to facilitate the evaluation by the reviewer. The statistical procedures and tests used to analyse the data should be described, so that the logic can be followed by independent statisticians if needed. Typically statisticians and reviewers will pay attention to raw data from control and treated groups and any subsequent data transform applied if needed, the mean or the median, the standard error, the pertinence of the statistical test used and the statistical difference the test can detect, and whether a pattern in the dose/concentration-response exists. For *in vitro* methods that make use of a prediction model in the interpretation of data, reviewers will be interested in the model and how it enables to predict conclusions; also the consistent treatment of equivocal results is important in building confidence in the test method. Data owners are free to publish the outcome of the validation exercise in the scientific peer-reviewed literature. However, the Working Group of the National Coordinators of the Test Guidelines Programme is keen on having access to a stand-alone validation report that usually provides more details that an article in a peer-reviewed journal. The Working Group of the National Coordinators of the Test Guidelines Programme, or a sub-group of it, also reviews the validation report and can ask for more details if needed, in particular if such information is critical to further acceptance of the Test Guideline.

At the OECD, various approaches to the review and peer-review of validation studies have been used, and are accepted by member countries, provided transparency and clarity are guaranteed. Experience shows it is not easy to find truly independent experts who have never been involved in discussions about a specific test method for a given area of hazard assessment and have no interest. Equally important to the Working Group of the National Coordinators of the Test Guidelines Programme is the transparency at all stages of the validation, and the clarity in opinions expressed in the review, including possible interest or conflict of interest of the expert providing his/her views. At the end of the review or peer-review process, the WNT takes a decision based on mutual agreement or consensus on the regulatory acceptance as an OECD Test Guideline.

Several formal peer-reviews have been organised successfully by validation centres in the United States, Japan and in Europe, in particular for *in vitro* methods. Experts from various countries participate in the peer-review panels and have dedicated meetings to discuss whether the validation has been successful in demonstrating the relevance and reliability of the method, and give an opinion about the scientific validity of the test method. The questions addressed by the peer-review panel generally mirror the validation principles and may address additional questions on specific aspects of the validation. The validation report and subsequent peer-review report or recommendations are then brought to the attention of the Working Group of the National Coordinators of the Test Guidelines Programme who approve, or not, the Test Guideline, in the light of all information available. This approach to peer-review is the most formal one, but cannot always be implemented given the resources involved, unless a country or region is paying for it. Occasionally, it has been used for the peer-review of endocrine disruption-related test methods or very new test methods (OECD 2007, 2011). Alternatively, the Working Group of the National Coordinators of the Test Guidelines Programme considers that the outcome of the validation may also be reviewed by existing

OECD expert groups who have the opportunity to discuss issues on the performance of the method and propose solutions that may facilitate regulatory acceptance. A number of validation reports supporting the development of Test Guidelines, reviewed by an Expert Group, are being endorsed by the Working Group of the National Coordinators of the Test Guidelines Programme, and published in the OECD Series on Testing and Assessment. In this way, reports supporting the validation status are referenced in the Test Guidelines and remain accessible to the public.

3.9 Conclusions

The above-mentioned validation principles are generally applicable across the range of test methods entering a validation study. The process used for the validation should remain flexible and adaptable (or modular), taking into account pre-existing information on the status of a test method, experience in performing the method, the intended purpose and use/place of the method (i.e. stand-alone replacement method, alternative method, part of a battery of assays, etc.). These preliminary considerations are useful in determining the extent of validation remaining, either prospectively or retrospectively, depending on the quality data available. Although various approaches are possible, it is important that decisions on how to conduct a validation study be guided by clear purposes for each phase of the validation. In a prospective validation study, not all purposes and questions can be addressed at once, and several phases may be necessary, typically 2–3 phases, depending on what supporting information already exist that determine the objectives of a given phase of validation. From experience at OECD with the validation of endocrine disruption-related assays, separate portions of the validation programme have been organised to address different purposes: the demonstration of inter-laboratory variability, the demonstration of the relevance of the assay for the detection of a range of chemical activities, and the blind-testing of some chemicals. Not all laboratories were necessarily involved in all portions of the validation programme, and these portions have either been performed one after the other, or in combinations. The commonality between most validation studies is the availability of a management group that defines together with the lead laboratory the gaps and the specific objectives to be addressed in a validation programme composed of several types of studies.

Over the last 20 years, validation studies have been performed by individual member countries, by test method developers, by established validation centres in countries/region or under the auspices of OECD. Most of the validation programmes have resulted in the adoption of an OECD Test Guideline, with a few exceptions. There are now several examples that can illustrate different situations, to name a few in the area of alternative test methods:

- ICCVAM evaluation of the Human Skin Corrosion test (TG 431),
- ECVAM validation of the Fish Embryo Toxicity Test (TG 236),

- JP METI validation of the Estrogen receptor-stably transfected transactivation assay (ER-STTA), for the screening of agonistic activity of chemicals (TG 455)
- US EPA validation of the Steroidogenesis Assay (TG 456).

Each of these cases has resulted in the regulatory acceptance and adoption of an OECD Test Guideline. Also, for each test method under consideration, the project to develop an internationally agreed Test Guideline can be proposed to the OECD at various stages of development. Most importantly, sponsors of test methods need to be engaged in discussions with international experts and regulators at an early stage of method development, at the OECD or elsewhere in meetings of scientific societies or expert networks and fora. These early discussions represent an important step to gauge interest from peers and regulators, to get feedback from the regulatory community on important issues to be addressed in the validation and information needed by regulators to make decisions. Each project and test method will have its specificities in terms of validation needs. Especially nowadays for some hazard areas, a number of internationally agreed Test Guidelines already exist, so the requirements for new test methods for the same hazard endpoint will be different (e.g. they will need to demonstrate superiority compared to other methods to be accepted by regulators who do not want a plethora of similar methods doing the same thing).

4 Other Elements Influencing Regulatory Acceptance

The test methods adopted as OECD Test Guidelines are intended to generate valid and high quality data to support chemical safety regulations in member countries. Advances in life sciences allow the continuous development of new and improved alternative testing methods. The regulatory acceptance of validated alternative testing methods at the OECD level represents the ultimate step leading to their regulatory implementation. Other upstream factors come into play, namely an enabling policy environment for the development of the alternative methods. In Europe, the regulatory framework for cosmetic products aims to strike the right balance among policy mechanisms that facilitate the regulatory acceptance of non-animal methods. In the last two decades, investments in research enabled the emergence of a wealth of candidate methods in particular for topical toxicity testing. The recent ban of animal testing for cosmetics in the European Union has pushed further the development of non-animal methods, while setting time pressure to get valid and acceptable test methods. The validation process has been applied to filter methods of sufficient relevance, reliability and predictive capacity. In recent examples of OECD Test Guidelines, the regulatory acceptance has only been possible when protection of human health was not jeopardized. The interpretation of negative results and the potential for a test method to generate false negatives remain difficult issues for regulators. The acceptance of negative results from a given source or test is generally better accepted when the results are interpreted in a framework where other available sources of information and possible alternative tests are also integrated and show concordant results. This means that test methods tend to be less and less regarded as stand-alone: the context, the purpose, the mechanistic understanding, the predictive capacity (including the rate of false negatives) and possible agreed and harmonised testing strategies to combine sources of information are key elements for further regulatory acceptance.

For more complex endpoints such as reproductive toxicity and carcinogenicity, the tasks remaining to be undertaken are daunting. Further efforts to understand toxicity pathways and to build integrated approaches to testing and assessment (IATA) are needed, in supplement to rigorous validation of individual methods. These efforts will set the scientific basis and provide the context under which alternatives to animal testing can be consistently and safely applied by regulatory bodies. Recently the Syrian Hamster Embryo Cell Transformation Assay, went through a validation programme, but it still not currently accepted as an OECD Test Guideline. One apparent reason was the lack of framework to guide regulators on how this assay could be used in the regulatory assessment of substances, mainly non-genotoxic carcinogens. Also, the limitation in the predictive capacity and the limited understanding of the underlying mechanisms of action appeared to hamper its regulatory acceptance. There are certainly lessons to be learnt for future test methods aiming to address complex hazard endpoints. In particular there is a need for a careful consideration of the regulatory context, possible purpose and use of a test method and data generated, and its applicability domain.

5 Challenges Ahead

5.1 Complex Endpoints Need Integrated Approaches to Testing and Assessment (IATA)

For more complex endpoints, alternative methods have to be combined in some ways to provide meaningful predictions. If possible, mode(s) of action will have to be known or postulated for hypotheses and toxicity pathways to be elaborated. The validation will then be very useful to demonstrate the relevance of the method and its utility for a given regulatory purpose. Obviously, the reliability of the method will also have to be established, but provided the procedures are well described the reproducibility of assays tends to be more straightforward with improved techniques, properly calibrated equipment and standardised practices.

Having a clear scope and realistic goal for use of a test method also facilitates its future regulatory acceptance. Experience at the OECD with some assays has shown that regulatory acceptance is hampered by unrealistic or changing objective over the course of validation and Test Guideline development. Also, when the expected need and possible of the assay are not well defined, the selection of reference chemicals cannot be optimal and this can cause problems in meeting the objectives of a validation study.

It is illusory to believe that all mechanisms of action will soon be discovered and that future alternative test methods will all be mechanistically-based. Although this is a wishful thinking for what the future should be, test methods developers and regulators will continue to rely on alternative methods where mechanisms are yet unknown, but utility and predictive capacity of the method are experimentally established for a given hazard endpoint. Beyond experimental validation, given the number of alternative methods addressing the same endpoint, developers and regulators should endeavour to talk together to agree on frameworks of application, describing how they complement each other and which test should be used under what circumstances. These frameworks, also called integrated approaches to testing and assessment (IATA) can be articulated or not around modes of action/adverse outcome pathways. Frameworks can also be developed in the absence of complete knowledge on mode of action or adverse outcome pathway, as long as it proposes a harmonised, meaningful and efficient use of methods to reach a conclusion. At OECD, these IATAs are the best place where methods and approaches to testing can be explained, with their advantages and limitations, and where harmonised testing strategies can be proposed (OECD 2014). For regulators, it is also re-assuring that despite a choice of possible alternative methods, they are guided by an agreed framework for their application.

5.2 High Throughput Screening (HTS) Assays May Need a Streamlined Validation Process

High-throughput screening techniques are more and more used beyond research and development, and experience gained to date raise expectations that results of these techniques may be used for screening and priority setting of chemicals in regulatory programmes. From a manual method to the equivalent (ultra) high-throughput methods, the principle of the test (e.g. binding to a receptor) and material used (e.g. transformed cell line) may not differ substantially, thus the relevance of the test in itself remains the same regardless of the throughput level. The main hurdles in the validation may be of a technical nature, and validation principles may need to be revisited and adapted to these new techniques.

For HTS assays that have an equivalent *in vitro* manual assay validated, the validation is greatly facilitated by having a well-defined list of reference chemicals that has been used in the validation of the equivalent manual method and possibly other similar methods. An adequate calibration of equipment used, a sufficient number of internal, positive, negative controls, are important to the success of the reliability of the test system. Given the small volumes of individual test chambers (i.e. microwells on the plates) and the high degree of robotisation, an issue could be the higher variability impacting accuracy of the results to predict a biological response. This may potentially be compensated by a higher level of standardisation and precision enabled by the robotisation of procedures and lesser human handling. For HTS assays that have no equivalent manual *in vitro* assay validated, one consideration might be whether validation of the manual method is a pre-requisite, whether it will facilitate the validation of the HTS method, or whether it is superfluous and not needed for further regulatory acceptance.

Furthermore, there is a limited number of testing facilities around the world equipped to perform these advanced techniques, given costly investments implied. These facilities are expected to represent highly performing laboratories where all calibration and quality control procedures are in place and working well. Provided this assumption is correct, the reproducibility of HTS screening techniques across laboratories should not be the main challenge of the validation process. As these tools will be used on large numbers of chemicals for screening purposes, it will be important to have assays with large applicability domains or multiple assays that together cover a large applicability domain, to build confidence of regulators that the test system does not miss positive effects. As a consequence, a streamlined approach to validation may be needed to address these relevant aspects. Testing more chemicals in fewer laboratories would make sense for an efficient use of resources (Judson et al. 2013). Other principles of the validation, such as having a detailed protocol and a description of the relationship between the test methods endpoint(s) and the biological phenomenon of interest, certainly remain important pre-requisites for any future acceptance of methods as OECD Test Guidelines, if such methods are expected to be covered by the Mutual Acceptance of Data.

6 Conclusions and Concepts to Preserve

Existing OECD-agreed validation principles will most likely generally remain relevant and applicable to address challenges associated with the validation of future test methods. Some adaptations may be needed, but demonstration of relevance and reliability will continue to play a central role as pre-requisite for the regulatory acceptance. Despite the fact that methods and techniques are getting more and more sophisticated and require a good level of proficiency, having harmonised standards for generating reliable results globally remain an important goal for the efficient use of resources. The Mutual Acceptance of Data among OECD member and partner countries is essential to maintain efficiency in testing and assessment of chemicals; trustable methods and harmonised approaches will continue to be needed. It is also important to continue to promote the OECD validation principles globally so that new techniques and assays emerging from science are supported by a good quality data generated using best practice to appraise their utility, potential for validation, and further regulatory acceptance.

References

Balls M et al (1990) Report and recommendations of the CAAT/ERGATT workshop on the validation of toxicity test procedures. ATLA 18:313–337

Balls M et al (1995) Practical aspects of the validation of toxicity test procedures: the report and recommendations of ECVAM workshop 5. ATLA 23:129–147

Fentem JH et al (1995) Validation, lessons learned from practical experience. Toxicol *In Vitro* 9(6):857–862
Fentem JH et al (1998) The ECVAM international validation study on *in vitro* tests for skin corrosivity. 2. Results and evaluation by the Management Team. Toxicol *In Vitro* 12:483–524
Judson R et al (2013) Perspectives on validation of high-throughput assays supporting 21st century toxicity testing. ALTEX 30(1):51–66
OECD (1981) Decision on the Mutual Acceptance of Data in the Assessment of Chemicals [C(81)30/Final]
OECD (1996) Report of the OECD workshop on "Harmonisation of validation and acceptance criteria for alternative toxicological test methods" (Solna report). OECD, Paris, 60 pp [ENV/MC/CHEM(96)9]
OECD (1997) Report of the final ring test of the Daphnia Magna Reproduction Test, OECD/GD(97)19. OECD, Paris
OECD (2005) Guidance Document on the Validation and Regulatory Acceptance of New and Updated Test Methods for Hazard Assessment, Series on Testing and Assessment, No. 34 [ENV/JM/MONO(2005)14]. OECD, Paris
OECD (2006) Guidance Document on the Development of OECD Guidelines for the Testing of Chemicals [ENV/JM/MONO(2006)20/REV1], Series on Testing and Assessment No. 1. OECD, Paris
OECD (2007) Report of the Validation Peer Review for the Hershberger Bioassay, and the Agreement of the Working Group of the National Coordinators of the Test Guidelines Programme on the Follow-up of this Report [ENV/JM/MONO(2007)34], Series on Testing and Assessment No. 85. OECD, Paris
OECD (2009a) Guidance Document for the Development of OECD Guidelines for Testing of Chemicals [ENV/JM/MONO(2006)20/REV1], Series on Testing and Assessment, No. 1. OECD, Paris
OECD (2009b) Report on Biostatistical Performance Assessment of the Draft TG 436 Acute Toxic Class Testing Method for Acute Inhalation Toxicity (ENV/JM/MONO(2009)9], Series on Testing ad Assessment No. 105. OECD, Paris
OECD (2009c) Guidance Document for Histologic Evaluation of Endocrine and Reproductive Tests in Rodents [ENV/JM/MONO(2009)11], Series on Testing and Assessment, No. 106. OECD, Paris
OECD (2010a) Report of the Validation of a Soil Bioaccumulation Test with Terrestrial Oligochaetes by an International ring test [ENV/JM/MONO(2010)33], Series on Testing and Assessment, No. 134. OECD, Paris
OECD (2010b) Guidance Document on the Diagnosis of Endocrine-related Histopathology in Fish Gonads [ENV/JM/MONO(2010)14], Series on Testing and Assessment No. 123. OECD, Paris
OECD (2010c) Guidance Document on Histopathology for Inhalation Toxicity Studies, Supporting TG 412 (Subacute Inhalation Toxicity: 28-day Study) and TG 413 (Sub-chronic Inhalation Toxicity: 90-day Study) [ENV/JM/MONO(2010)16], Series on Testing and Assessment No. 125. OECD, Paris
OECD (2011) Peer Review Report of the Validation of The Skin Irritation Test Using LabcyteEpi-Model 24 [ENV/JM/MONO(2011)144]. Series on Testing and Assessment, No. 155. OECD, Paris
OECD (2013) Summary Document on the Statistical Performance of Methods in OECD Test Guideline 431 for Sub-categorisation [ENV/JM/MONO(2013)14], Series on Testing and Assessment No. 190. OECD, Paris
OECD (2014) Guidance Document on Integrated Approaches to Testing and Assessment for Skin Irritation and Corrosion [ENV/JM/MONO(2014)19]. Series on Testing and Assessment, No. 203. OECD, Paris

Chapter 3
Regulatory Acceptance of Alternative Methods in the Development and Approval of Pharmaceuticals

Sonja Beken, Peter Kasper, and Jan-Willem van der Laan

Abstract Animal studies may be carried out to support first administration of a new medicinal product to either humans or the target animal species, or before performing clinical trials in even larger populations, or before marketing authorisation, or to control quality during production. Ethical and animal welfare considerations require that animal use is limited as much as possible. Directive 2010/63/EU on the protection of animals used for scientific purposes unambiguously fosters the application of the principle of the 3Rs when considering the choice of methods to be used.

As such, today, the 3Rs are embedded in the relevant regulatory guidance both at the European (European Medicines Agency (EMA)) and (Veterinary) International Conference on Harmonization ((V)ICH) levels. With respect to non-clinical testing requirements for human medicinal products, reduction and replacement of animal testing has been achieved by the regulatory acceptance of new *in vitro* methods, either as pivotal, supportive or exploratory mechanistic studies. Whilst replacement of animal studies remains the ultimate goal, approaches aimed at reducing or refining animal studies have also been routinely implemented in regulatory guidelines, where applicable. The chapter provides an overview of the implementation of 3Rs in the drafting of non-clinical testing guidelines for human medicinal products at the level of the ICH. In addition, the revision of the ICH S2 guideline on genotoxicity testing and data interpretation for pharmaceuticals intended for human use is discussed as a case study.

S. Beken (✉)
Division Evaluators, DG PRE Authorisation, Federal Agency for Medicines and Health Products (FAMHP), Victor Hortaplace 40/40, Brussels 1060, Belgium
e-mail: sonja.beken@fagg-afmps.be

P. Kasper
Federal Institute for Drugs and Medical Devices (BfArM), Kurt-Georg-Kiesinger Allee 3, Bonn 53175, Germany

J.-W. van der Laan
Pharmacology, Toxicology and Biotechnology Department, Medicines Evaluation Board (MEB), Graadt van Roggenweg 500, 3531 AH Utrecht, The Netherlands

C. Eskes, M. Whelan (eds.), *Validation of Alternative Methods for Toxicity Testing*, Advances in Experimental Medicine and Biology 856,
DOI 10.1007/978-3-319-33826-2_3

In October 2010, the EMA established a Joint *ad hoc* Expert Group (JEG 3Rs) with the mandate to improve and foster the application of 3Rs principles to the regulatory testing of medicinal products throughout their lifecycle. As such, a Guideline on regulatory acceptance of 3R testing approaches was drafted that defines regulatory acceptance and provides guidance on the scientific and technical criteria for regulatory acceptance of 3R testing approaches, including a process for collection of real-life data (safe harbour). Pathways for regulatory acceptance of 3R testing approaches are depicted and a new procedure for submission and evaluation of a proposal for regulatory acceptance of 3R testing approaches is described.

Keywords ICH • EMA • JEG 3Rs • Regulatory testing • Non-clinical • Genotoxicity • Pharmaceuticals • Reduction • Replacement • Refinement

1 Introduction

To comply with Directives 2001/83/EC (Directive 2001a) and 2001/82/EC (Directive 2001b) and their associated Guidelines, non-clinical[1] testing to support clinical trials as well as marketing authorisation of human and veterinary medicinal products often requires the use of laboratory animals. In addition, animal studies may be used to control quality during production of the medicinal product. Ethical and animal welfare considerations require that animal use is limited as much as possible.

In this respect, Directive 2010/63/EU (Directive 2010) on the protection of animals used for scientific purposes is fully applicable to regulatory testing of human and veterinary medicinal products.[2] Directive 2010/63/EU unambiguously fosters the application of the principle of the 3Rs (replacement, reduction and refinement) by stating in article 4 that:

1. Member States shall ensure that, wherever possible, a scientifically satisfactory method or testing strategy, not entailing the use of live animals, shall be used instead of a procedure.[3]
2. Member States shall ensure that the number of animals used in projects is reduced to a minimum without compromising the objectives of the project.

[1] Referred to as safety testing in marketing authorisation applications for veterinary medicinal products.

[2] With the exception of clinical trials for veterinary medicinal products, which are specifically excluded from the scope of the directive.

[3] A 'procedure' means any use, invasive or non-invasive, of an animal for experimental or other scientific purposes, with known or unknown outcome, or educational purposes, which may cause the animal a level of pain, suffering, distress or lasting harm equivalent to, or higher than, that caused by the introduction of a needle in accordance with the good veterinary practice (Directive 2010).

3. Member States shall ensure refinement of breeding, accommodation and care, and of methods used in procedures, eliminating or reducing to the minimum any possible pain, suffering, distress or lasting harm to the animals.

The choice of methods is to be implemented according to article 13 which states that:

1. Without prejudice to national legislation prohibiting certain types of methods, Member States shall ensure that a procedure is not carried out if another method or testing strategy for obtaining the result sought, not entailing the use of a live animal, is recognised under the legislation of the Union.
2. In choosing between procedures, those which to the greatest extent meet the following requirements shall be selected:

   (a) use the minimum number of animals;
   (b) involve animals with the lowest capacity to experience pain, suffering, distress or lasting harm;
   (c) cause the least pain, suffering, distress or lasting harm; and are most likely to provide satisfactory results.

The application of all 3Rs is currently embedded in the drafting process of regulatory guidance both at the European and at International Conference on Harmonisation ((V)ICH) level. With respect to non-clinical testing requirements for human medicinal products, over the past years, new *in vitro* methods have been accepted for regulatory use via multiple and flexible approaches, either as pivotal, supportive or as exploratory mechanistic studies, wherever applicable. Whilst replacement of animal studies remains the ultimate goal, the application of all 3Rs needs to be the focus. As such, approaches aiming at reducing or refining animal studies are and have been routinely implemented in regulatory guidelines, where applicable.

This chapter provides an overview of the implementation of 3Rs in the drafting of non-clinical testing guidelines for human medicinal products at the level of the ICH. The revision of the ICH S2 guideline on genotoxicity testing and data interpretation for pharmaceuticals intended for human use will be discussed in more detail as a case study. Finally, the approach from European Medicines Agency (EMA) to regulatory acceptance of 3Rs testing approaches is specifically highlighted.

2 Critical View on 3Rs at the Level of ICH

2.1 ICH and 3Rs

"*In Europe as in other parts of the world, you will be aware that real concern has been expressed regarding the testing of medicinal and many other products, on animals. … Certain indispensable testing procedures on animals must therefore be accepted. It is nevertheless absolutely clear that we should only tolerate testing of animals where it can be shown to be scientifically justified and of relevance to the marketing authorization decision.*"

This citation is taken from the third page of the opening speech by Dr. M. Bangemann, at that time vice-president of the European Commission (CEC) at the first ICH in Brussels in 1991 (Bangemann 1992). It highlights the interest in reducing the use of animals for toxicological testing of human pharmaceuticals from the very first beginning of the International Conference on Harmonisation of technical requirements for registration of pharmaceuticals for human use (abbreviated as ICH).

2.1.1 Start of ICH

When a delegation of the European Commission together with European Pharmaceutical Industry visited Japan, the history of ICH had its definite start. Differences in technical requirements for pharmaceuticals for human use were identified as being a stumbling block in the cooperation between the two economic parts in the world (Arnold 1992).

The participants started a discussion on such differences between regulatory agencies, because of the fact that the agencies in their own region had the same duty, i.e. ensuring the safety, quality and efficacy of the medicines for humans on their respective markets. By reducing these differences, the costs of developing promising new pharmaceuticals could go down.

The European Community elaborated the project further with US and its regulatory authority, the Food and Drug Administration (FDA) with its Center for Drug Evaluation and Research (CDER) and its Center for Biologics Evaluation and Research (CBER). In October 1989 in Paris, the project received the green light to proceed.

Dr. Bangemann clearly expressed that animal testing should be kept to a minimum. Animal testing is still needed to ensure safety of humans. Therefore, the ICH should strive to reduce the use of animals as much as possible. During the ICH-process it has been requested that each Expert Working Group reporting to the Steering Committee, gives explicit attention to this aspect.

At the first ICH in Brussels, Michael Perry identified four topics (Perry 1992).

- The Toxicity Testing Program: short and long term toxicity testing and carcinogenicity
- Reproductive toxicology
- Biotechnology
- The timing of toxicity studies in relation to the conduct of clinical trials

These topics have been discussed intensively during this first ICH-meeting, and recommendations have been given for the follow-up. In this chapter, these recommendations and their follow-up will be discussed in the sequence of the ICH numbering. We refer to the proceedings of the ICH-conferences. Five 2-yearly conferences have been held and the discussions minuted.

2.2 *Individual ICH Guidelines and Their Impact on 3Rs*

2.2.1 Acute Toxicity as Refinement and Reduction

The first toxicological issue at ICH1 in Brussels was "single dose toxicity". The discussions in ICH Expert Working Groups have not been reported in the proceedings of ICH1 in a large detail. Only the conclusions were presented. Apparently, the issue of single dose toxicity was well known at that time. Testing of single dose toxicity is usually causing severe pain and suffering to the animals and the usefulness of the data is low. Therefore, several authorities did not require at that time LD50s for the estimation of acute toxicity of a pharmaceutical for human use. Detailed information about the toxicity profile was (and still is) considered more important than a more or less precise estimate of the dose resulting in the death of 50 % of the animals.

The background paper in the ICH1-proceedings clearly states that "*the classic LD50 determination is no longer a formal requirement for single-dose toxicity testing in any of the three regions*". An increasing-dose tolerance study is recommended in two mammalian species (Perry 1992).

The Japanese authorities further reduced their requirements by explicitly indicating the number of rodent species could be reduced from 2 to 1, and only an approximation of the lethal dose was required. For non-rodents, toxicity features would be sufficient, and dosing up to lethality not needed. The latter measure would also enable the repeated use of an animal, as histopathology is not needed after a single dose (Ohno 1992). While these developments reflect the focus of ICH on 3R's, it has to be said that they cannot only be due to the existence of ICH, but should be seen in relation to developments that started already earlier in various Regulatory Authorities. However, these first statements on single dose toxicity help to further reflect on the emphasis of ICH on the 3Rs, especially focusing on Reduction and Refinement in the very first meetings of the ICH.

Recent Developments

Recently, the need for single dose toxicity studies has been discussed again and with the revision of the ICH M3 guideline in 2009, the general request for a single dose toxicity study was dropped from the list altogether. Acute toxicity information can be derived from appropriately conducted dose-escalation studies or short duration dose ranging studies defining a maximum tolerated dose (MTD) in an animal species. Single dose toxicity studies are only needed where there is no need for a repeated dose toxicity studies, e.g. with diagnostic drugs that are expected to be given only once clinically. Even non-GLP studies contribute to acute toxicity information if they are supported by data from other studies in compliance with GLP.

2.2.2 Carcinogenicity Testing

The classical approach of carcinogenicity testing requests the use of two species, rats and mice; and a carcinogenicity assay requires testing of four groups (control and three dosages up to the MTD) with at least 50 males and 50 females per group, tested for 18 months (mice) or 2 years (rats). In general, carcinogenicity testing leads to the use of around 1200–1600 animals (rats and mice together), and accounts for 40–50 % of the total number of animals used to characterize the safety of a new individual compound. This approach is based on the classical presumption that the development of cancer is a process of chance. The chance is maximal with a lifetime exposure at the MTD. However, the differentiation between genotoxic and non-genotoxic mechanisms of action has led to changes in the approaches to study carcinogenic potential, but it is not within the scope of this chapter to extend further on this issue.

The area of carcinogenicity testing focused rather on refinement and reduction than on replacement.

An important aspect was the dose-selection in the study design. The resistance against a rigid application of the MTD became a driving force for a separate guideline in this field, which became clear when the first S1-Guideline was nr. S1C Dose selection. Discussions in Brussels were on "*high-dose selection*", and on "*survival*" of the animals.

High-dose selection. In the same period as developing new ICH guidelines on carcinogenicity testing, another topic was the development of guidelines for toxicokinetics (ICH S3) (see Sect. 2.2.3). Analytical assays became more sensitive and generally applicable. Companies were therefore required to conduct the determination of exposure, e.g. via plasma or serum levels of the compound or its metabolite(s), and not to rely only on a theoretical dose extrapolation based on the velocity of the basal metabolism.

The ICH Expert Working Group on Carcinogenicity finally proposed several endpoints to determine the maximum dose in a carcinogenicity study, with the main criterion being the pharmacokinetics, i.e. a 25-fold ratio of the AUC in humans at the intended therapeutic dose. Decisions on the details of the study design of a carcinogenicity study would be taken at a stage that these pharmacokinetic data from humans should be known, namely at the end of Phase 2 of development. Contrera et al. (1995) showed that applying an MTD approach led to very high exposure ratios as compared to human exposure in approximately 30 % of the cases. The limit of 25-fold the human AUC would therefore lead to reduction of animal exposure, and improvement of animal welfare (Refinement). It should be kept in mind, however, that this approach is applicable to only a small part of the carcinogenicity studies. From the dataset of Contrera et al. (1995) it is clear that in many cases the exposure in animals might not be as high as compared to the intended therapeutic exposure. Safety margins cannot always be established, and the clinical "tolerance" is leading more than a toxicological approach in setting safe doses.

Other endpoints are e.g. saturation of exposure, and pharmacodynamic response.

Need for carcinogenicity studies. Another possibility to reduce the number of animals used for carcinogenicity was to agree on a better definition on the need for carcinogenicity studies, with an emphasis on when a study would not contribute to further risk assessment. In the ICH S1A Guideline it was defined that in case of unequivocal genotoxicity this risk could be sufficiently determined by the short-term genotoxicity assays as explained in ICH S2. A full dataset on carcinogenicity based on 2-year studies in two species would not add any value to establish the risk for such a compound, and these 2-year studies should therefore not be conducted. Again, the emphasis of the ICH was on reduction of the use of animals.

Species selection. An important discussion in this area was the need for rats and mice for testing of carcinogenic properties. An evaluation was started of the history of carcinogenicity studies with human pharmaceuticals (Van Oosterhout et al. 1997; Contrera et al. 1997). Important data became available in this respect. The EU EMA (Safety Working Party—SWP) concluded that the outcome of mouse carcinogenicity studies did not contribute to the weight of evidence of carcinogenicity assessment of human pharmaceuticals. Mechanistic studies in rats were seen as more important than additional data from mice. Therefore, the EU proposed to skip the mouse as a second species.

However, the FDA could not accept this proposal, as a few compounds would exist for which mouse data could not be dismissed in carcinogenicity assessment, causing uncertainty about the irrelevance of the mouse study. Because of this, the position to skip the mouse as a testing species could not be maintained by the EU in the negotiations with US FDA and Japanese MHLW (Van der Laan 2013).

In the ICH S1B guideline (ICH 1997) a compromise was formulated indicating that in the testing strategy for carcinogenicity, the rat is the preferred species, with a second study using either normal mice (with a 2-year study) or transgenic mice with a knock-out p53 gene (tumor-suppressor gene) or a knock-in RasH2 gene (oncogene).

From a 3Rs point of view the introduction of these genetically modified animals was interesting. As genetically modified animals already carry an induced mutation the induction of specific tumours is supposed to occur earlier in life (6–12 months) and in certain organs/tissues only. Therefore, the use of less animals per dosing group should be possible to obtain a statistically significant result. Initially groups of 15 animals were used, but for screening of unknown compounds, groups of 25 animals per dose are recommended. It allowed sponsors to use a maximum of 160–200 animals for 6 months instead of 400–500 animals required for a 2-year study.

The S1B guideline came into force in 1997/8 (see Table 3.1), and suggested the use of these models, but at that time the models were not evaluated yet. Under auspices of the ILSI-HESI Alternatives to Carcinogenicity Testing Technical Committee (ACT-TC), the use of these mouse strains has been evaluated rather than validated, based on an agreed set of compounds (Robinson and MacDonald 2001). FDA and EU have explicitly accepted the use of the heterozygous p53 mice, as well as the TGRasH2 mice for genotoxic and non-genotoxic compounds (for review see Nambiar and Morton 2013).

Table 3.1 List of ICH Safety Guidelines developed up to 2014

Topic	Reference number	Publication date	Effective date
S 1 Regulatory notice on changes to core guideline on rodent carcinogenicity testing of pharmaceuticals	EMA/CHMP/ ICH/752486/2012	Sept 2013	Sept 2013
S 1 A The need for carcinogenicity studies of pharmaceuticals	CPMP/ICH/140/95	Dec 1995	July 1996
S 1 B Testing for carcinogenicity of pharmaceuticals	CPMP/ICH/299/95	Sept 1997	March 1998
S 1 C (R2) Dose selection for carcinogenicity studies of pharmaceuticals	CPMP/ICH/383/95	April 2008	Oct 2008
S 2 (R1) Guidance on genotoxicity testing and data interpretation for pharmaceuticals intended for human use	CHMP/ICH/126642/08	Dec 2011	June 2012
S 3 A Toxicokinetics: A guidance for assessing systemic exposure in toxicology studies	CPMP/ICH/384/95	Nov 1994	June 1995
S 3 B Pharmacokinetics: Guidance for repeated-dose tissue-distribution studies	CPMP/ICH/385/95	Nov 1994	June 1995
S 4 Duration of chronic toxicity testing in animals (rodent and non-rodent toxicity testing)	CPMP/ICH/300/95	Nov 1998	May 1999
S 5 (R2) Detection of toxicity to reproduction for medicinal products and toxicity to male fertility	CPMP/ICH/386/95	Sept 1993	March 1994
S 6 (R1) Preclinical safety evaluation of biotechnology-derived pharmaceuticals	CHMP/ICH/731268/1998	July 2011	Dec 2011
S 7 A Safety pharmacology studies for human pharmaceuticals	CPMP/ICH/539/00	Nov 2000	June 2001
S 7 B The non-clinical evaluation of the potential for delayed ventricular repolarisation (QT interval prolongation) by human pharmaceuticals	CPMP/ICH/423/02	May 2005	Nov 2005
S 8 Immunotoxicity studies for human pharmaceuticals	CHMP/ICH/167235/04	Oct 2005	May 2006
S 9 Non-clinical evaluation for anticancer pharmaceuticals	CHMP/ICH/646107/08	Dec 2009	May 2010
S 10 Guidance on photosafety evaluation of pharmaceuticals	CHMP/ICH/752211/2012	January 2014	June 2014

Till now, surprisingly, most companies are still conducting a 2-year mouse study, and have not included genetically modified mice in their testing strategy. Some experts indicate that this could be due to the uncertainty about the outcome of these alternatives. Unexpected findings in normal mice can be readily explained, but in case of unexpected findings in transgenic mice for which no reasonable explanation can be found, this might be the death of a compound.

Recent Developments

A new process is ongoing in this area, again with a focus on reduction, rather than on replacement. A consortium of 13 pharmaceutical industries has compiled data from 182 compounds on 2 year carcinogenicity studies in the rat compared to 6 months repeat dose toxicity studies in the same rat strains. Negative predictivity of the 6 months data was defined as the absence of signs in this time period (hyperplasia, hypertrophy, hormonal effect) associated with the absence of tumours in 2-year studies of the same rat strain (Sistare et al. 2011). This absence of tumours did occur in 80 % of all compounds, which were negative after 6 months.

PhRMA suggested that conducting a full-term carcinogenicity study of 2-years duration does not add value when the prediction of absence of tumours would be so high. The organization raised a plea to the regulators to take this on board as a way to reduce the use of animals.

In the EU, the pharmacological data received attention, especially because of the false negatives in the Sistare paper (Sistare et al. 2011) (i.e. those compounds negative after 6 months, but inducing tumours after 2 years). What would be the cause for the tumours that showed up in those cases? As such, another strategy was introduced based on the pharmacology of the compounds. Evaluation of the pharmacology of the compounds in relation with the tumours induced in specific organs gave the confirmation that in nearly all cases, in rats, the tumours are related to their pharmacological action (Van der Laan et al., manuscript in preparation). Starting from this viewpoint not only a negative (Sistare et al. 2011) but also a positive prediction is expected to be possible. In a rather unique regulatory experiment, the Regulatory Authorities involved in ICH are now working together with industry to evaluate virtual waiver requests for carcinogenicity assays. Companies are expected to write a Carcinogenicity Assessment Document (CAD) to support a potential waiver request (or a justification why a study should be conducted), and Regulatory Authorities are evaluating these as if they were real requests for waiving a 2-year rat study. The virtual waivers will then be compared with the outcome of the study afterwards. It is the intention to evaluate around 50 of such cases, and then to conclude whether a revision of the S1 Guidelines would be possible, to allow waivers of life time studies to be granted in real time (ICH 2013).

2.2.3 Genetic Toxicity Testing

Genotoxicity testing has always been an example of how *in vitro* approaches can be included as an important part of the testing strategy. The most well-known test for genotoxicity is the Ames-test, a bacterial mutagenicity assay.

The discussions in ICH were on the so-called testing battery approach. However, it took several years to come to agreement on which assays should populate the battery itself. This was based on the choice of predictive endpoints, and the extent of testing with positive and negative compounds. Are all negative cases really negative (otherwise compounds with a risk would not be stopped), and vice versa, are all positive compounds really carrying a risk (otherwise compounds will be stopped unnecessarily in their development)?

Therefore, in the first ICH S2 guideline (S2A) only details of testing procedures have been discussed, while only in a second stage (1997) a standard battery has been defined as consisting of 2 *in vitro* assays (Ames test and mammalian cell assay), and an *in vivo* assay at an adequate dose (Müller et al. 2013). The *in vivo* assay has the status of confirmation of the *in vitro* data, which are more sensitive, but sometimes oversensitive. In the early ICH days the discussion was whether one or two mammalian cell assays should be included, especially to have a mammalian mutation endpoint, but the *in vivo* test was included without debate.

In relation to the clinical trial stage the *in vitro* approaches have more or less a stand-alone status. If both are negative then one is able to proceed with the first-into-human clinical trial. An *in vivo* study is still required in the next stage of clinical development.

Also in the recent revision of the ICH S2 guidelines (bringing them together into one guideline) the *in vivo* assay remains important, and in fact an option was introduced to skip one mammalian cell assay. From a 3Rs point of view, however, the revised guidance emphasises that the *in vivo* genotoxicity endpoint can also be included in a repeated dose toxicity study, allowing a combination of these studies and as a consequence a reduction of the number of animals used.

For a detailed discussion see Sect. 3.

2.2.4 Toxicokinetic Testing

The ICH S3 Guideline started to emerge after the first ICH. It was at that time a breakthrough in the thinking on risk assessment, not only focusing on dose, but rather on exposure, leading to more insight in the extrapolation from animal species to humans. Not only exposure, but also the role of metabolites became clearer. The emphasis on toxicokinetics was rather unique in the regulatory field, and was possible because of the intended use of the compounds, i.e. for therapeutic purposes. The knowledge of human exposure and pharmacokinetics is an enormous advantage in the risk assessment of pharmaceuticals as compared with other chemicals, such as pesticides where human exposure should be avoided. The intended use of pharmaceuticals in humans allows the design of Phase 1 studies with a first estimate of human pharmacokinetics.

However, from the viewpoint of the 3Rs, this guideline was a disaster. The need for frequent blood sampling was not compatible with the principles of toxicological screening. Repeated blood sampling of volumes of 500 μL would impact on the health of the animal irrespective of the exposure to the compounds. Determination of the toxicokinetics is important to estimate the exposure during the course of the study, especially in chronic studies, where induction of metabolism and accumulation of metabolites might occur.

Adding satellite groups to the various dose groups was adding high numbers of animals to the study design, especially in case of rats and mice. For larger animals such as dogs and monkeys this could be handled usually with the same animals.

However, there was also an awareness of unnecessary use of animals, i.e. in the Japanese requirement for repeated dose tissue distribution studies. Ohno (2013) has highlighted the role of 3Rs in the Japan-driven input in ICH. In this respect he refers to the ICH 3B, the need for repeated dose tissue distribution studies because of the risk of accumulation. These studies were emphasized in Japan, but not in EU and US. Through the ICH process it was better defined in which cases the sponsor should conduct a repeated dose tissue distribution study using radiolabelled material, i.e. when the half-life is longer than two times the dosing period. This ICH-driven guideline therefore resulted in a reduction of the use of animals for this type of studies.

Recent Developments

Recently, the technique of microsampling has become more common, and the ICH has decided in 2014 to include this approach in a Q&A document belonging to S3. This may lead to important reductions in animal use.

Microsampling might be applied even in the case of monoclonal antibody administration in mice (Marsden, personal communication).

2.2.5 Repeated Dose Toxicity Testing

Redundancy of studies and increasing animal welfare (or reducing animal suffering) was the topic for S4. The focus was on the duration of the so-called chronic toxicity studies, in the area of human pharmaceuticals; i.e. the request for chronic rodent studies, 6 and 12 months, and the duration of chronic non-rodent studies, 6 versus 12 months.

For the rodent studies (usually the rat) there was a quick win: a 12 months-study could easily be dropped from the list of requirements, as it was clear that for compounds warranting a 12 months study also a 24 months was required in the framework of testing the carcinogenic potential. With having toxicological endpoints from a 24-months study, especially detailed histopathology of non-cancerous tissue, data from a 12 months study duration would be redundant.

With respect to the studies in non-rodents, a more difficult discussion took place. The US FDA wanted to maintain their requirement for a 12 months study in non-rodents, while in the EU, the duration of the study (6 months) was included in the legislation i.e. in then applicable EU Directive 75/318 (more stringent than in a guideline). In case of an agreement on a duration longer than 6 months, the EU legislation should be changed. A number of 18 cases were identified which from the viewpoint of the US FDA led to labelling because of possible human relevance of observed toxicity occurring beyond 6 months in non-rodents. These cases have been discussed in detail to see what the added value of the 2nd period of 6 months would be (Contrera et al. 1993; DeGeorge et al. 1999). However, the industry viewpoint was very different after in depth evaluation (Van et al. 2000).

Finally the focus was on a few cases in which additional (potential clinically relevant) toxicity was observed later than 6 months. A compromise of 9 months duration was a general solution that is followed now for small molecules. When revising the pharmaceutical legislation, i.e. the EU-directive in 2001 the wording of this requirement has been changed to now read, "to be specified by appropriate guidelines" (Directive 2001a).

As a whole, the issue of repeated dose toxicity studies has resulted in a reduction in the use of rodents previously needed for a 12 months study, and in a refinement by lowering the stress load on non-rodents by reducing the study duration from 12 months to 9 months. A reduction would be likely too, as many companies conducted a 6 months as well as a 12 months non-rodent study in order to fulfil both EU and US requirements, respectively. It would be important to estimate the impact of this measure on the reduction of the number of animals.

Recent Developments

A similar discussion could be expected about the duration of chronic repeated dose toxicity studies for biopharmaceuticals. Indeed, a huge number of companies apply this figure of 9 months to this type of product. Regulatory Authorities, which are approached for approval of clinical trial applications, or for scientific advice, may also stimulate this. When writing the addendum to ICH S6 (see below) the line was drawn again at 6 months. All Regulatory Authorities agreed upon this aspect.

2.2.6 Reproductive Toxicity Testing

Reproductive toxicity testing was introduced more stringently after the thalidomide disaster (1957–1961). Various Regulatory Authorities created different schemes in the 1960s, leading to different testing practices all over the world. Pharmaceutical industries had to duplicate studies in order to fulfil the requirements from the various regions. However, even within regions (e.g. EU) different requirements still existed.

Rolf Bass (Bass et al. 2013) nicely showed that the ICH starting in early 1990s could take advantage from an existing network of experts in the ICH-areas, which

Table 3.2 Design of Segment 2 studies before 1990 (taken from Omori (1992) with permission)

	USA	EU	Japan
Title	Teratology study	Embryotoxicity studies	Study on administration of drug during the period of organogenesis
Species	Two species, one rodent (rat, mouse) and one non-rodent (rabbit)	Two species, one of which should be other than a rodent	At least two species, one rodent such as rat or mouse and one non-rodent, e.g. rabbit
Rodents			
No. of animals	20 female rodents/group	20 pregnant rodents/group	30 female rodents/group
Dosing period	Rat and mouse: day 6–15	Throughout the period of embryogenesis (organogenesis)	Rat: day 7–17 Mouse: day 6–15
Dose levels	At least three dose levels	Normally three dose levels	At least three dose levels

was a 'well-oiled machinery' enhancing the efficiency of the negotiations. In his chapter of the book on the background of ICH Guidelines (Bass et al. 2013), Bass clearly showed the pre-ICH activities in his own network. The draft S5 Guideline used during the Brussels meeting in 1990 could carry, therefore, the follow-up number 12 of a series of drafts starting long before the ICH had begun (Bass et al. 1991).

The strength of the approach chosen to come to harmonization lay in the choice for a scientific fundamental and systematic discussion regarding all possible stages and effects of pharmaceuticals during reproduction starting from fertility to postnatal development. It was this systematic approach that led to the flexibility in the assessment applied by the Regulatory Authorities (Sullivan et al. 1993).

The reproductive cycle could be covered by three studies, i.e. Segment 1 (prior to and in the early stages of pregnancy), Segment 2 (organogenesis and embryofoetal development) and Segment 3 (throughout pregnancy and lactation, up to fertility in pups).

The various ICH regions had different requirements for these three segments, often leading to duplication of these studies when fulfilling these specific criteria, such as dosing scheme, number of animals, endpoints and species selection (Omori 1992). In Table 3.2 the differences are clearly spelled out for Segment 2 studies. The differences in dosing period and number of females can easily lead to unnecessary repetition if a formal approach is followed in the design of the safety assessment strategy for a new pharmaceutical.

The three ICH Regulatory Authorities first discussed the possibility of mutual acceptance before coming to a harmonized concept of studies. This was because of the urgency felt in this area. Takayama (1992) described this need for mutual acceptance as follows; from the number of products marketed between 1980 and 1990 (489) less than 10 % showed adverse effects on reproductive and developmental function. The sensitivity of the approach was judged to be sufficient to cover the risk for new teratogenic drugs. The new guideline became available in 1994 and this has reduced the number of animals used in this area.

FDA-representative Judy Weissinger (1992) confirmed the acceptability of these Japanese proposals, suggesting to come to a new reproductive and developmental toxicity screen. As 75–80 % of the drugs do not demonstrate this type of toxicity, further testing would not be needed when passing this test. For the remaining 20–25 % a more focused approach can be applied.

Recent Developments

In vitro approaches in reproductive toxicity testing started to emerge already in the 1970s and 1980s, with rat whole embryo cultures and mouse embryonic stem cells (Spielmann et al. 1997). Lots of money has been invested since then, but so far no Regulatory Authorities have accepted *in vitro* tests for risk assessment purposes.

The EURL ECVAM has published a preliminary validation of the mouse Embryonic Stem Cell test, but further work was recommended (Marx-Stoelting et al. 2009).

In June 2006, a first brainstorming session within ICH was organised in Yokohama under the chairmanship of EU EMA and the Japanese MHLW, where new topics for ICH Guidelines have been listed, as well the possibility to revise the first guidelines. It was decided not to include these *in vitro* approaches in ICH S5 as the state of the art of the *in vitro* alternatives was not mature.

The European Commission sponsored a research project under the Framework Programme 6, named Reprotect (Hareng et al. 2005), focusing on *in vitro* testing for developmental and reproductive toxicity, which was finalised around 2009. A brainstorming workshop was organized under the auspices of the safety group of ICH in Tallinn in June 2010, to discuss the further steps needed in follow up of this project. Reduction of the need for two species in the embryofoetal toxicity test (now rodent and non-rodent, resp. rats and rabbits) was intended, but the question remained which species should be considered as the most sensitive, and/the most predictive.

This question appeared to be the most urgent and since then work has been started to build a database on embryofoetal development testing in rats and rabbits (Theunissen et al. 2014). Workshops have been held in Leiden (October 2011) and Washington (April 2012) to support the process (Van der Laan et al. 2012; Brown et al. 2012).

In 2014, the start of the ICH process revising the S5 document was marked by the establishment of an Informal Working Group in Minneapolis (US).

2.2.7 Safety Testing of Biotechnology-Derived Proteins

During the first ICH in Brussels, the unique position of biotechnology-derived proteins has been emphasized. This has led to the writing of a guideline, ICH S6, advocating a flexible approach dependent on the identity of the product (ICH 1996).

Table 3.3 Number of non-human primates used to support marketing authorisation of biotechnology-derived proteins in Europe (1988–2002)

	Recombinant proteins	Monoclonal antibodies
Number of studies (average)	6.4	7.8
Number of animals/study	12.8	11.0
Number of animals/product	81.8	86.0
Highest number of animals/study	64	60
Lowest number of animals/study	3	5
Highest number of animals/compound	269	308
Lowest number of animals/compound	6	8

Data derived from Tessa van der Valk, Van der Laan, Moors, Schellekens (2002, unpublished) and Nicole van de Griend, Van der Laan, Moors, Schellekens (2003, unpublished)

In this way an important issue became clear for this type of product, i.e. the responsiveness of the test animal for the product. Biotechnology-derived proteins are usually derived from endogenous proteins to a certain extent, and their effectiveness is dependent on the existence of targets in humans and in the animal species to be used for safety testing. Issues related to species specificity for these types of proteins, i.e. large molecules built up with high numbers of amino acids, led to serious constrains with respect to species selection.

The nature of the products, i.e. being protein products, led to a more predictable scheme of metabolism, via the common pathways of protein breakdown. Species differences because of differences in metabolism became much less important, and allowed for single species testing under the condition of having a pharmacodynamic response to the product.

The higher species specificity led unavoidably to the use of non-human primates, as the species most alike to humans. The specificity for the receptor is usually the same, although the affinity for the target might be lower in the non-human primate. See Table 3.3 for an overview of the number of non-human primates to support European marketing authorisation applications for biotechnology-derived proteins.

The use of monkeys was still low for development of small peptide molecules, such as insulin, human Granulocyte Stimulating Factor (GCSF), and even with haemophilia factors, such as factor VII and factor VIII. Rats, and rabbits are responsive to human GCSF and insulin whereas dogs were the preferred species for the coagulation factors, also in view of spontaneous disease models in this species. The use of monkeys was important in the development of interferons, in fact leading to the use of chimpanzees. The use of chimpanzees for regulatory safety testing was, however, forbidden globally around 2003.

For monoclonal antibodies only for the very first products, safety was tested without monkeys. This was due to the fact that the epitope was lacking in monkeys, and the compounds were used for diagnostic purposes. Therapeutic proteins could be developed only when humanized proteins could be produced thereby reducing the risk for immunogenicity. Later on, with new *in vitro* culture techniques and more advanced recombinant techniques, there was a development in the direction of humanized proteins (Van Meer et al. 2013). For this type of product, there is a huge experience in monkeys. A few products are human-specific, e.g. efalizumab and infliximab, and testing of these molecules have been conducted in chimpanzees. Due to the design of the protocols, the added value of these studies was low, however (Van Meer et al. 2013).

Already at the time of the start of ICH it was generally accepted that biotechnology-derived proteins exert highly specific effects, and toxicological effects are rather characterized as exaggerated pharmacology then as off-target phenomena. Despite this recognition, toxicological testing has been required up to the recent writing of the addendum to the ICH S6.

Recent Developments

The discussion on the use on monkeys has been stimulated in various ways.

Van Meer et al. (2013) published a review on the marketed monoclonal antibodies in Europe since the start of the EMA, extending and updating the work listed in Table 3.3, and showing that toxicity phenomena associated with administration of monoclonal antibodies in chronic studies in monkeys are indeed phenomena of exaggerated pharmacology. The approach can be criticized as molecules dropped during development are not included in the review, but recent cases were presented in a workshop in Berlin (Baumann et al. 2014). The authors conclude: "*While effects were mostly pharmacologically mediated, they were not necessarily always predicted, and identified previously unknown consequences of the respective pharmacologies*".

Discussions can be started with the question what the minimum package should be to detect this type of rare off-target effects, while most other effects of monoclonal antibodies can be readily predicted from *in vitro* studies by receptor- or ligand-binding and by binding to FcR and FcRn receptors (van der Laan 2013).

2.2.8 Safety Pharmacology

Where as the ICH S7A is focusing on the investigation of potential undesirable pharmacodynamic effects of human pharmaceuticals on physiological function at therapeutic dose ranges and above, the second guideline, ICH S7B, focuses on the investigation of the potential for QT interval prolongation. The latter guideline specifically refers to an *in vitro* approach to assist in the detection of QT-prolonging properties of human pharmaceuticals as part of an integrated assessment strategy.

The outcome of the *in vitro* approach is to be complemented by an *in vivo* study, but such an assessment could be conducted in the same animals (non-rodents) as in the repeated dose toxicity study, as is recommended in the ICH M3(R2) and ICH S9 guidelines. It could be defended that even such an *in vivo* study is not needed, as this testing can be done in a so-called clinical Thorough-QT-(TQT) study in healthy volunteers. The compound is administered under controlled conditions and the Electrocardiogram (ECG) is performed (see ICH E14 Guideline). However, as exceeding the therapeutic dose might present unacceptable safety risks for healthy volunteers, the conduct of a TQT study might be ethically unacceptable.

2.2.9 Immunotoxicity

The focus on the discussion on Immunotoxicity testing of human pharmaceuticals was on the routine need for additional data on immune functioning. When special immunotoxicity testing was introduced in the EU EMA Repeated Dose Guideline, it was expected that this testing would be conducted for each compound. The US FDA and the Japanese MHLW were not happy with this high level of concern with respect to immunotoxicity, and therefore the topic became an issue at the ICH level (Putman et al. 2003).

Potential reduction of animal use was rather a co-effect in the discussion about ICH S8, but can also be taken as evidence that within ICH reduction of animal use is an important driving force.

Before agreeing on the development on an ICH Guideline on this topic, a survey has been conducted to learn in how many cases a specific immunotoxic compound would have been missed based on routine toxicity studies only. Eventually, 45 compounds were eligible in this dataset, and only 6 compounds were characterized as immunotoxic based on specific immunotoxicity studies (Weaver et al. 2005). Because of this low number, the EU EMA changed their mind and dropped the request for routine testing of immunotoxicity by specific studies.

The discussion on immunotoxicity did not specifically address the 3Rs as a purpose.

2.2.10 Safety Testing of Anticancer Products

The focus of ICH S9 is on reduction and refinement rather than on replacement of animal studies. If the disease is life threatening, then a higher risk can be accepted, which can also be seen as a high level of uncertainty. However, in practice it might have led also to an increase in animal use in certain aspects. Previously anticancer products such as cytostatics were developed on a large scale and tested with *in vitro* methodology, and safety testing was restricted by conducting a mouse lethality test. The LD10 in mice was taken as the starting point to calculate a safe starting dose in humans.

In the present S9 Guideline, the package is much more defined in line with the development of other small molecules. When compared with the package needed for a small molecule for the treatment of e.g. diabetes, the number of studies is small, but the data requirements are certainly more than only an LD10 determination in mice. Furthermore, anticancer drugs are no longer restricted to cytostatics, but are covering now also targeted therapies such as tyrosine kinase inhibitors.

2.2.11 Photosafety Testing

The most recent guideline is the ICH S10 focusing on photosafety testing of human medicinal products. The guideline is rather unique from the viewpoint of 3Rs, as it explicitly indicates that several *in vivo* animal models are hardly relevant to the human situation, and are not predictive. Their use should therefore be discouraged.

During the development of the guideline not only an existing *in vitro* approach, i.e. the use of 3T3-Neutral Red Uptake Phototoxicity Test (NRU-PT), but also the Reactive Oxygen Species (ROS) assay was accepted as a reasonable approach to screen for phototoxic properties.

2.3 ICH and 3Rs, Future Perspectives

Reduction of animal use is one of the driving forces in the ICH process as can be derived from the discussion of all the safety guidelines agreed upon under ICH auspices. Reduction of animal use is, however, not the primary purpose for the ICH-process on nonclinical safety testing, which is focused on reducing the trade barriers between the economic areas involved.

Furthermore, it has to be admitted that in certain cases drafting of ICH Guidance may have led to additional animal use, as explained above for S3 Toxicokinetics. Indeed, enhancing the technical abilities has led to an increased need to enhance knowledge about the fate of pharmaceutical products in animals, this to refine the interpretation of toxicity studies, and to use information on animal exposure for risk assessment by comparing this to the exposure in humans.

For several topics, especially those discussed early in the ICH process, new developments might exist that could further reduce animal experimentation. It needs to be understood that changing ICH-guidelines cannot be done at a regional level. This emphasizes that changing guidelines at a global level might need a long way of negotiations. The initiative to come to revisions of S1 Carcinogenicity is driven by the general agreement that using a huge number of animals in a life-time carcinogenicity study in case of a high predictability of the outcome, is over the top in the ethical use of animals for safety testing.

3 The Revision of ICH S2 Genotoxicity Testing Guideline, a Case Study

The implementation of new 3Rs testing approaches in regulatory testing of pharmaceuticals through the ICH process occurs by either drafting a new guideline or updating an existing one. New testing approaches in this context include both, newly developed assays and changes to existing assays with an impact on 3Rs aspects such as protocol modifications or improved advices in data interpretation and follow-up strategies. As an example of how new 3Rs testing approaches have achieved regulatory acceptance the revision of the ICH Guidelines on genotoxicity testing of pharmaceuticals will be described in this section.

The original ICH guidelines on genotoxicity testing, designated ICH S2A and S2B, were finalized in 1996–1997 and recommended a battery of two *in vitro* and one *in vivo* genotoxicity tests (Müller et al. 1999). The *in vitro* assays were a bacterial mutagenicity assay (Ames test) and an *in vitro* mammalian cell assay, either a metaphase chromosome aberration (CAb) assay in cultured cells or an assay for mutations at the *tk* locus in L5178Y mouse lymphoma cells (MLA). The *in vivo* assay was an assay for chromosome damage; the most widely used assay for this purpose is the rodent bone marrow micronucleus assay. For follow-up testing when a positive result was found in an *in vitro* mammalian cell assay, additional *in vivo* testing was recommended, and while a range of possible assays was mentioned, emphasis was placed on the UDS (unscheduled DNA synthesis assay) in rat liver.

In 2006 an ICH Concept Paper was published with a proposal for revision of the ICH S2 guidelines on genotoxicity testing. One of the major goals for the revision was to "*reduce the numbers of animals used in routine testing by improving the current procedures and clarifying the follow-up testing in case of positive findings*" (Final concept paper ICH S2(R1) 2006). A six-party Expert Working Group (EWG) was set up in charge of developing a scientific consensus of the revised guideline elements. As a result of the revision process the two original ICH Guidelines S2A and S2B were merged into one document titled "ICH S2R1 Guidance on Genotoxicity Testing and Data Interpretation for Pharmaceuticals intended for Human Use" which was adopted in 2011 (ICH guideline S2(R1) 2012).

In the absence of formal validation studies and/or OECD guidelines or when proposed ICH guideline recommendations differed from existing OECD guidelines the decisions of the EWG for revision of the ICH guideline was mainly made based on thorough review of both industry-held and published data and recommendations from international expert workshops. Of particular importance in this context are the International Workshops on Genotoxicity Testing (IWGT). The IWGT process is implemented through working groups of recognized international experts from industry, academia and the regulatory sectors (Kirkland et al 2007a). The remit of each specific topic group is to derive consensus recommendations based on data, and not on unsupported opinion or anecdotal information. In 2002, the International Association of Environmental Mutagen Societies (IAEMS) formalized these workshops under IAEMS umbrella and agreed that they would be held on a continuing basis in conjunction with the International Conferences on Environmental Mutagens (ICEM) that are held every 4 years. The Sixth IWGT Workshop was recently held in Foz do Iguassu, Brazil, as a satellite to the 2013 ICEM. IWGT recommendations have been seen as state-of-the-art and have high credibility. These recommendations serve as important supplements to establish regulatory guidelines and provide a sound basis for those guidelines as the state of science advances (Kirkland et al 2007a).

The ICH S2 EWG identified mainly two areas with opportunities for reduction of use of animals considered as essential for incorporation in the revision: (1) enhance the performance of *in vitro* mammalian cell assays to reduce the need for *in vivo* follow-up testing of irrelevant/false-positive *in vitro* findings and (2) animal reduction opportunities in genotoxicity *in vivo* testing.

3.1 Enhance Performance of In Vitro Mammalian Cell Assays by Lowering the Test Concentrations

The genotoxicity assays are used early in pharmaceutical development to determine whether drug candidates are safe enough to be given to human volunteers, and then later in development to patients. However, the ICH S2A/B testing battery did not serve well in this respect because of the poor specificity that is associated with the high sensitivity of *in vitro* mammalian cell assays. The traditional justification of the use of such tests has been "hazard identification". Based on considerable experience that is now available in the pharmaceutical sector it can be concluded that in the vast majority of cases of *in vitro* positive genotoxicity results, no genotoxicity is detectable *in vivo*, and in still many further cases there is no evidence for induction of tumours in rodents that are thought to reflect a genotoxic mechanism (Kirkland et al. 2005, Matthews et al. 2006). Also, there is now a large body of experience that demonstrates that positive results in the existing *in vitro* regulatory tests, the CAb assay and MLA, at high concentrations and/or associated with toxicity, most often do not reflect DNA damaging capability of the test compound, but are a secondary response to perturbation of cell physiology (Kirkland et al. 2007b). Since the genotoxicity seen under these *in vitro* conditions occurs through mechanisms that are not operating at lower doses, there is a threshold, not a linear dose relation, and the results are not likely to be informative about risk in the human therapeutic context, that is they are considered "non-relevant" to *in vivo* conditions. Nevertheless, the occurrence of such *in vitro* findings required follow-up testing in a second animal study, usually a rat liver UDS test.

Based on these identified shortcomings the EWG agreed that there is a need to reduce the reliance of testing on *in vitro* assays carried out under such extreme conditions on the principle of hazard identification. A refinement of the test strategy with the use of tests/protocols that identify potential genotoxic effects under more realistic conditions would be required. One approach to achieve more realistic testing conditions and thus improve test specificity (reduce number of irrelevant positives) is to challenge the need for testing to such high concentrations to which an article should be tested in *in vitro* mammalian cell assays. The original ICH S2 guideline as well as the respective OECD guidelines recommended that the top concentration when toxicity is not limiting is 10 mM or 5 mg/mL, whichever is lower. The ICH S2R1 EWG proposed a reduction from 10 to 1 mM as the top concentration in the revised ICH S2 guidance. Two types of information were considered in justifying this recommendation.

First, the question was addressed whether a 10-fold reduction of the used top concentration would maintain sufficient sensitivity of the *in vitro* assays in detecting *in vivo* relevant genotoxicants. Sets of data were reviewed on Ames test negative compounds that were positive in the CAb assay/MLA only above 1 mM. Experience in the pharmaceutical industry from Japan, Europe and the US showed that generally compounds that were positive at >1 mM only lacked genotoxic potential *in vivo*. However, the data set is incomplete as many results lacked follow-up studies, but overall it contributed to supporting a 1 mM upper limit. Similarly, one of the data sets used initially in adopting the 10 mM limit (Scott et al. 1991) was re-examined, and it was noted that all *in vivo* positive chemicals were detected in the Ames test or *in vitro* in mammalian cell assays below 1 mM.

Second, an important factor in the ICH recommendation of 1 mM as a top concentration for *in vitro* mammalian cell assays is that it would be a high multiple of the known human exposures to most pharmaceuticals. The pharmacologically active concentrations for drugs, and the optimal substrate concentrations for many enzymes including P450s, are typically below 10 μg/mL (equivalent to 20 μM for average molecular weight of 500). For defining a top concentration, it is useful to consider high dose pharmaceuticals with high bioavailability, and those that are known to accumulate in the body (mostly in specific cell types) after repeat dosing. Data on human exposure levels of 313 marketed pharmaceuticals (Goodman and Gilman 2001) were reviewed by the EWG. Peak exposure to pharmaceuticals is generally below 10–50 μM. Only 53 out of 313 marketed pharmaceuticals had a C_{max} of more than 10 μg/mL (or about 20 μM). These included antibiotics, anti-tumour and antiviral drugs. Of the others, ibuprofen, acetaminophen and clofibrate are examples of higher exposures, with C_{max} of 300, 130 and 450 μM respectively. There are examples of drugs that accumulate extensively in tissues; an example of a lipophilic drug with a long half-life is fluoxetine, a cationic amphiphilic drug with an elimination half-life of 1–3 days and an active metabolite with an elimination half-life of 4–16 days. Fluoxetine accumulates in the brain to ~10 μg/mL (~35 μM) with a brain/plasma ratio of 20:1. However, no example of a drug was found for which there were both high plasma levels and a high (10–20 fold) accumulation in tissue. Thus, a top concentration of 1 mM would capture low potency drugs and other high dose drugs including cases of extensive tissue accumulation.

The proposed recommendation of the ICH S2R1 EWG to reduce the top concentration in *in vitro* mammalian cell assays from 10 to 1 mM triggered further data reviews and discussions on this topic in the scientific community such as IWGT (Galloway et al. 2011; Moore et al. 2011). A review of existing databases and published literature to determine the concentrations at which rodent carcinogens induced damage in three mammalian cell assays was also supported by EURL ECVAM (Parry et al. 2010). An analysis of this review confirmed the ICH S2R1 EWG view that testing above 1 mM was unnecessary to identify genotoxic compounds, provided the compound was tested and found negative using a bacterial gene mutation (Ames) assay (Kirkland and Fowler 2010).

3.2 *Animal Reduction Opportunities in Genotoxicity In Vivo Testing: Integration in Standard Toxicity Studies and Combination of Endpoints*

For pharmaceuticals the assessment of genotoxicity *in vivo* is an essential part of the standard battery. The most widely used assay for this purpose has been the rodent bone marrow micronucleus assay (MN). The use of peripheral blood is an alternative approach for both mice and rats (when the youngest fraction of reticulocytes are sampled) which provides equivalent data to the bone marrow assay and is technically less demanding. More recently, the use of flow cytometry scoring methods has increased further the efficiency and reproducibility of the rat peripheral blood assay. This technique requires only microliter quantities of blood which can be obtained without sacrificing the animals and can therefore be easily integrated into routine toxicology and pharmacokinetic studies. The available data were reviewed by groups of experts at the 4th IWGT in 2005 in San Francisco and the 5th IWGT 2009 in Basle and it was concluded that the use of rat peripheral blood reticulocytes and its integration into repeat-dose toxicity (RDT) studies is scientifically acceptable and ready for regulatory use (Hayashi et al. 2007; Rothfuss et al. 2011). The ICH EWG agreed to these conclusions and revised the S2 Guideline accordingly. The revised ICH S2R1 guideline now accepts the use of rat peripheral blood reticulocytes as target cells for the MN assay and also the use of flow cytometry scoring methods. Moreover, the revised ICH guideline encourages integration of *in vivo* MN analysis into RDT studies so that it is not necessary any more to conduct an independent study for this purpose.

For pharmaceuticals with positive results in genotoxicity tests *in vitro* follow-up testing *in vivo* should be done in two tissues. In practice, this has been usually done by performing a MN test in rodent bone marrow erythrocytes and a rat liver UDS test. As part of the ICH S2 revision the EWG agreed to replace the rat liver UDS test by a rat liver Comet assay, mainly due to the identified insensitivity of the UDS (Kirkland and Speit 2008). Although there was no formal validation of the *in vivo* comet assay and no OECD guideline at the time the ICH S2 revision process was ongoing the EWG considered this test method as suitable for routine use. This decision was mainly based on the available knowledge regarding adequate testing protocols and data interpretation provided by IWGT expert working groups and other international workshops (Tice et al. 2000; Hartmann et al. 2003; Burlinson et al. 2007).

In addition, at the instigation of the ICH EWG, a collaborative trial was run using 15 compounds in 13 laboratories in Europe, Japan and the US, to examine the performance of the liver Comet assay especially as a complement to the *in vivo* micronucleus assay and potential replacement for the liver UDS assay. The results supported the EWG proposal that the Comet assay was suitable for routine use (Rothfuss et al. 2010).

Besides the higher sensitivity of the liver comet assay compared to the UDS test another advantage is the ease with which the comet assay can be combined with the *in vivo* micronucleus test. The above mentioned collaborative trial therefore also investigated the sensitivity and practicality of integrating the liver Comet assay into acute and 2- or 4-week repeat-dose rat toxicity studies (Rothfuss et al. 2010). These and other data were considered by the Basle 2009 IGWT working groups as part of their recommendations on *in vivo* genotoxicity testing, in particular the suitability of tests with multiple genotoxicity endpoints integrated into acute or RDT studies. The combination of the acute *in vivo* MN and Comet assays was considered by the working group to represent a technically feasible and scientifically acceptable alternative to conducting independent assays. For the integration of Comet assays into RDT studies, the working group reached the consensus that, based upon the limited amount of data available, integration is scientifically acceptable and that the liver Comet assay can complement the MN assay in blood or bone marrow in detecting *in vivo* genotoxins. Practical issues need to be considered when conducting an integrated Comet assay study (Rothfuss et al. 2010).

The ICH EWG has adopted these advices for combining different endpoints into one study in the revised ICH S2 guideline. When *in vivo* assessment of genotoxicity with two tissues is required the guideline encourages the incorporation of two genotoxicity assays in one study using the same animals, e.g. bone marrow micronucleus test and liver DNA strand breakage assay.

Recent study report submissions to regulatory health authorities clearly indicate that the new options offered by the ICH S2R1 are increasingly utilized by pharmaceutical industry. In particular, integration of MN analysis in RDT studies in cases where *in vitro* testing is negative and acute combined *in vivo* MN/comet assay to follow-up positive findings in *in vitro* mammalian cell tests are the preferred options in recently performed genotoxicity testing programs. It can thus be roughly estimated that the number of animals used for *in vivo* genotoxicity assessment in new drug development may be decreased by nearly 50 % as a result of the revisions of the ICH S2 guideline.

In summary, the experiences with the process leading to revision of the ICH S2 guideline clearly show that formal validation is not a necessary prerequisite for regulatory acceptance of new 3R testing approaches. Instead, the scientific credibility of new methodologies and its use for regulatory purposes is pragmatically assessed in a formalized ICH procedure by a working group of recognized industry and regulatory experts in the field based on high quality data from different sources.

4 The EMA Approach to Regulatory Acceptance of 3R Test Methods

4.1 The JEG 3Rs

To demonstrate its commitment to the application of replacement, reduction and refinement (the 3Rs) of animal testing as detailed in Directive 2010/63/EU (Directive 2010), in October 2010, the EMA endorsed the establishment of a Joint *ad hoc* Expert Group on the application of 3Rs in the development of medicinal products

(JEG 3Rs). The JEG 3Rs has as a mandate to improve and foster the application of 3Rs in the regulatory testing of medicinal products throughout their lifecycle. Moreover, this group provides advice and recommendations to the Committees (i.e. CHMP and CVMP) on all matters related to the use of animals in regulatory testing of medicinal products.

The core of the JEG 3Rs consists of experts from CVMP and CHMP and their working parties for which animal testing is relevant, and can be complemented, as necessary, by specific experts. As such, all relevant disciplines (i.e. quality, safety and, in the case of veterinary medicinal products, efficacy) are represented, for both pharmaceutical and biological/immunological products.

As it is recognised that much work is already being done in the 3Rs area by other European Commission bodies and consequently it is considered that duplication of efforts should be avoided, the JEG 3Rs works in close cooperation with EURL ECVAM (European Union Reference Laboratory for Alternatives to Animal Testing) and EDQM (European Directorate for the Quality of Medicines and Healthcare).

Amongst the achievements of the JEG 3Rs so far are the efforts to ensure compliance of existing regulatory guidance with the 3Rs. As such, a concept paper announcing a review of existing EMA guidance to ensure compliance with best 3Rs practice was published in February 2014 (Concept paper on review and update of European Medicines Agency Guidelines 2014) and publication of the first amended guidelines will ensue. Eight guidelines have so far been identified for revision but the review is not yet complete. In addition, a draft guideline on regulatory acceptance of 3Rs testing approaches has been published for a 3-month period of consultation (Draft Guideline on regulatory acceptance of 3R 2014). This guideline describes the process for submission and evaluation of a proposal for regulatory acceptance of 3R (Replacement, Reduction and Refinement) testing approaches for use in the development and quality control during production of human and veterinary medicinal products. Furthermore, scientific and technical criteria for validation of 3R testing approaches are presented and pathways for regulatory acceptance of 3R testing approaches are described.

Another work stream of the JEG 3Rs is related to the implementation of the 3Rs in batch release testing of human and veterinary medicinal products. To this end, a statement was published highlighting the need for marketing authorization holders (MAHs) to ensure that batch safety and potency tests comply with 3Rs options available in the Ph Eur (Recommendation to marketing authorisation holders 2012). A statement was published highlighting the need for MAHs for veterinary vaccines to update MAs to remove the target animal batch safety test following removal of the requirement from the Ph Eur ("Recommendation to marketing authorisation holders" 2013). At their July 2014 meetings CXMP adopted a concept paper for publication for consultation, announcing the development of guidance on transferring validated quality control methods to a product/laboratory specific context (Concept paper on tranferring quality control methods 2014). Work on developing the guideline is now underway. In addition, review of batch release testing for human and veterinary vaccines is conducted and follow-up communications to MAHs ensured.

JEG 3Rs coordinates responses to requests from the European Union Reference Laboratory for Alternatives to Animal Testing (EURL ECVAM) for preliminary analysis of regulatory relevance of new alternative methods (PARERE).

The existence of the JEG 3Rs provides a strong signal indicating that EMA takes 3Rs issues seriously. A clear statement of the EMA on the application of the 3Rs in regulatory testing of human and veterinary medicinal products was published in the EMA website at the very beginning of the JEG 3Rs' first mandate (Statement of the EMA position on the application of the 3Rs 2011). This reads as follows:

The European Medicines Agency (EMA) commits to the application of replacement, reduction and refinement (the 3Rs) of animal testing as detailed in Directive 2010/63/EU1. To this end, a Joint ad hoc Expert Group (the JEG 3Rs) has been created in order to promote best practice in the implementation of the 3Rs in regulatory testing of medicinal products and to facilitate full and active cooperation with other European groups working in the 3Rs area.

While significant progress has been made in relation to regulatory testing involving animals it remains the case that certain types of data can only be generated by means of animal studies. Where such studies are needed they should be selected and conducted in strict adherence to the 3Rs principles.

As a European body with responsibility for developing harmonised European regulatory requirements for human and veterinary medicinal products the EMA has and will continue to play a key role in eliminating repetitious and unnecessary animal testing in the European Economic Area (EEA), in collaboration with other European organisations such as EDQM. Through its active participation and collaboration in the work of other multinational organisations such as the ICH and the VICH, the EMA contributes to the application of the 3Rs in the development of globally harmonised requirements, the implementation of which contributes to the elimination of unnecessary animal testing.

The JEG 3Rs is now recognised at international level and is often cited as an example of how regulatory agencies should tackle 3Rs issues whilst providing a clear entry point for questions or comments in this area. The JEG 3Rs mandate was renewed in October 2014 for the second time, thus allowing this group to continue its work in order to achieve progress in the field of 3Rs, an area for which Europe is clearly a global frontrunner.

4.2 Draft Guideline on Regulatory Acceptance of 3R Testing Approaches

4.2.1 Ontogeny

The application of the 3Rs were already highlighted in the Position on the Replacement of Animal studies by *in vitro* models (Replacment of animal studies by *in vitro* models 1997), adopted by the then called EMA Committee on Proprietary Medicinal Products (CPMP) at its meeting in February 1997. This Position Paper addressed the feasibility of replacing *in vivo* animal studies by *in vitro* investigations in the preclinical development of medicinal products. In addition, considerations regarding validation procedures for *in vitro* methods and their incorporation into CPMP Notes for Guidance were presented.

Whilst replacement of animal studies remains the ultimate goal, the application of all 3Rs needs to be the focus. As exemplified by the ICH regulatory safety guidelines described earlier, approaches aiming at reducing or refining animal studies are being routinely implemented in regulatory guidelines, where applicable. At the same time, over the past years, new *in vitro* methods have been accepted for regulatory use via multiple and flexible approaches, either as pivotal, supportive or as exploratory mechanistic studies, wherever applicable. As such, although regulatory acceptance of 3Rs testing approaches is currently possible, a formal regulatory acceptance process has been lacking and implementation of new test methods in routine regulatory testing has sometimes proven problematic.

Consequently, a review of the position paper focusing primarily on Replacement was needed. As such, on March 11th 2011 a Concept Paper on the Need for Revision of the Position on the Replacement of Animal studies by *in vitro* models (Concept paper on the Need for Revision of the Position on the Replacement of Animal Studies 2011) was drafted by the CHMP Safety Working Party and published on the EMA website. Herein, aside from the extended focus to all 3Rs principles, the revision intended to describe a clear process for regulatory acceptance of 3Rs testing approaches, to discuss qualification criteria and bring the requirements in line with Directive 2010/63/EC.

As 3Rs principles apply to all regulatory testing requirements involving animal use for both human and veterinary medicinal products, a multidisciplinary drafting group was set up under the JEG 3Rs to develop the draft Guideline for Regulatory acceptance of 3Rs testing approaches (Draft Guideline on regulatory acceptance of 3R testing approaches 2014). Concomitantly the JEG 3Rs started a thorough review of the current regulatory testing requirements for human and veterinary medicinal products and identification of opportunities for implementation of the 3Rs.

The Draft guideline was forwarded to the relevant EMA Working Parties and Committees for comments on the 17th of March 2014 and the final draft was launched for public consultation on the 3rd of October 2014. Comments received are being considered and an update of the guideline is currently under way.

4.2.2 Draft Guideline for Regulatory Acceptance of 3Rs Testing Approaches

This guideline only applies to testing approaches that are subject to regulatory guidance for human and veterinary medicinal products. More specifically, those that are used to support regulatory applications, such as clinical trial and marketing authorisation applications. The process of uptake of 3Rs testing methods in the Ph. Eur. Monographs is excluded.

In line with the above, regulatory acceptance of a new 3Rs testing approach is defined by its incorporation into a regulatory testing guideline. However, on a case-by-case basis, it is also seen as the acceptance by Regulatory Authorities of new approaches not (yet) incorporated in testing guidelines but used for regulatory decision making.

The modification of existing testing approaches to achieve refinement, reduction and replacement of laboratory animal use and, if possible, at the same time increase predictive power of regulatory testing is expected to occur at different levels. These levels range from discrete modifications of existing testing approaches (eg reduction of the top concentration used in *in vitro* genotoxicity testing in ICH S2R, see Sect. 3) to the implementation of a completely new approach in regulatory toxicology (e.g. Toxicity Testing in the twenty-first century; (Committee on Toxicity Testing and Assessment of Environmental Agents 2007)).

The draft guideline clearly lists a number of criteria that need to be fulfilled before a 3Rs testing approach can be considered for regulatory acceptance, namely:

1. Demonstration of method validation. This implies that there is a defined test methodology/standard protocol with clear defined/scientifically sound endpoints and demonstration of reliability and relevance. The amount of information needed and the criteria applied to a new method will depend on the regulatory and scientific rationale for the use of the method, the type of method (e.g. existing test, new method), the proposed uses (e.g. mechanistic, total or partial replacement, as part of a testing strategy), the mechanistic basis for the test and its relationship to the effect(s) of concern, and the history of use of the test method, if any, within the scientific and regulatory communities. The draft guideline clearly indicates the acceptability of different routes of method validation. This includes formal validation by recognised institutions such as the VAMs (Balls et al. 1995; Balls and Karcher 1995; NIH 1997, 1999; OECD 2005; Hartung et al. 2004) and EDQM but also allows for the acceptance of 3R testing approaches that have sufficient demonstration of scientific validity but have not been assessed in a formal validation process. The latter case implies that the relevant Working Parties, Expert Working Groups or National Control Authorities will conduct data evaluation.
2. Demonstration that the new or substitute method or testing strategy provides either new data that fill a recognised gap or data that are at least as useful as, and preferably better than those obtained using existing methods.
3. On a case-by case basis, demonstration of adequate testing of medicinal products under real-life conditions (human and veterinary). This can be achieved by voluntary submission of data obtained by using a new 3Rs testing approach in parallel with data generated using existing methods under a safe harbour. This implies that data generated with the new 3Rs testing approaches are not to be used for regulatory decision but need to be evaluated independently for the purpose of decision making on the regulatory acceptability.

Finally, the new draft guideline now unambiguously anchors the process for submission and evaluation of proposals for regulatory acceptance of 3R testing approaches for human medicinal products to the EMA procedure on Qualification of Novel Methodologies for Drug Development ("Qualification of novel methodologies for drug development" 2014; Manolis et al. 2011). This voluntary procedure was established by the EMA in 2008 under the auspices of the Scientific Advice Working Party (SAWP) of the CHMP. This procedure is innovative as it can be independent of a specific medicinal product and can include a formal assessment of submitted data by the SAWP itself. Typically, the outcomes are either a CHMP Qualification Advice on future protocols and methods for further development of the new method towards qualification for regulatory use, based on the evaluation of the scientific rationale and on preliminary data submitted. On the other hand, there can also be the formulation of a CHMP Qualification Opinion on the acceptability of a specific use for the proposed method in a research and development context (non-clinical studies), based on the assessment of submitted data.

With respect to 3Rs testing approaches for veterinary medicinal products only, proposal submission is to be in accordance with existing scientific CVMP guidance for companies requesting scientific advice (Guidance for companies requesting scientific advice 2012). The actual assessment of the new 3R testing approaches will be performed in collaboration with the relevant 3Rs experts from CHMP and/or CVMP working parties.

One could reflect on the added benefit of having a specific process for regulatory acceptance at the EU level, especially taking into account the regulatory guidance issued by ICH and VICH. Indeed, although major topics are governed by ICH or VICH, this does not represent the totality of the regulatory realm and EMA guidelines necessitating 3Rs improvements could benefit from EMA qualified 3Rs testing approaches. Moreover, the existence of such a regional process can thoroughly prepare global harmonization efforts.

References

Arnold (1992) Objectives and preparation of the conference and the role of workshops. In: D'Arcy PF, Harron DWG (eds) Proceedings of the second international conference on harmonisation, Brussels 1991. Queen's University Belfast, 1992, pp 7–11

Balls M, Karcher W (1995) The validation of alternative test methods. ATLA 23:884–886

Balls M, Blaauboer BJ, Fentem JH, Bruner L, Combes RD, Ekwall B, Fielder RJ, Guillouzo A, Lewis RW, Lovell DP, Reinhardt CA, Repetto G, Sladowski D, Spielmann H, Zucco F (1995) Practical aspects of the validation of toxicity test procedures. The report and recommendations of ECVAM workshop 5. ATLA 23:129–147

Bangemann M (1992) Welcome address. In: D'Arcy PF, Harron DWG (eds) Proceedings of the second international conference on harmonisation, Brussels 1991. Queen's University, Belfast, pp 1–5

Bass R, Ulbrich B, Hildebrandt AG, Weissinger J, Doi O, Balder C, Fumero S, Harada Y, Lehman H, Manson J, Neubert D, Omori Y, Palmer A, Sullivan F, Takayama S, Tanimura T (1991) Guidelines on detection of toxicity to reproduction for medicinal products (Draft nr 12). Adverse Drug React Toxicol Rev 10:143–154

Bass R, Ohno Y, Ulbrich B (2013) Why and how did reproductive toxicity testing make its early entry into and rapid success in ICH? In: Van der Laan JW, DeGeorge JJ (eds) Global approach in safety testing. Advances in the pharmaceutical sciences series, vol 5, pp 37–75

Baumann A, Flagella K, Forster R, De Haan L, Kronenberg S, Locher M, Richter WF, Theil FP, Todd M (2014) New challenges and opportunities in nonclinical safety testing of biologics. Regul Toxicol Pharmacol 69:226–233

Brown ES, Jacobs A, Fitzpatrick S (2012) Reproductive and developmental toxicity testing: from *in vivo* to *in vitro*. ALTEX 29(3):333–339

Burlinson B, Tice RR, Speit G, Agurell E, Brendler-Schwaab SY, Collins AR, Escobar P, Honma M, Kumaravel TS, Nakajima M, Sasaki YF, Thybaud V, Uno Y, Vasquez M, Hartmann A (2007) Fourth international workgroup on genotoxicity testing: results of the in vivo Comet assay workgroup. Mutat Res 627:31–35

Committee on Toxicity Testing and Assessment of Environmental Agents, National Research Council (2007) Toxicity testing in the 21st century: a vision and a strategy. The National Academies Press, USA

Concept paper on review and update of European Medicines Agency Guidelines to implement best practice with regards to 3Rs (replacement, reduction and refinement) in regulatory testing of medicinal products (EMA/CHMP/CVMP/JEG-3Rs/704685/2012) (2014) http://www.ema.europa.eu/docs/en_GB/document_library/Scientific_guideline/2014/02/WC500161024.pdf

Concept paper on the Need for Revision of the Position on the Replacement of Animal Studies by *in vitro* Models (CPMP/SWP/728/95) (2011) http://www.ema.europa.eu/docs/en_GB/document_library/Scientific_guideline/2011/04/WC500105110.pdf

Concept paper on transferring quality control methods validated in collaborative trials to a product/laboratory specific context (CHMP/CVMP/JEG-3Rs/94304/2014) (2014) http://www.ema.europa.eu/docs/en_GB/document_library/Scientific_guideline/2014/07/WC500169977.pdf

Contrera JF, Aub D, Barbehenn E, Belair E, Chen C, Evoniuk G, Mainigi K, Mielach F, Sancilio L (1993) A retrospective comparison of the results of 6 and 12 months non-rodent studies. Adverse Drug React Toxicol Rev 12:63–76

Contrera JF, Jacobs AC, Prasanna HR, Mehta M, Schmidt WJ, DeGeorge JJ (1995) A systemic exposure-based alternative to the maximum tolerated dose for carcinogenicity studies of human therapeutics. J Am Coll Toxicol 14:1–10

Contrera JF, Jacobs AC, DeGeorge JJ (1997) Carcinogenicity testing and the evaluation of regulatory requirements for pharmaceuticals. Regul Toxicol Pharmacol 25:130–145

DeGeorge JJ, Meyers LL, Takahashi M, Contrera JF (1999) The duration of non-rodent toxicity studies for pharmaceuticals. Toxicol Sci 49:143–155

Directive 2001/83/EC of the European Parliament and of the Council of 6 November 2001 on the Community code relating to medicinal products for human use (consolidated version: 05/10/2009)

Directive 2001/82/EC of the European Parliament and of the Council of 6 November 2001 on the Community code relating to veterinary medicinal products. Official J L311:1–66. 28/11/2001 (consolidated version: 18/7/2009)

Directive 2010/63/EU of the European Parliament and of the Council of 22 September 2010 on the protection of animals used for scientific purposes. Official J L 276/33

Draft Guideline on regulatory acceptance of 3R (replacement, reduction, refinement) testing approaches (EMA/CHMP/CVMP/JEG-3Rs/450091/2012) (2014) http://www.ema.europa.eu/docs/en_GB/document_library/Scientific_guideline/2014/10/WC500174977.pdf

Final concept paper ICH S2(R1) (2006) guidance on genotoxicity testing and data interpretation for pharmaceuticals intended for human use. http://www.ich.org/fileadmin/Public_Web_Site/ICH_Products/Guidelines/Safety/S2_R1/Concept_papers/S2_R1__Concept_Paper.pdf

Galloway S, Lorge E, Aardema MJ, Eastmond D, Fellow M, Heflich R, Kirkland D, Levy DD, Lynch AM, Marzin D, Morita T, Schuler M, Speit G (2011) Workshop summary: top concentration for *in vitro* mammalian cell genotoxicity assays; and report from working group on toxicity measures and top concentration for *in vitro* cytogenetics assays (chromosome aberrations and micronucleus). Mutat Res 723:77–83

Goodman & Gilman (August 13, 2001) In: Hardman JG, Limbird LE, Gilman AG (eds) The pharmacological basis of therapeutics, 10th edn. McGraw-Hill Professional, New York

Guidance for companies requesting scientific advice (EMEA/CVMP/172329/2004-Rev.3) (2012) http://www.ema.europa.eu/docs/en_GB/document_library/Regulatory_and_procedural_guideline/2009/10/WC500004147.pdf

Hareng L, Pellizzer C, Bremer S, Schwarz M, Hartung T (2005) The integrated project ReProTect: a novel approach in reproductive toxicity hazard assessment. Reprod Toxicol 20(3):441–452
Hartmann A, Agurell E, Beevers C, Brendler-Schwaab S, Burlinson B, Clay P, Collins A, Smith A, Speit G, Thybaud V, Tice RR (2003) Recommendations for conducting the *in vivo* alkaline Comet assay. Mutagenesis 18:45–51
Hartung T, Bremer S, Casati S, Coecke S, Corvi R, Fortaner S, Gribaldo L, Halder M, Hoffmann S, JanuschRoi A, Prieto P, Sabbioni E, Scott L, Worth A, Zuang V (2004) A modular approach to the ECVAM principles on test validity. ATLA 32:467–472
Hayashi M, MacGregor JT, Gatehouse DG, Blakey DH, Dertinger SD, Abramsson-Zetterberg L, Krishna G, Morita T, Russo A, Asano N, Suzuki H, Ohyama W, Gibson D (2007) *In vivo* erythrocyte micronucleus assay. III. Validation and regulatory acceptance of automated scoring and the use of rat peripheral blood reticulocytes, with discussion of non-hematopoietic target cells and a single dose-level limit test. Mutat Res 627:10–30
ICH (1996) In: D'Arcy PF, Harron DWG (eds) Proceedings of the third international conference on harmonisation, Yokohama 1995. Queen's University, Belfast, 998p
ICH (1997) S1B testing for carcinogenicity of pharmaceuticals. http://www.ich.org/fileadmin/Public_Web_Site/ICH_Products/Guidelines/Safety/S1B/Step4/S1B_Guideline.pdf
ICH (2013) ICH guideline S1, Regulatory notice on changes to core guideline on rodent carcinogenicity testing of pharmaceuticals
ICH guideline S2 (R1) on genotoxicity testing and data interpretation for pharmaceuticals intended for human use, Step 5 (2012) http://www.ema.europa.eu/docs/en_GB/document_library/Scientific_guideline/2011/12/WC500119604.pdf
Kirkland D, Fowler P (2010) Further analysis of Ames-negative rodent carcinogens that are only genotoxic in mammalian cells *in vitro* at concentrations exceeding 1 mM, including retesting of compounds of concern. Mutagenesis 25:539–553
Kirkland D, Speit G (2008) Evaluation of the ability of a battery of three *in vitro* genotoxicity tests to discriminate rodent carcinogens and non-carcinogens. III. Appropriate follow-up testing *in vivo*. Mutat Res 654:114–132
Kirkland D, Aardema M, Henderson L, Müller L (2005) Evaluation of the ability of a battery of three *in vitro* genotoxicity tests to discriminate rodent carcinogens and non-carcinogens. I. Sensitivity, specificity and relative predictivity. Mutat Res 584:1–256
Kirkland D, Hayashi M, Jacobson-Kram D, Kasper P, MacGregor JT, Müller L, Uno Y (2007a) The international workshops on genotoxicity testing (IWGT): history and achievements. Mutat Res 627:1–4
Kirkland D, Pfuhler S, Tweats D, Aardema M, Corvi R, Darroudi F, Elhajouji A, Glatt H, Hastwell P, Hayashi M, Kasper P, Kirchner S, Lynch A, Marzin D, Maurici D, Meunier J-R, Muller L, Nohynek G, Parry J, Parry E, Thybaud V, Tice R, van Benthem J, Vanparys P, White P (2007b) How to reduce false positive results when undertaking *in vitro* genotoxicity testing and thus avoid unnecessary follow-up animals tests: report of an ECVAM workshop. Mutat Res 628:31–55
Manolis E, Vamvakas S, Isaac M (2011) New pathway for qualification of novel methodologies in the European medicines agency. Proteomics Clin Appl 5:248–255
Marx-Stoelting P, Adriaens E, Ahr HJ, Bremer S, Garthoff B, Gelbke HP, Piersma A, Pellizzer C, Reuter U, Rogiers V, Schenk B, Schwengberg S, Seiler A, Spielmann H, Steemans M, Stedman DB, Vanparys P, Vericat JA, Verwei M, van der Water F, Weimer M, Schwarz M (2009) A review of the implementation of the embryonic stem cell test (EST). The report and recommendations of an ECVAM/ReProTect workshop. Altern Lab Anim 37(3):313–328
Matthews EJ, Kruhlak NL, Cimino MC, Benz RD, Contrera JF (2006) An analysis of genetic toxicity, reproductive and developmental toxicity and carcinogenicity data. I. Identification of carcinogens using surrogate endpoints. Regul Toxicol Pharmacol 44:83–96
Moore MM, Honma M, Clements J, Awogi T, Douglas GR, van Goethem F, Gollapudi B, Kimura A, Muster W, O'Donavan M, Schoeny R, Wakuri S (2011) Suitable top concentration for tests with mammalian cells: mouse lymphoma assay workgroup. Mutat Res 723:84–86
Müller L, Choi E, Yamasaki E et al (1999) ICH-harmonized guidances on genotoxicity testing of pharmaceuticals. Evolution, reasoning and impact. Mutat Res 436:195–225

Müller L, Tweats D, Galloway S, Hayashi M (2013) The evolution, scientific reasoning and use of ICH S2 guidelines for genotoxicity testing of pharmaceuticals. In: Van der Laan JW, DeGeorge JJ (eds) Global approach in safety testing. Advances in the pharmaceutical sciences series, vol 5, pp 37–75

Nambiar PR, Morton D (2013) The rasH2 mouse model for assessing carcinogenic potential of pharmaceuticals. Toxicol Pathol 41:1058–1067

NIH (1997) Validation and regulatory acceptance of toxicological test methods. A report of the *ad hoc* interagency coordinating committee on the validation of alternative methods. NIH Publication 97-3981. NIEHS, Research Triangle Park, NC, USA, 105 pp

NIH (1999) Evaluation of the validation status of toxicological methods: general guidelines for submissions to ICCVAM (revised, October 1999). NIH Publication 99-4496. NIEHS, Research Triangle Park, NC, USA, 44 pp

OECD (2005) Guidance document on the validation and international acceptance of new or updated test methods for hazard assessment. OECD Testing Series and Assessment Number 34. ENV/JM/MONO(2005)14. OECD, Paris, France, pp 96

Ohno (1992) Toxicity testing: regulatory perspectives. In: D'Arcy PF, Harron DWG (eds) Proceedings of the second international conference on harmonisation, Brussels 1991. Queen's University, Belfast, pp 186–188

Ohno Y (2013) A Japanese perspective on implementation of the three Rs: incorporating best scientific practices into regulatory process. In: Van der Laan JW, DeGeorge JJ (eds) Global approach in safety testing. Advances in the pharmaceutical sciences series, vol 5, pp 37–75

Omori Y (1992) Principles and guidelines—a review of recommendations (on detection of toxicity) in the three regions. In: D'Arcy PF, Harron DWG (eds) Proceedings of the first international conference on harmonisation, Brussels 1991. Queen's University Belfast, pp 256–266

Parry JM, Parry E, Phrakonkham P, Corvi R (2010) Analysis of published data for top concentration considerations in mammalian cell genotoxicity testing. Mutagenesis 25:531–538

Perry (1992) Toxicity testing programme. Background paper. In: D'Arcy PF, Harron DWG (eds) Proceedings of the second international conference on harmonisation, Brussels 1991. Queen's University, Belfast, pp 183–186

Putman E, Van der Laan JW, Van Loveren H (2003) Assessing immunotoxicity: guidelines. Fundam Clin Pharmacol 17:615–626

Qualification of novel methodologies for drug development: guidance to applicants (EMA/CHMP/SAWP/72894/2008) (2014) http://www.ema.europa.eu/docs/en_GB/document_library/Regulatory_and_procedural_guideline/2009/10/WC500004201.pdf

Recommendation to marketing authorisation holders for veterinary vaccines, highlighting the need to update marketing authorisations to remove the target animal batch safety test (TABST) following removal of the requirement from the European Pharmacopoeia monographs (EMA/CHMP/CVMP/JEG-3Rs/746429/2012) (2013) http://www.ema.europa.eu/docs/en_GB/document_library/Scientific_guideline/2013/06/WC500144488.pdf

Recommendation to marketing authorisation holders, highlighting the need to ensure compliance with 3Rs methods described in the European Pharmacopoeia (EMA/CHMP/CVMP/JEG-3Rs/252137/2012) (2012) http://www.ema.europa.eu/docs/en_GB/document_library/Scientific_guideline/2012/07/WC500130369.pdf

Replacement of animal studies by *in vitro* models (Position adopted by the CPMP on 19 February 1997) (CPMP/SWP/728/95) (1997) http://www.ema.europa.eu/docs/en_GB/document_library/Scientific_guideline/2009/09/WC500003354.pdf

Robinson DE, MacDonald JS (2001) Background and framework for ILSI's collaborative evaluation program on alternative models for carcinogenicity assessment. International Life Sciences Institute. Toxicol Pathol 29(Suppl):13–19

Rothfuss A, O'Donovan M, De BM, Brault D, Czich A, Custer L, Hamada S, Plappert-Helbig U, Hayashi M, Howe J, Kraynak AR, van der Leede BJ, Nakajima M, Priestley C, Thybaud V, Saigo K, Sawant S, Shi J, Storer R, Struwe M, Vock E, Galloway S (2010) Collaborative study on fifteen compounds in the rat-liver Comet assay integrated into 2- and 4-week repeat-dose studies. Mutat Res 702:40–69

Rothfuss A, Honma M, Czich A, Aardema MJ, Burlinson B, Galloway S, Hamada S, Kirkland D, Heflich RH, Howe J, Nakajima M, O'Donovan M, Plappert-Helbig U, Priestley C, Recio L, Schuler M, Uno Y, Martus HJ (2011) Improvement of *in vivo* genotoxicity assessment: combination of acute tests and integration into standard toxicity testing. Mutat Res 723:108–120

Scott D, Galloway SM, Marshall RR, Ishidate M, Brusick D, Ashby J, Myhr BC (1991) Genotoxicity under extreme culture conditions, a report from ICPEMC Task Group 9. Mutat Res 257:147–204

Sistare FD, Morton D, Alden C, Christensen J, Keller D et al (2011) An analysis of pharmaceutical experience with decades of rat carcinogenicity testing: support for a proposal to modify current regulatory guidelines. Toxicol Pathol 39:716–744

Spielmann H, Pohl I, Döring B, Liebsch M, Moldenhauer F (1997) The embryonic stem cell test (EST), an *in vitro* embryotoxicity test using two permanent mouse cell lines: 3T3 fibroblasts and embryonic stem cells. *In Vitro* Toxicol 10:119–127

Statement of the EMA position on the application of the 3Rs (replacement, reduction and refinement) in the regulatory testing of human and veterinary medicinal products (EMA/470807/2011) (2011) http://www.ema.europa.eu/docs/en_GB/document_library/Other/2011/10/WC500115625.pdf

Sullivan, FM, Watkins, WJ, van der Venne, MTh (1993) The toxicology of chemicals—series two: reproductive toxicology, EUR 12029 EN 14991

Takayama S (1992) Proposal for mutual acceptance of studies. In: D'Arcy PF, Harron DWG (eds) Proceedings of the first international conference on harmonisation, Brussels 1991. Queen's University Belfast, pp 266–269

Theunissen PT, Beken S, Cappon GD, Chen C, Hoberman AM, Van der Laan JW, Stewart J, Piersma AH (2014) Toward a comparative retrospective analysis of rat and rabbit developmental toxicity studies for pharmaceutical compounds. Reprod Toxicol 47:27–32

Tice RR, Agurell E, Anderson D, Burlinson B, Hartmann A, Kobayashi H, Miyamae Y, Rojas E, Ryu JC, Sasaki YF (2000) Single cell gel/comet assay: guidelines for *in vitro* and *in vivo* genetic toxicology testing. Environ Mol Mutagen 35:206–221

van der Laan JW, Herberts CA, Jones DJ, Thorpe S, Stebbings R, Thorpe R. The nonclinical evaluation of biotechnology-derived pharmaceuticals, moving on after the TeGenero case. In: Corsini E, van Loveren H (eds) Molecular immunotoxicology. Wiley-VCH Verlag, pp 189–207

Van der Laan JW, Chapin RE, Haenen B, Jacobs AC, Piersma AH (2012) Testing strategies for embryo-fetal toxicity of human pharmaceuticals. Animal models vs *in vitro* approaches. A workshop report. Regul Toxicol Pharmacol 63:115–123

Van der Laan JW, DeGeorge JJ, Sistare F, Moggs J (2013) Toward more scientific relevance in carcinogenicity testing. In: Van der Laan JW, DeGeorge JJ (eds) Global approach in safety testing. Advances in the pharmaceutical sciences series, vol 5, pp 37–75

Van Meer PJ, Kooijman M, van der Laan JW, Moors EH, Schellekens H (2013) The value of non-human primates in the development of monoclonal antibodies. Nat Biotechnol 31(10): 882–883

Van Oosterhout JPJ, Van der Laan JW, De Waal EJ, Olejniczak K, Hilgenfeld M, Schmidt V, Bass R (1997) The Utility of two rodent species in carcinogenic risk assessment of pharmaceuticals in Europe. Regul Toxicol Pharmacol 25:6–17

Van Cauteren, Bentley P, Bode G, Cordier A, Coussement W, Heining P, Sims J (2000) The industry view on long-term toxicology testing in drug development of human pharmaceuticals. Pharmacol Toxicol 86(Suppl I):1–5

Weaver J, Tsutsui N, Hisada S, Vidal J-M, Spanhaak S, Sawada J-I, Hastings KL, Van der Laan JW, Van Loveren H, Kawabata TT, Sims J, Durham SK, Fueki O, Matula T, Kusunoki H, Ulrich P, Nakamura K (2005) Development of the ICH guidelines on immunotoxicology. evaluation of pharmaceuticals using a survey of industry practices. J Immunotoxicol 2:171–180

Weissinger J (1992) Commentary on proposal for mutual acceptance and proposed alternative approaches. In: D'Arcy PF, Harron DWG (eds) Proceedings of the first international conference on harmonisation, Brussels 1991. Queen's University, Belfast, pp 183–186

Chapter 4
Validation of Alternative *In Vitro* Methods to Animal Testing: Concepts, Challenges, Processes and Tools

Claudius Griesinger, Bertrand Desprez, Sandra Coecke, Warren Casey and Valérie Zuang

Abstract This chapter explores the concepts, processes, tools and challenges relating to the validation of alternative methods for toxicity and safety testing. In general terms, validation is the process of assessing the appropriateness and usefulness of a tool for its intended purpose. Validation is routinely used in various contexts in science, technology, the manufacturing and services sectors. It serves to assess the fitness-for-purpose of devices, systems, software up to entire methodologies. In the area of toxicity testing, validation plays an indispensable role: "alternative approaches" are increasingly replacing animal models as predictive tools and it needs to be demonstrated that these novel methods are fit for purpose. Alternative approaches include *in vitro* test methods, non-testing approaches such as predictive computer models up to entire testing and assessment strategies composed of method suites, data sources and decision-aiding tools. Data generated with alternative approaches are ultimately used for decision-making on public health and the protection of the environment. It is therefore essential that the underlying methods and methodologies are thoroughly characterised, assessed and transparently documented through validation studies involving impartial actors. Importantly, validation serves as a filter to ensure that only test methods able to produce data that help to address legislative requirements (e.g. EU's REACH legislation) are accepted as official testing tools and, owing to the globalisation of markets, recognised on international level (e.g. through inclusion in OECD test guidelines). Since validation creates a credible and transparent evidence base on test methods, it provides a quality stamp, supporting companies developing and marketing alternative methods and creating considerable business opportunities.

C. Griesinger • B. Desprez • S. Coecke • V. Zuang (✉)
European Commission, Joint Research Centre (JRC), Ispra, Italy
e-mail: Valerie.ZUANG@ec.europa.eu

W. Casey
Division of the National Toxicology Program, National Institute of Environmental Health Sciences, Research Triangle Park, NC, USA

Interagency Coordinating Committee on the Validation of Alternative Methods (ICCVAM), Washington, DC, USA

C. Eskes, M. Whelan (eds.), *Validation of Alternative Methods for Toxicity Testing*, Advances in Experimental Medicine and Biology 856,
DOI 10.1007/978-3-319-33826-2_4

Validation of alternative methods is conducted through scientific studies assessing two key hypotheses, reliability and relevance of the test method for a given purpose. Relevance encapsulates the scientific basis of the test method, its capacity to predict adverse effects in the "target system" (i.e. human health or the environment) as well as its applicability for the intended purpose. In this chapter we focus on the validation of non-animal *in vitro* alternative testing methods and review the concepts, challenges, processes and tools fundamental to the validation of *in vitro* methods intended for hazard testing of chemicals. We explore major challenges and peculiarities of validation in this area. Based on the notion that validation per se is a scientific endeavour that needs to adhere to key scientific principles, namely objectivity and appropriate choice of methodology, we examine basic aspects of study design and management, and provide illustrations of statistical approaches to describe predictive performance of validated test methods as well as their reliability.

1 Introduction

What is validation and why do we need it? Validation of alternative methods has been defined as the process by which the reliability and relevance of a particular method is established for a defined purpose (Balls et al. 1990a, b, c, 1995a, b; OECD 2005). This definition has then later been extended to alternative approaches in the wider sense, i.e. not only covering individual methods but also combinations thereof, including strategies for data generation and integration. The reliability relates to the within- and between-laboratory reproducibility as well as to the transferability of the method or approach in different laboratories, whereas relevance relates mainly to its predictive capacity and, importantly, to the biological/mechanistic relevance, traditionally subsumed as "scientific basis". Judging the overall relevance however also includes aspects of applicability domain and even the level of reliability required in view of the purpose of the method. The defined purpose can be various and range from full replacement of a regulatory test to the generation of mechanistic information relevant to the type and extent of toxic effects which might be caused by a particular chemical (Frazier 1994).

In regulatory toxicity testing, validation is placed between research/development and regulatory acceptance and aims at the characterisation of an *in vitro* test method under controlled conditions which in turn leads to the standardisation of the test method protocol. This aspect of test method development has been summarised in Coecke et al. (2014). Validation generally facilitates and/or accelerates the international (regulatory) acceptance of alternative test methods. In fact, the regulatory acceptance of tests that have not been subjected to prevailing validation processes is discouraged by international bodies (OECD 2005). This is true not only for alternative methods but also for tests conducted in animals. The term "regulatory acceptance" of an *in vitro* test method relates to the formal acceptance of the method by regulatory authorities indicating that the test method may be used as an official tool to provide information to meet a specific regulatory requirement. This includes, but is not limited to, a formal adoption of a test method by EU and/or OECD as an EU test method and included in the EU Test Methods Regulation and/or as an OECD Test Guideline, respectively. Standardisation and international adoption of testing approaches supports worldwide acceptance of data. Under the OECD Test Guideline Programme this is known as Mutual Acceptance of Data (MAD). MAD saves every year an appreciable number of animals and other resources as it avoids duplicate testing.

Three main types of validation processes have been defined: prospective, retrospective and performance standards-based validation—the latter being a form of prospective validation. Prospective validation relates to an approach to validation when some or all information necessary to assess the validity of a test is not available, and therefore new experimental work is required (OECD 2005). Retrospective validation relates to an assessment of the validation status of a test method carried out by considering all available information, either as available in the published literature or from other sources (e.g. data generated during previous validation studies (OECD 2005) or in-house testing data from industry). Validation based on Performance Standards relates to a validation study for a test method that is structurally and functionally similar to a previously validated and accepted reference test method. The candidate test method should incorporate the essential test method components included in Performance Standards developed for the reference test method, and should have comparable performance when evaluated using the reference chemicals provided in the Performance Standards (OECD 2005).

The European Reference Laboratory for Alternatives to Animal Testing (EURL ECVAM) [formerly known as the European Centre for the Validation of Alternative Methods, ECVAM] and its international collaborators published recommendations concerning the practical and logistical aspects of validating alternative test methods in prospective studies (Balls et al. 1995a, b). These criteria were subsequently endorsed by and mirrored in the procedures of the US Interagency Coordinating Committee on the Validation of Alternative Methods (ICCVAM 1997), and later internationally, summarised in the "Guidance Document 34" of the Organisation for Economic Cooperation and Development (OECD) (OECD 2005).

In 2004, ECVAM proposed a modular approach to the validation of alternative methods (Hartung et al. 2004), according to which the various information requirements for peer-review and as generated during the validation process are broken down into seven independent modules. According to this modular approach, the information requirements can be fulfilled by using data obtained from a prospective study, by a retrospective evaluation of already existing data/information, or by a combination of both.

More recently, the concepts of weight of evidence validation/evaluation (Balls et al. 2005) and evidence-based validation (Hartung 2010) have been introduced; Weight of evidence validation involves the careful analysis and "weighing" of data with regard to their quality, plausibility, etc. in view of concluding whether it supports one or the other side of an argument, in this context whether or not a particular method is useful for a specific purpose. Evidence-based validation essentially refers to the use of tools from evidence-based medicine for purposes of alternative method validation. These may range from systematic reviews (e.g. to determine reference data or analyse a set of existing data) over data grouping and meta-analysis to more probabilistic descriptors of test method performance as are used in medicine, for instance to describe the performance and usefulness of diagnostic tests.

This chapter explores the fundamental concepts behind validation, the hypotheses assessed and information generated, outlines specific challenges of alternative methods validation that relate to the nature of test methods being reductionist proxies for the human situation and provides a detailed discussion of the practical aspects of organising, designing, planning and conducting a validation study and analysing the data generated by appropriate statistical analyses (see also Chap. 5).

2 Validation: Principles, Hypotheses Assessed and Information Generated

This section examines fundamental principles of validation and explores the hypotheses and information generated by validation studies of alternative methods conducted in the context of their envisaged use for the safety assessment of specific test materials such as chemicals (of various chemical and/or use categories) and their integration in integrative approaches (e.g. Integrated Testing Strategies, ITS or Integrated Approaches to Testing and Assessment, IATA). Instead of simply recapitulating commonly accepted concepts of alternative method validation described in OECD guidance document Nr 34 (OECD 2005), we unfold this topic in the following way:

- First we will consider a series of fundamental issues that are necessary for the understanding of some unique features of the validation of alternative approaches.
- Second, we will examine three key concepts and explain their meaning in view of avoiding confusion regarding terminology. These are (a) *validation workflow*, (b) *validation study type (or validation process)* and (c) *the validation information generated through dedicated studies*. These three are often subsumed under the term "validation" but it is important to understand them as separate categories.
- Third, we will discuss the broader concept of 'validation' in view of deducing the central hypotheses assessed by alternative method validation. This will serve to understand the commonalities between validation in general and validation of alternative methods, and sculpt out some specific characteristics of the latter, in particular those constituting major challenges. These challenges include (a) finding appropriate reference data for *in vitro* test method development ("calibration") and validation and (b) the identification of mechanisms that are causative for downstream (i.e. more complex) events and hence should be modelled in reductionist and mechanistically-based alternative methods.
- Finally, we will discuss in more detail the information that needs to be satisfied in order to consider an alternative method valid for a specific purpose. We will put a particular emphasis on the composite nature of judging the overall relevance of alternative methods. This discussion will then lead over to section three and

four that explore the management, planning, design and conduct of validation studies in a manner so as to satisfy these information requirements. Details on EURL ECVAM's specific approach regarding multi-laboratory trials can be found in Chap. 5.

2.1 Fundamental Considerations

2.1.1 Validation in the Current Context Relates to *In Vitro* Methods Used for Toxicity Testing

Validation, i.e. the process of establishing the usefulness and appropriateness of a method for a given purpose, is applicable to a wide range of biological and analytical methods, e.g. in diagnostic medicine, food safety, etc. In the current context, we focus on the validation of biological *in vitro* test methods for toxicity testing of chemicals and safety testing of biologicals (Hendriksen et al. 1998). Therefore, chemicals (or biologicals) are the basic entities used to study and report the performance, utility and applicability of alternative method during validation. Consequently, selecting an appropriate set of test chemicals is of key importance when planning a validation study (see Sect. 4.2). One should however not lose sight of the fact that a mere summary and analysis of testing data would not yet make a complete validation study: a fundamental aspect to consider relates to the biological and physiological processes modelled by the test method and thought to be relevant for the chemicals' adverse effects. This is called the "scientific basis" of a test method and needs to be properly described. This helps judging the plausibility of results obtained with a given test method and supports the assessment of its relevance for a given purpose (see Sect. 2.3.3).

For more than a quarter century validation studies have been conducted in the areas of safety testing of chemicals and biologicals (e.g. vaccines) as well as ecotoxicological toxicity testing. Considerable efforts have been invested in developing internationally agreed validation frameworks, notably by the European Commission's EURL ECVAM, the Centre for Alternative to Animal Testing (CAAT), the US Interagency Coordinating Committee on the Validation of Alternative Methods (ICCVAM) as well as many individual scientists in academia, industry, government and international organisations (Scala 1987; Balls et al. 1990a, b, c, 1995a, b; Frazier 1990a, b; Frazier 1992; Green 1993; Balls 1994; Walum et al. 1994; Fentem et al. 1995; Balls and Karcher 1995; Goldberg et al. 1995; Bruner et al. 1996). This led to the development of reports and guidance documents adopted on international level, such as the OECD report on the harmonisation of validation and acceptance criteria for alternative toxicological test methods (OECD 1996, updated in 2009) which later gave rise to the more complete OECD guidance document on the validation and international acceptance of new or updated test methods for hazard assessment (OECD 2005). These documents reflect the status of international agreement at the beginning of the millennium.

Validation has played and is continuing to play a key role in toxicity testing because of its confidence- and trust-building role. Validation, overseen by impartial actors and subjected to scientific peer review leads to the comprehensive characterisation and transparent documentation of novel test methods. It is an important prerequisite for the international recognition and regulatory acceptance of test methods, e.g. through uptake into relevant legislations outlining official test methods recognised for use in a specific jurisdiction. Examples are the EU's Test Method Regulation EC 440/2008 and, on a global level, internationally accepted guidelines such as OECD's Test Guidelines (TGs). Although formally not relating to legislation per se, data produced in agreement with TGs are binding for OECD member countries due to the OECD agreement on Mutual Acceptance of Data (MAD) dating back to 1981. This stipulates that *"data generated in the testing of chemicals in an OECD Member country in accordance with OECD Test Guidelines and OECD Principles of Good Laboratory Practice shall be accepted in other Member countries for purposes of assessment and other uses relating to the protection of man and the environment.*" Notably, combinations of methods as described in OECD guidance documents on "Integrated Approaches on Testing and Assessment" (the OECD term for Integrated Testing Strategies) are not covered by the MAD agreement at present. We will focus in this chapter mainly on validation as a means of characterising and assessing alternative approaches in view of their fitness for regulatory acceptance.

2.1.2 Validation Has Largely Focused on Hazard Testing So Far

Importantly, validation of alternative methods for toxicity testing has mainly focused on predicting potential hazards of chemicals, that is, their intrinsic potential to cause adverse effects in a particular test system (i.e. an animal, cell type, etc.), without providing much information on the potency. Potency relates to the doses required to provoke adverse effects in a whole organism and is key information for a complete risk assessment of chemicals. What are the major bottlenecks concerning methods addressing potency? First, the concentrations that a given cellular population in a human body is exposed to following systemic exposure through the environment are typically not known: hence it is difficult to define appropriate concentrations of test chemicals that should be used in *in vitro* systems—including when validating these. More effort needs to be invested in approaches (including *in vitro* systems) for assessing toxicokinetic processes, i.e. the absorption, distribution, metabolism and excretion (ADME) of chemicals. Reliable data and/or simulations would assist in defining the appropriate range of chemical concentrations to be used in alternative approaches. This has been already highlighted by Balls and colleagues in 1995 (Balls et al. 1995a, b). Second, due to the reductionist nature of alternatives (see below), processes that may influence the human *in vivo* potency (including ADME) are present only to a limited extent in alternative approaches. Hence there is considerable uncertainty regarding the use of concentration-response information from an artificially reduced test system (e.g. a confluent layer of hepatocytes) for predicting potential dose-response relationships (e.g. for hepatotoxicity) via *in vitro–in vivo* extrapolation (IVIVE), even if rooted in mechanistically informed physiologically based kinetic modelling (PBK).

2.1.3 Definition of "Alternative Approaches"

We refer to the definition of "alternative approaches" as suggested by Smythe (1978), i.e. alternatives to *established scientific procedures* which can lead to the *replacement*, the *reduction* or the *refinement* of animal experimentation, thus addressing the 3Rs principle as established by Russell and Burch (1959). Alternative approaches in this sense cover individual test methods, test batteries, strategic combinations of test methods (testing strategies) as well as holistic approaches towards data generation, evaluation and integration. These have been termed "Integrated Testing Strategies (ITS)" or, more recently, "Integrated Approaches to Testing and Assessment (IATA)" and can be composed of testing and non-testing methods. Validation can in principle extend to the assessment of integrated approaches (Kinsner-Ovaskainen et al. 2012). A surprisingly common misunderstanding regarding validation is that it is focusing on one-to-one replacements, i.e. one single alternative that replaces one single traditional animal test. This is however not the case, validation is context-dependent and purpose driven and includes all sorts of assays, also those that address initial mechanisms of action, intermediate effects, pathways of toxicity or modes of action. Further, the term 'alternative method' can relate to *empirical testing methods* (often *in vitro* methods) or methodologies that are not based on empirical testing and therefore referred to as "non-testing methods".

Non-testing methods are essentially approaches employing basic logical and plausibility reasoning or sophisticated mathematical approaches. Examples of non-testing methods include grouping of substances, read-across from one substance to another on the basis of properties such as chemical structure or biological mechanisms, structure-activity relationships (SARs) and quantitative SARs (QSARs). It also includes, in the wider sense, biological modelling approaches including modelling the kinetics of xenobiotics such as physiologically based pharmacokinetic (PBPK) modelling and its applications in toxicokinetics.

With *test methods* we refer to a scientific methodology based on a biological test system (e.g. a cell population, a reconstructed tissue or an excised organ) as well as provisions for handling this system and performing measurements following exposure to chemicals (i.e. the test method's procedure, normally captured in Standard Operating Procedure(s), SOP(s), outlining the related life science or analytical measurement techniques), as well as those relating to data analysis, processing and interpretation.

All alternative methods will need to process the raw data and translate them into toxicologically meaningful information, i.e. the actual results of the test method. This process is often referred to as *data analysis*. The results can then further be converted into predictions of the toxic effects of interest. This is achieved by so-called *prediction models* (Archer et al. 1997; OECD 2005), a description of how to interpret the data or measurements in view of obtaining categorical predictions. This often takes the form of a mathematical function or algorithm The predictions can stretch the entire spectrum of biological organisation, from molecular interactions over mechanisms on organelle or cell level (e.g. signalling pathways) up to mechanisms of cell ensembles, tissues, organs up to the entire organism or (sub)populations. A prediction model can be generically phrased as

$$P = f(x)$$

with P the prediction, f the mathematical transformation of the measured data x. Prediction models can be very simple, for instance in case of *in vitro* skin irritation test methods based on reconstructed human epidermis, a 50 % cell viability of the exposed skin equivalent is taken as a cut-off for ascribing either irritant or non-irritant properties to the test chemical. Notably, not all alternative test methods do feature prediction models, e.g. ecotoxicological assays.

The key components of alternative test methods are schematically summarised in Fig. 4.1.

2.1.4 Alternative Methods are Proxies and Reductionist Models

Typically, life science research is conducted on model systems, which can be further separated into (a) "*proxy*" (*or* "*surrogate*") *systems* and (b) *reductionist systems*, with possible overlap between the two (see below).

First, proxy systems are entities used to study properties of another system: a substantial amount of basic research in biology and biomedicine is conducted on

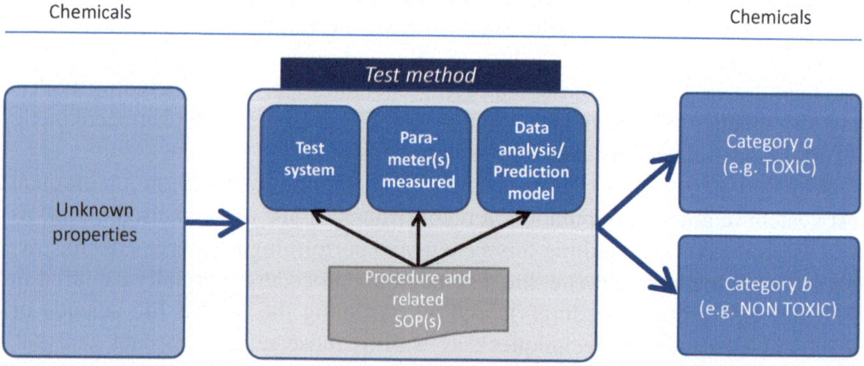

Fig. 4.1 An alternative test method and its main constituting elements: the biological 'test system', the biological parameters measured in response to exposure of the test system to chemicals and the element of data interpretation and analysis that translate data into toxicologically useful information. This may (but does not need to) include a 'prediction model', i.e. a prescriptive procedure of how to translate the measurements obtained in the test system into categorical predictions. Chemicals with unknown properties can be tested by the method and, using the prediction model or 'classifier', can be assigned to specific categories that can relate to any property ranging from activation of a cellular pathway to a downstream human health effect. Test methods can therefore be seen as "sorting machines" that allow to allocate chemicals with initially unknown properties to distinct categorical classes with defined properties normally relating to the presence or absence of the capacity to trigger a specific biological mechanism related to toxicity

animals (so-called "animal models") with a view to extrapolate the obtained results on animal genetics, biology and physiology to the human situation. Animals are used as proxies for humans with the basic assumption that, with increasing phylogenetic proximity, the results obtained in the proxies are considered more relevant, accurate and less uncertain with respect to reflecting the situation in the "target system", i.e. the human. In many cases, this has proven a successful approach in life science. For example, understanding the dopaminergic system and its involvement in Parkinson's disease has been largely obtained through animal experimentation. Despite these successes, there are limitations with regard to the use of proxies, probably due to phylogenetic differences, including at the level of gene expression, physiological mechanisms, metabolism, etc. that add uncertainty to results from animal studies as models for the human situation. Recent examples from pharmacological preclinical safety trials include the "tegenero incident" (Horvath and Milton 2009; Attarwala 2010) and the unexpected (hepato)toxicity of the antibiotic trovafloxacin (Borlak 2009; Gregory 2014). Secondly, there are reductionist models, such as brain slice cultures, dissociated primary cells and cell lines which are used to study specific physiological processes which recapitulate, at a highly reduced level of complexity, specific mechanisms, structures or other properties of the target system.

Alternative *in vitro* methods represent an interesting blend of these two concepts: they are proxies inasmuch as they are used for human safety testing *in lieu* of humans but they are also highly reductionist methods, since they are modelling only aspects of the target system (e.g. a complete organism or an organ, etc.) and are used to predict the properties of the target system or some of its constituting parts. Consider a barrier model composed of confluent polarised epithelial cells from the gut used to study uptake of substances through this epithelium or the use of monocytic cell lines for studying markers of epidermal inflammation and immune cell activation in the context of skin sensitisation leading to the clinical manifestations of allergic contact dermatitis. Both are highly reduced systems that model key properties thought to underlie higher level ("downstream/apical") effects in the system of interest, the 'target system'. As a major consequence of these facts, both test method development and validation are typically undertaken in relation to proxies or surrogate systems (i.e. animal data) and not the species of interest (Fig. 4.2).

Epistemologically, *in vitro* alternatives used to predict behaviour of chemicals in more complex systems up to the entire organism can be seen as a variety of *explanatory reduction* (Weber 2005). This type of reductionism can be seen as based on the identification of a "difference-making principle" (Waters 1990, 2007) assumed to be a causative (and/or explanatory) factor that is sufficient for studying and explaining features that are emergent at a higher level of organisation. Reductionist systems used to predict such higher level (downstream) properties need to model this "principle" (here: a physiological mechanism; in genetics: the concept of the gene) in order to study potential consequences in the complex target system (e.g. toxicity in humans and human (sub)populations). In the current context, finding the difference making principle is equal to the identification of physiological mechanisms believed to underlie a response (i.e. an adverse effect) in the target system.

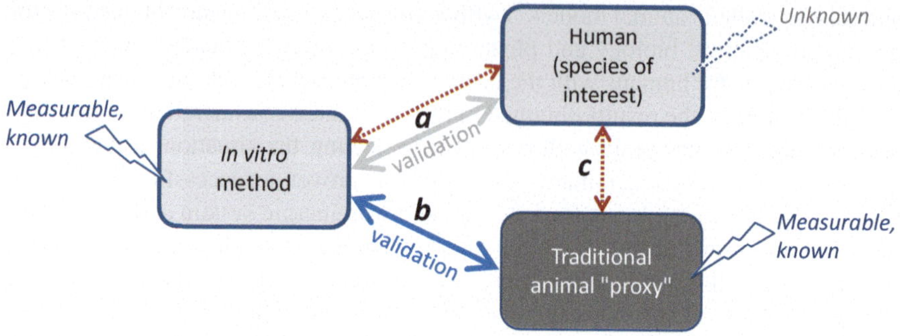

Fig. 4.2 Peculiarities and obstacles specific to validation of alternative methods. The effects of chemicals on the target system of interests (humans) are normally not known (*dotted flash*) or only known to a very limited extent and associated with much uncertainty. (**a**) Alternatives can therefore not be readily validated in relation to the target system (*grey arrow*). (**b**) Since traditionally animals have been used as proxies for humans, a lot of data is available concerning chemical effects on whole animal systems (*blue outlined flash*). In the absence of standardised human data, these data can be used as reference data for validation (*blue arrow*) and related to the measured data—also for developing a prediction model that translates the measured data *in vitro* into a prediction of measured effects in the animal proxy. (**c**) Importantly, the true relationships of the effects measured in the animal proxy and in the *in vitro* method to the target system are often not known (*red dotted arrows*)

In the context of validation, the entire mapping of biological and mechanistic information has been referred to as the "scientific basis" of alternative test methods. This relates to their biological and mechanistic design in terms of recapitulating mechanisms of toxicity (e.g. cellular pathways) or any relevant disturbances of structure or function at the cellular level (mode of action concept) or throughout the different layers of biological organisation. This means that responses from a reduced system need to be extrapolated to a more complex system.

Identifying the key causative factors or events that underlie specific adverse outcomes and which allow predicting these outcomes with sufficient accuracy and reliability is one of the biggest challenges of modern toxicology. The basic assumptions are that it is (a) unnecessary and (b) practically impossible to model all potential mechanisms. Finding those that truly make a difference in view of tilting the homeostatic balance and driving adverse effects is pivotal for developing relevant test methods. While a thorough understanding and description of the scientific basis of alternative test methods has always been part of validation, there are increasing efforts to organise the existing scientific knowledge in a consistent manner so as to improve interactions between various actors within the community (e.g. scientists, test method developers, validators, regulators, legislators, test method users). Identifying and describing physiological key events that can be perturbed by toxicants will allow adjusting chemical design of new substances so as to avoid the interference of substances with known "toxicity pathways" but will allow also help tailoring the scientific development of new test methods and informing their validation. The OECD guidance of describing "Adverse Outcome Pathways" (AOP)

(OECD 2013a, b) is a recent effort to describe the biological events leading to toxicity in humans in a concise but consistent manner. The AOP concept foresees the structuring of key events in relation to the biological level of complexity on which they occur and arranging the different events in causality chains, starting from a molecular initiating event and describing the causal relationships between one key event and another. This approach could improve the identification and description of such key factors and might thus support the development and validation of test methods that map/recapitulate mechanisms and pathways that underlie downstream events or higher-level features of the system. In addition, it has been proposed that evidence-based methods such as systematic reviews could help identifying key causative events triggering the development of biologically relevant test methods (Guzelian et al. 2005; Hoffmann and Hartung 2006b).

2.1.5 Reductionism: Consequences for Test Development and Validation

Above considerations show that the usefulness of alternative test methods needs to be assessed in relation to the target system: *reference points* (=data) need to be derived ideally from the target system that the alternative approach is modelling. It is not sufficient to validate *in vitro* methods in relation to other *in vitro* systems (Goldberg et al. 1995). This is in contrast to many forms of validation where the usefulness of systems is judged in relation to reference points relating to performance of similar systems (e.g. diagnostic tests). Finding accurate reference data (Fig. 4.2) of the actually "true" effect of a given chemical on the species of interest (humans) would be the obviously the ideal approach for assessing the usefulness ("relevance") of alternative methods. However, human data relating to chemical effects (Fig. 4.2) are normally not readily available or need at least to be derived from highly uncertain information (e.g. epidemiological data), involving moreover expert judgement.

This absence of human data makes it very difficult to "calibrate" alternative methods during test method development against the target system whose properties the alternative is intended to predict. This "calibration" typically consists of developing a data analysis procedure for processing the raw measurements into toxicologically meaningful results. This can include a prediction model that translates the measurements obtained in the alternative methods into categorical predictions, either relating to a category system used for hazard labelling relating to adverse health effects (e.g. UN GHS categories) or to a specific mechanism of action or toxicity pathway. To overcome this issue of non-availability of human data, reference data are traditionally taken from proxies or "surrogates", i.e. animal models that have been used in toxicology for many decades (consider for instance the Draize eye and skin tests in rabbits dating to 1944) although these animal models have never been validated themselves (i.e. how well they model or predict the effects in humans or how reliable/repeatable they are).

Moreover, there may be cases where no *reference method* is available, for instance when a method for a new purpose needs to be developed. This would mean that there are no *reference data* at all for the development and validation of an *in vitro* method. The statistical tool of "latent class analysis" (LCA) may be a viable approach to estimate assay performance parameters even when the true state of nature is not known or has not been observed (=is 'latent') (Hothorn 2002; Hoffmann et al. 2008). Yet another situation is found in environmental toxicity assessment which uses very few surrogate species/proxies for judging the impact of chemicals on a specific environment, habitat or ecosystem. An example is the use of daphnia or fish as surrogates (i.e. two of thousands of species!) for judging the potential impact on the aquatic ecosystem. For more details on reference points in validation of alternatives see the ECVAM workshop report by Hoffmann et al. (2008).

2.1.6 Modelling the Mechanism Is Necessary But Not Necessarily Sufficient

As outlined above, alternatives that are based on modelling biological events that are assumed to be causative for adverse effects in the species of interest are more credible and useful than methods that show only correlative results with the target system, i.e. without modelling relevant biological mechanisms. This has been pointed out already in early publications on validation (Goldberg et al. 1995; Bruner et al. 1996). It is thus tempting to assume that methods which model such results should quasi automatically produce results that are informative and relevant for downstream health effects. This is however not necessarily the case: biological systems have a great capacity to repair and reset their properties once disturbed (homeostasis). Reduced test methods typically do not model all those homeostatic mechanisms and hence the results can be of limited relevance, especially for health effects that depend on repeated exposure and a variety of stressors (e.g. epigenetic changes involved in cancer). Hence, the modelling of mechanisms in alternative test methods is a necessary precondition of robust and relevant predictions, but it is not necessarily sufficient with respect to the accuracy of such predictions.

2.1.7 Reductionism Requires Integration at Later Stages

A consequence of the fact that alternative *in vitro* methods are *reductionist* models is that in most cases no single method will suffice to describe the properties of the higher-level target system with its complex anatomical and physiological organisation. Consider the health effect of reproductive toxicity: several organs and complex hormonal feedback loops are involved which cannot be modelled by one single reductionist system. Instead, the lack of complexity at the level of individual test methods is sought to be compensated by using a suite of test methods and other information sources that each address different properties of the target system. The complexity of the target system is basically dissected into aspects that can be modelled and experimentally manageable in several reductionist systems. Information from such groups of methods (including also non-testing methods) then needs to be integrated through strategic combinations of test methods in holistic data gathering and evaluation schemes. These, initially, have been termed tier-testing strategies or testing strategies (Balls et al. 1995b) and later referred to as "Integrated Testing Strategies", implemented in the REACH guidance published on ECHA's website from 2007 onwards and including also elements of data collection and evaluation (see also Kinsner-Ovaskainen et al. 2012; Balls et al. 2012). Subsequently, the OECD has introduced the term "Integrated Approaches to Testing and Assessment" (IATA) (OECD 2008; OECD 2014a, b). The communality is that data from various sources, irrespective of whether already available or to be generated, are integrated in order to yield conclusions on whether specific chemicals trigger a particular property of the target system.

This need for data integration has important consequences for validation. Back in the early 90s validation of alternative methods initially intended to establish single replacement methods for addressing an entire health effect (e.g. EC/HO validation study on alternatives to the Draize eye irritation test). This has worked to some extent in topical toxicology where the Draize test for skin corrosion and irritation could be successfully addressed by two sets of *in vitro* test methods, both based on Reconstructed human Epidermis (RhE) (additional methods are available for skin corrosion assessment). However, it is plausible that other health effects of a more systemic nature (often referred to as "complex endpoints" or "systemic toxicity") will require a strategic combination of alternative methods and this has to be taken into account already when validating the individual "building blocks" of such strategies (Bouvier d'Yvoire et al. 2012). This has been discussed in a joint EURL ECVAM/EPAA workshop report (Kinsner-Ovaskainen et al. 2012). We will return to this aspect later in the context of the requirements in terms of chemical number for assessing reliability versus predictive capacity/applicability domain.

The later use of an alternative test method within larger integrative approaches has impacts on validating such a method and the study design needs to take this into account. For example, if a method is used in (strategic) combination with other assays, it is conceivable that the requirements regarding predictive capacity and even reliability are different as opposed to situations where a method would be used as a stand-alone test. The same holds true for a screening assay versus one used to address regulatory requirements as observed by Green (1993).

2.2 *Validation: Basic Terminology*

Traditionally validation has been seen as a process of assessing the scientific validity of an alternative method. While this is still correct, validation carries additional meanings: for example, when scientists talk about a test method as "being validated" they rather refer to whether or not a method has been shown to be reliable and relevant, i.e. whether the hypotheses that modelling a specific mechanism of action in such a reductionist model indeed picks up, during validation testing, chemicals known to cause specific adverse effects. Therefore, the term validation in the area of toxicity texting incorporates at least three aspects: (a) the formal process of validation or validation workflow; (b) the validation study type (or "validation process") and (c) information generated during validation or the hypotheses assessed during validation.

2.2.1 Validation Workflow

Validation traditionally has been overseen by independent and impartial organisations ('validation bodies') that do not have vested interests. Examples are EURL ECVAM in the EU, NICEATM/ICCVAM in the US, JaCVAM in Japan, KoCVAM in South-Korea, Health Canada and BraCVAM in Brazil. This impartiality is important for the validation of alternative methods that are intended for regulatory acceptance: it ensures that the characterisation and confirmation of validity of test methods is done on the basis of scientific considerations only and independent of specific/vested interests (financial, etc.) of test method submitters. It thus guarantees impartiality, scientific rigour and consistency of approach. All validation organisations follow a practical workflow or process for prioritising test methods for validation, for conducting studies, for subsequent independent peer review and for organising and communicating their main conclusions and recommendations. The above mentioned (supra)national validation bodies in the EU, Japan, Canada, South-Korea and the US work together within the ICATM framework (ICATM = International Cooperation on Alternative Test Methods) and have recently attempted to align and streamline their workflow (see Chap. 14 on international collaboration). A generic validation body workflow comprising four basic steps is shown in Fig. 4.3 and explained below.

Step 1: Evaluation of *in vitro* test methods: Importantly, not all *in vitro* methods that are developed by test method developers will be necessarily validated by validation bodies. Before being able to enter validation, proposed *in vitro* test methods will need to be evaluated against a catalogue of criteria such as: is the test method sufficiently developed to enter validation, in particular is there a "mature" protocol/SOP available and are there some initial data on within-laboratory repeatability and reproducibility (for details see, Sect. 2.3.2)? Does it produce information that could be useful for the intended application, in particular regulatory decision making? Once these criteria have been confirmed, a test method might be considered for validation, a process that involves the use of considerable funding and resources, typically of public funds. Only methods that promise to generate useful information and, in particular, address toxicity effects for which there is no 'alternative coverage' yet, will merit such investment. Essentially, at this step, validation organisations will conduct a cost/benefit analysis in view of prioritisation.

Step 2: Designing and conducting a validation study: Validation involves dedicated scientific studies to determine whether the alternative method appropriately models and, if applicable, predicts the properties of the target system. We will discuss

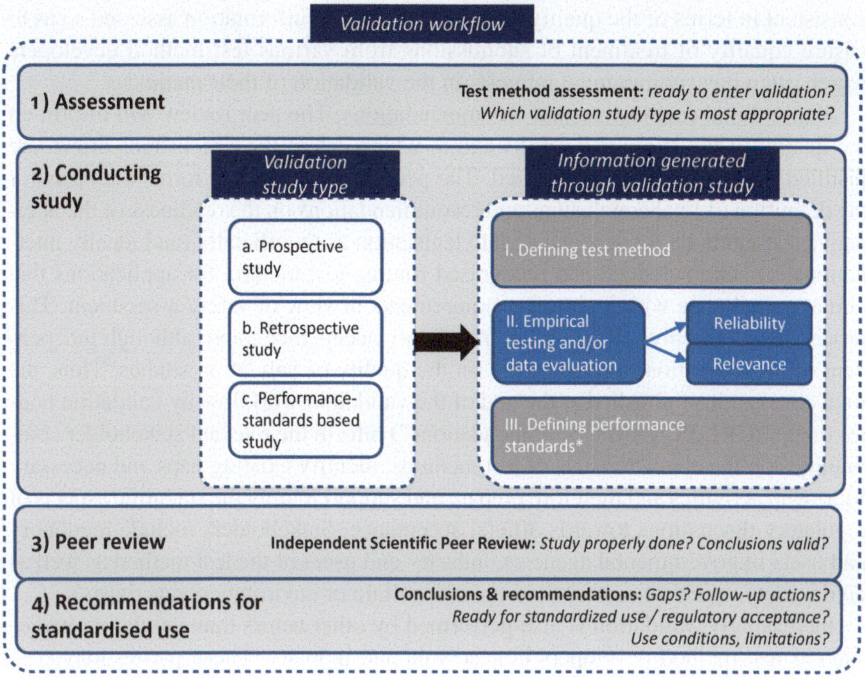

Fig. 4.3 Schematic outline of the overall validation workflow of independent validation organisations

the various validation study types in more detail in Sect. 2.2.2 below. Irrespective of the validation study type, the information generated relates to three aspects: (i) a better characterisation and definition of the *in vitro* test method (as a result of validation), (ii) the assessment of the key hypotheses of reliability and relevance (which we will explore below in more detail below, Sect. 2.3) as well as (iii) the setting of performance standards and operational criteria (e.g. refinement of test acceptance criteria) that will guide development and validation of future test methods based on similar principles.

<u>Step 3: Independent scientific peer review.</u> Since validation is achieved through scientific experimental studies which, as all scientific endeavours, contain elements of data interpretation and inference to reach conclusions, an essential element of validation is the assessment of the results obtained and the conclusions drawn by an independent group of knowledgeable scientists. This peer review assesses whether key scientific principles such as objectivity and appropriateness of methodology have been observed and will to this end evaluate managerial and study design aspects, ranging from the choice of specific readouts, over composing the data matrix to statistical tools used for analysing the data. In contrast to the peer review of scientific manuscripts undertaken by individual scientists with normally relatively little guidance from the journals/publishers, the peer review of alternative methods, especially when conducted by public validation bodies, needs to be highly

consistent in terms of the quality criteria used and the information assessed so as to ensure equality of treatment of submissions from various test method developers which often have commercial interests in the validation of their methods.

Step 4: Final conclusions and recommendations: The peer review will inform on the quality of the study and results and to which extent the conclusions drawn are justified by the data/results obtained. The peer review typically forms the basis for the definition of final conclusions and recommendations on the readiness of the alternative test methods for acceptance into legislation as an officially (and ideally internationally) acknowledged and recognised routine test method for applications that aim at compliance with legislative requirements in view of safety assessment. This process is commonly referred to as 'regulatory acceptance', and, although independent of the validation process, relies on the quality of validation studies. Thus, the final conclusions published at the end of the validation workflow by validation bodies (e.g. "EURL ECVAM Recommendations") inform the relevant stakeholder communities on the characteristics of test methods, identify existing gaps and necessary follow-up activities and therefore prepare and support mainly the scientific aspects of regulatory discussions towards official acceptance. Stakeholders include regulatory end users in governmental agencies, industry end users of the test method as well as civil society organisations (such as animal welfare or environmental activists).

Increasingly, validation is also performed by other actors than validation bodies such as test method developers in academia and industry. These parties may seek independent and impartial evaluation and peer review of their studies by validation bodies that are neutral with respect to the assay (i.e. do not have vested interests). For instance, EURL ECVAM is regularly evaluating 'external' validation studies and having them reviewed by its independent EURL ECVAM Scientific Advisory Committee (ESAC).

2.2.2 Validation Study Types

Validation of alternative methods for toxicity testing is centred on the analysis of testing data relating to a relevant set of chemicals (the so-called "test chemicals"). Testing is normally carried out through formal validation studies that should follow scientific principles and good scientific practice with regard to study design and conduct (see Chap. 5), in particular relating to chemical selection, the statistical planning (e.g. to calculate the power required to derive dependable point estimates such as sensitivity and specificity), but also with regard to the statistical analysis of the study data themselves. This will be explored in detail in Sect. 3.

There are different types of validation studies conceivable that vary in their design. A useful distinction is based on whether the chemicals testing data need to be generated *de novo* (so called prospective studies) or whether they are already existing and are analysed in view of a defined purpose (retrospective studies). Studies can of course also contain both prospective and retrospective elements: make use of newly generated data as well as existing data (see Sect. 2.4 on modular approach).

Prospective Studies

(a) Prevalidation studies
Prevalidation studies are studies conducted in view of assessing whether a test method and associated SOP is ready to merit potential further full validation (Curren et al. 1995) and robust enough to merit the considerable expense of such a study. Prevalidation studies focus on the aspect of transferring the SOP/test method from an experienced laboratory (e.g. test method developer) to naïve laboratories. These studies allow optimising further the SOP based on the experiences during such transfer. Prevalidation studies thus help minimising the risk of transfer problems during full prospective validation studies. Transfer problems due to shortcomings of the SOP or training protocols only create unnecessary cost during full validation studies without contributing to the core goal of a full validation study, i.e. test method characterisation in view of a purpose. Transfer(ability) is assessed through testing a small but conscientiously selected set of chemicals with also challenging properties, such as chemicals that are at the border of the prediction model cut-off (see also Sect. 4.7.2). A major benefit of conducting prevalidation studies is that they produce limited but quality controlled data sets on within-laboratory reproducibility (WLR), between-laboratory reproducibility (BLR) and predictive capacity which may inform about the possible overall *performance* to be expected of a specific test method (see also Sect. 4.6). Like in other validation contexts (e.g. analytical method validation) such *a priori* knowledge and other historical data may support the realistic setting of goals and objectives of subsequent full prospective validation studies, including potential validation acceptance criteria where useful (i.e. if the precise use of the method in a regulatory setting is already known). Although the term prevalidation is not any longer frequently used, there are still studies conducted that adhere to the principles of prevalidation, namely a first check of transferability of SOP from one laboratory to another, identification of pitfalls and improvement of SOP and/or training, if necessary, before embarking on a costly multi-laboratory ring trial.
(b) Full prospective validation studies
These are large-scale studies involving the testing of a sufficient sample of chemicals for characterising a test method in terms of WLR, BLR and predictive capacity and for characterising, with some confidence, its applicability domain and potential limitations. Adaptions of the design of such studies have been suggested and will be discussed below. Such studies create confidence and trust in alternative novel test approaches for regulatory applications when involving a sufficiently large set of test chemicals. When considering the size of the chemical testing set, it is important to separate what would be statistically desirable (in terms of chemical sample size) from what is realistically doable taking also considerations of cost and availability of test materials into account.
(c) Performance standards-based (PS-Based) validation studies
PS-based studies are conducted in relation to a set of predefined "standards", including biological criteria and reference chemicals, as a means of

efficiently assessing test methods considered to be sufficiently similar to a previously-validated one. This concept has been initially proposal by Balls (1997). PS are typically defined upon completion of a full validation study. However assessment criteria (factually standards) can also be defined for test method development already and carried over to test method evaluation/validation.

Normally PS-based studies are used to validate, through a smaller scale study involving significantly less chemicals, test methods that are scientifically sufficiently similar to the previously validated "reference methods". The rationale of these studies is that similar biological and operational characteristics will most likely mean that the general performance of a similar method can be assumed to be equivalent to the validated reference method and that it is therefore justifiable to test a smaller set of chemicals instead of repeating a full scale validation exercise. Performance standards typically are composed of three elements: (i) The essential test methods components, defining the test methods and key operational parameters, (ii) a set of "Reference Chemicals" that need to be tested (typically in the range of 20 or so), (iii) target performance values in term of WLR, BLR and predictive capacity. Importantly both the reference chemicals and the target values are defined on the basis of the full validation and the parameters achieved: thus, these values map at a reduced scale the chemical, toxicological and functional spectrum of test chemicals and the values attained during validation. A significant drawback of this study type is that the test chemicals are known beforehand and can be used for test method development.

Retrospective Studies

These studies use existing testing information that can be analysed through data grouping and meta-analysis tools. For a short introduction to meta-analysis see Mayer (2004). Retrospective validation may sometimes be conducted through a 'systematic review'; this term however rather relates to the methodology used. As long as the goal of the systematic review is to characterise a method, and through this, assess its validity for a purpose, it technically constitutes a validation exercise. Retrospective studies require particular attention with respect to the selection of the data through using pre-defined search and selection criteria.

Weight of Evidence Validation and Evidence-Based Validation

While the terms prospective and retrospective validation relate to the temporality of data generation, the concept of weight of evidence (WoE) validation relates to the tools for evaluating data sets relevant for a given validation study. Weight of evidence generally relates to the considerations made in a situation where there is uncertainty and which are used to ascertain whether the information/evidence at hand support one or the opposite side of an argument or a conclusion. In the context of validation, WoE considerations can be useful in situations where there is uncertainty regarding the available reference data or in case different and opposing findings from reference methods are available. WoE judgment may of course also be used in case there are several contradictory testing results in retrospective data sets. Possible principles of WoE validation have been summarised by Balls et al. (2005). However, WoE approaches can be tailored to individual needs as long as they are underpinned by the consistent use of a predefined set of criteria relating to quality, relevance, plausibility, etc. of the data. The second element, integrating this information in view of arriving at a final judgement, may depend on the specific case.

Evidence-based validation (Hartung 2010) is a term suggested for validation studies that make full use of data assembly and analysis tools as well as advanced statistical tools as used in (evidence-based) medicine (Mayer 2004). This includes data grouping, the concepts and techniques of meta-analysis as well as the use of likelihood ratios to summarise predictive performance and the consideration of prevalence (Hoffmann and Hartung 2005). This should be seen in the wider context of introducing evidence-based methods from medical research (including systematic reviews) also in toxicology (Hoffmann and Hartung 2006b; Griesinger et al. 2009; Guzelian et al. 2009; Stephens et al. 2013) in order to address in particular issues of variability and uncertainty (Aggett et al. 2007) and make remaining uncertainties transparent (Guzelian et al. 2005). While the evidence-based approaches from medicine can to some extent be used also in toxicology (Neugebauer 2009), there are however important differences between medicine and toxicology which need to be taken into account (Griesinger 2009) (e.g. the focus of prevention in toxicology versus prevention and cure in medicine or the differences of the entities studied through test methods: chemical properties in toxicology and diseased patients in medicine).

2.3 Validation: Hypotheses Assessed and Information Generated

Having outlined fundamental concepts relating to the assessment of alternative methods and validation workflow, we explore the term validation in more detail in the following.

2.3.1 The General Concept of Validation

Validation aims to show whether or not something is valid. "Valid" is rooted in the latin verb *valere*—to be (of) worth. This shows the core goal of any validation: assessing whether something has (some) worth or usefulness. From this, two key characteristics of all validation exercises can be deduced:

- First, the terms "worth" or "valid" are highly context-dependent: something is of "worth" in relation to something or for a specific use, application or performance. Thus, validation always relates to a *specific context* or *purpose*. This purpose may change over time, requiring revisiting or re-conducting validations of systems that have been previously validated in relation to a different purpose. In the context of alternative methods validation, this purpose-oriented aspect is described by the term "**relevance**". Relevance has been described as the *usefulness* and *meaningfulness* of the results of an alternative method (Balls et al. 1990a, b, c; Frazier 1990a, b). We would like to emphasize that it is this rather broad understanding of relevance (Bruner et al. 1996; OECD guidance document Nr. 34, glossary) that we are using here. Unfortunately, relevance has sometimes been reduced to mere aspects of predictive capacity and applicability of an assay. However, judging the overall relevance requires the integration of many types of information and requires also scientific judgement: relevance is a *composite measure* and involves also the biological/mechanistic relevance ("scientific basis") and may also include considerations of reliability of a test method (Bruner et al. 1996). We will discuss this in more detail in Sect. 2.3.3.

- Secondly, a system/method or process is only then fully relevant for an application or purpose, if it is reliable: if it performs in the same manner each time it is applied, irrespective of the operator and in a reasonable independence of the setting within which it is used (e.g. a computer programme should not only work on the developer's computer, but on those of millions of users). This is described by the term "**reliability**". It is immediately intuitive that a test method that is unreliable cannot be relevant for its purpose. Inversely, the purpose of a method will have an influence on the reliability that is requested from of a given test method. For some purposes (e.g. when combining test methods in a battery) a lower reliability may be acceptable than when using an alternative method as a stand-alone replacement test. Thus, reliability may need to be taken into account when judging the overall relevance of a test method for a purpose.

Based on these brief considerations, one can frame the key characteristics of any validation exercise including alternative method validation:

1. Validation is the *process* required to assess/confirm or assess validity for purpose as described under (2)
2. The validation process concerns the *assessment of the value* (*validity*) of a system *within a specific context and/or for a specific purpose*, typically a use scenario or a specific application by examining whether the system reliably **fulfils the requirements** of that specific purpose in a **reliable** manner and is **relevant** for the intended purpose ("fitness for purpose") or application.
3. The validation process is a **scientific endeavour a**nd as such needs to adhere to principles of objectivity and appropriate methodology (study design). Accepting that validation studies are of a scientific nature means that they should be described in terms of assessing clearly described *hypotheses*. These hypotheses include (1) the reliability of an assay when performed on the basis of a prescriptive protocol, (2) the mechanistic or biological relevance of the effects recapitulated. This is measured through testing, during validation, a wide array of chemicals with known properties regarding an adverse health effect: if the modelled mechanism is relevant, this will be reflected in the accuracy of the

predictions or measurements. This will also show whether there are specific chemical classes or other properties for which no accurate predictions can be obtained (applicability/limitations); (3) the predictive relevance, i.e. the appropriateness of the prediction model developed typically on a small set. Obviously, hypotheses 1–3 are related. For practical purposes, they are grouped in *reliability* and *relevance*.

Typically, validation has assessed this "fitness for purpose" outlined in the three hypotheses above by studying (i) whether or to which extent the system fulfils *predefined specifications* relating to performance (for instance sensitivity and specificity of predictions made), (ii) the reliability (and operability) deemed necessary to satisfy the intended purpose as well as (iii) robustness, which is measured *inter alia* through the ease of transferring a method from one to another laboratory which is typically done in prevalidation studies (Curren et al. 1995). Points of reference or predefined standards for predictive capacity and reliability therefore play a key role in validation (Hoffmann et al. 2008). Importantly, the process of validation will inevitably lead to the **characterisation** of the system's performance and, if applicable, its operability, generating useful information even in case the validation goal/objective is not met, the method is not (yet) found fit for purpose or "scientifically valid". Test method validation, should therefore also be seen as a way of characterising a system for future improvement and adaptation. It is this general concept of validation that underlies also the validation of alternative approaches.

2.3.2 Validation of Alternative Methods: Reliability and Relevance

As outlined above, the theoretical basis of alternative method validation can be readily deduced from the general concept of validation: the two key hypotheses assessed by alternative method validation are **reliability** and (overall) **relevance** incorporating biological relevance, relevance (concordance) of predictions for various chemicals (applicability domain) and, at times, reliability.

This definition goes back to discussions at a workshop in Amden, Switzerland in 1990 conducted by the *Centre for Alternatives to Animal Testing* (*CAAT*), USA and the *European Research Group for Alternatives in Toxicity Testing* (*ERGATT*) whose results have been published as CAAT/ERGATT workshop report on the validation of toxicity test procedures (Balls et al. 1990a, b, c). Being sufficiently general, the original definition relating to relevance and reliability provides an appropriate framework for validation of alternatives still today.

In the following we would like to explore how these two hypotheses are addressed in validation studies in more detail:

First, an alternative test method can only be considered useful if it shows ***reliability***, i.e. if it provides the same results or shows the same performance characteristics over time and under identical as well as different conditions (e.g. operators, laboratories, different equipment, cell batches, etc.). In the context of validation studies, reliability has been defined as assessing the (***intra-laboratory***) ***repeatability*** and the reproducibility of results ***within*** **and** ***between laboratories*** over time (Balls et al. 1990a, b, c, 1995a, b; OECD 2005). Repeatability relates to the agreement of results within one laboratory when the procedure is conducted *under identical conditions* (OECD 2005), while reproducibility relates to the agreement of results using the same procedure but not necessarily under identical conditions (e.g. different operators in one laboratory or different laboratories).[1] Reliability assessment is important in view of assessing the performance of methods in their final use scenario, i.e. employed in laboratories across the world. Assessment of within- and between-laboratory reproducibility is often done by means of measuring *concordance* of (i.e. agreement between) predictions obtained with the prediction model. This has the advantage that the reliability is measured on the basis of the **intended results** or **output** generated by the test method, i.e. again under final use conditions. However, it is also important to describe, using appropriate statistical methods, the **intrinsic variability of the parameter**(s) **measured** (see also Sect. 4.7.2) in the test method (e.g. cell viability, fluorescence as a result of the expression of a reporter gene, etc.). This will allow producing data on reproducibility (or inversely variability) independent of the prediction model and therefore closer to the actual data produced. Such data may be useful in case the prediction model is changed due to post hoc analyses. A *post-hoc* improvement of the prediction model has recently been done on the basis of *in vitro* skin corrosion methods (Desprez et al. 2015). In addition, the transferability of a method is an aspect that needs attention during validation: it relates to how easily a method can be transferred from one experienced laboratory (e.g. test method developer) to naïve laboratories that may have relevant experience with alternative methods but are, at least, inexperienced with the particular SOP associated with the test method (Fig. 4.1). Transferability relates to both the reliability but also the "robustness" of a test method: the more sensitive a method is to slight variations of equipment and operators, the less robust it is. Robustness is important when considering a test method for standardised routine use. A practical way of gauging robustness at early stages is through checking the ease with which a test method can be transferred from one to another laboratory (e.g. in the context of a prevalidation study). Robustness however will also be reflected in the levels of repeatability, and within- and between laboratory reproducibility obtained during validation.

Second, in view of ensuring that an alternative test method is fit for a specific purpose (i.e. the reliable generation of data on the properties of test chemicals) its ***relevance*** *for this purpose* needs to be assessed. This requires that the *purpose is clearly defined* before validation. A surprisingly common shortcoming of validation exercises is that the intended purpose of the test method and, therefore, the goal and

[1] *Repeatability* has been defined as "*the agreement of test results obtained within a single laboratory when the procedure is performed on the same substance and under identical conditions*" (OECD 2005) i.e. the same operator and equipment.

Reproducibility has been defined as "*the agreement of test results obtained from testing the same substance using the same protocol*" (OECD 2005), but not necessarily under identical conditions (i.e. different operators and equipment).

objectives of a validation study are not defined with sufficient precision. This has been already remarked on by Balls and colleagues in 1995 (Balls et al. 1995a, b). Inversely, over-ambitious goals are sometimes set, including the specification of target performance values (e.g. for specificity and sensitivity) which are not sufficiently backed by prior data. Lack of goal setting or defining objectives has a negative impact on the clarity of study design (see Sect. 4): as for all scientific experiments, the objectives of a study will determine the necessary design. Thus, study design is not a 'one-size fits all' issue, but depends on the specifics of the study. With regard to validation of alternatives, relevance for a particular purpose has been defined as assessing the scientific *meaningfulness* and *usefulness* of results from alternative methods (Balls et al. 1990a, b, c, 1995a, b; Frazier 1990a, b). Meaningfulness in this context is crucial and relates to the plausibility of data or predictions and how convincing they are on the basis of a variety of considerations. As observed by Goldberg et al. (1995) and Bruner et al. (1996), hazard predictions from alternative methods that address a specific known mechanism of action or because they closely model a specific tissue are scientifically more credible and are probably more likely to be correct than predictions from a test methods that that does provide correct predictions but does *not* model the biology of the target system or whose relationship with the latter are at least unknown (such assays could be called "correlative methods"). Thus, when judging the overall relevance of a test method, also biological or mechanistic relevance needs to be taken into consideration, i.e. to which extent the alternative model recapitulates key aspects of biology, physiology and toxicity that need to be assessed. This aspect has traditionally been referred to as the "**scientific basis**" of a test method.

2.3.3 Key Information for Relevance: Scientific Basis, Predictive Capacity, Applicability Domain and Also Reliability

As indicated above, relevance is a rather broad term and judgement of relevance is to some extent a subjective process that relies on the evaluation and integration of scientific data. To assess or establish the relevance of a method for a defined purpose requires considering the method's predictive capacity, its applicability domain and limitations, its reliability and, at a more fundamental level, its scientific basis: the biological and/or mechanistic relevance of the test method in view of it being considered a suitable proxy or surrogate for the target system and a model of key causative elements that are involved in emergent properties of the target system (see discussion on explanatory reductionism Sect. 2.1 subpoint 3). Figure 4.4 schematically summarises the information taken into account for judging the overall relevance against the defined purpose.

The four aspects for judging relevance of a method are elaborated in the following:

(a) **Scientific basis** relating to the biological or mechanistic relevance of a test method and its underlying test system. Does it recapitulate a specific tissue

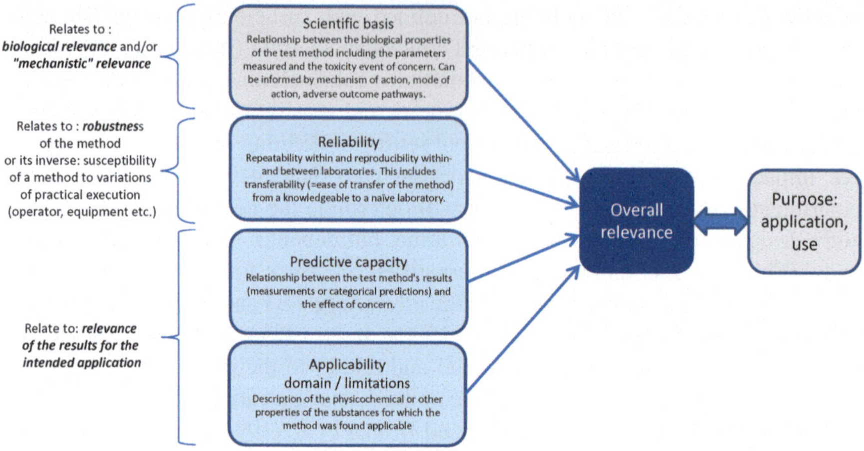

Fig. 4.4 Judging the overall relevance of a method against a specified purpose upon completion of a validation study requires information on the biological and mechanistic relevance (scientific basis) of a test method, its reliability, its predictive capacity and applicability domain. Note that the scientific basis of a method should be defined on the outset of a study (*light grey*) and is not based on empirical testing generated during the study, while information on reliability, predictive capacity and applicability are assessed through the data generated during validation (*boxes in light blue*). Empirical data on the relevance of the results (e.g. an IC_{50} measurement) or categorical predictions (="predictive capacity") in regard of the effect of concern allow falsifying or "verifying" the hypothesis that a particular scientific basis is relevant for predicting an adverse effect. The scientific basis hence is the foundation of a test method. Its description is informed by considerations of mechanisms of action (MOA, relating to the specific biochemical interaction by which a drug/toxin acts on the target system), mode of action (MoA, relating to functional or anatomical changes correlated with the toxicity effect) and adverse outcome pathways (AOP, relating to descriptions of sequences of biological key events that lead from initial molecular interactions of the toxin with the system to downstream adverse health effects of individuals or populations)

architecture, mechanism or action or biological/toxicological pathway? We provide a few examples to illustrate this point:

Reconstructed human epidermis used for skin irritation testing has a high biological relevance for the intended application (prediction of the irritancy potential of chemicals) as it models the upper part of the human skin and is based on human keratinocytes. The predominant readout used for skin irritation testing is cell viability which has some relation to the toxicity mechanisms: it models cell and tissue trauma which is a key event for triggering an inflammatory response in skin leading to the clinical symptoms of irritation (redness, swelling, warmth) (Griesinger et al. 2009, 2014). However, more specific markers that directly probe inflammatory processes would be closer to the toxicity event from a mechanistic point of view (draft AOP in Griesinger et al. 2014).

As another example, transactivation assays for measuring the potential of chemicals to act as (ant)agonists on endocrine receptors (e.g. estrogen, androgen receptors) typically are based on cell lines intrinsically expressing these receptors. Such assays have a high mechanistic relevance as they directly model the mode of action. However, depending on the test system used and the degree of reduction applied (i.e. cell line versus tissue), they have a reduced biological relevance.

(b) **Predictive capacity: The relationship between the measurements obtained with the alternative method and the effects in the biological system that the alternative method is supposed to model.** Typically this relationship is captured through assessing the capacity of the alternative method to provide accurate predictions of specific effects in the biological target system. This is called a test method's "predictive capacity". The effects predicted typically relate to distinct categories and constitute "classifiers" (in standard scientific terms one could say that the continuum of effects from non-toxic to highly toxic has undergone a binning procedure; the basis for this binning often relate to decision-rules that relate to regulatory traditions of categorising health effects). These classifiers normally relate to predictions of downstream adverse health effect ("apical endpoint" such as skin or eye irritation and their respective classification and labelling categories), but they may also relate to a specific cellular mechanism involved in toxicogenesis ('toxicity pathway'), to an organ-level effect, etc.

An example of predictive capacity of a health endpoint is *in vitro* skin irritation: skin equivalent models based on human keratinocytes that grow into epidermis-like tissue equivalents in the dish are used to predict the skin irritation effect of chemicals in humans (OECD TG 439 2010; Griesinger et al. 2009). The capacity to predict skin irritation is characterised through an evaluation of test chemicals with known reference properties in the target (or surrogate) system. Here they relate to irritants as defined by classification and labelling schemes such as GHS versus 'non-classified'. The predictive capacity is described by standard statistical measures used for analysing diagnostic or predictive test methods, as long as these methods aim at making categorical predictions of the sort "positive" versus "negative" (=true presence or absence of a property). These are mainly sensitivity (=true positive rate), specificity (=true negative rate) and accuracy (sum of true negatives and true positives over all predictions made); see Fig. 4.5. Importantly, these are all statistical point estimates and they are independent of the balance between positives and negatives in the reference data. Often positive and negative predictive values (PPV, NPV) are also used to characterise the performance of alternatives. However, these values are dependent on the prevalence of positives amongst the test chemicals (see Fig. 4.5) and care needs to be taken when using these descriptors for predictive capacity of test methods after validation studies where normally the balance is 50:50 % (i.e. there is a 50 % prevalence). NPV and PPV only provide meaningful information when either the prevalence of the test chemicals during validation matches the prevalence in the real situation or by taking the prevalence into account when calculating NPV and PPV on the basis of the sensitivity/specificity values obtained during validation using a balanced set (50:50 %). Analogies between the assessment of test methods for chemical safety assessment and those for diagnosing diseases are tempting and hold true

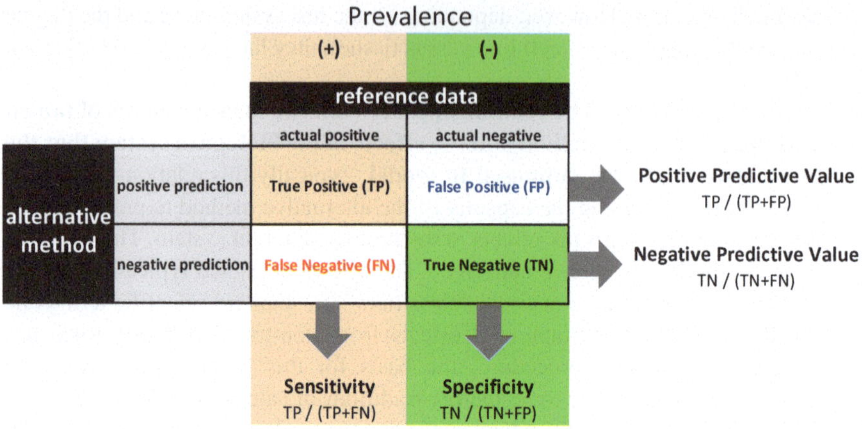

Fig. 4.5 Predictive capacity of a test method is described by assessing its ability to yield correct predictions for classes of properties described by reference data. In the example below, a classical contingency table, there are two categories of the reference data: actual positive and actual negative. The prevalence of chemicals that are ascribed these properties has impact on the statistical analyses and the parameters that are useful. The alternative test method has a prediction model that allows binary classification, either "positive" or "negative". Comparing the results of the alternative method with the reference data allows ascribing to the results of the alternative method the arguments True Positive (TP), False Positive (FP), False Negative (FN) and True Negative (TN). Note that the sensitivity (=true positive rate) and specificity (=true negative rate) are independent of the prevalence of actual negatives/positives. In contrast, both positive and negative predictive values are dependent on this balance

for the most of the statistical issues (Hoffmann and Hartung 2005), but should be used with some care due to obvious differences of the entities examined (diseased people versus chemicals causing adverse effects) (Griesinger 2009) and some issues related to prevalence: while the prevalence of a given disease (e.g. type II diabetes) may be grounded on solid evidence, establishing the 'prevalence' of toxic chemicals with regard to a specific health effect can be challenging. One approach used in the past was to assess the number of entries in chemical registries (e.g. the EU new chemicals database). However it should be noted that the chemicals listed there have already undergone safety assessments and the real prevalence of chemicals when they are being subjected to test methods may be different. Further, other measures in addition to NPV and PPV may be useful when expressing the quality of binary classifications, in particular in cases when actual positives and negatives are highly unbalanced. This includes the "Matthews Correlation Coefficient" (MCC) (Matthews 1975) that indicates the correlation between predictions and observations (actual negatives/positives) on a scale of −1 (no correlation whatsoever) over 0 (random) to 1 (fully correlated).

Assessing the predictive capacity of a test method requires the availability of **reference data** that are used to "calibrate" the prediction model of the method and to assess its predictive capacity during validation. These reference data are often from animal studies and relate to categorical values such as "actual positive" and "actual negative" ascribed to a set of test chemicals. Notably, reference data already carry a considerable degree of simplification due to the reduction of a much more complex reality of a continuum of physiological events into a binary (or other) classification. Reference data therefore need to be used with care, especially when derived from surrogate/proxy animal models, i.e. not the species of interest as is typically the case in toxicology.

(c) **The reliability** of a test method also may influence judgements on its overall relevance. Consider for instance the impact of the practical use scenario of a test method on its relevance judgment: test methods that will be used on their own (stand-alone replacements) will have to show a high degree of reproducibility in order to be judged relevant for the purpose of effectively replacing a traditional animal test. For example, reliability thresholds for single replacement test methods such as skin corrosion and skin irritation are very high. Other test methods on the other hand will be used in conjunction with others, either in parallel, assessing the frequency/mode of predictions obtained from such a "battery" or through strategic step-wise combinations of test methods.[2] In such use cases, test methods with reproducibility performances lower than those of single replacement methods may be nevertheless useful and judged relevant, for instance when used in weight-of-evidence approaches to support plausibility reasoning such as read-across of properties from one chemical substance to another. The relationship between intended use and requirements in terms of accuracy and also reliability was first noted by Green (1993).

Figure 4.6 schematically summarises the three main aspects covered for judging relevance: the scientific basis (triangle or circle) of the alternative test method, that is the mechanism or property recapitulated or modelled by the method and thought to be causally related to an adverse effect in the target system (triangle in target system), the reliability and the accuracy (predictive capacity) of the measurements made in the alternative method with respect to the prediction of properties in the target system. Test methods (a)–(c) have a strong scientific basis since they model mechanism *p* (white triangle) that is either underlying or correlating with property *P* in that system: the predictive capacity shows to which extent the method is able to identify chemicals that activate *p* and which is thought to lead to *P* in the target system. Test methods (d) and (e) have a weaker scientific basis: they do not model mechanism *p* but another one *q*, indicated by a white circle. With regard to the overall relevance of the methods (a)–(e) the following can be said:

Method (a) is highly reliably (=always yields the same results) and scientifically relevant, but it is not accurate with respect to the predictions made: for

[2] Such strategic combinations have been proposed in the context of "Integrated Testing Strategies" that were proposed during the implementation of the REACH legislation in the EU (2006–2007) and consisted of steps of data gathering, evaluations and empirical (strategic) testing using several data sources. Later the concept of ITS has been further promoted under the term "Integrated Approaches to Assessment and Testing (IATA) by the OECD (OECD 2008).

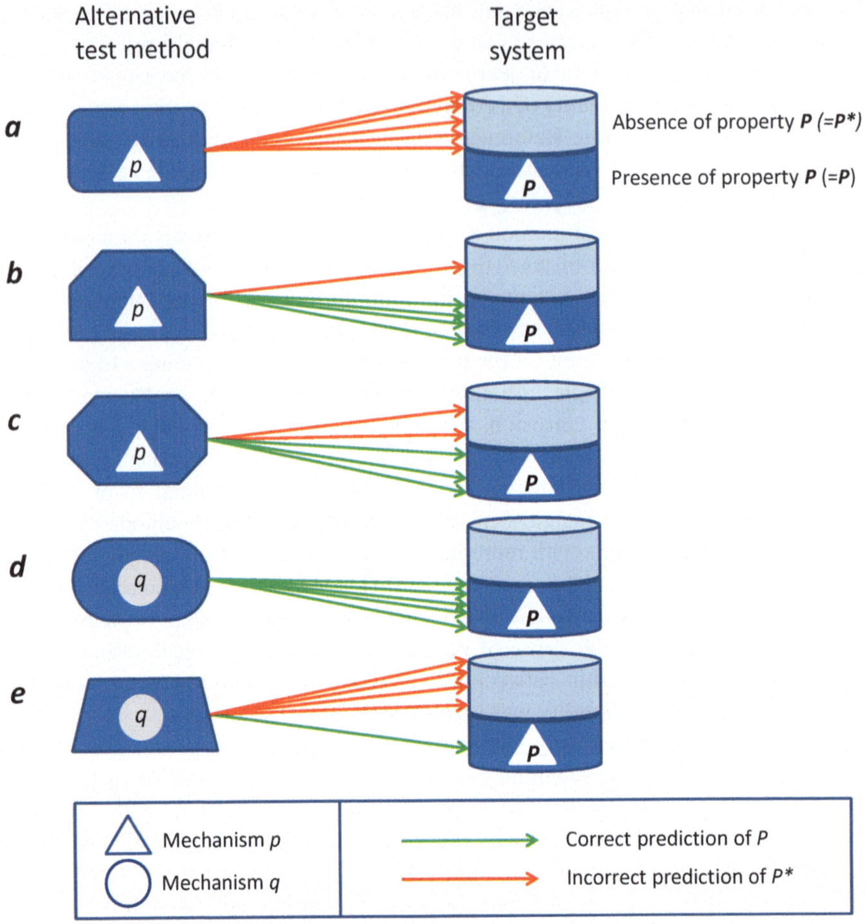

Fig. 4.6 Schematic representation of the main aspects impacting on the overall relevance of a test method, i.e. the meaningfulness and usefulness of its data. *Arrows* represent test results from five repeated experiments of the same test chemical. Correct predictions in *green*, incorrect predictions in *red*. The test method's purpose is to predict the presence of property **P** in the target system (e.g. a toxicity pathway). Reference data for the target system are available that have been simplified in two categories: chemicals that trigger P and others that do not trigger P. Thus, the alternative method needs to provide accurate predictions on absence (**P***) or presence (**P**) of property P. Some test methods (**a–c**) model the mechanism **p** thought to underlie property **P** (*white triangle*). Other test methods (**d–e**) do not model mechanism **p**, but **q**, which is not thought to be causative for **P**. Detailed explanations in the text

chemicals known to activate p, it predicts (P^*)=absence of property (P). These wrong predictions are indicated by red arrows. Its overall relevance therefore is very low. **Method (b)** has a strong scientific basis, is reliable and accurate. Its overall relevance is high. **Method (c)** is neither reliable nor accurate, although its scientific basis is relevant. Its overall relevance is low. **Method (d)** is reliable, but its results are more uncertain than those of method (b) since (d) does not model the mechanism of action *p* thought to be related to the occurrence of P in the target system. Thus, although (d) is accurate, its results correlate with rather than predict the adverse effect. **Method** (e) is reliable but inaccurate and has a week scientific basis. Its overall relevance is rather low.

(d) Applicability domain and limitations

An additional important aspect for judging the relevance of alternative test methods is applicability. Since test methods are used to assess chemicals, it is the applicability of a test method *to chemicals* that has been traditionally considered under the term "applicability domain". This would cover physicochemical properties, structural groups "chemical categories" or also sectorial use groups (e.g. biocides, pesticides, industrial chemicals, etc.) and such like. The applicability domain cannot be fully defined during validation but only be outlined based on the test chemicals used during validation. The wider the applicability domain, the more useful and hence more relevant is a method.

However, instead of restricting applicability domain only to aspects of chemical structure or physicochemical properties, it is useful to think of the applicability as a multidimensional space that is set up by as many descriptors as needed to describe how a method can be applied (Fig. 4.7). Notably, OECD guidance document 34 goes beyond the mere aspect of *chemical* applicability when defining applicability domain: "*a description of the physicochemical or other properties of the substances for which a test method is applicable for use*" (OECD 2005). Other properties (or descriptors) that may be useful for describing applicability are test outcomes (e.g. only applicable to positives), specific biological mechanisms of action/toxicity pathways.

It is obvious that 'applicability domain' in the above sense always refers to a positive description of what a method is applicable to. Inversely, the term "limitations" can be understood as a negative delineation of applicability, i.e. of "non-applicability". However, in practice, limitations more often relate to simple *technical limitations* and *exclusions* due to technical/procedural incompatibility of test items with a test method. Consider for instance a test methods based on measuring the cell viability using a colorimetric assay: test chemicals that are coloured may interfere with the readout and thus constitute a technical limitation due to incompatibility with the readout. Another example is the use of cells as a test system kept in submerged culture: this will result in a restriction to chemicals that can be dissolved in cell culture medium acting as a vehicle; the limitation would thus relate to insoluble substances such as some waxes or gels.

Thus, while applicability and limitation can be thought of as complementary terms, in reality, it is much easier to describe the limitations of a test method (especially technical limitations relating to compatibility with the test system) than to describe the applicability at the stage of validation. The reason is simply that during a validation exercise, for practical and economic reasons, only a limited number of test chemicals can be assessed: each chemical can be seen as probing with one single entity into the chemical universe composed of a vast space of hundreds of thousands of manufactured and natural chemicals. From each substance one can extrapolate to neighbouring substances within the chemical space (similar structure) or the biological space (similar mechanism of action).

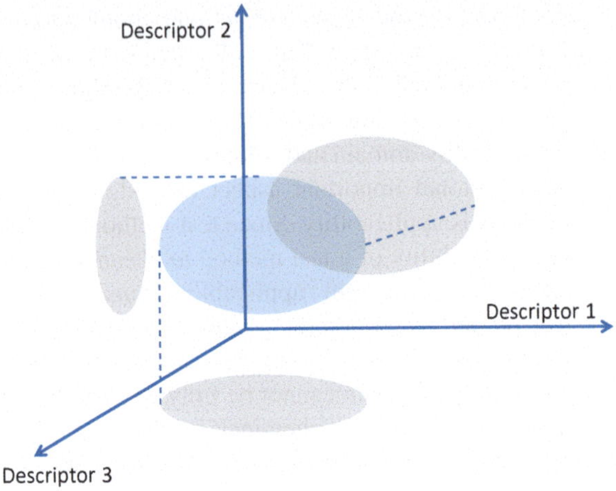

Fig. 4.7 The applicability domain of an alternative method can be seen as the space occupied by the method in a multidimensional coordinate space set up by various descriptors such as chemical structure, biological action, predictive parameters (applicable to negatives or positives only), etc. The space is indicated in *blue* and is a function of the relationships between the various descriptors

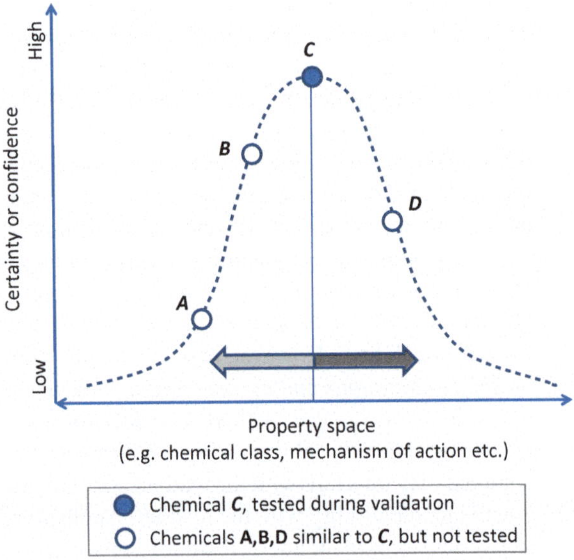

Fig. 4.8 For practical and economic reasons, validation studies can only empirically test a small sample of the chemical population. From these testing data, inferences can be made on substances with similar properties, e.g. relating to chemical structure or biological activity. Notably the certainty or confidence of these inferences decrease with increasing distance of these chemicals (*A*, *B*, *D*) from the chemical with empirical data (*C*)

The further one moves away from the substance with empirical data, the more uncertain this extrapolation gets (Fig. 4.8). It is clear that it is simply not feasible during a single scientific study to comprehensively delineate the entire space of applicability by testing, so extrapolation and "read across" of results will remain a key aspect of describing the applicability domain. To improve the description of applicability and limitations beyond the scope of validation studies, mechanisms of post-validation surveillance through which end users can report the successful use of test method to new substances as well as report problems, should be used in a more consistent manner and appropriate tools would need to be set up for such reporting.

Finally, since applicability can only be assessed or proven by testing or evaluating existing testing information, the certainty with which the applicability domain is determined is strongly correlated with the number of chemicals that has been assessed. Similarly, the certainty with respect to the predictive capacity is depending on the number of chemicals and minimum requirements in terms of sample size and power calculations for assessing for instance a dichotomous prediction model can be precisely calculated. However, for applicability and predictive capacity one could state that "the more chemicals, the better", i.e. increasing the chemical number will always increase the sharpness and accuracy with which both applicability and predictivity are defined and therefore increase the trust and confidence in the method.

In contrast, this "open-ended" approach regarding chemical number does not hold for reliability assessment: while there is a minimum number of substances statistically required for reliability assessment which can be calculated through statistical methods (sample size/power calculations), this number can be much lower than that required for a more robust description of predictive capacity and applicability domain. In contrast, to the assessment of applicability and predictive capacity, there is no substantial benefit in increasing the number of chemicals for reliability assessment. The different requirements regarding chemical number are schematically depicted in Fig. 4.9. These differences should be kept in mind when

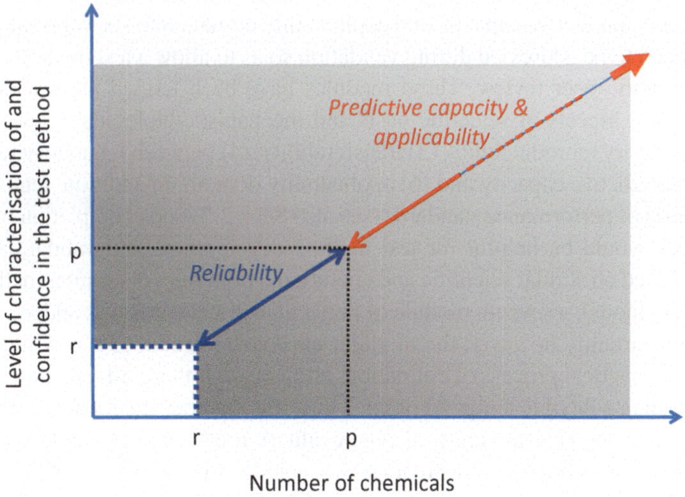

Fig. 4.9 The minimum requirements in terms of chemical number for assessing reliability on the one hand and predictive capacity and applicability on the other are different. There is a minimum number of chemicals that is required for reliability assessment in view of satisfying statistical needs (*blue dotted lines*, r). There is, however, no real need to go beyond a certain number of chemicals as defined by statistics to determine reliability since certainty will not increase to a substantial degree. In contrast, while there is also a minimum number of chemicals required for assessing predictive capacity (depending on the number of classifiers used) and applicability (p), the certainty with which these two can be considered characterised will always increase with increasing the numbers of chemicals assessed (*big arrowhead*)

discussing the potential adaptations to validation in terms of "lean processes" (see also Sect. 4.5.1): it is obvious from the above that the different requirements can be exploited in terms of adapting the data matrix generated during validation by dissecting the chemical testing set that has been traditionally assessed for all information requirements (reliability, predictive capacity and applicability domain) into two sets: one for assessing the reliability and a larger one for assessing predictive capacity. We will discuss this in more detail in Sect. 4.5.

2.4 Supporting the Practice of Validation: The Modular Approach

In 2004 EURL ECVAM proposed the "modular approach" to validation (Hartung et al. 2004) that has proven to be a very useful tool for adapting the validation study design not only to the intended purpose but also to the available information. Importantly, this modular approach should not be confused with the one proposed by Goldberg and colleagues in 1995 which relates to validation of *in vitro* methods on the basis of one defined readout against concurrent human data where possible (Goldberg et al. 1995).

Starting from the observation that validation until then had emphasized the process rather than the information requirements, the modular approach suggests to structure the information of scientific basis, within- and between laboratory reproducibility, transferability, predictive capacity and applicability domain into six information modules that need to be addressed during validation so as to allow a test methods to progress to scientific peer review. These modules have been termed (1) test definition (encapsulating aspects of scientific basis and mechanistic/biological relevance), (2) within laboratory reproducibility, (3) transferability, (4) between laboratory reproducibility, (5) predictive capacity and (6) applicability domain. In addition, realising that the definition of performance standards (see also Sect. 2.2) upon completion of validation studies would be helpful for test method development and validation of test methods, based on similar scientific and operational principles (="similar methods" or "me-too" methods), a seventh module of performance standards was added.

Most importantly however, the modular approach introduced a new philosophy towards the practical process of validation, allowing that these information modules be addressed in a flexible temporal order. Thus, test methods do not necessarily have to run through the typical ring-trial type evaluation of classical validation studies but need to address only the information that is judged to be missing. This information can then either be produced prospectively through dedicated new testing or retrospectively through analysis of existing information. For instance, for a specific test method, there may be ample information on predictive capacity, so that validation can focus on defining the test method and assessing mainly the reliability. EURL ECVAM has in recent years conducted several modular studies (see EURL ECVAM webpage, section "EURL ECVAM Recommendations", EURL ECVAM 2012 onwards), notably in the area of skin sensitisation. EURL ECVAM exploited the fact that, for some well-established methods (e.g. Direct Peptide Reactivity Assay, DPRA), there was a wealth of publicly available information on predictive capacity and applicability from user laboratories. This allowed to focus the design of the validation studies on protocol transferability and reliability (within and between laboratories) in order to complete these information modules.

3 Validation Study Management

3.1 Generic Design of a Validation Study

As outlined in Sect. 2.2.2, there are various types of validation studies in terms of the scientific design to assess reliability and relevance. Here we provide a brief outline on the managerial aspects of validation studies (Fig. 4.10).

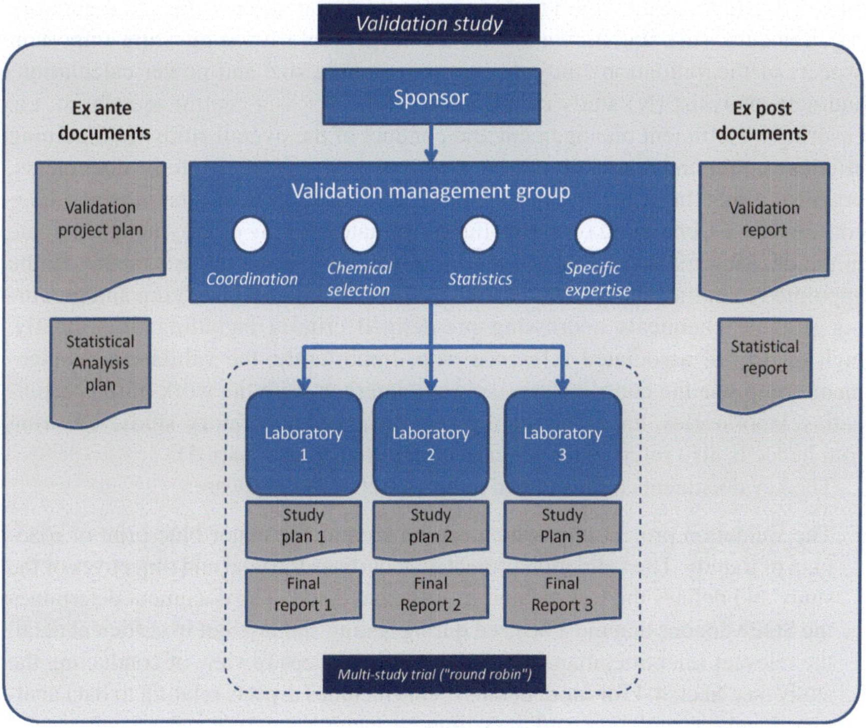

Fig. 4.10 Generic outline of the overall organisation of a prospective validation study: main actors, key documents produced at the outset (ex ante), during testing and upon completion of a study (ex post). Main actors are (1) the sponsor or sponsor consortium, initiating and normally financing the study, (2) the validation management group that is set up by the sponsor in view of managing the science and logistics of the study and composed of experts with different roles and expertise including coordinators, statisticians, chemists and regulators for selecting chemicals and other experts (e.g. in validation, the test method under scrutiny, etc.), (3) the participating laboratories conducting the testing within a round robin or 'multi study trial'. In case of retrospective studies, the design would be the same, without however the participating labs

3.2 Roles and Responsibilities of Actors

Validation studies are typically initiated by a sponsor or sponsor consortium. The sponsor has an interest in validating the method either because of economic interests and/or in view of legislative requirements necessitating a particular validated alternative method for routine use. The sponsor typically appoints a validation management group to oversee the entire study, i.e. to decide on study design, to manage and coordinate the study execution phase (involving dedicated chemicals testing in case of prospective studies, to analysing the results and concluding on and reporting the main outcomes by writing up the final validation report. The validation management group is composed in view of gathering the expertise needed to conduct the specific study in question. This includes (i) a Chair who is moderating meetings, teleconferences as well as discussions and the decision-making process related to all VMG decisions; (ii) experts with knowledge in the test method under scrutiny and related scientific and regulatory requirements; (iii) statistician(s) that are responsible for suggesting important aspects of the validation study design (e.g. sample size and power calculation) and data analysis; (iv) study coordinator(s) who act as a central secretariat, i.e. ensuring the efficient management and conduct of the overall study (maintaining efficient communication, preparing drafts of key validation study documents, organising meetings, recording key decisions and reports of meetings and teleconferences). Depending on study, the coordinator(s) may or may not participate in the decision making of the group. Finally, among these experts, some can be appointed to define and perform the chemicals selection: identifying and procuring suitable chemicals addressing pre-defined criteria including, importantly, high quality of associated reference data. Importantly, the validation management group, via the coordinator(s), closely interacts with the work of the participating laboratories, each conducting one dedicated laboratory study. The ring trial hence is also referred to as "multi study trial" (see Chap. 5).

The key documents to be defined at the outset of the study are:

- The validation project plan which can be seen as the major blue-print or roadmap of a study. The validation project plan outlines the goal and objectives of the study and defines the test method in sufficient detail. The document determines the SOP versions that must be used during testing and lays out in sufficient detail the relevant scientific, managerial and logistical steps in view of conducting the study (see Sect. 4.4 for more details). This includes aspects relating to data analysis, handling problems and deviations. It includes contributions from specific experts of the management group, e.g. from the chemical selection committee which will outline the test chemicals to be studied and their associated reference data or from the statistician, describing the sample size calculations conducted in view of addressing the study goal and objectives).
- **The statistical analysis plan**, outlining the data handling, analysis, interpretation and reporting. This plan can be part of the project plan or a stand-alone document.

Key documents during the validation study are:

- The laboratory **study plans** and **final reports** (requirements under GLP) that outline all the relevant SOPs required (not only that of the test method, but also those relating to equipment and other issues of the local laboratory) and that define how the testing data will be reported in agreement with the quality assurance measures in operation at the laboratory.

Key documents upon completion of a validation study are:

- The **statistical report** summarising the analysis of the data and the statistical findings. This report can be a stand-alone document or be part of the validation report. Important is that the statistical analysis and its conclusions are not influenced by the VMG (who may be biased with respect to the decisions it took during the study) and is conducted solely on the basis of the data available.
- The **validation report** that summarising the entire validation study (referring where necessary to other documents, e.g. the statistical report), the problems encountered and which has to clearly outline results obtained, the conclusions drawn and take clear position with respect to whether or not the study goal has been achieved.

4 Validation Study Design

Having discussed the key actors, the key documents and the generic organisation of a validation study in Sect. 3, we now explore the most important elements to be addressed during validation study design. These typically would be captured in a validation project plan (see Fig. 4.9).

4.1 Number of Chemicals, Sample Size and Power Considerations

Conclusions drawn on the basis of empirical testing can be considered solid scientific insight only if they can be generalised beyond the *single experimental result*. The assessment of the capacity of an alternative test method in view of obtaining predictions on the effects of chemicals cannot be done on an infinite number of chemicals, but, for practical and economic reasons, on the basis of a restricted number. This should however be sufficient to allow such generalisations, taking also into account the restricted reproducibility of scientific experimentation. Thus, empirical testing will be restricted to a *sample* of the population (chemical substances). In the following we discuss this 'sample size' problem, that is, the problem of concluding from the relative frequency of events *in a sample* to the relative frequency *in the entire population*. We equate here sample size with number of chemicals since the goal of validation is to make inferences on the ability of a test method to predict the properties of chemicals. It is however noted that the term may also reflect the sample size of two or more distinct populations or simply to the number of observations or replicates.

The number of chemicals used for the validation study needs to be determined by statistical means so as to allow adequate quantitative metrics in view of the validation study goal and objectives. The quantitative metrics relate to mainly the within-laboratory and between-laboratory reproducibility (WLR and BLR) and predictive capacity; for the latter the number of categories predicted (dichotomous/binary or more; see Fig. 4.1) will be an important factor influencing the sample size/power calculations.

The sample size, here the number of chemicals, should be large enough to represent sufficient statistical power for comparing two (or more) populations by a statistical test on the basis of a measured parameter; the latter can be a *mean* or a *proportion*. Two types of errors can be encountered, type 1 and type 2. Both types are taken into consideration for the sample size calculation:

- The type 1-error is the error that consists in rejecting the null hypothesis H_0 of equality of the parameter when H_0 is true. It represents therefore the false positive cases. The probability that this type of error occurs is usually denoted by α.
- The type 2-error is the error that consists in not rejecting the null hypothesis H_0, i.e. accepting H_0, when H_0 is false. This type 2-error represents therefore the false negative prediction. The probability that this type of error occurs is usually denoted by β. The power of the statistical comparison is defined by $1-\beta$.

In the case of *in vitro* test methods, predictions typically consist of categorical outcomes relating to specific mechanisms (e.g. activating estrogen receptors) or entire health outcomes (e.g. in Skin Corrosion Tests, Category 1A, Category 1B/1C, and Non-Corrosive). The value of WLR is typically obtained by calculating the proportion (i.e., fraction in percentages) of chemicals that have concordant predictions throughout the runs used in one laboratory. The test chemicals represent the population for which the calculation of the sample size is required. This WLR is the measured parameter over the population of chemicals. For defining the sample required, the expected values (target value, here relating to WLR) is an important aspect to be defined prior to testing. The target value should be based on prior testing of a small set of chemicals (e.g. in the context of a so-called "prevalidation" study) or can be derived from other historical information. The formula to be used, for calculating the sample size, is the one based on proportions and will include this target values as well as α and 1- β values.

The following equation shows the advantage of simultaneously taking into account the targeted WLR value and the lower limit of this value (i.e. WLR should not go below this value). The target value is represented by π, the error by δ, the lower limit by π-δ.

Z_α and Z_β respectively represent the Z distribution values for the probabilities α and β. This formula was proposed by Flahault et al. (2005) and can be derived by the one presented by Lachin (1981).

$$\textit{Number of chemicals} = \frac{\left(Z_\alpha\sqrt{(\pi-\delta)(1-\pi+\delta)}+Z_\beta\sqrt{\pi(1-\pi)}\right)^2}{\delta^2}$$

Such calculation of the number of chemicals needed, prior to testing, plays an important role in the validation study as this sets up the level of confidence—and conversely deals with the uncertainty towards the obtained values of WLR.

Statistical considerations also apply to the calculation of the number of chemicals needed to reach target values of BLR. Similarly to WLR, BLR is a proportion—the fraction of chemicals for which concordant predictions have been made over the participating laboratories. The difference δ accepted for the target value π plays a critical role in the formula: when δ decreases, n increases according to the inverse of δ square root. For instance, a target value of WLR of 90 %, i.e. $\pi=0.9$ with a power of 80 %, i.e. $1-\beta=0.8$ ($Z_\beta=0.842$), and a risk $\alpha=0.05$ ($Z_\alpha=1.645$) will result in a different sample size whether the value of δ is 0.1 or 0.2. If $\delta=0.2$ the total number of chemicals needed is 25; if $\delta=0.1$ the total number of chemicals needed is 83 and therefore much higher.

Therefore, the assumptions (or the certainty of preliminary target values) play a critical role for calculating the number of chemicals to be assessed in a validation study. These assumptions cover not only the target values of WLR or BLR, but also the underlying statistical formulae used for the calculation (normal approximation to the binomial law).

4.2 Selection of Test Chemicals and Associated Reference Data

For above said reasons, the selection of chemicals used in validation studies is critical and the success or failure of a validation study may largely depend on it. This includes issues of both *number* and *nature* of chemicals selected. Ideally, a high number of chemicals should be selected to represent different chemical classes and, depending on the purpose of the validation study, also different chemical use categories, such as e.g. industrial chemicals, food additives, pharmaceuticals, cosmetic ingredients, pesticides, etc. Ideally, the following information on the selected chemicals should be known and compiled: use applications, *in vivo* data sources, substance supply, chemical classes, physical chemical properties and GHS classifications (if applicable).

Chemical selection has traditionally focused on mono-constituent substances of high purity, ensuring correspondence of documented *in vivo* data to sample material acquired for *in vitro* testing. Nevertheless, acknowledging the REACH definitions, 'pure mixtures' (multi-constituent substances with negligible impurities) have also been admitted, provided composition was reported quantitatively and consistent with the material used for the *in vivo* study.

In general, a principal requirement for chemical selection is the availability of complete and quality assured supporting *reference* data sets, for comparative evaluation of *in vitro* mechanistic relevance and/or method predictive capacity. These reference data are typically from surrogate animal studies ("*in vivo* data"), but can also be derived from other sources. In areas where the mechanisms of action is not preserved across species, (e.g. metabolism, CYP induction), the availability of human reference data for the mechanism studied is essential. Human toxicity data however are often problematic with respect to their availability and their quality (see Sect. 2.1).

The availability of human reference data for many areas in toxicokinetics and toxicodynamics is often limited to pharmaceuticals since this is the only sector where testing is performed in humans after pre-clinical toxicological testing. Human data from the pharmaceutical and other sectors can also be obtained from selected scientific references and poison control centres. In such registries, human data derived from clinical case studies, hospital admissions, and emergency department visits can be found. Although this information is not acquired systematically, it represents a potential source of human toxicity and toxicokinetic data available for commonly encountered chemicals. Thus, the clinical information is used as a basis for comparison with *in vitro* values.

Another source of more reliable human toxicological data may be obtained through the testing on human volunteers for some areas of local toxicity, such as skin and eye irritation. Human volunteers for skin irritation testing produce concentration-effect curves for fixed endpoints, while in the case of eye irritation, testing is, for ethical reasons, limited to minimal mild effects (redness, itchiness). A more recent technology to obtain human data is the human microdosing. This technology seems promising for obtaining human toxicity data as only extremely low amounts of chemical need to be given to the human volunteers. These external amounts could well remain below current threshold of toxicological concern (TTC) values. However this area needs to be further explored and it is stressed that all experiments with human volunteers need to be carefully considered for their ethical implications before being conducted.

In general, the selected chemicals should be (1) commercially available, (2) stable after fresh preparation of a stock solution, (3) soluble in saline, or in solvents that are used in concentrations not affecting the mechanism of interest and (4) not precipitate for defined time frames when used under standard operating procedures.

Experience has shown that all laboratories should use the same solvent and the same non-cytotoxic highest concentration of the test item over a defined period as defined in the standard operating procedure during a validation study.

Another prerequisite is to use defined chemicals (that is by their Chemical Abstracts Service (CAS) formulas or their generic names) rather than proprietary mixtures or coded industrial products. Studies performed with defined chemicals allow for between-laboratory testing and clear definition of critical components of the validation study.

4.3 *Defining the Data Matrix Required*

Once the sample size has been determined, it is advisable to determine the precise data matrix that would be required for a statistically appropriate analysis of the performance characteristics of the test method during validation. By data matrix we simply mean the number of data points required for each test chemical in view of characterising the performance of the method. Aspects of lean design (see Sect. 4.5.1) can be taken into account when defining the data matrix.

A typical example of a data matrix can be defined by

- X number of laboratories testing…
- Y number of chemicals in …
- Z number of experiments

The terms "experiment" and "run" or "complete run" are sometimes used interchangeably. Importantly, these terms usually relate to all the measurements and data processing/analysis required to generate a *final result* for a given test item (either a toxicological measure or a categorical prediction). Thus, runs or experiments relate to the intended application of the test method in practice when routinely assessing test items.

In the following section we present some illustrative examples from test methods in the area of topical toxicity testing (mainly skin irritation) as summarised in a background document on *in vitro* skin irritation testing (Griesinger et al. 2009), see Fig. 4.11:

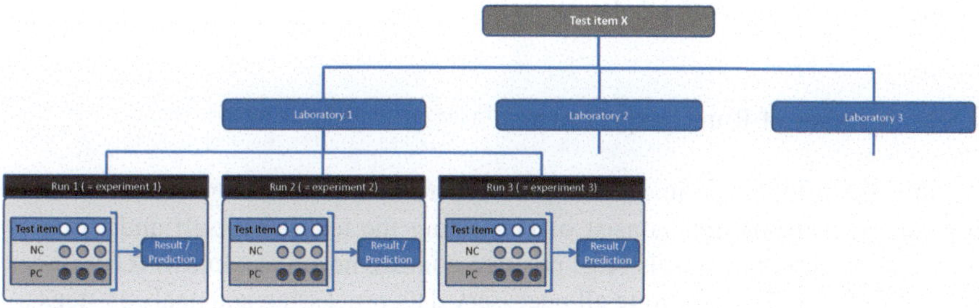

Fig. 4.11 Schematic depiction of a possible data matrix for one given test item (X) in the context of a validation study. The example is based on *in vitro* skin irritation testing. Each test item is tested in three laboratories. In each laboratory, three experiments (=runs) are being conducted. A run is the experiment that will yield the final result of the test method as intended in practice, i.e. either a final toxicological measure or a categorical prediction. Thus, a run incorporates all steps necessary to produce this information and thus includes the testing of the test item, the controls, as well as all data analysis as required. This can include conversion of the result into categorical predictions by means of a prediction model. The three runs conducted in each laboratory can be used to assess the within-laboratory reproducibility (e.g. by assessing concordance of run predictions). Runs are based on several replicate measurements (circles) whose results normally are being averaged and analysed for variability as a measure for the quality of the data underlying the experiment or run. Variability measures such as Standard Deviation (SD) or Coefficient of Variation (CV) can be used to define "Test Acceptance Criteria", i.e. quality criteria for accepting or rejecting an experiment based on replicate measures

4.3.1 Number of Replicates

The replicates are the repeated individual measurements of the parameter of interest, for a given test chemical, and together with other relevant measurements (e.g. controls) constitute the data underlying a run, i.e. the actual result of the test method when used in practice. Each replicate measures the parameter of interest (Griesinger et al. 2009). Replicate measurements can be used to calculate mean and standard deviation (SD) values. The SD value can be used to further calculate the coefficient of variation (CV) defined, in percentages, by $CV=(SD/Mean)\times 100$. SD and CV are quantitative indicators of variability. Measures from all replicates are usually averaged to derive the final result or prediction for the test item tested. Importantly, the use of replicate measurements allows assessing the quality of the experiment: the variability of these replicate measurements should be below a pre-defined threshold (e.g. a SD value), otherwise the run result is considered invalid (or "non-qualified"). The SD thus serves as a tool for applying a "Test Acceptance Criterion) (TAC)". In the example of skin irritation testing, the SD derived from three tissue replicates must be equal or below 18 %. Importantly, the TAC must be set on the basis of a sufficiently large set of historical testing data and the number of replicates required to assess within-experimental variability should also be based on sufficient previous data. Typically, during a validation study, the number of "replicates" will follow the provisions of the test method protocol intended for later application. However, when defining the validation data matrix, it should be carefully assessed whether the number of replicates can be reduced ("lean design"), e.g. by analysing historical data sets and assessing the impact of such reduction. Importantly, the number of replicates is specific to each test method and, unlike for the number of runs or laboratories, no general recommendations can be made.

4.3.2 Number of Runs (Experiments)

A run is the actual experiment that provides a final result on a given test item. A run (or experiment) thus consist of (1) testing the test item itself and, concurrently, all necessary controls (e.g. positive control, negative control) (Griesinger et al. 2009) and (2) performing all necessary data processing and analysis steps to generate a final results for the test item. This may, where applicable, include the conversion of the toxicological result into categorical predictions by means of a prediction model.

In a validation study, typically three runs (or experiments) are performed in each laboratory. Since each run provides a final prediction, the between run-concordance (=agreement between) such predictions can be used to assess the within a laboratory repeatability and within-laboratory reproducibility of the test method.

Predictions at run level may also be used for deriving a final prediction per chemical in one laboratory. This has typically been done by simply determining the "mode" of predictions and settling unequivocally on a final prediction per chemical. If this approach is used, the number of runs needs to be an odd number (e.g. three runs).

4.3.3 Number of Laboratories

For the same considerations as described above for the number of runs, three laboratories are usually participating in a validation study. The involvement of several laboratories allows evaluating the reproducibility of the test method between laboratories. The between-laboratory reproducibility can be calculated as described in Sect. 4.7.2.

4.4 Validation Project Plan

The validation project plan serves as a driver and a reference for the conduct of the validation study. It covers an extensive range of topics relevant for conscientiously planning the scientific and managerial aspects of the validation study. It takes into account logistical and practical considerations and sets up timelines. The project plan defines the test methods under validation, the goal and objectives of the study, it describes the actors involved and their respective roles and responsibilities, and defines specific stages/timelines of the study.

A typical project plan can include the following main sections:

1. Definitions: this section provides definitions of the test methods studied during validation, outlining (1) the test systems (e.g. reconstructed human tissue of multi-layered epithelium) used as well as (2) determining the associated protocols/SOPs and the precise version numbers to be used during the study.
2. Validation study goal and objectives: goal and objectives of the study should be clearly outlined. Typically the goal of a study corresponds to a regulatory requirement and often to the prediction of specific hazard classes or categories of chemicals (e.g. Category 2 of eye irritant in the United Nations Global Harmonized Systems for classification and labelling, UN GHS). Therefore, this section should explicitly mention the name of the regulation addressed. If several regulations are concerned it should be specified how the study will relate to these. The objectives would be more detailed aims, such as validation for identification of negatives or for a specific class of chemicals in view of filling an existing methodological gap, etc.
3. *In vitro* test methods: this section provides a detailed scientific characterisation of the *in vitro* test methods undergoing validation. This relates to the scientific basis, the test method's mechanistic and biological relevance, as well as historical aspects relating to test method development (test method development, optimisation, previous assessments including prevalidation studies, etc.).
4. Validation management group (VMG): the VMG is the body that oversees and manages the validation study (see Sect. 3.2). The validation project plan should outline the expertise required in view of ensuring an efficient conduct of the study. Typically a VMG consists of (i) a Chair responsible for chairing meetings, facilitating decision making and representing the VMG; (ii) relevant

experts with specific expertise required for the study; (iii) statistician(s); (iv) study coordinator(s) acting as focal contact point and running the study secretariat. Moreover, depending on study, a VMG subgroup dedicated to selection of test items and associated reference data, Moreover, observers or liaisons may participate (e.g. representing other validation bodies). Also, representatives of the laboratories can be involved for specific agenda items of VMG meetings related to technical and/or experimental issues. The specific role of each of the above mentioned categories of participants and the way they interact together should be clearly explained and may be supported by a schematic figure. In order to maintain an impartial and unbiased study, the VMG must not include members directly involved in the development of the methods undergoing the validation process. However, the VMG may consult the test method developer if necessary.

5. Validation study coordination and sponsorship: this part of the validation project plan defines sponsors of the study as well as the activities that should be covered by the study coordinators, including logistical aspects (e.g. coding and distribution of chemicals), communication (e.g. frequency, means), organisation of VMG meetings, teleconferences, minutes, etc. This section should also describe the allocation of financial resources, e.g. purchasing of test chemicals and other relevant service contracts (e.g. statistical support).
6. Chemicals selection: The process and criteria for selecting test chemicals should be detailed in this section. Chemical selection can be done by *ad-hoc* experts or by a dedicated VMG chemical selection group (CSG). Experts can include members of the validation study coordination, independent scientists, liaisons and representatives of the competent authorities. Moreover, since *in vitro* methods will be evaluated against reference data, this section should also stipulate criteria for the selection of such data associate with the test chemicals. To this end, the type of reference data and the sources of these data (e.g. databanks, literature, etc.) are specified. Eligible chemicals are usually compiled in table format (e.g. classification of selected chemicals according to the UN GHS for skin corrosion). Number of chemicals needed for the validation study, obtained from sample size calculation (see paragraph 4.1), will be mentioned as well as the proportions of distinct classes/categories (e.g. negative vs positive, solids vs liquids, etc.). In terms of procedure, the CSG proposes the list of eligible chemicals to the VMG. This latter may also take into account the availability of the chemicals to be tested, especially those commercially available versus the proprietary ones as well as other practical factors such as potential health effects of test chemicals: since validation studies are conducted under blind conditions, substances with specifically high risks can be excluded (e.g. "CMR substances" with carcinogenic, mutagenic and reproductive toxicity effects) as long as these are not related to the health effect of concern to the study.
7. Chemical acquisition, coding and distribution: This section should outline the provisions regarding acquisition, coding and distribution of the test chemicals. This should be accomplished by a person affiliated to a certified ISO 9001/GLP

structure. Individuals involved in this process must be independent from those conducting the testing. The process should foresee a purity analysis of the chemicals and the provision expiry dates. In laboratories testing different versions of one protocol (e.g. separate protocols for testing solid and liquid chemicals), codes of chemicals will be different for each version.

8. Receipt and handling of chemicals: this part of the validation project plan tackles the shipping of the coded chemicals, the storage time and conditions as well as health and safety measures related to their handling.
9. Participating laboratories: This section should outline the requirements of the participating laboratory, e.g. study director, quality assurance officer/unit, study personnel and a safety officer. This section also includes a description of how laboratories, within a group, may communicate together and when the VMG should be involved in these discussions. For instance, during the testing phase, the participating laboratories must not contact each other without approval of the VMG.
10. Laboratory staff: the validation project plan specifies the roles of the study directors, the quality assurance officers/unit, the study personnel and the safety officers. The study director should be an experienced scientist in the field and acts as main contact point of the VMG. He/she is responsible for preparing each necessary report. The quality assurance officers will assure that compliance with any quality requirements (e.g. GLP) is respected. The quality officer needs to be independent from the study director direction and from the study personnel conducting the experiments. The experimental team will perform the testing. It should be trained, experienced and competent for the specific techniques. The safety officer is in charge of receiving the coded chemicals and transmitting them to the responsible person of the laboratory. He/she is in charge of the sealed material data sheets (MSDs) corresponding to the test chemicals and their codes. These will be disclosed only in case of accident.
11. Validation study design: this section of the project plan includes details on each type of assay taking part in the validation study. For instance, number of chemicals, runs and replicates should be clearly defined. Specific technical aspects of the test methods are tackled. For instance, if there are two different protocols for a given test method with different exposure times, those will be mentioned.
12. Data collection, handling and analysis: this part of the validation project plan describes how final reports and the reported data are forwarded to the biostatistician. He/she will decode the chemicals and proceed to the analysis (see paragraph 4.6, Statistical analysis plan) and produce a biostatistical report to the VMG. This report should present the results (predictive capacity, within- and between laboratory reproducibility, quality criteria) as well as how data were analysed and the statistical tools used. Data analysis strategy should be developed, before the end of the experimental phase, by the biostatistician in a statistical analyses and reporting plan. This latter will be submitted to the VMG for approval.

13. Quality assurance good laboratory practices: it is usually desirable that the validation study complies with OECD good laboratory practices (GLP) in order to facilitate international acceptance of the validation study and its outcomes (OECD GD 34 2005). This allows full traceability of the study at all levels of its experimental phases.
14. Health and safety: the laboratories should comply with applicable (and required) health and safety statutes. The safety officer of each laboratory is designated as the contact point for these questions.
15. Records and archives: provisions should be made for the appropriate archiving of raw data, interim and final reports of the validation study (where, how many copies, by which means) as well as for the management of the archiving.
16. Timelines: defines the critical timelines that should be respected. Timelines are established for each critical phase of the validation study (e.g. chemical eligibility, approval of the validation project plan, approval of the validation study design, dates of testing, etc.).
17. Documents and data fate: proprietary questions in relation with the documents and data generated are described. This also covers the confidentiality of these elements and whether and to which extent information can be disclosed.

Finally the validation project plan should also make provisions for retesting in case of non-qualified (invalid) runs so that this can be implemented in the study plans for the laboratories under supervision of the individual study directors. In particular this should address how often experiments relating to one chemical can be repeated, i.e. how many retesting runs are permissible. Typically, the validation coordinator prepares an example of a study plan that can be adapted by the laboratories in compliance with their own specific laboratory procedures (see Chap. 5).

4.5 Adaptations of Validation Processes

The modular approach (Sect. 2.4) can be regarded as an important adaptation of the classical validation approach. Traditionally information on reliability and the judgement of relevance followed a rather rigid sequence towards producing a comprehensive data matrix. The modular approach introduced a significant degree of flexibility with regard to the generation of the information. Two further adaptations have been under discussion recently namely approaches to reduce the data matrix without compromising the adequacy of the validation study ("lean design") and, secondly, the use of automated equipment (e.g. automated platforms, medium- and high-throughput platforms) for generating empirical testing data. Third, some methods used for prioritisation have been developed on custom-made automated platforms and some aspects of validation cannot be always applied to such assays (e.g. transferability assessment). These three adaptations are briefly discussed below.

4.5.1 Lean Design of Validation Studies

As discussed in Sect. 2.3.3(d), the requirements in terms of sample size for assessing reliability and for assessing predictive capacity and applicability domain are different. This can potentially be used in view of adapting the data matrix in order to reduce both cost for test chemicals, test systems and the labour involved. As a general consideration, it is conceivable to assess the reliability of a test method using a small set but statistically sufficient set of chemicals in three laboratories, while assessing the predictive capacity (e.g. in terms of a dichotomous prediction model requiring a higher sample size) with more chemicals but only in one laboratory or by testing subsets of this larger set in various laboratories. A feasibility study of this approach has been conducted by Hoffmann and Hartung (2006a, b) using the data set of the EURL ECVAM skin corrosion validation study (Barratt et al. 1998; Fentem et al. 1998). Using resampling techniques it was shown that the number of test runs could be reduced by up to 60 % without compromising significantly the level of confidence. While this result is promising it should be noted that the reproducibility of these methods was very high and this has probably led to the remarkable reduction rates of the data matrix that were possible. It still needs to be evaluated to which extent the lean design can also be useful for other test methods and other use scenarios in particular.

4.5.2 Automated Testing as a Data Generation Tool for Validation

Validation studies normally assess test methods on the basis of manually executed SOPs. This ensures that validated test methods and their associated protocols are universally usable, also by laboratories that do not have automated platforms at their disposal. This however does not mean that automated methodology (e.g. relating to liquid handling steps in a manual method) could not be used during validation. Automated or robotic platforms can greatly accelerate the generation of testing data and allow the economical testing of a larger numbers of test items in shorter a time. This supports a more complete characterisation of the predictive capacity and applicability (see Sect. 2.3.3) of a test method (Bouhifd et al. 2012). An important prerequisite to use automated approaches for validation is to ensure that the automated protocol is equivalent to the manual one in terms of the results and/or predictions it generates. There may be variations that need to be assessed with attention (e.g. smaller exposure volumes, slightly different application regimes with regard to the test chemicals etc). When used for additional data generation during validation, automated testing represents rather a technical than a conceptual adaptation of the validation process.

4.5.3 High-Throughput Assays for Chemicals Prioritisation

In the context of alternative *in vitro* testing methods, high-throughput assays (HTAs) are those using automated protocols to test large chemical libraries over a range of concentrations. Chemical prioritization is often the objective when using HTAs which aims to identify those chemicals in large libraries that may exert a specific mechanism of action with the potential to lead to particular adverse effects. While these HTAs are not intended for global use by end users (e.g., via OECD test guidelines), data generated via HTAs may be used by regional agencies and international organizations to inform regulatory decision-making, especially as part of a weight-of-evidence approach. Consequently, it is important to consider whether adaptations of standard validation approaches may be appropriate for use with HTAs.

The principals of validation outlined in Sect. 2 are applicable to all alternative methods, including HTAs. However, the unique nature of the automated assays and the resulting volume of data generated using HTAs differ significantly from traditional "manual" methods, and these aspects need to be taken into account during the validation process.

Most HTAs are performed using highly automated processes developed on custom-built robotic platforms and are therefore not amenable to traditional "ring-trial" studies used to demonstrate transferability of the method. Transferability, one of the assessments of reliability along with inter-laboratory repeatability, is important because (i) it provides independent verification of results obtained using the same method in another laboratory and (ii) it allows a statistical assessment of between laboratory reproducibility (BLR, see Sect. 4.7) that can be used in an overall assessment of how robust the protocol is when used in different laboratories. The statistical characterization of method transfer is generally not germane to HTAs due to the highly customized and unique nature of these assays, Judson et al. (2013). However, the ability to confirm independently the results of the HTAs remains an extremely important aspect of method validation and deserves careful consideration. Since many HTAs are adapted from previously existing low-throughput methods (i.e. manual protocols), the most straightforward approach to confirm results from HTAs is *via* use of performance standards developed for mechanistically and procedurally similar assays (see Sect. 2), the latter without regard of the equipment used to execute specific procedures (i.e. protocol steps), i.e. manual or automated.

In the event that the HTA is measuring a unique event or utilizing a proprietary technology, data generated in other assays measuring activity in the same biological pathway may be useful in confirming or at least supporting results of the HTA assay undergoing validation. If a number of chemicals produce consistent results across several different key events in a given biological pathway, then the activity of those chemicals may be able to serve as a reference for other (new) assays that target key events in the same pathway. For example, if the HTA undergoing validation measures one key event in a signaling pathway (estrogen receptor dimerization, for example), then data generated in other assays measuring different key events in the same pathway (e.g., ligand binding, DNA binding, mRNA production, protein production, cellular proliferation) may potentially be used to establish confidence in the HTA data.

Another critical aspect to consider when validating HTAs is the volume of data generated by these methods, which necessitates increased reliance on laboratory information management systems (LIMS) and automated algorithms for data analysis. Although data management and statistical analysis (see Sect. 4.7) are important components of all validation studies, the large amount of data associated with HTAs often results in analysts being "disconnected" from the data, which has the potential to lead to wide-scale misinterpretation of the results. With this in mind, the validation of data management tools and the statistical approaches employed become paramount.

4.6 Ex Ante *Criteria for Test Method Performance*

Clear criteria relating to desired or expected performance defined at the outset of validation (before data generation) can support an objective evaluation of the results and conclusions of a validation study and in particular to which extent its goals have been met. These criteria can be fixed values or ranges relating to specificity, sensitivity and within- and between-laboratory reproducibility. They should be based on reliable empirical data from prevalidation or derived from other relevant data sets such as in-house (non-blinded) testing in the test developer's laboratory. Importantly, the performance criteria should relate to the intended purpose of the test method, i.e. its practical application, e.g. whether the test will be used in pre-regulatory screening or for the generation of data for regulatory dossiers in response to legislative requirements (Green 1993). Moreover, the use scenario is a key factor to be considered: for instance, will the method be a stand-alone or be merely part of an integrative approach? Ex ante performance criteria have been used by EURL ECVAM when validating *in vitro* skin corrosion methods (Fentem et al. 1998), using ranges of sensitivity and specificity that were subdivided in bands of acceptability. This approach was recently used again by EURL ECVAM when validating *in vitro* methods for eye irritation testing (EURL ECVAM 2014).

4.7 Statistical Analysis Plan

The statistical analysis plan includes a series of calculations that aim to demonstrate two main features of the test method to be validated. The first one deals with the reliability of the method and covers two main parameters: the within-laboratory reproducibility and the between-laboratory reproducibility. This second feature is the predictive capacity of the method. Below we outline the basic statistical approaches that can be used to describe these. Most of the relevant literature to describe predictive capacity deals with evaluations of diagnostic tests during clinical trials (i.e. versus a gold standard test). Most of the concepts and tools can be applied also to predictive toxicity tests, although there are important differences with regard to the entities tested and the nature of predictions obtained (see Sect. 2.1.4). An overview of statistical evaluations of test methods can be found in Pepe (2003).

4.7.1 Statistical Evaluation of the Information Provided by Alternative Test Methods

Fundamental Considerations

Two basic groups of test methods can be distinguished with regard to the results they provide: Test methods that provide meaningful toxicological information without transforming these into categorical predictions and those that convert measurements into distinct categorical predictions by means of a prediction model.

(1) Results are measures of some sort but no categorical predictions: Examples include assays that provide *in vitro* concentration-response curves and thus information about *in-vitro* potency. Generally, ecotoxicological test methods provide results that are not in form of categorical predictions. An example is the Fish Embryo Toxicity Test (FET) which yields an LC_{50} value (concentration that leads in 50 % of the animals in the observation group to lethality).
(2) Results are categorical predictions: The final measurements are converted into categorical predictions. These, in most cases, are dichotomous (or binary) predictions of the general form "toxic" versus "non-toxic". Test methods used for hazard identification in relation to categorical systems such as the United Nations Globally Harmonised System (UN GHS) for classification and labelling (C&L) of chemicals will need to produce categorical predictions to be useful in practice. The categories in this case relate to downstream ("apical") health effects such as skin corrosion, acute oral toxicity, etc. However, categorical predictions do not necessarily need to be tied to C&L classes or apical health effects. Categories can in principle relate to events at any level of biological organisation (e.g. activation of a given pathway). When considering and using categorical information from any toxicological test method (irrespective of whether it is a traditional animal test or an alternative method) one should keep in mind that the distinct categories (as defined for purposes of C&L) have been set as an arbitrary convention to simplify risk management and transport of chemicals. Unlike other testable properties that may come in two classes (e.g. absence or presence of a disease marker), toxicity and hazard are continuous events and categorical differences do not exist in reality. This is especially important when considering data close to the cut-off of a prediction model (see Fig. 4.13, Sect. 4.7.2). Chemicals close to the cut-off can lead to an apparent high variability (or low reproducibility) of the test system and impact on the predictive capacity. It can be useful to consider such data close to the cut-off as "inconclusive" results which need to be further processed by expert judgement (i.e. ascribing one of the two categories). This judgement can be aided by additional statistical measures (e.g. Confidence Intervals) and/or other sources of toxicological information (read-across, QSAR, etc.).

   In this chapter we will focus on statistical measures of predictive capacity of categorical predictions. Statistical analyses of the results from non-categorical methods need to be defined on a case-by-case basis. To return to the example of the Fish Embryo Toxicity test: in this case the predictive relationship between

LC_{50} values of embryonic fish and LC_{50} values from adult and juvenile fish was assessed by means of orthogonal regression (Belanger et al. 2013) providing information on slope, intercept and range of concentrations over which the correlation held.

Predictive Capacity (PC)

The predictive capacity of tests that provide categorical predictions informs about test method performance in terms of correct and incorrect predictions in comparison to pre-selected reference data that are considered "true" and referred to a "actual positives" and "actual negatives". These data can be derived from the species of interest (Goldberg et al. 1995) or from other reference methods, typically surrogate animal methods. The latter has been, for reasons outlined in Sect. 2.1.4, typical practice during validation. The predictive capacity gives quantitative information on test method performance in terms of translating the actual measurements obtained (e.g. cell viability, quantified gene expression) into predictions of a defined effect (e.g. a pathway or a downstream health effect). The predictive capacity therefore reflects the final outcomes of the test method when applied as intended.

The predictive capacity serves as a tool for policy makers and regulators to evaluate to which extent the test method considered is likely to accurately predict the biological effect(s) of interest. Based on the predictive capacity and duly considering its intended use scenario, regulators can decide whether or not a given method is ready to be implemented in regulation as a routine tool for contributing to risk assessment. Due to the fact that alternative methods have primarily focused so far on hazard identification (Sect. 2.1), the predictions often relate to categories of classification and labelling as defined in international classification systems such the United Nations (UN) Global Harmonized System for Classification and Labelling (GHS).

For calculating the predictive capacity, the final outcomes/predictions of the test method are compared to those from a reference method or to other reference data. The reference method is usually an *in vivo* test method (see Sect. 2.1.4), but comparison can also be performed against human data if available.

Sensitivity and Specificity

Typically, test methods provide binary outcomes (see Fig. 4.1). This is true for most diagnostic tests in medicine but also for most alternative methods. Binary (=dichotomous) predictions here relate to diagnostic results of yes/no (absence or presence of a property) or predictions on causative properties of test items in the system of interest, i.e. "positive" = causing a toxicity effect or negative = not causing this effect (or at least at a threshold below concern = "cut-off").

To characterise the diagnostic or predictive capacity of methods with binary outcomes, the sensitivity (Se) and specificity (Sp) of the test method is calculated. To this end, the binary predictions of the alternative test method are compared to binary predictions obtained from the reference data, typically the *in vivo* test method, for the same set of test chemicals. Predictions from the reference method are considered as actual positive or actual negatives.

As defined by OECD Guidance Document No. 34, the *sensitivity* is the proportion of positive chemicals for the endpoint considered that are correctly identified by the test method (true positive predictions) as compared to the actual positives of the reference method; conversely, the proportion of positive chemicals wrongly predicted as negative corresponds to the false negatives. The *specificity* is the proportion of negative chemicals that are correctly identified by the test method (or true negative predictions); conversely the proportion of negative chemicals wrongly predicted as positive is the false negative rate. Additionally, the accuracy of the test method is the proportion of correct predictions made—in comparison to the reference data—over all predictions obtained.

Two-by-two contingency tables are useful tools for summarising the outcomes of test methods in relation to the actual positives and actual negatives of the reference data. These tables show the frequency of each type of prediction: a positive prediction of the test method for a test item considered an actual positive is termed "true positive" (a). Accordingly the outcomes "false negative" (c), "true negative" (d) and "false positive" (b) are determined. Additionally the number of chemicals assessed is shown (see Table 4.1 and Fig. 4.5).

The fraction (P) of chemicals that produce a positive result in the reference method, over the total number (N) of chemicals is often named 'prevalence'. Conversely the fraction of chemicals that produce a negative result in the reference method is $(1-P)$. Therefore, the number of chemicals producing a positive result in the reference method is $P \times N$ and the number of chemicals producing a negative result in the reference method is $(1-P) \times N$ (Table 4.1). Denoting the reference method by R, for which the outcome can be positive or negative (respectively R+ and R–) and the test method by T, for which the outcome can be positive or negative as well (respectively T+ and T–), the prevalence P can be expressed as the probability in the reference data set that the outcome is positive and formulated as $P = P(R^+)$ and $1 - P = P(R^-)$.

The calculation of sensitivity and specificity can be formulated with the use of Bayes' formulas as follows:

$$Se = P\left(T^+ \mid R^+\right) = \frac{P\left(T^+ \cap R^+\right)}{P\left(R^+\right)} \Leftrightarrow Se = \frac{a}{a+c} \tag{4.1}$$

$$Sp = P\left(T^- \mid R^-\right) = \frac{P\left(T^- \cap R^-\right)}{P\left(R^-\right)} \Leftrightarrow Sp = \frac{d}{b+d} \tag{4.2}$$

Those equations show that the proportion of actual negatives and actual positives in the sample do not influence the calculations of Se and Sp. One can also say that both are independent on the prevalence (number of actual positives) in the sample. That means that Se and Sp are indicators directly related to the intrinsic features of the test method.

Table 4.1 Two-by-two contingency table for binary outcomes, providing types of predictions and their respective proportions

	Reference + (actual positive)	Reference − (actual negative)	
Test+ (positive prediction)	a = P × N × Se	b = (1 − P) × N × (1 − Sp)	a + b
	True Positive prediction	False Positive prediction	
Test− (negative prediction)	c = P × N × (1 − Se)	d = (1 − P) × N × Sp	c + d
	False Negative prediction	True Negative prediction	
	a + c = P × N	b + d = (1 − P) × N	a + c + b + d = N

× = multiplication sign

Table 4.2 Three-by-three contingency table for three possible outcomes, providing types of predictions and their respective proportions

	Reference Category 1	Reference Category 2	Reference Category 3
Test	**a**	**b**	**c**
Category 1	Correct prediction as Category 1	Underprediction as Category 2	Underprediction as Category 3
	Rate = $(a/n_1) \times 100$	Rate = $(b/n_2) \times 100$	Rate = (c/n3) × 100
Test	**d**	**e**	**f**
Category 2	Overprediction as Category 1	Correct prediction as Category 2	Underprediction as Category 3
	Rate = $(d/n_1) \times 100$	Rate = $(e/n_2) \times 100$	Rate = (f/n3) × 100
Test	**g**	**h**	**i**
Category 3	Overprediction as Category 1	Overprediction as Category 2	Correct prediction as Category 3
	Rate = $(e/n_1) \times 100$	Rate = $(h/n_2) \times 100$	Rate = (i/n3) × 100
	$a + d + g = n_1$	$b + e + h = n_2$	$c + f + i = n_3$

× = multiplication sign

Positive and Negative Predictive Values

Apart of sensitivity and specificity, two other quantitative indicators can be calculated: Positive Predictive Value (PPV) and Negative Predictive Value (NPV). They correspond to the *probability* that a chemical produces a positive result in the reference method when the outcome of the test method is positive (PPV), and the probability that a chemical produces a negative result in the reference method when the outcome of the test method is negative (NPV). Using Bayes' formulas, they are respectively calculated as:

$$PPV = P\left(R^{+} \mid T^{+}\right) = \frac{P\left(T^{+} \cap R^{+}\right)}{P\left(T^{+}\right)} \tag{4.3}$$

$$NPV = P\left(R^- | T^-\right) = \frac{P\left(T^- \cap R^-\right)}{P\left(T^-\right)} \tag{4.4}$$

They respectively result in:

$$\begin{aligned} PPV = P\left(R^+ | T^+\right) &= \frac{P\left(T^+ \cap R^+\right)}{P\left(T^+\right)} = \frac{P\left(T^+ \cap R^+\right)}{P\left(T^+ \cap R^+\right) + P\left(T^+ \cap R^-\right)} \\ &= \frac{\left(P \cdot N \cdot Se\right) / N}{\left(\left(P \cdot N \cdot Se\right) / N\right) + \left(\left(1 - P\right) \cdot N \cdot \left(1 - Sp\right) / \mathrm{N}\right)} \\ &= \frac{P \cdot Se}{P \cdot Se + \left(1 - P\right) \cdot \left(1 - Sp\right)} \end{aligned} \tag{4.5}$$

$$\begin{aligned} NPV = P\left(R^- | T^-\right) &= \frac{P\left(T^- \cap R^-\right)}{P\left(T^-\right)} = \frac{P\left(T^- \cap R^-\right)}{P\left(T^- \cap R^-\right) + P\left(T^- \cap R^+\right)} \\ &= \frac{\left(\left(1 - P\right) \cdot N \cdot Sp\right) / N}{\left(\left(\left(1 - P\right) \cdot N \cdot Sp\right) / N\right) + \left(\left(P \cdot N \cdot \left(1 - Se\right)\right) / \mathrm{N}\right)} = \frac{\left(1 - P\right) \cdot Sp}{\left(1 - P\right) \cdot Sp + P \cdot \left(1 - Se\right)} \end{aligned} \tag{4.6}$$

It is obvious that both positive predictive value (Eq. (4.5)) and negative predicted value (Eq. (4.6)) depend on the prevalence (P), unlike sensitivity and specificity.

Therefore PPV and NPV calculations do represent the performance of the test method per se but for a specific set of chemicals in terms of the relative proportion of actual negatives and actual positives. They give only post-testing indications on how predictions were made for the set of chemicals that has been used; those indications would be different with another set of chemicals (e.g. where the prevalence of positive chemicals would be different—see Sect. 2.3.3). In contrast, the calculations of Sensitivity and Specificity are representative of the intrinsic test method performance, independent of the prevalence, i.e. the fraction of chemicals producing positive results. The examination of both sensitivity and specificity allows capturing the test method performance. This simultaneous evaluation of sensitivity and specificity can also be done when performing Receiver Operating Characteristic (ROC) analysis, as described below.

Considerations for More Than Binary Outcomes

When the possible outcomes of a test method are not binary and thus provide more than two types of prediction, sensitivity and specificity *sensu stricto* are not used but similar calculations are performed. For instance, when the prediction model yields three (sub-)categories, the resulting contingency table is therefore a three-by-three table, covering nine possible predictions. Still, predictions performed by the *in vitro* method are compared to those from the reference data (e.g. the *in vivo* reference method or other relevant data relating to the toxicity event). Consider a situation with three categories, category 1 relating to the most severe effect, category 2 to intermediate effects and category 3 to the least severe effects. For predictions regarding category 1, the three possible outcomes are: correct predictions into category 1, under-prediction into category 2, and under-prediction into category 3. For category 3, the three possible outcomes are: correct predictions into category 3, over-prediction into category 2, and over-prediction into category 1. For the middle category 2, the three possible outcomes are: correct predictions into category 2, under-predictions into category 3, and over-prediction into category 1. For each of these nine predictions it is possible to calculate their respective rates in percentages within the category predicted by the reference method.

Accuracy

Whether the outcome is binary or not, the accuracy of the test method—also referred to as 'overall accuracy'—can additionally being calculated. It is defined by the total number of correct predictions divided by the total number of predictions performed.

When examining the most common case of binary outcome (see Table 4.1), the overall accuracy (OA) is also related to the Prevalence (P) by the following formula:

$$
\begin{aligned}
OA &= P\left(\left(T^+ \cap R^+\right) | \left(T^- \cap R^-\right)\right) = P\left(T^+ \cap R^+\right) + P\left(T^- \cap R^-\right) \\
&= P\left(T^+ | R^+\right) \cdot P\left(R^+\right) + P\left(T^- | R^-\right) \cdot P\left(R^-\right) = Se \cdot P + Sp \cdot (1-P)
\end{aligned}
\tag{4.7}
$$

The same result is obtained when calculating the OA using the expressions in Table 4.1 cells.

$$OA = \frac{a+d}{N} = P.Se + (1-P).Sp = P(Se - Sp) + Sp \tag{4.8}$$

If Se>Sp, then from the above formula (Eq. 4.8) it follows necessarily that Se>OA>Sp

If Se<Sp, then it is derived from the same formula that necessarily that Sp>OA>Se

Additionally, when Se>Sp and if P increases, the OA increases as well; if P decreases, the OA decreases. When Se<Sp, if P increases, the OA decreases; if P decreases, the OA increases.

In other words, the OA is influenced by the prevalence P and always takes values that are necessarily between Se and Sp, whatever the value of P is—except for the particular case of Se=Sp, then OA=Sp=Se. During the validation process, the OA is sometimes used and reported. However using the OA does not reflect the intrinsic performance of the test method—in contrast to Se and Sp—as it depends on the prevalence P. or instance, an overall accuracy of OA=0.78 could correspond to three different cases such as: {Se=0.9; Sp=0.7; P=0.4} or {Se=0.9; Sp=0.5;

P=0.7} or {Se=0.9; Sp=0.3; P=0.8}. Therefore, the single use of OA is not a very useful tool to describe the concordance of a test method against a reference method (or reference data).

Likelihood Ratios

As demonstrated above, typical measures characterising test method performance relate to the prevalence-independent measures of sensitivity, specificity and overall accuracy taking into consideration the number of chemicals tested. However, likelihood ratios can be useful for assessing and reporting test method performance. For binary tests, one distinguishes likelihood ratio positive (LR^+) from likelihood ratio negative (LR^-). Likelihood ratios are routinely used in medicine in the context of describing how informative diagnostic tests are. However they have not been used much in toxicity for describing how informative a particular test result is from a specific test method.

Positive and negative likelihood ratio are defined as follows:

$$LR^+ = \frac{P(T^+ \mid R^+)}{P(T^+ \mid R^-)} = \frac{P(T^+ \mid R^+)}{1 - P(T^- \mid R^-)} = \frac{Se}{1 - Sp} \tag{4.9}$$

$$LR^- = \frac{P(T^- \mid R^+)}{P(T^- \mid R^-)} = \frac{1 - P(T^+ \mid R^+)}{P(T^- \mid R^-)} = \frac{1 - Se}{Sp} \tag{4.10}$$

In the expressions of LR^+ and LR^- (Eqs. (4.9) and (4.10)) the prevalence P is absent. That means that both likelihood ratios are not influenced by the prevalence P. In that sense, they are not mere ratios (re-)using sensitivity and specificity values. They represent probabilistic indicators reflecting how likely it is that a type of prediction is true. The LR^+ indicates the probability of a positive result being a true positive. In terms of performance, it is desirable that LR^+ is as high as possible which corresponds to high rate of true positive and/or low rate of false positive. Conversely, the LR^- should be as low as possible which corresponds to high rate of true negative and/or low rate of false negative. In medicine, likelihood ratios are often translated into qualitative stratified ratings ("qualitative strength") using cut-offs of LR's. These ratings aid the communication of test method strength. Mayer (2004) for instance lists four categories corresponding to "excellent", "very good", "fair" and "useless".

ROC as Means of Evaluating Optimal Cut-Off

Variations of the cut-off value are usually examined at the stage of test method development, but can be taken into account at any point in time. Desprez et al. (2015), have recently provided an example of a post hoc analysis of prediction models used for skin corrosion sub-categorisation and, on the basis of the analysis, proposed improved prediction models. For prediction models using cut-off values for assigning the predictions "negative" or "positive", any variation of the cut-off value will result in changes of the Se and Sp, in opposite directions. Thus, depending on the intended application it is possible to set a cut-off (i.e. within the prediction model) so that it optimises either sensitivity or specificity. To systematically assess the impact of shifting the cut-off, a useful approach consists in obtaining a Receiver Operating Characteristic (ROC) curve which provides quantitative indications of the predictive capacity.

A ROC curve is a graphical representation of test method performance: the x-axis represents values of (1-Specificity) and the y-axis represents values of the Sensitivity when monotonic variation of the cut-off value is applied for binary predictions (Fig. 4.12). The best theoretical performance of the method is obtained when both Se and Sp are close to 1 i.e., when Se is close to 1 and $1-\text{Sp}$ close to 0. The area under the ROC curve is necessarily between 0 and 1, and the best performance of the method is obtained when this area is close to 1. In contrast to the simple use of single values of Se and Sp, the ROC curve represents all possible values of Se and 1-Sp for all possible cut-offs. The ROC analysis will thus consist of finding the cut-off that will maximize the value of Se and minimize the value of $1-\text{Sp}$ (i.e. maximize the value of Sp). Usually the diagonal line—defined by the points (0; 0) to (1; 1)—is also represented. The shape of the ROC curve gives also an indication of the test method performance; it should have a hyperbolic shape, that is it should be as far away as possible from the midline and follow as closely as possible a curve that would link the points (0; 0) to (0; 1) and (0;1) to (1; 1). Such a curve would lead to an area under ROC close to 1, i.e. the best possible result.

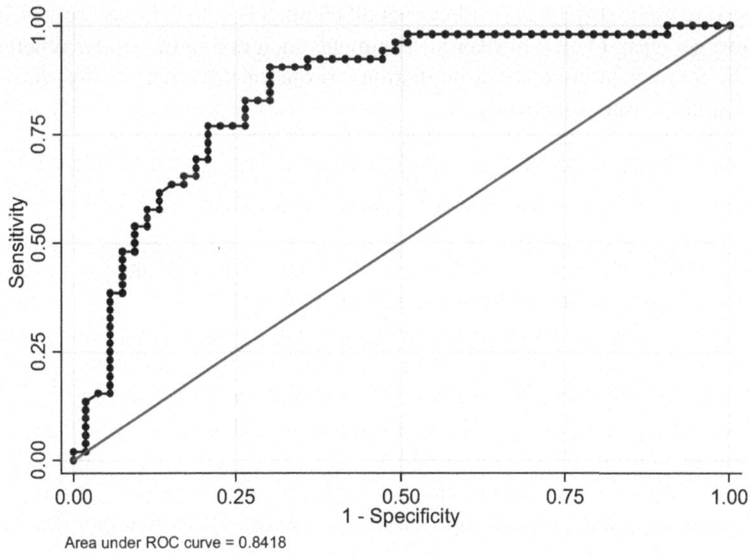

Fig. 4.12 Theoretical example of receiver operating characteristic (ROC) curve

4.7.2 Statistical Evaluation of the Within- and Between Laboratory Reproducibility

Within laboratory reproducibility (WLR) (or intra-laboratory reproducibility) gives information on the extent to which a test provides the same results over time when conducted in the same laboratory (OECD 2005), while the between-laboratory reproducibility (BLR) addresses this question for different laboratories (OECD 2005). In a more general manner, WLR and BLR may not only focus on the obtained prediction but may also examine the variability (e.g. standard deviation) of the measured endpoint of the test method.

Reproducibility and Variability Within One Laboratory

Within-Laboratory Reproducibility

The OECD Guidance Document No. 34, on the validation of new test methods (OECD 2005), provides a definition of the "within laboratory reproducibility" (WLR) or "intra-laboratory reproducibility". The WLR aims to determine the "extent that qualified people within the same laboratory can successfully replicate results using a specific protocol at different times". Typically, the experiment is performed over several runs that are independent and the WLR is assessed considering the agreement between the predictive results of these runs. The WLR is part of the indicators that measure the test method reliability (together with the between laboratory reproducibility, see below).

Several ways can be used to assess the WLR. Classically, the WLR is calculated on the basis of the fraction of chemicals (in %) for which concordant predictions in all runs were made (Eq. (4.11)) either over all chemicals with valid test results in the laboratory (see Eq. (4.11)) or over all chemicals included in the study. Whether to relate the number of concordant predictions to one or the other ideally should be defined at the outset of the study.

$$WLR = \frac{\textit{Number of chemicals for which concordant predictions are made in all runs}}{\textit{Total number of chemicals used for these runs}} \tag{4.11}$$

The advantage of this type of calculation is that it uses the final outcome or result of the method as used in practice and is easy to perform.

However, it should be kept in mind that an analysis of the reproducibility (or inversely variability) of the underlying measurement (e.g. normalised cell viability) allows assessing reproducibility without potential misleading results of substances close to the cut-off of the prediction model: obviously, measures that are close to the cut-off value defined for deriving predictions may show low variability between each other and yet result in different predictions which would be interpreted as "non-reproducibility" (Fig. 4.13). Notably, the closer

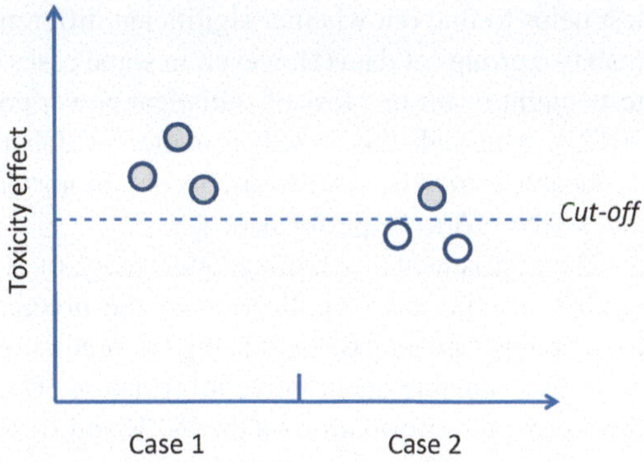

Fig. 4.13 Vicinity of measurements to an arbitrary categorisation cut-off may lead to non-concordant predictions that are interpreted as non-reproducibility although the dispersion between the individual runs (*circles*) is very similar in case 1 versus case 2

measurements are to the cut-off, the greater the influence of random variations that tilt results in one or the other direction (i.e. positive or negative prediction). In these borderline cases, the assessment of WLR based on concordant predictions may not capture accurately the reproducibility of the test method and when interpreting reproducibility via concordance of predictions the vicinity of values to arbitrarily fixed cut-offs needs to be taken into account. It is therefore also useful to assess and quantify the variability (dispersion) of the actual measurements before application of the prediction model.

Variability

In addition to assessing the agreement of predictions it is useful to study the variability of the measurements obtained, e.g. over runs. Variability can be studied by examining medians, means, as well as standard deviation (SD) values, and coefficient of variations (CV) of the measured parameters. Observation of the SD value helps establishing a threshold: data points for which the SD values are below this threshold have a low variability and are considered concordant. Analysis of variance (ANOVA) can further be performed and would compare the variability of the parameter over the runs. However before performing an ANOVA, some conditions regarding the data should be verified first. These conditions are that (i) the groups of comparison (i.e. the runs) are independent, (ii) the distribution of the data is normal, and (iii) the equality of variance in the compared groups is verified. This ANOVA can be combined with pairwise comparisons that help determining which pairs of runs are eventually significantly different (e.g. if four runs were performed, six pair comparisons should be done between runs 1 and 2; 1 and 3; 1 and 4; 2 and 3; 2 and 4; 3 and 4).

When the conditions of the ANOVA are not verified, the analysis can be performed on the basis of non-parametric statistical tests, such as the Kruskal-Wallis or Mann–Whitney tests as those statistical tests are based on the ranks of the parameter (Van Hecke 2012). For instance, when three runs are performed, the Kruskal-Wallis test helps to find out whether significant differences are globally observed on more than two groups of data (However, in some cases the performance of non-parametric tests might result in a loss of statistical power compared to ANOVA (Ferreira et al. 2012)). Although this is still a matter of debate, the data transformation—when applicable—may be worthwhile to obtain normally distributed data, and therefore allow ANOVA to be performed.

The assessments of the reproducibility of a test method based on concordant predictions (i.e. after application of the prediction model) and variability of the measured parameter (without using the prediction model) are complementary. They both give valuable quantitative information. The assessment of concordant predictions provides information on the WLR and BLR of the test method for its intended use and is therefore necessary for regulatory purposes. The assessment of the measured parameter is also necessary to capture variations of this parameter over runs, especially to identify borderline cases (when the measured parameter approaches the cut-off value) and therefore helps in understanding how predictions were performed and may help identifying chemicals for which predictions have been problematic. Moreover, defining variability independent of the prediction models may support later adaptation of the prediction model if necessary (Desprez et al. 2015).

Between Laboratory Reproducibility

The between laboratory reproducibility (BLR) is also called inter-laboratory reproducibility and has been also defined in the OECD Guidance Document No. 34. The BLR provides information on the reproducibility of a test method in different laboratories, i.e. under slightly different conditions. Together with within-laboratory reproducibility and the ease of transferring a method from one experienced to less experience laboratories ("transferability"), BLR informs on the robustness of a test method, i.e. its "resilience" towards minor variations in terms of equipment, operator and aspects such as shipment of cells, etc.

The way to assess BLR is similar to the one for assessing WLR and it can therefore be based on the fraction of chemicals that led to concordant predictions in all different labs (see Eq. (4.12)) either over all chemicals with valid test results in the laboratories (see Eq. (4.12)) or over all chemicals included in the study. Whether to relate the number of concordant prediction to the one or the other ideally should be defined at the outset of the study.

$$BLR = \frac{Number\ of\ chemicals\ for\ which\ concordant\ predictions\ are\ made\ in\ all\ laboratories}{Total\ number\ of\ chemicals\ used\ for\ these\ laboratories} \tag{4.12}$$

Similarly to what has been said before, it can be useful to assess also the variability between laboratories using medians, means, standard deviations and coefficient of variations of the measured parameter.

4.7.3 Providing Confidence Intervals Instead of a Single Point Estimates

The use of a single value (i.e. point estimate) does not entirely capture the uncertainty related to the use of a test method and its predictions. The key values of Sensitivity, Specificity, WLR and BLR may be given within a confidence interval (CI), for example at 95 % (CI_{95}). Calculating and reporting CIs takes this into account and communicates the uncertainty associated with a point estimate, thus improving the description of test method performance.

Consider the use of CI in the following example of WLR: In theory, if the whole population of chemicals would be tested, the obtained WLR would be the exact WLR (WLR_{ex}). However, for validation studies only a very limited number of chemicals(= a representative sample of chemicals) is used. In terms of statistics, this set of test chemicals is a sample of the entire population of existing chemicals. Therefore the WLR value (WLR_{est}), obtained with this set of test chemicals, is an estimated value of the exact one (WLR_{ex}). The CI_{95} represents the range of WLR values for which the probability to find the exact one is 95 %, that also means that the probability of not including the exact value in this interval is 5 %. Any value included in the CI has the same probability than any other to occur, including also the mean (see Fig. 4.14). Obviously, the sample size plays a critical role. The greater the sample, the narrower the confidence interval.

For instance, a test that classifies 20 chemicals and for which 17 out of 20 chemicals are concordantly predicted has, according to previous definition, a WLR rate of $(17/20) \times 100 = 85\,\%$. The CI_{95} for this value is [62.1–96.8 %], following binomial distribution. If we now consider a set of 60 test chemicals, for which the WLR rate is also 85 % i.e., 51 out of 60 chemicals have concordant prediction, then the CI_{95} is [73.4–92.9 %]. This CI is therefore much narrower than the previous one that has the same mean value of WLR.

Any value of this CI has the same probability to occur and the mean value of 85 % is included in this interval. If the whole population of chemicals was tested

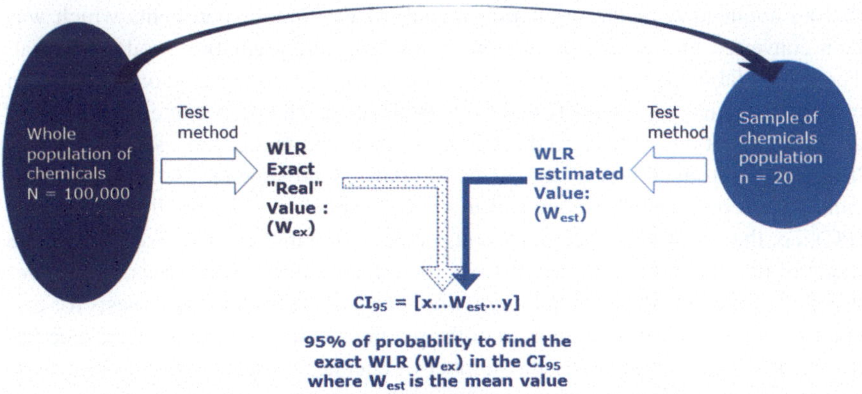

Fig. 4.14 WLR estimation with confidence intervals

then we would get the exact value of WLR. The CI_{95} represents the range of value of WLR for which there is 95 % of chances to find the exact WLR value.

In the examples above, where WLR = 85 %, the use of the CI helps to quantify the uncertainty of this value with an error risk of 5 %. When comparing the CI for 20 test chemicals, which is [62.1–96.8 %], and the one for 60 test chemicals, which is [73.4–92.9 %] it becomes clear that the extent of the first one is much greater than the second: the larger the CI, the higher the uncertainty. In other words, the central value of 85 % in the first CI is framed by a much more extended range of values compared to the second one.

4.7.4 Using All Experimental Observations for PC, WLR and BLR

Up to now we have presented analyses that are based on the assumption that, for BLR analysis, there is *one final prediction per laboratory* that can be used to determine concordance of predictions between laboratories (Eq. (4.12)) and that there is one final prediction per chemical so as to calculate the predictive capacity of the assay for the sample tested (Table 4.1, Fig. 4.5). However, this is normally not the case since during validation studies, experiments are typically conducted in triplicate in each of the laboratories which creates nine available experimental predictions for each chemical. The reason for this data-rich matrix is the need to assess within-laboratory reproducibility of the predictions of experiments, i.e. the final outcome of the test method as used in practical application. For BLR and PC however, the data matrix results in the following problems: which of the three from each laboratory should be used for determining BLR and PC within the laboratory, and which of the nine predictions available should be used to determine PC for the assay? To address these issues, it has been common practice in many validation studies to derive "final calls" (i.e. a final prediction). For BLR per one final call per laboratory was derived either by (a) calculating the mode of predictions of the three experiments (hence per laboratory normally an odd number of experiments is performed, e.g. three) or (b) by calculating an average value of the final measurement (before application of the prediction model) of the three experiments which was then converted into a final prediction by applying the prediction model as usual. These final laboratory predictions were then analysed for their concordance, i.e. in exactly the same way as WLR had been established. The percentage value of concordant predictions *between* laboratories is then communicated as the between-laboratory reproducibility of the assay. Similarly, final singular predictions ("final calls") were produced per chemical in view of calculating the predictive capacity (PC), i.e. the sensitivity, specificity and accuracy of a test method. To this end, the mode of the final laboratory predictions (determined for BLR analysis, see above) was determined yielding one *final call per chemical*. This created the basis for calculating one point estimate for each of the predictive indicators sensitivity, specificity, accuracy based on exactly the number of chemicals analysed during the study. Although this analysis may appear a straightforward way of simplifying the artificially inflated validation data matrix, mentioned approach has a fundamental disadvantage: instead of using the data from the test methods as it would be used in practice, results from experiments are artificially "aggregated" or "condensed" by means of averaging or, basically, majority voting (mode of predictions). Moreover, this approach leads to a loss of information on the experimental level.

This problem is of course not specific to toxicological data sets but is encountered in many disciplines including biology, medicine (e.g. evaluation of diagnostic test methods, clinical trials), epidemiology, etc. where large sets of (non-independent or not fully independent) observations are available. Standard statistical literature has been cautions with regard to fully using all observations (Colton 1974), mainly because this may be misleading with regard to the actual sample size that should be reflected in the analysis (Colton gives an example of 800 blood pressure measurement in a drug study which were however based on 10 measurements weekly over an 8-week treatment course in only ten patients, which would be the actual sample). In our example, using all observations would mean calculating the sensitivity and specificity on the basis of *all* predications generated during the study, i.e. nine observations per chemical times the sample of chemicals tested. So if 100 chemicals have been tested, the sample would appear to consist of 900 and not the 100 that have been tested in reality. Thus the actual sample is overstated and it also misleadingly narrows the confidence intervals. More recently more publications have addressed the issue of using how all observations can be used or, in particular, to which extent each observation contributes statistically to the overall analysis in such cases. The statistical technique of *Generalized Estimating Equations* (GEE) (Hanley et al. 2003) can be used in such situations and its applicability to validation data sets should be considered. Another way to estimate the WLR, BLR and predictive capacity is the *bootstrapping* technique (Holzhütter et al 1996; Hoffmann and Hartung 2006a). The data from all experiments performed during the validation study are resampled over a large number of times e.g., 1000 times, and on the basis of the resampling the studied parameter is estimated. The idea is that the entire population of chemicals cannot be studied and the sample size is deemed to be limited (e.g. 20 chemicals) for the estimation of the parameter. Therefore the performance of resampling on the sample itself and repeated many of times may better capture the variability of the parameter than the approach based on a single value. For instance, for the WLR, the resampling can be performed at the level of the different runs of a given laboratory. Then WLR is calculated on the basis of concordant results obtained in this resampling. The resampling procedure can also be done at the level of the chemicals. For the BLR, the principle would be the same, e.g. resampling over the results obtained in all laboratories.

5 Conclusions

In this chapter we have explored the fundamental concepts underlying the validation of *in vitro* test methods for hazard/safety assessment of chemicals or biologicals and have summarised the major challenges as well as established processes and tools for validation. Validation sits between the development of novel test methods and their routine use for safety assessment by industry and regulators. The aim of validation is to provide a robust, transparent and trustworthy scientific basis regarding the characterisation of a test method in view of its application for a particular purpose ("fitness for purpose"). From this it follows that there can never be a final or ultimate validation of a test method: validation is context-dependent. Validation studies and subsequent recommendations support regulators, policy makers and stakeholders when considering whether or not to formally adopt (i.e. into legislation) a given test method for a specific use in relation to legislation that aims to protect human health (e.g. workers, consumers) and the environment.

We stress that the term validation incorporates various meanings: it relates to the formal process of assessing and establishing fitness-for-purpose of a test method (often conducted by impartial governmental or supranational organisations), but also to the scientific study type for achieving this and, last, to the testing of a set of hypotheses. These are: (1) that the scientific basis of a test method is relevant for an adverse outcome or toxicity pathway in a target system, (2) that a given protocol associated with the test method allows its reproducible use and (3) that a given prediction model allows making sufficiently accurate predictions on adverse outcomes. All of these hypotheses are assessed through empirical testing of test chemicals (prospective studies) or the evaluation of existing information (retrospective studies). Consequently validation studies are scientific endeavours that need to be conducted in agreement with key scientific principles such as objectivity and appropriateness of methodology. These relate to statistically informed sample size calculations, conscientious selection of test chemicals, ex ante criteria for test method performance and the independence and/or impartiality of some of the actors (at least the scientific peer review).

We have discussed some major challenges of validated alternative test methods, mainly relating to the fact that these are proxy systems and highly reductionist models. We have also discussed the basic design of test methods. These are based on specific test systems (e.g. a specific cell line or tissue), the measurement of specific parameters as well as a prediction model. These elements of a test method are normally described in the procedure associated with the test method. Prediction models are of key importance for the validation of test methods as they are used to derive a performance characterisation in terms of predictive capacity and applicability. Prediction models are functions that convert the measured parameters into categorical predictions relating to any classification that is relevant of the purpose. Classifications can relate to chemicals being sorted according to their intrinsic potential to activate a toxicity pathway or to downstream (apical) health effects/adverse outcomes. Using the terminology of adverse outcome pathways (AOP), classifiers of alternative test methods can relate to everything from molecular initiating events to adverse effects on population level. Typically however, these relate to categories as defined by classification and labelling systems, for instance the United Nations Global Harmonized Systems for Classification and Labelling (UN GHS). Therefore the conversion of the measured parameter into classes/categorical variables, by means of the prediction model, is a simplification process that renders the outcome of the test more comprehensive but loses resolution with regard to the reality of a continuum of toxicity effects from non-toxic to highly toxic.

The validation process also encompasses the careful examination of the regulatory context. Due to the reductionist nature of alternative methods (i.e. modelling only small aspects of a more complex system), validated methods will be increasingly used in integrated testing strategies (ITS) or integrated approaches on testing and assessment (IATA), bringing together data from a variety of sources. The concept of adverse outcome pathways (AOPs) supports consensus-building on what may be the most important toxicological events leading to a final adverse effect: AOPs provide a description of these so-called "key events" and, to the extent possible, their causal links. In that sense the AOP concept promises to contribute to the identification of knowledge gaps and is expected to expedite the development of new test methods that model upstream mechanisms relevant for the downstream (apical) adverse effect of concern. The AOP concept thus also supports the validation of alternative methods of greater mechanistic and biological relevance and, it is hoped, greater predictive power and overall relevance.

The validation workflow typically includes four steps: assessment of test methods, conduct of validation studies, independent scientific peer review and final conclusions and recommendations. Regarding the practice of validation, the so-called "modular approach" has proven extremely useful: the information generated during the validation studies is systematically assessed through several information modules that all need to be sufficiently satisfied in view of scientific peer review of the validity status of a test method—notably, what constitutes "sufficient" depends on the purpose. The modules include the test definition (i.e. a description of the scientific basis of the method, within- and between-laboratory reproducibility, transferability, predictive capacity, applicability domain and performance standards, defined upon completion of a validation study. All of these modules are informed by testing data on chemicals. Thus, the number of chemicals tested influences the certainty of the data obtained. Therefore calculation of sample size, prior to the conduct of the study, is a prerequisite for enabling the generation of a sufficient amount of data. This relates to the statistical power and the target values defined for the study (e.g. target values of within-laboratory reproducibility or sensitivity). The reliability relies to the reproducibility of the method within a given laboratory, so called within laboratory reproducibility (WLR), the reproducibility over several laboratories, or between laboratory reproducibility (BLR) as well as the ease with which methods are amenable to transfer from one to another laboratory ("transferability"). WLR and BLR are assessed by the proportion of concordant predictions obtained. However this may not capture all the variability observed when using the method and other quantitative tools for assessing data variability before application of the prediction model are useful. The predictive relevance relies to how useful the obtained predictions are for the intended regulatory use. This is quantitatively assessed by the predictive capacity of the method. The predictive capacity uses accuracy values, such as sensitivity and specificity. Reporting confidence intervals helps capturing the uncertainty on the values obtained. ROC curves are another useful tool for assessing in a systematic manner the performance of the test method as a function of variations of the cut-off value of the prediction model.

In summary, validation is a multidisciplinary scientific exercise requiring expertise in a wide range of disciplines and areas, including biology, physiology, chemistry, statistics and regulatory frameworks. All these aspects are necessary for as complete a characterisation of a test method as possible through validation: This will help to understand and describe the extent of certainty and confidence in a test method and the remaining level of uncertainty. Validation will therefore play an ever greater role as new tools and more probabilistic approaches emerge in risk assessment wherein alternative methods are likely to play a central role.

References

Aggett P et al (2007) Variability and uncertainty in toxicology of chemicals in food, consumer products and the environment. Report by the Committee on toxicity of chemicals in food, consumer products and the environment. http://cot.food.gov.uk/sites/default/files/cot/vutreport-march2007.pdf

Archer G, Balls M, Bruner LH, Curren RD, Fentem JH, Holzhütter H-G, Liebsch M, Lovell DP, Southee JA (1997) ATLA 25:505–5016

Attarwala H (2010) TGN1412: from discovery to disaster. J Young Pharm 2(3):332–336

Balls M (1994) Replacement of animal procedures: alternatives in research, education and testing. Lab Anim 28:193–211

Balls M (1997) Defined structural and performance criteria would facilitate the validation and acceptance of alternative test procedures. ATLA 25:483–484

Balls M, Karcher W (1995) Comment: the validation of alternative test methods. ATLA 23:884–886

Balls M, Blaauboer B, Brusick D, Frazier J, Lamb D, Pemberton M, Reinhardt C, Roberfroid M, Rosenkranz H, Schmid B, Spielmann H, Stammati AL, Walum E (1990a) Report and recommendations of the CAAT/ERGAAT workshop on the validation of toxicity test procedures. ATLA 18:313–337 ("Amden I report")

Balls M, Botham P, Cordier A, Fumero S, Kayser D, Koeter H, Koundakjian P, Lindquist NG, Meyer O, Pioda L, Reinhardt C, Rozemond H, Smyrniotis T, Spielmann H, Van Looy H, van der Venne MT, Walum E (1990b) Report and recommendations of an international workshop on promotion of the regulatory acceptance of validated non-animal toxicity test procedures. ATLA 18:339–344 ("Vouliagmeni report")

Balls M, Bridges J, Southee J (1990c) Animals and alternatives in toxicology: present status and future prospects. VCH Publishers, New York

Balls M, Blaauboer BJ, Fentem JH, Bruner L, Combes RD, Ekwall B, Fielder RJ, Guillouzo A, Lewis RW, Lovell DP, Reinhardt CA, Repetto G, Sladowski D, Spielmann H, Zucco F (1995a) Practical aspects of the validation of toxicity test procedures. ECVAM workshop report 5. ATLA 23:129-147 ("Amden II report")

Balls M, De Klerck W, Baker F, van Beek M, Bouillon C, Bruner L, Carstensen J, Chamberlain M, Cottin M, Curren R, Dupuis J, Fairweather F, Faure U, Fentem J, Fisher C, Calli C, Kemper F, Knaap A, Langley G, Loprieno G, Loprieno N, Pape W, Pechovitch G, Spielmann H., Ungar K, White I, Zuang V (1995b) Development and validation of non-animal tests and testing strategies: the identification of a coordinated response to the challenge and the opportunity presented by the sixth amendment to the Cosmetics Directive (76/768/EEC). ATLA 23:398–409

Balls M, Amcoff P, Bremer S, Casati S, Coecke S, Clothier R, Combes R, Corvi R, Curren R, Eskes C, Fentem J, Gribaldo L, Halder M, Hartung T, Hoffmann S, Schechtman L, Laurie Scott L, Spielmann H, Stokes W, Tice R, Wagner D, Zuang V (2005) The principles of weight of evidence validation of test methods and testing strategies. the report and recommendations of ECVAM workshop 58. ATLA 34:603–620

Balls M, Combes RD, Bhogal N (2012) The use of integrated and intelligent testing strategies in the prediction of toxic hazard and in risk assessment. Adv Exp Med Biol 745:221–253

Barratt MD, Brantom PG, Fentem JH, Gerner I, Walker AP, Worth AP (1998) The ECVAM international validation study for *in vitro* tests for skin corrosivity. 1. Selection and distribution of the test chemicals. Toxicol *In Vitro* 12:471–482

Belanger SE, Rawlings JM, Carr GJ (2013) Use of fish embryo toxicity tests for the prediction of acute fish toxicity to chemicals. Environ Toxicol Chem 32(8):1768–1783

Borlak J (2009) Trovafloxacin: a case study of idiosyncratic or iatrogenic liver toxicity—molecular mechanisms and lessons for pharmacotoxicity. In: Hayes W, Griesinger C, Guzelian (eds) Proceedings of the 1st international forum towards evidence-based toxicology. Hum Exp Toxicol 28:119–212

Bouhifd M, Bories G, Casado J, Coecke S, Norlén H, Parissis N, Rodrigues RM, Whelan MP (2012) Automation of an *in vitro* cytotoxicity assay used to estimate starting doses in acute oral systemic toxicity tests. Food Chem Toxicol 50(6):2084–2096

Bouvier d'Yvoire M, Bremer S, Casati S, Ceridono M, Coecke S, Corvi R, Eskes, C, Gribaldo L, Griesinger C, Knaut H, Linge JP, Roi A, Zuang V (2012) ECVAM and new technologies for toxicity testing. In: Balls M, Combes RD, Bhogal N (eds) New technologies for toxicity testing. Springer series "Advances in experimental medicine and biology." Springer/Landes Bioscience 745:154–180

Bruner LH, Carr GJ, Chamberlain M, Curren RD (1996) Validation of alternative methods for toxicity testing. Toxicol *In Vitro* 10:479–501

Coecke S, Bowe G, Millcamps A, Bernasconi C, Bostroen AC, Bories G, Fortaner S, Gineste Jm, Gouliarmou V, Langezaal I, Liska R, Mendoza E, Morath S, Reina V, Wilk-Zasadna I, Whelan M (2014) In: Jennings P, Price A (eds) Considerations in the development of *in vitro* toxicity testing methods intended for regulatory use, Coecke S. et al. 2014. *In vitro* Toxicology Systems series: methods in pharmacology and toxicology. Springer Science+Business Media, LLC, pp 551–569

Colton T (1974) Statistics in medicine. Little, Brown and Company, Boston

Curren RD, Southee JA, Spielmann H, Liebsch M, Fentem JH, Balls M (1995) The role of prevalidation in the development, validation and acceptance of alternative methods. ATLA 23:211–217

Desprez B, Barroso J, Griesinger C, Kandárová H, Alépée N, Fuchs H (2015) Two novel prediction models improve predictions of skin corrosive sub-categories by test methods of OECD Test Guideline No. 431. Toxicol *In Vitro* 29(2015):2055–2080

Draize JH, Woodard G, Calvery HO (1944) Methods for the study of irritation and toxicity of substances applied topically to the skin and mucous membranes. J Pharmacol Exp Ther 82:377–390

EURL ECVAM (2014) The EURL ECVAM—Cosmetics Europe prospective validation study of reconstructed human tissue-based test methods for identifying chemicals not requiring classification for serious eye damage/eye irritation testing

EURL ECVAM, European Commission, Joint Research Centre (2012 onwards) Website on EURL ECVAM Recommendations on validated test methods; the site is continuously updated. https://eurl-ecvam.jrc.ec.europa.eu/eurl-ecvam-recommendations

Fentem JH, Prinsen MK, Spielmann H, Walum E, Botham PA (1995) Validation—lessons learned from practical experience. Toxicol *In Vitro* 9:857–862

Fentem JH, Archer GE, Balls M, Botham PA, Curren RD, Earl LK, Esdaile DJ, Holzhütter HG, Liebsch M (1998) The ECVAM International validation study on *in vitro* tests for skin corrosivity. 2. Results and evaluation by the management team. Toxicol *In Vitro* 12(4):483–524

Ferreira E, Rocha M, Mequelino D (2012) Sigmae Alfenas 1(1):126–139. http://www.google.it/url?sa=t&rct=j&q=&esrc=s&frm=1&source=web&cd=1&ved=0CCUQFjAA&url=http%3A%2F%2Fpublicacoes.unifal-mg.edu.br%2Frevistas%2Findex.php%2Fsigmae%2Farticle%2Fdownload%2F99%2Fpdf&ei=4hK5VJS3H8i7ygPf0wI&usg=AFQjCNHE4phiKyiXay4fNnpctO489uddBQ&sig2=tNR25ijrcheBiS87zWLcrA&bvm=bv.83829542,d.bGQ

Flahault A, Cadilhac M, Thomas G (2005) Sample size calculation should be performed for design accuracy in diagnostic studies. J Clin Epidemiol 58:859–862

Frazier JM (1990a) Scientific criteria for validation of *in vitro* toxicity tests. Environment Monographs no. 36. Paris: Organization for Economic Co-Operation and Development; 62 pp. Not available any longer. The document gave rise to the OECD Solna report (OECD, 1996)
Frazier JM (1990b) Validation of *in vitro* models. J Am Coll Toxicol 9:355–359
Frazier JM (1992) Validation of *in vitro* toxicity tests. In: Frazier JM (ed) *In vitro* toxicity testing: applications to safety evaluations. Marcel Dekker, New York, pp 245–252
Frazier JF (1994) The role of mechanistic toxicology in test method validation. Toxicology *In Vitro* 8:787–791
Goldberg, AM, Epstein, LD, Zurlo J (1995) A modular approach to validation—a work in progress. In: Salem H, Katz SA (eds) Advances in animal alternatives for safety and efficacy testing. Taylor & Francis, WA, pp 303–308. Expanded edition of: Animal test alternatives. Marcel Dekker, New York
Green S (1993) Regulatory agency considerations and requirements for validation of toxicity test alternatives. Toxicol Lett 68:119–123
Gregory CD (2014) Cell biology: the disassembly of death. Nature 507:312–313
Griesinger C (2009) Comparing medicine with toxicology— a mapping of knowledge creation, concepts and basic epistemology. In: Hayes W, Griesinger C, Guzelian P (eds) Proceedings of the 1st international forum towards evidence-based toxicology. Hum Exp Toxicol 28(2–3):101–104
Griesinger C, Hoffmann S, Kinsner A, Coecke S, Hartung T (2009) Special issue: evidence-based toxicology (EBT). Preface. Hum Exp Toxicol. In: Hayes W, Griesinger C, Guzelian P (eds) Proceedings of the 1st international forum towards evidence-based toxicology. Hum Exp Toxicol 28(2–3):83–86
Griesinger C, Schäffer M, Worth A, Zuang V (2014) Skin corrosion and irritation. In: Alternative methods for regulatory toxicology—a state-of-the-art review. JRC science & policy report. Publications Office of the European Union. http://bookshop.europa.eu/is-bin/INTERSHOP.enfinity/WFS/EU-Bookshop-Site/en_GB/-/EUR/ViewParametricSearch-Dispatch
Guzelian PS, Victoroff MS, Halmes NC, James RC, Guzelian CP (2005) Evidence-based toxicology: a comprehensive framework for causation. Hum Exp Toxicol 24(4):161–201
Guzelian PS, Victoroff MS, Halmes C, James RC (2009) Clear path: towards an evidence-based toxicology (EBT). In: Hayes W, Griesinger C, Guzelian P (eds) Proceedings of the 1st international forum towards evidence-based toxicology. Hum Exp Toxicol 28(2–3):71–79
Hanley JA, Negassa A, Edwardes MD, Forrester JE (2003) Statistical analysis of correlated data using generalized estimating equations: an orientation. Am J Epidemiol 157(4):364–375
Hartung T (2010) Evidence-based toxicology—the toolbox of validation for the 21st century? ALTEX 27(4):253–263
Hartung T, Bremer S, Casati S, Coecke S, Corvi R, Fortaner S, Gribaldo L, Halder M, Hoffmann S, Roi AJ, Prieto P, Sabbioni E, Scott L, Worth A, Zuang V (2004) A modular approach to the ECVAM principles on test validity. Altern Lab Anim 32(5):467–472
Hendriksen C, Spieser J-M, Akkermans A, Balls M, Bruckner L, Cussler K, Daas A, Descamps J, Dobbelaer R, Fentem J, Halder M, van der Kamp M, Lucken R, Milstien J, Sesardic D, Straughan D, Valadares A (1998) Validation of alternative methods for the potency testing of vaccines. ATLA 26:747–761
Hoffmann S, Hartung T (2005) Diagnosis: toxic!—trying to apply approaches of clinical diagnostics and prevalence in toxicology considerations. Toxicol Sci 85(1):422–428
Hoffmann S, Hartung T (2006a) Designing validation studies more efficiently according to the modular approach: retrospective analysis of the EPISKIN test for skin corrosion. Altern Lab Anim 34(2):177–191
Hoffmann S, Hartung T (2006b) Toward an evidence-based toxicology. Hum Exp Toxicol 25(9):497–513
Hoffmann S, Edler L, Gardner I, Gribaldo L, Hartung T, Klein C, Liebsch M, Sauerland S, Schechtman L, Stammati A, Nikolaidis E (2008) Points of reference in the validation process: the report and recommendations of ECVAM Workshop 66. Altern Lab Anim 36(3):343–352

Holzhütter HG, Archer G, Dami N, Lovell DP, Saltelli A, Sjöström M (1996) Recommendation for the application of biostatistical methods during the development and validation of alternative methods. ATLA 24:511–530

Horvath CJ, Milton MN (2009) The TeGenero incident and the Duff Report conclusions: a series of unfortunate events or an avoidable event? Toxicol Pathol 37(3):372–383

Hothorn LA (2002) Selected biostatistical aspects of the validation of *in vitro* toxicological assays. ATLA 30(2):93–98

ICCVAM (1997) Validation and regulatory acceptance of toxicological test methods: a report of the *ad hoc* interagency coordinating committee on the validation of alternative methods. National Institute of Environmental Health Sciences (NIEHS), Research Triangle Park, NC, USA, p 105. NIH Publication No: 97-3981. http://iccvam.niehs.nih.gov/docs/guidelines/validate.pdf

Judson R, Kavlock R, Martin M, Reif D, Houck K, Knudsen T, Richard A, Tice RR, Whelan M, Xia M, Huang R, Austin C, Daston G, Hartung T, Fowle JR 3rd, Wooge W, Tong W, Dix D (2013) Perspectives on validation of high-throughput assays supporting 21st century toxicity testing. ALTEX 30(1):51–56

Kinsner-Ovaskainen A, Maxwell G, Kreysa J, Barroso J, Adriaens E, Alépée N, Berg N, Bremer S, Coecke S, Comenges JZ, Corvi R, Casati S, Dal Negro G, Marrec-Fairley M, Griesinger C, Halder M, Heisler E, Hirmann D, Kleensang A, Kopp-Schneider A, Lapenna S, Munn S, Prieto P, Schechtman L, Schultz T, Vidal JM, Worth A, Zuang V (2012) Report of the EPAA-ECVAM workshop on the validation of Integrated Testing Strategies (ITS). Altern Lab Anim 40(3):175–181

Lachin JM (1981) Introduction to sample size determination and power analysis for clinical trials. Control Clin Trials 2:93–113

Matthews BW (1975) Comparison of the predicted and observed secondary structure of T4 phage lysozyme. Biochim Biophys Acta 405(2):442–451

Mayer D (2004) Essential evidence-based medicine. Cambridge University Press, Cambridge

Neugebauer EA (2009) Evidence-based medicine—a possible model for evidence-based toxicology? In: Hayes W, Griesinger C, Guzelian P (eds) Proceedings of the 1st international forum towards evidence-based toxicology. Hum Exp Toxicol 28(2–3):105–107.

OECD (1996, updated in 2009) Final report of the OECD workshop on the harmonisation of validation and acceptance criteria for alternative toxicological test methods. "Solna Report". http://www.oecd.org/officialdocuments/publicdisplaydocumentpdf/?cote=env/mc/chem/tg(96)9&doclanguage=en

OECD (2005) Guidance document on the validation and international acceptance of new or updated test methods for hazard assessment. OECD series on testing and assessment, document No. 34. http://www.oecd.org/officialdocuments/publicdisplaydocumentpdf/?cote=env/jm/mono%282005%2914&doclanguage=en

OECD (2008) Workshop on integrated approaches to testing and assessment. Series on testing and assessment No. 88. http://www.oecd.org/officialdocuments/publicdisplaydocumentpdf/?cote=env/jm/mono(2008)10&doclanguage=en

OECD (2013 (rev.), adopted 2010) Test guideline No. 439. *In vitro* skin irritation: Reconstructed human Epidermis test methods. http://www.oecdilibrary.org/docserver/download/9713241e.pdf?expires=1417441398&id=id&accname=guest&checksum=DB5C47DAF73F635E918A565FCCABCA01

OECD (2013) Guidance document on developing and assessing adverse outcome pathways. Series on Testing and Assessment No. 184. http://www.oecd.org/officialdocuments/publicdisplaydocumentpdf/?cote=env/jm/mono(2013)6&doclanguage=en

OECD (2014 (2nd rev.), adopted 2004) Test guideline No. 431. *In vitro* skin corrosion: Reconstructed human Epidermis test methods. http://www.oecdilibrary.org/docserver/download/9714521e.pdf?expires=1417442281&id=id&accname=guest&checksum=BA3044BD757F25BF4559D9D83E8E83A6

OECD (2014) New guidance on an integrated approach on testing and assessment (IATA) for skin corrosion and irritation. Series on testing and assessment No. 203. http://www.oecd.org/officialdocuments/publicdisplaydocumentpdf/?cote=env/jm/mono(2014)19&doclanguage=en

Pepe M (2003) The statistical evaluation of medical tests for classification and prediction. Oxford statistical science series. Oxford University Press, Oxford

Russell WMS, Burch RL (1959) The principles of humane experimental technique. Methuen, London

Scala RA (1987) Theoretical approaches to validation. In: Goldberg AM (ed) *In Vitro* toxicology: approaches to validation. vol 5, Alternative methods in toxicology. Mary Ann Liebert, New York, pp 1–9

Smythe DH (1978) Alternatives to animal experiments. Scolar Press for the Research Defence Society, London

Stephens ML, Andersen M, Becker RA, Betts K, Boekelheide K, Carney E, Chapin R, Devlin D, Fitzpatrick S, Fowle JR 3rd, Harlow P, Hartung T, Hoffmann S, Holsapple M, Jacobs A, Judson R, Naidenko O, Pastoor T, Patlewicz G, Rowan A, Scherer R, Shaikh R, Simon T, Wolf D, Zurlo J (2013) Evidence-based toxicology for the 21st century: opportunities and challenges. ALTEX 30(1):74–103

Van Hecke T (2012) J Stat Manage Syst 15(2–3). http://www.tandfonline.com/doi/abs/10.1080/09720510.2012.10701623?journalCode=tsms20#.VH7SxRAhB8E

Walum E, Clemedson C, Ekwall B (1994) Principles for the validation of *in vitro* toxicology test methods. Toxicol *In Vitro* 8:807–812

Waters CK (1990) Why the antireductionist consensus won't survive the case of classical Mendelian genetics. In: Fine A, Forbes M, Wessels L (eds) Proceedings of the biennial meeting of the Philosophy of Science Association, vol 1. Philosophy of Science Association, East Lansing, pp 125–139

Waters CK (2007) Causes that make a difference. J Philos 104:551–579

Weber M (2005) Philosophy of experimental biology. Cambridge University Press, New York

Chapter 5
Practical Aspects of Designing and Conducting Validation Studies Involving Multi-study Trials

Sandra Coecke, Camilla Bernasconi, Gerard Bowe, Ann-Charlotte Bostroem, Julien Burton, Thomas Cole, Salvador Fortaner, Varvara Gouliarmou, Andrew Gray, Claudius Griesinger, Susanna Louhimies, Emilio Mendoza-de Gyves, Elisabeth Joossens, Maurits-Jan Prinz, Anne Milcamps, Nicholaos Parissis, Iwona Wilk-Zasadna, João Barroso, Bertrand Desprez, Ingrid Langezaal, Roman Liska, Siegfried Morath, Vittorio Reina, Chiara Zorzoli and Valérie Zuang

Abstract This chapter focuses on practical aspects of conducting prospective *in vitro* validation studies, and in particular, by laboratories that are members of the European Union Network of Laboratories for the Validation of Alternative Methods (EU-NETVAL) that is coordinated by the EU Reference Laboratory for Alternatives to Animal Testing (EURL ECVAM). Prospective validation studies involving EU-NETVAL, comprising a multi-study trial involving several laboratories or "test facilities", typically consist of two main steps: (1) the design of the validation study by EURL ECVAM and (2) the execution of the multi-study trial by a number of qualified laboratories within EU-NETVAL, coordinated and supported by EURL ECVAM. The approach adopted in the conduct of these validation studies adheres

S. Coecke (✉) • C. Bernasconi • G. Bowe • A.-C. Bostroem • J. Burton • T. Cole • S. Fortaner
V. Gouliarmou • C. Griesinger • E. Mendoza-de Gyves • E. Joossens • A. Milcamps
N. Parissis • I. Wilk-Zasadna • J. Barroso • B. Desprez • I. Langezaal • R. Liska • S. Morath
V. Reina • C. Zorzoli • V. Zuang
European Commission, Joint Research Centre (JRC), Ispra, Italy
e-mail: Sandra.COECKE@ec.europa.eu

A. Gray
UK GLP Monitoring Authority, MHRA, London, UK

S. Louhimies
Directorate General for Environment, European Commission, Brussels, Belgium

M.-J. Prinz
Directorate General for Internal Market, Industry, Entrepreneurship and SMEs, European Commission, Brussels, Belgium

C. Eskes, M. Whelan (eds.), *Validation of Alternative Methods for Toxicity Testing*, Advances in Experimental Medicine and Biology 856,
DOI 10.1007/978-3-319-33826-2_5

to the principles described in the OECD Guidance Document on the Validation and International Acceptance of new or updated test methods for Hazard Assessment No. 34 (OECD 2005). The context and scope of conducting prospective *in vitro* validation studies is dealt with in Chap. 4. Here we focus mainly on the processes followed to carry out a prospective validation of *in vitro* methods involving different laboratories with the ultimate aim of generating a dataset that can support a decision in relation to the possible development of an international test guideline (e.g. by the OECD) or the establishment of performance standards.

Abbreviations

AOP	Adverse Outcome Pathway
CAS	Chemical Abstracts Service
CYP	Cytochrome P450
DMSO	Dimethyl sulfoxide
ECHA	European Chemicals Agency
EMA	European Medicines Agency
EFSA	European Food Safety Authority
EFTA	European Free Trade Association
ESAC	EURL ECVAM's Scientific Advisory Committee
ESTAF	EURL ECVAM Stakeholder Forum
EU	European Union
EU-NETVAL	European Union Network of Laboratories for the Validation of Alternative Methods
GIVIMP	Good *In Vitro* Method Practice (Guidance)
GLP	Good Laboratory Practice
IATA	Integrated Approach for Testing and Assessment
ICCVAM	Interagency Coordinating Committee on the Validation of Alternative Methods
ISO	International Organization for Standardization
ICATM	International Collaboration on Alternative Test Methods
JaCVAM	Japanese Center for the Validation of Alternative Methods
LIMS	Laboratory Information Management Systems
MAD	Mutual Acceptance of Data
MSDS	Material Safety Data Sheet
NICEATM	NTP (National Toxicology Program) Interagency Center for the Evaluation of Alternative Toxicological Methods
OECD	Organisation for Economic Co-operation and Development
PARERE	Preliminary Assessment of Regulatory Relevance
PBTG	Performance-based OECD test guidelines
(Q)SAR	(Quantitative) Structure-Activity Relationship
REACH	Registration Evaluation, Authorisation and Restriction of Chemicals
SOP	Standard Operating Procedure

1 Introduction

Before moving to the stage of conducting a prospective validation study, *in vitro* methods need to undergo thorough assessment. Experience has shown that development, validation and regulatory acceptance of a new *in vitro* method at international level is challenging as regulatory discussions relating to the integrative use of information from various *in vitro* methods and non-testing methods become increasingly complex. Therefore, it is of critical importance to consider the potential contribution or added-value of a particular *in vitro* method during the development stage, i.e. its contribution to a mechanistic understanding (e.g. in context of Adverse Outcome Pathways—AOP), its potential contribution to an Integrated Approach for Testing and Assessment (IATA) as described in Chap. 13, and hence its potential future use and usefulness (Bouvier d'Yvoire et al. 2012; Coecke et al. 2014a, b).

Generally, the development of an *in vitro* method purely for screening purposes may not require regulatory acceptance. However, if a method is intended for regulatory applications, this requires a high level of scientific description and characterisation as well as compliance with quality systems that need to be respected once such *in vitro* methods are routinely used. Regulators commonly consider "good" *in vitro* methods as those that are relevant for their specific regulatory purpose, are scientifically sound, applicable to the substances they are interested in, and demonstrated to be technically reproducible within and between different laboratories. This generally means that the *in vitro* method employs human cells or tissues in order to ensure biological relevance to the species of interest and that it measures endpoint(s) relevant to humans and underlying human Mechanisms of Action and/or Mode of Action. The selection of test chemicals as adequate positive and negative control items or, reference items is of critical importance.

1.1 Validation

Validation has been defined as a process to characterise the reliability and relevance of methods in view of a particular pre-defined purpose. Validation is an essential step towards ensuring that an *in vitro* method (1) is sufficiently *reliable* when used under standardised conditions on a routine basis and (2) produces data that are *relevant* in view of its envisaged purpose and application. Validation is a pivotal step towards the regulatory acceptance and the international recognition of *in vitro* methods for a range of scientific purposes by a variety of end-users (European Parliament and Council 2010). It is a prerequisite for the development of international standards and test guidelines that underpin regulatory decision-making and, due to agreements such as OECD's Mutual Acceptance of Data (MAD), supports global trade. Internationally accepted validation principles for alternative approaches are described in the OECD no. 34 "Guidance document on the validation and international acceptance of new or updated test methods for hazard assessment"

(OECD 2005), which outlines the essential considerations and steps to assess the reliability and relevance of an *in vitro* method.

A prospective validation study can be divided into two parts: (1) designing a validation study and (2) conducting a validation study. Key decisions should be made at the stage of trial design, including the establishment of the appropriate standard operating procedure (SOP) versions to be used during the validation study. In contrast, the phase of conducting a validation study mainly relates to the logistical and organisational aspects relating to the practical testing as well as the management of potential deviations.

2 Submission and Evaluation of *In Vitro* Methods for Validation Studies

The validation process starts with a theoretical, paper-based evaluation of the *in vitro* method submission or submissions covering a specific class of *in vitro* methods (Fig. 5.1). Once the theoretical assessment of the proposed (submitted) *in vitro* method has been completed, in some cases, the method is evaluated at the EURL ECVAM GLP test facility for technical and usability aspects. When *in vitro* methods covering a class of assays are submitted, one or more representative methods can be defined aiming to obtain, for this class of assays, the corresponding and desirable performance standards. The experimental evaluation of the submitted *in vitro* method or representative method(s) is performed by carrying out at least two studies, depending on the complexity of the method. One of these studies is performed as a GLP study, as OECD guidance states that it is preferred that validation studies are performed and reported in accordance with GLP (OECD 2005).

Before introducing the proposed *in vitro* method into the EURL ECVAM test facility, personnel (study director and study personnel) are trained adequately on the method, either in the test submitter's facility or at the EURL ECVAM GLP test facility. The training is usually carried out by the test developer. After completion of the training phase, the method is transferred to the EURL ECVAM GLP test facility. During this step, data analysis templates and raw data forms are assessed for completeness, usability and possible calculation errors. If no analysis templates or raw data forms are available, new templates or forms may be drawn up if deemed necessary. Any issues found during this *in vitro* method transfer phase will be communicated to the test submitter who shall address the findings before the method is moved into the next phase. If the method is substantially altered, it may need to be re-validated in-house. The last step in the evaluation process is to perform a GLP study within the EURL ECVAM GLP test facility, usually with more than one test chemical. All data analysis templates need to be validated according to the facility's SOPs on validation of electronic spreadsheets before they are used in a GLP study. The results of the GLP study, i.e. the final report, will be provided to the sponsor, allowing EURL ECVAM to take the necessary follow-up actions (Fig. 5.1).

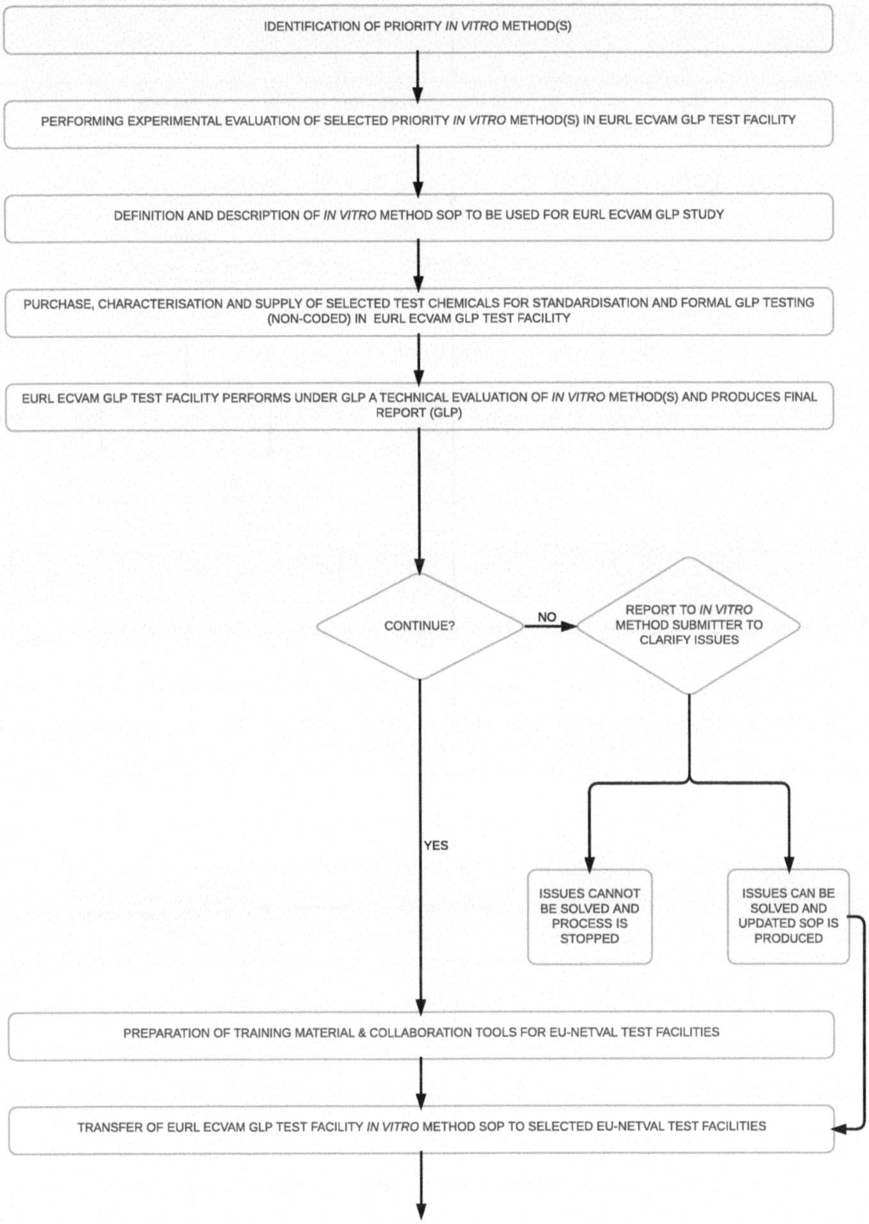

Fig. 5.1 Process flow in a validation study

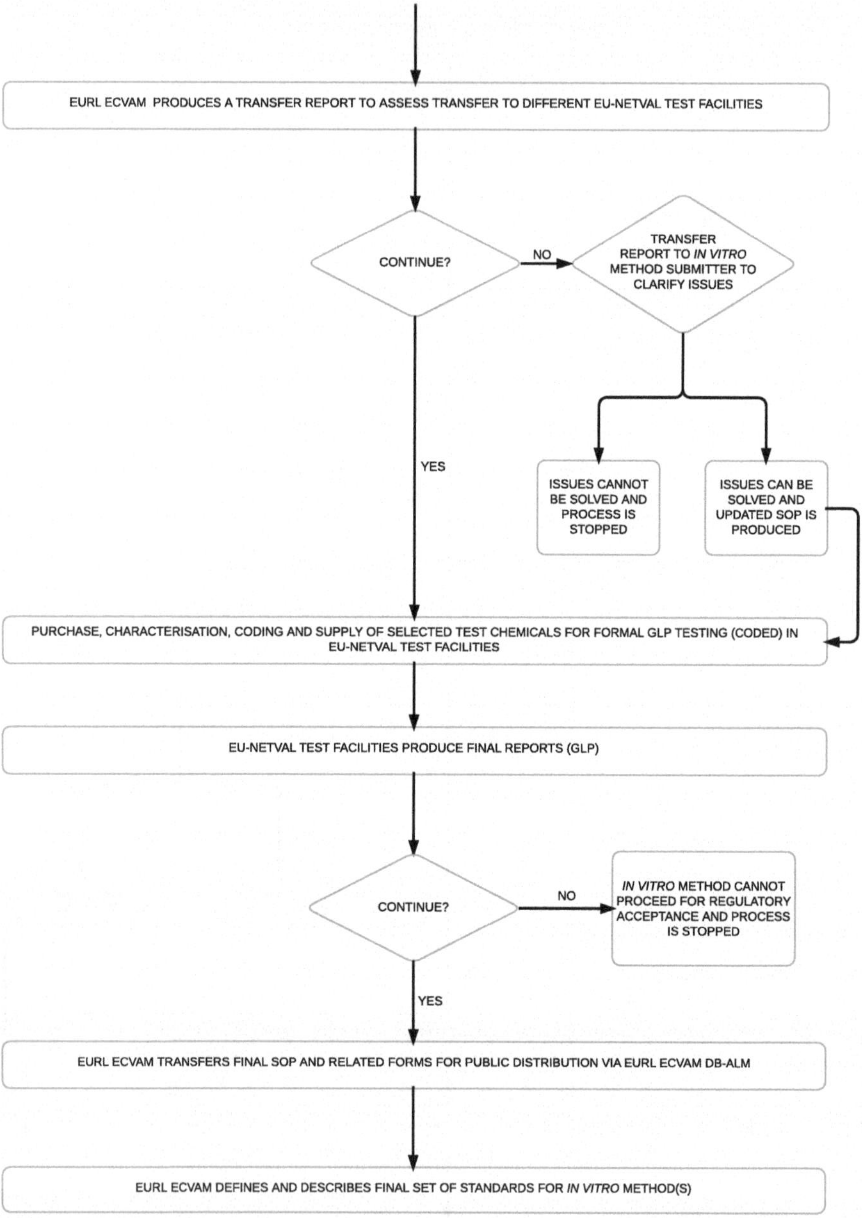

Fig. 5.1 (continued)

3 Designing a Prospective Validation Study

This is the part of the prospective validation process which, in practice, may require the most time and resources. Therefore, the better planned and thought through it is, the higher the probability that the validation study will succeed. In principle, the design phase starts already from the *in vitro* method development phase, since, as it will be explained further, a well-designed and robust *in vitro* method is of paramount importance. All the important aspects that should be taken into consideration during the design of the prospective validation study are described in the following sections.

3.1 Roles and Responsibilities

3.1.1 Role of EURL ECVAM as an EU Reference Laboratory

If necessary, the EURL ECVAM GLP test facility improves, in close collaboration with the *in vitro* method submitter, the *in vitro* method procedure in terms of completeness, clarity and test definition and test description. The *in vitro* method procedure should be standardised to accelerate its translation into internationally recognised test guidelines, including performance based test guidelines, and to ensure its acceptance for regulatory use. Then the EURL ECVAM GLP test facility generates GLP compliant test data using the *in vitro* method under validation. This serves as a preparatory step towards the design and execution of studies carried out by the EU-NETVAL laboratories. Thus EURL ECVAM acts as the interface between the *in vitro* method submitter and the EU-NETVAL test facilities.

EURL ECVAM coordinates the validation study according to a validation project plan which includes aspects such as finalisation of test definitions and SOPs, training and support of EU-NETVAL test facilities participating in the multi-study trial, provision of materials and test chemicals, collection and statistical analysis of test data, and the preparation of the validation report for formal ESAC peer review. Thus EURL ECVAM facilitates the international harmonisation and standardisation of validated *in vitro* methods to aid their translation into internationally recognised standards and test guidelines and to ensure their acceptance for regulatory use. It takes the lead in the development of guidance documents and training materials covering various technical aspects of good *in vitro* method development and practices. As such, EURL ECVAM strives to sustain a high level of efficiency and effectiveness of the network in its support of validation studies, and to expand its capacity and expertise in order to keep pace with technological and methodological developments that are reflected in methods submitted for validation.

3.1.2 Role of EU-NETVAL Laboratories

EU-NETVAL laboratories contribute to validation studies through the execution of one or more specific tasks including conducting single study elements of a multi-study trial. The support sought from members varies in scope depending on the task and the capacities and the areas of expertise of the laboratories. Tasks constituting this support address the particular data and information requirements of one or more validation modules applicable to the study. Tasks of the EU-NETVAL laboratories include:

i. **Definition and description of *in vitro* methods**
   Support the definition of *in vitro* method procedures including technical assessment (non-experimental or experimental) in terms of (a) clarity of the description of the scientific basis of the *in vitro* method, (b) completeness of the SOP, (c) SOP's overall clarity, (d) putative robustness and suitability for implementation within a GLP environment. This includes reflecting the definition of an *in vitro* method in a suitably elaborated method description, prepared in a format fit for public dissemination through EURL ECVAM's database on alternative methods, DB-ALM (see http://ecvam-dbalm.jrc.ec.europa.eu/).
ii. **Transfer of *in vitro* methods between laboratories**
   Support the demonstration and assessment of the transferability of *in vitro* methods between laboratories. This includes the preparation of technical training courses and related training materials on the method undergoing validation to aid in the transfer process.
iii. **Assessment of the reproducibility of *in vitro* methods**
   Support the generation of datasets on selected reference chemicals for the assessment of within-laboratory and between-laboratory reproducibility of *in vitro* methods. This may include participation in a multi-study trial and acting as a lead laboratory for such trials, if appropriate.
iv. **Assessment of the predictive capacity and applicability domain of *in vitro* methods**
   Support the generation of datasets on selected reference chemicals for the performance assessment of an *in vitro* method in relation to its predictive capacity into its intended purpose and/or its contribution to an integrated testing strategy or testing battery. This will also include the assessment of the mechanistic, chemical, physico-chemical, sectorial and regulatory applicability domains of *in vitro* methods, and the generation of datasets suitable for the establishment of performance standards for particular classes of *in vitro* method.
v. **Guidance documents and training materials supporting validation**
   Support the development of guidance documents and training materials covering various technical aspects of good *in vitro* method development and practices in order to sustain a high level of efficiency and effectiveness of the network in supporting validation studies, and to expand its capacity and expertise in order to keep pace with technological and methodological developments that are reflected in methods submitted for validation.

vi. **Surveillance of uptake and use of validated *in vitro* methods**
Support the surveillance of the uptake and use of *in vitro* methods that have undergone validation to assess in-field post-validation performance against the originally intended purpose and to exploit data generated by end-users to further refine the method description and review the applicability domain.

3.1.3 Role of the Validation Management Group

Following the principles for *in vitro* method validation, a Validation Management Group (VMG) is established by EURL ECVAM (Balls 1995). According to OECD Guidance document no. 34 (OECD 2005), the VMG, also called Validation Management Team, is an independent oversight group, consisting of individuals with experience with the types of assays being performed, biostatisticians, and others with knowledge of the purpose of the validation study. The main roles of the VMG are: (1) review and approval of the validation project plan, (2) progress monitoring, (3) management of deviations, (4) trouble shooting, (5) interpretation of results and drawing of conclusions, (6) assistance, review and approval of the validation report, (7) consultation after the validation study. Representatives of other international validation organisations, such as ICCVAM and NICEATM (USA) and JaCVAM (Japan) may participate as observers also. A chemical selection group shall be responsible for the selection of the test chemicals to be used in the validation study. The statistical analysis of the *in vitro* data is the responsibility of an independent biostatistician. The biostatistician is independent from the *in vitro* method submitter and from the EU-NETVAL laboratories involved in the validation study.

3.2 *Quality and Good Scientific Practices During Method Development*

It is very important to properly define and control the essential components of the *in vitro* method, including the exposure regime to the test chemicals, the *in vitro* biological models (test systems), the analytical or life science measurement techniques used and the experimental design. Therefore, EURL ECVAM coordinates efforts related to Good *In Vitro* Method Practice (GIVIMP) to provide detailed updates on today's state-of-the-art of good practices when applying *in vitro* methods in regulatory human safety assessment for various kinds of chemicals. Well-designed, relevant and reliable *in vitro* methods that can run in a GLP environment are becoming more and more instrumental for supporting regulatory decisions. GIVIMP contributes to increase standardisation and international harmonisation in the generation of *in vitro* information on test item safety and will give guidance to obtain high quality data based on sound scientific principles to support regulatory human safety assessment of chemicals using *in vitro* methods. GIVIMP also facilitates the application of the OECD Mutual Acceptance of Data agreement (MAD) for data generated by *in vitro* methods avoiding as such unnecessary duplication of testing by MAD adherent countries.

3.3 The Importance of the Standard Operating Procedure (SOP)

3.3.1 Need to Define Minimum SOP Requirements in View of Reporting Features

An *in vitro* method should be robust and transferable and allow for standardisation. Therefore, the first step in developing a new *in vitro* method is achieving a well-defined *in vitro* method. This should be done in the format of the SOP, a detailed, written instruction to achieve uniformity of the performance of a specific *in vitro* method. The SOP should be written in a concise, step-by-step, easy-to-read format and should be able to be executed in a GLP environment. The information presented should be complete and unambiguous. The document should not be too wordy, contain redundant information, or be too long. Any specialised or unusual terms (e.g. acronyms or abbreviations) should be defined either in a separate definition section or in the appropriate discussion section. Moreover, it is also important that the SOP states the required personnel qualifications and the related roles and responsibilities, together with safety considerations.

The *in vitro* method definition process begins with defining a clear and concise title and the chemical, biochemical or biological basis of the method. This includes a rationale for the relevance of the results produced such as the endpoints to be measured and a rationale or decision criteria for how the results are to be interpreted and used. The title should be clearly worded to be readily understandable by a person knowledgeable with the general concept of the procedure. The *in vitro* method definition has to be carried out in an orderly, logical, or systematic way and result from existing instructions, inquiries, experimental and non-experimental investigations and studies. The *in vitro* method description is more related to the presentation of the method and to the evaluation of what is the best way to describe a method. The content should be described in such a way that it is meeting the standards required to enable an end-user of the method with the necessary technical and scientific *in vitro* method competence to properly carry out the procedure. A numbered list is useful when explaining instructions that need to be performed in sequence. Instead of burying the key points inside large blocks of text, the significant parts should be pulled out, so readers can see with a glance what the most important parts are. Some complex key experimental steps can also be described by audio-visual tool such as short video clips.

The *in vitro* method SOP should avoid an overly complex structure by breaking the information into a series of logical steps or headings. Over use of multiple headings should be avoided. Some of the questions that should be asked by the authors of *in vitro* methods are:

- Is the description and the definition for the intended use of the *in vitro* method complete?
- Is the mechanistic basis of the *in vitro* method adequately described?
- Are the reference and control items clearly identified in the *in vitro* method?
- Are the acceptance criteria clearly defined and are they based on solid and verified experimental data?
- Does the solvent interfere with the results? What type of solvents may be used?
- Are the limits of the *in vitro* method adequately characterised?
- Are there gaps or missing information in the overview of how the *in vitro* method should be conducted?
- Is it possible for a person with the adequate technical background to reproduce the method based only on the information included in the SOP?
- Is a dose-range finding procedure established that allows the selection of a meaningful dose-range for the test item?

3.3.2 Limitations and Applicability of the *In Vitro* Method

It is important to clearly describe in the SOP the applicability domain of the *in vitro* method, as well as any limitations or exceptions. For instance, some *in vitro* methods will only be compatible for technical reasons with liquid chemicals but not with solids or other physic-chemical states. This will enable the proper application of the method and will avoid the generation of misleading data. For example, limitations in terms of applicability could stem from already known limits of use for a particular class of method, difficult chemical types (e.g. volatiles), lack of metabolic competence (biotransformation) of the test system or an absence of critical transporters.

3.3.3 Apparatus, Reagents and Special Consumables

A brief description of the essential requirements of the apparatus (analytical and life science measurement techniques) required for the *in vitro* method should be included into the SOP. Trademarks should be avoided, unless a specific manufacturer's product is required for a well-defined reason. If special types of plastic ware are required, then the significant characteristic desired should be clearly stated. Reagents and materials required for each procedure should be listed as a separate section under each subdivision, including purity (if applicable), CAS or identification number. Finally, it is important to describe any specific requirements in case a complex apparatus is used (e.g. precision, detection limit, limit of quantitation), or in case a critical reagent is used (e.g. purity or special handling etc.).

3.3.4 Establishing Acceptance and Decision Criteria

Acceptance criteria should be established and described in the SOP for each of the following:

Test system—performance (assessed by the response of positive and negative controls items), growth rate/curves, contamination free (e.g. mycoplasma), passage number range, cell recovery.

Test chemicals—performance (reference items, negative and positive control items), solubility, cytotoxicity, dose response, etc.

Analytical or life science measurement technique—linearity, accuracy, sensitivity, reproducibility, performance (reference item(s), internal controls, standards, quantitation limit).

Data analysis—number of runs (SD, % CV, etc.), acceptable failure rate, statistics to be employed (e.g. curve fitting), curve fitting acceptance.

Data analysis templates (spreadsheets)—ensure they are a true reflection of the SOP (i.e. are all data analysis described in SOP)

Reporting—outcome to be reported and units of reporting

Additionally, decision criteria to decide whether to accept or reject a test run should be described in detail in the SOP. These decision criteria should be realistic and take into account:

- What response of the biological test system can be achieved?
- What definitive activities or SOP steps must occur?
- What tools will be needed to execute the SOP steps? How one knows that the SOP steps are successfully completed?
- What processing apparatus is involved or will impact the SOP steps execution?
- Who is responsible for executing the SOP step(s) or following the procedure?
- In what order must these SOP steps occur in order to succeed?
- How should the steps be executed and the response achieved should be documented?

All appropriate Quality Assurance (QA) and Quality Control (QC) activities for that SOP should be described with significant references.

3.4 Selection of Test Chemicals

The selection of test chemicals for validation studies is a complex process and the following considerations should be addressed during the selection process:

1. Data on chemical activity/toxicity ideally in the species of interest (e.g. humans) should be available so that the relevance and reliability of the *in vitro* method can be assessed. A small set of chemicals to further test the limits of the method may be included;

2. The selected total pool of chemicals should be diverse in terms of their structure and biological effect (potency) to ensure the robustness of the method. Although not easy to practically implement, the known applicability domain of the method is extended if validated with diverse chemicals and further confidence can be achieved if the *in vitro* method has also been assessed with a small but representative set of chemicals of particular interest such as nanomaterials and mixtures;
3. Test chemicals should be suitable for testing and comply with practical constraints such as solubility, chemical stability, commercial availability and cost. Typically 10–50 test chemicals, depending on the availability of high quality reference data, are selected for a multi-study trial involving the EU-NETVAL test facilities. Besides the availability of high quality human (or other target species) reference data, above mentioned practical constraints as well as processing time, handling, storage and safety requirements, limit the number of test items.

To achieve such a complex task, manual selection based solely on expert knowledge becomes more difficult and computational tools may facilitate decision making.

3.4.1 Collection of Reference Data Associated with the Test Chemicals

The most reliable way to assess the biological behaviour of selected test chemicals is to explore the existing literature and documentation. Chemistry-based systems, such as PubChem or Scifinder allow querying by molecule (e.g. name, CAS number, structure) and retrieving a significant part of the scientific literature concerning the query chemical. Specific databases, depending on the biological mechanism studied, are more suitable. It is also acceptable to cite the Toxcast/Tox21 initiative that generated *in vitro* responses for a large set of chemicals and assays (US EPA 2010a, b; Tice et al. 2013). In the end, it is recommendable to collect a state-of-the-art of all (or most of) the chemical substances based on human exposure data, or on similar biological *in vitro* test systems with same or similar characteristics and conditions. It is then feasible to assign for each chemical substance a confidence score defined by the relevance of the data collected and the frequency of the behaviour observed.

Literature findings can be completed and/or confirmed by *in silico* predictions. For example, many reliable Quantitative Structure-Activity Relationship (QSAR) or docking models exist for receptor binding measurements and can be a good supplement to fill in gaps in datasets or reinforce a given observation. Phenomena involving very specific chemical mechanisms (e.g. protein or DNA binding) are tightly connected to well-known chemical reactivity and reliable models have been produced. The known space of biological response will define the chemical selection of a validation study. Exploratory compounds can nevertheless be added to the selection set for the sake of new data generation or mechanistic elucidation; but it remains quite minor compared to the whole set of selected chemicals. In order to have a better picture of the selected test items, it is often possible to collect *in vivo* data related to the tested system. However, in toxicology, the tested system may only be a part of a broader toxicity pathway (or Adverse Outcome Pathway, AOP) and finding a direct correlation between the tested *in vitro* method (often accounting for an effect at receptor or cellular level) and an underlying adverse effect (at the organism level) is not always easy. For example, when testing an endocrine-related receptor (oestrogen or androgen receptors), it is not obvious to relate the tested phenomenon (most probably agonism or antagonism) to an actual *in vivo* observation such as developmental, reproductive, or carcinogenic effects.

3.4.2 The Diversity Issue

When validating an *in vitro* method, the applicability domain is of major importance since it will define the chemical classes for which the method is able to give reliable measurements. The applicability domain is defined by the chemical substances tested during validation and in principle the method should only be reliable for similar chemicals by interpolation or, to some extent, with a slight extrapolation. It is then advisable to maximise the diversity of the chemicals selected for validation in order to extend the applicability domain of the tested method.

From a chemistry point of view, it is desirable to cover a wide chemical space, meaning a large number of different structures. The exercise of covering chemical space is a trade-off between the number of chemical classes covered (the "area") and the number of representatives of each class (the "density"). It should be kept in mind that the chemical selection is always limited by the number of compounds that can be used for validation. Therefore, it is not possible to cover all chemical classes while maintaining a high density. Depending on the aim of the validation study, one wants to focus on the area covered or the density. A large area with low density will ensure a large applicability domain of the tested system but with a loss in reliability. On the other hand, a smaller area covered more densely will allow more reliable responses but will only be applicable for a limited count of chemical classes.

The structural diversity can be evaluated with structural descriptors like structural keys or fragment fingerprints. Similarity measures (such as the Tanimoto index) are able to quantify the diversity (*i.e.* a low average similarity between compounds) of a dataset. Strategies based on chemo-informatics methods are used to pick up a restricted, yet diverse, subset out of a source dataset. The sphere exclusion (or cell-based) method and the clustering method, which are illustrated in Fig. 5.2, are examples of such strategies. In the sphere exclusion, chemicals are projected in a multidimensional space made of user defined descriptors (either structural or physical/chemical properties). A compound is randomly picked as the seed and selected. All the compounds in the neighbourhood (defined by a threshold) are excluded and another molecule is then picked up outside the excluded sphere. Iteratively, compounds are picked-up and their closest neighbours excluded until there are no more compounds left to select or until the target number of chemicals is reached. It is also possible to use clustering methods. The dataset is clustered

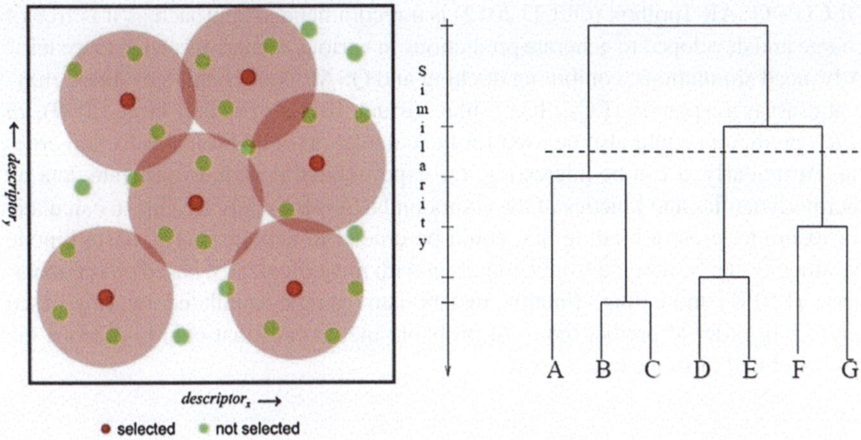

Fig. 5.2 Example of application of two diversity selection methods: sphere exclusion and cluster selection

according to the inter-compound similarity (again, either based on structural or chemical descriptors). A threshold is set to define *n* clusters (*n* being the target number of chemicals used for the validation study) and one compound is selected (randomly or according to other criteria) inside each cluster giving eventually a selection of *n* compounds.

Following the same reasoning, diversity of biological response should not be neglected. As chemical diversity has to be maximised to ensure a large applicability domain, similarly the potency spectrum should be adequately covered. The data compiled in the data collection step help at ranking compounds, triggering high, medium or low responses from the system. The final selection should then include compounds covering a wide range of biological response to be sure that the system is tested in all possible values of the spectrum. Obviously, it concerns only the positive response (triggering an observed effect) while the negatives are considered as zeros or close to it. At this stage, QSAR models could also help at evaluating or confirming the expected behaviour of the compounds. Note that, due to the nature of the biology and the specificity of the system tested, positive compounds typically cover a smaller structural space than the negative ones. This is especially true for specific receptor-based assays for which only very fine structural details may trigger a response (agonist or antagonist).

3.4.3 Property Predictions

Existing computational models can help evaluate the biological response of the tested system. An extensive number of quantitative (QSAR) and qualitative (structural alerts, decision trees, profiler) models are available nowadays covering a large landscape of biological effects, from receptor binding to systemic toxicity. The

OECD's QSAR Toolbox (OECD 2012) is a recommended approach as it is free of charge and developed to generate predictions in various scenarios related to toxicity. Advanced simulations, combining docking and QSAR could help to evaluate enzymatic assay responses (EC_{50}, IC_{50}), like Virtual ToxLab (Vedani et al. 2009). *In silico* predictors could also be used for the calculation of additional selection criteria. Particularly, it can be interesting, for experimental design, to generate data on thermodynamics and kinetics of the compounds. Some models are able to calculate, for example, protein binding that could be crucial in certain assays. Also, kinetic parameters can be derived from simulation with physiologically-based pharmacokinetic (PBPK) modelling. Finally, metabolism can be simulated by rule-based engines in order to predict the most probable metabolites that could influence the results of cell or tissue based assay.

3.4.4 Practical Examples

For the validation of an androgen receptor transactivation assay (namely, AR-CALUX) at EURL ECVAM, the test item selection relied mainly on collection of available knowledge from various sources. We gathered expert opinions, published literature, collaborators' results as well as in-house results in a comprehensive data table for about 80 chemicals. They were evaluated based on a score accounting for the number of observations of three possible behaviours in the assay (agonist, antagonist, negative). Ranked by confidence (higher scores), the top chemicals were selected to obtain a balanced set of 45 (15 of each of the 3 classes). Available potency data were also gathered from the same data sources (combined with in-house experiment on solubility of compounds) in order to set up experimental protocols (e.g. test doses range).

In conclusion, in the context of a prospective validation study, chemical selection is often totally based on theoretical work. That is why *in silico* tools can be used appropriately at several stages of the chemical selection. Ideally, data on the potential chemicals should be collected through dedicated databases or inventories. The selection should maximize the chemical diversity covered with the help of chemoinformatics that provides similarity-based tools. Also, the biological response should be thoroughly explored by selecting compounds with different profiles (determined by literature findings that could be supported with predictions). Finally, theoretical models are able to predict a whole range of properties that can help at fine-tuning the chemical selection taking into account the experimental design and the underlying limitations.

3.5 Test Chemical Management

Solubility and stability testing of test chemicals (test, reference and control items) is part of the test chemical management activities. The process starts with their acquisition and ends with their shipment to the EU-NETVAL test facilities for the

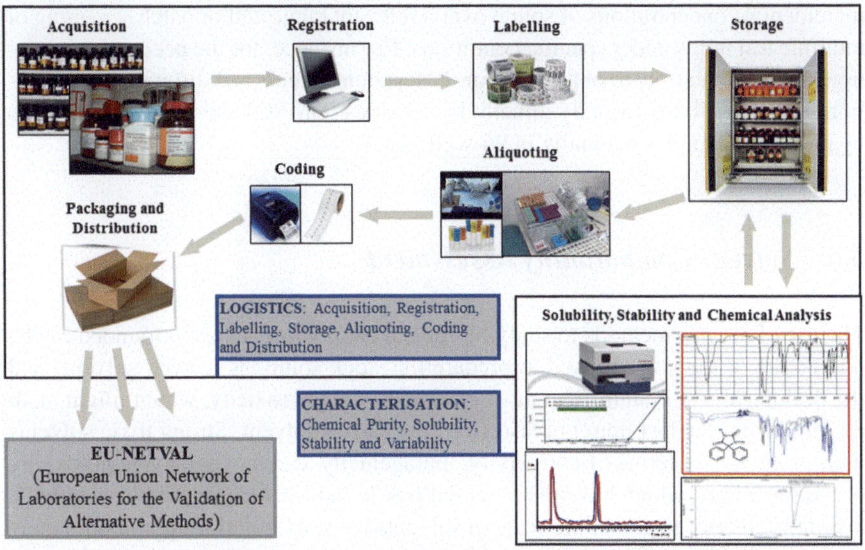

Fig. 5.3 Test chemical management: from acquisition to distribution

multi-study trial. The entire process includes first checking on the availability of the test chemicals, the acquisition, the registration, the labelling for their identification and the control of the correct storage conditions in order to trace and guarantee the chain of custody. The aliquoting, coding (when required), and packaging of the test items for the appropriate distribution is also an important test chemical management activity during a validation study. Tests for solubility, stability and variability between batches are performed at EURL ECVAM test facility. Figure 5.3 illustrates all these steps.

Solubility verification ensures compatibility of the test chemical with the *in vitro* method under evaluation. Before selecting a test chemical for a validation study, it is important to check if the test chemical is and remains soluble both in the stock solution and in the *in vitro* method incubation medium (working solution) at the desired concentration and under the *in vitro* method experimental conditions.

After a preliminary phase, theoretical *in vitro* method performance evaluation and theoretical SOP optimisation, solubility testing is performed in standard throughput mode to generate sufficient data in a reasonable time. The traditional solubility testing approach by visual inspection of the solution is subjective. A more objective method that has been used in EURL ECVAM validation studies is using the Tyndall-based nephelometer method. Nephelometry provides an objective indicator of solubility, differentiating trace dispersions from limpid solutions by relative turbidity, detected as Tyndall-effect light scattering of a transmitted laser beam. The method is relevant and accurate at threshold concentration ranges of solubility in contrast to the visual inspection. Automated instrumentation for multi-well microplates allows rapid serial measurement of sample aliquots, applicable to incremental concentrations of solute over a relevant range and/or batch screening of multiple test items under specific conditions. For instance, for the needs of a validation study it is feasible to obtain data on 12 test items/week, at 3 different concentrations, in pure solvent (usually dimethyl sulfoxide—DMSO) and in culture medium by preparing samples manually in 96-well plates.

3.6 Solvent Compatibility Assessment

The use of *in vitro* methods to study biological endpoints can be confounded by the interaction of the solvent used to prepare the stock solutions (carrier solvent) and the biological test system. The most common effect is toxicity, which might stem not only from the test item but also from the carrier solvent. Strong toxic solvents with properties in terms of corrosivity, mutagenicity, carcinogenicity, genotoxicity or teratogenicity, which have the potential risk to induce adverse effects, should be avoided and only a compatible scale of solvents for stock solutions preparation may be considered. In some cases, carrier solvent might not cause toxicity but it might interfere with the test system in such a way that it masks the *in vitro* response. For instance, some solvents might interact with Cytochrome P450 (CYP) enzymes and thus interfere with the results of *in vitro* CYP induction or *in vitro* metabolic clearance methods. Also in such cases care should be taken to use appropriate solvent and at acceptable final concentration in the incubation mixture of the *in vitro* method. Commonly, a sequence of compatible solvents can be resumed as: water, DMSO ethanol and methanol. In general DMSO is an appropriate solvent for organic test items such as pharmaceuticals. Alternatively water would be suitable for inorganic compounds. In conclusion, solubility testing must be restricted to carrier solvents that are compatible with the *in vitro* biological system employed in the *in vitro* method.

3.7 Test Chemical Purchase and Distribution

In practice, chemical substances are produced for laboratory research and development and are available from retail suppliers who provide a certificate of analysis. The facility responsible for test chemical recipient should ensure the chemicals are stored at the recommended temperature (refrigerator, ambient, freezer) with attention to any additional conditions (i.e., inert gas for air sensitivity, and darkness for light sensitivity). Unless otherwise indicated, chemicals are allocated an expiry date 2 years from acquisition. This practice should be handled with care and identity and stability verifications are highly recommended.

Distribution of the test items used for the validation should be in compliance with International Air Transport Association regulations, potentially hazardous goods require declaration by UN number, if applicable, indicated in the Material

Safety Data Sheet (MSDS). The shipment package should include the MSDS of each chemical, with the corresponding test item code, enclosed in a sealed envelope labelled *for customs use only* (to be discarded unopened, on arrival at the participating test facilities). Relevant to remedial procedures at a participating test facility in case of accident or emergency only, an additional MSDS for each test item should also be included, sealed individually in opaque envelopes identified only by code. The MSDS envelopes should remain sealed during the study, with any recourse for consultation reported to the VMG.

Each shipment should be addressed to a nominated responsible person at the test facility. Shipments should be made early in the week, avoiding potential delays due to week-ends. On arrival, recipients should confirm delivery of complete and intact test items, including integrity of the sealed MSDS envelopes.

Following the modular approach to validation studies, only certain modules would normally require coding of test items. In general the training and transfer phases would not involve coded test items, while the EU-NETVAL multi-study trial testing for reproducibility and predictive capacity may justify test item coding to ensure full independence of the experimental work by the participating test facilities.

Coding of test chemicals (chemical aliquots) for distribution to participants in a validation study should be systematic, to facilitate logistical management, but also unique, to ensure identity encryption.

3.8 Good Experimental Design and Data Interpretation Based on Statistical Analysis

Drawing conclusions on the basis of the data obtained from *in vitro* methods depends a lot on the correct interpretation of the information obtained and their statistical analysis. For a validation study a first statistical analysis should be done on data provided by the *in vitro* method submitter. EURL ECVAM assesses if acceptance criteria can be met by the EURL ECVAM GLP test facility. It is of crucial importance that the *in vitro* method submitter sets its acceptance criteria on historical data and uses the necessary statistical tool to set the acceptance window. During the validation study, a global statistical analysis is conducted centrally by EURL ECVAM.

The main analysis should reflect the purpose of the validation study. It is therefore needed to extract all of the useful information and present the data in a way that it can be interpreted, taking account of biological variability and measurement error. The methods applied should be of such a kind that any knowledgeable reader with access to the original data can verify the results. For a validation study it is therefore necessary to perform a statistical analysis to describe the within-laboratory reproducibility, the transferability (including goodness of fit and robustness) and the between-laboratory reproducibility. The statistical analysis also needs to assess the predictive capacity and the applicability domain of the proposed *in vitro* method. The predictive capacity will be influenced by the number and range of chemical substances and the quality of the reference chemical substances and should maximise the specificity (also called the negative rate) and sensitivity (or the true positive rate).

Often users of *in vitro* methods are confronted with the fact that the received *in vitro* method and its original method description and definition should be optimised for a variety of different reasons such as the need for the method to become a formal test guideline, the adaptation of the method for specific user-requirements, etc.. A systematic approach should be taken when standardising the *in vitro* method. In general, the parts that are to be standardised and improved should be clearly identified including problem formulation, and indicate how the standardisation and implementation will be measured (what parameters are to be optimised and how these optimised parameters will be measured e.g. using standard deviations, % CV, etc.).

Factorial design can be used as an experimental approach to obtain a well-defined *in vitro* method and can be subsequently used as a tool for any further standardisation and improvement needs as it permits simultaneous evaluation of multiple factors that might influence the performance of the *in vitro* method. Specific questions that can be asked are: Are the numbers of replicate and/or repeat experiments appropriate for each experimental step in the *in vitro* method? And is a good *in vitro* method experimental design of those steps used in the SOP where such design is critical in terms of adequate placing of test, reference and control items, generating enough data to draw conclusions, etc.?

A good statistical practice is a necessary element in the experimental design of *in vitro* methods and facilitate the method definition and optimisation and also, if applicable, the subsequent validation steps. If there is need to optimise an *in vitro* method, the parameters to be optimised should be well described, an objective function should be established and results should be supported by sound data. However, *in vitro* method developers should pay careful attention to ensure that the statistical practices are implemented correctly and the results are appropriately interpreted.

3.9 Importance of Good Data Management

When data is not well-defined there is the risk that they are misused and wrongly interpreted which could lead to false conclusions. In order to avoid this, good data management should be followed throughout the whole data lifecycle i.e. from the planning to the creation of data up to the storage and eventual when the data can be considered obsolete to the deletion of the data. Tools to obtain well-documented data are pre-defined guidance documents proposing terminologies and a fixed data reporting format. When eventually data is produced, it should be reported in the correct format and accompanied with a detailed, well-documented, description of the procedure followed (including also background literature used). Finally, clearly described instructions of acceptance criteria and data transformation should be provided in order to be able to draw correct conclusions from data. At the end of a validation study all data and documentation should be stored in such a way that any subsequent regulatory question that might come up in the future can be addressed.

After data recording of *in vitro* methods, data analysis is performed. Spreadsheet forms are used for such data analysis. Such forms belong to the category of computerised systems. Indeed, computerised systems can vary in type (e.g. hardware, software), in complexity and in dimension. Some examples of computerised systems are: programmable analytical instruments, personal computers and laboratory information management systems (LIMS), but also electronic spreadsheets used for the storage, processing and reporting of data. Spreadsheets (e.g. MS Excel) are widely used for data analysis and storage of electronic data.

Their complexity varies enormously depending on the actual *in vitro* method being performed. When using spreadsheets for performing routine data handling, these spreadsheets should be considered as part of the SOP. The design and validation of spreadsheets when used in a quality environment such as GLP has been addressed in specific guidelines for the development and validation of spreadsheets. When developing spreadsheet forms or other applications (e.g. database), both their design and validation should be planned and documented. For complex applications, their use should be documented either in the *in vitro* method SOP or in separated SOPs. The statistical method required for data treatment and analysis should be documented in the *in vitro* method SOP, including a description how to interpret the final results.

The use of computerised systems by test facilities for the generation, measurement or assessment of data is nowadays consolidated and computerised systems are fully integrated into the *in vitro* method. It is essential that for regulatory applications, computerised systems are developed, validated, operated and maintained in accordance with the OECD Consensus document No. 10 on "The application of the principles of GLP to computerised systems" (OECD 1995, undergoing a revision in 2014 and specific guidance documents (PIC/S 2007; AGIT 2007). All statistical methods and calculations to be used should be described in the method. Some checks can be done when evaluating the completeness of the SOP for the statistical aspects such as:

- A clear description and definition of the statistical or non-statistical methods used to analyse the resulting data are provided.
- A clear description and definition of the decision criteria and the basis for the prediction model used to evaluate the test item in relation to its required response are provided.
- Control if the relevance (e.g. accuracy/concordance, sensitivity, specificity, positive, and negative predictivity, false positive and false negative rates) of the *in vitro* method is adequately described.
- Assess if specific measures of variability are adequately included.
- Evaluate if the acceptance criteria are based on historical experimental data.

4 Conducting a Validation Study

The main players in this phase of the process are the EURL ECVAM GLP test facility and the EU-NETVAL laboratories (Figs. 5.1 and 5.4). This part of the process relies heavily on several important factors including the use of good scientific and quality practices when developing, optimising and standardising *in vitro* methods, having measures in place to control for biological test system quality, and using of good test chemical purchase and distribution practices.

4.1 Importance of the Good In Vitro Methods Practice

When developing and implementing *in vitro* methods intended for regulatory purposes, good practices, e.g. good scientific practices and good quality practices, are a critical prerequisite. GIVIMP is critical throughout the validation study (Rispin et al. 2004; Gupta et al. 2005; Coecke et al. 2014a, b). The aims and topics covered by GIVIMP include:

(1) Ensuring the use of good scientific and good quality practices since the validated *in vitro* methods target the area of regulatory human safety assessment;
(2) Ensuring that the Standard Operating Procedures (SOPs) of the *in vitro* methods under validation are well-designed, robust, well-defined and described and can run in a GLP environment, which is essential for regulatory use;
(3) To provide guidance on minimum SOP requirements and reporting features to strive for more harmonised approaches for today's regulatory needs in the field of human safety assessment (see Sect. 5 below);
(4) To describe the key importance of applying Good Cell Culture Practice (GCCP), essential in the identification, authentication and characterisation of the *in vitro* biological model (e.g. test systems such as cell lines, stem cells, primary cells and tissues) used in *in vitro* methods;
(5) To describe the key importance of applying good test item handling procedures and clarify the importance of a clear definition of the *in vitro* environment that hosts the *in vitro* test system, which is essential for the correct dosing of the test system, and for the assessment of test item compatibility with the specific *in vitro* environment;
(6) To describe the key importance of applying good experimental design, establishing acceptance criteria for *in vitro* methods, describing equipment requirements (including also those based on new technologies and any scientific progress in the field of detection methods) and performance standards based on scientific evidence from the generated *in vitro* data sets;
(7) To describe how international collaborations and networks can help in disseminating GIVIMP and the use of the generated data sets for specific regulatory applications. GIVIMP will contribute to the use of *in vitro* method data to support regulatory human and environmental safety assessment of chemicals by striving that such data are being generated in compliance with and based on current good scientific practices.

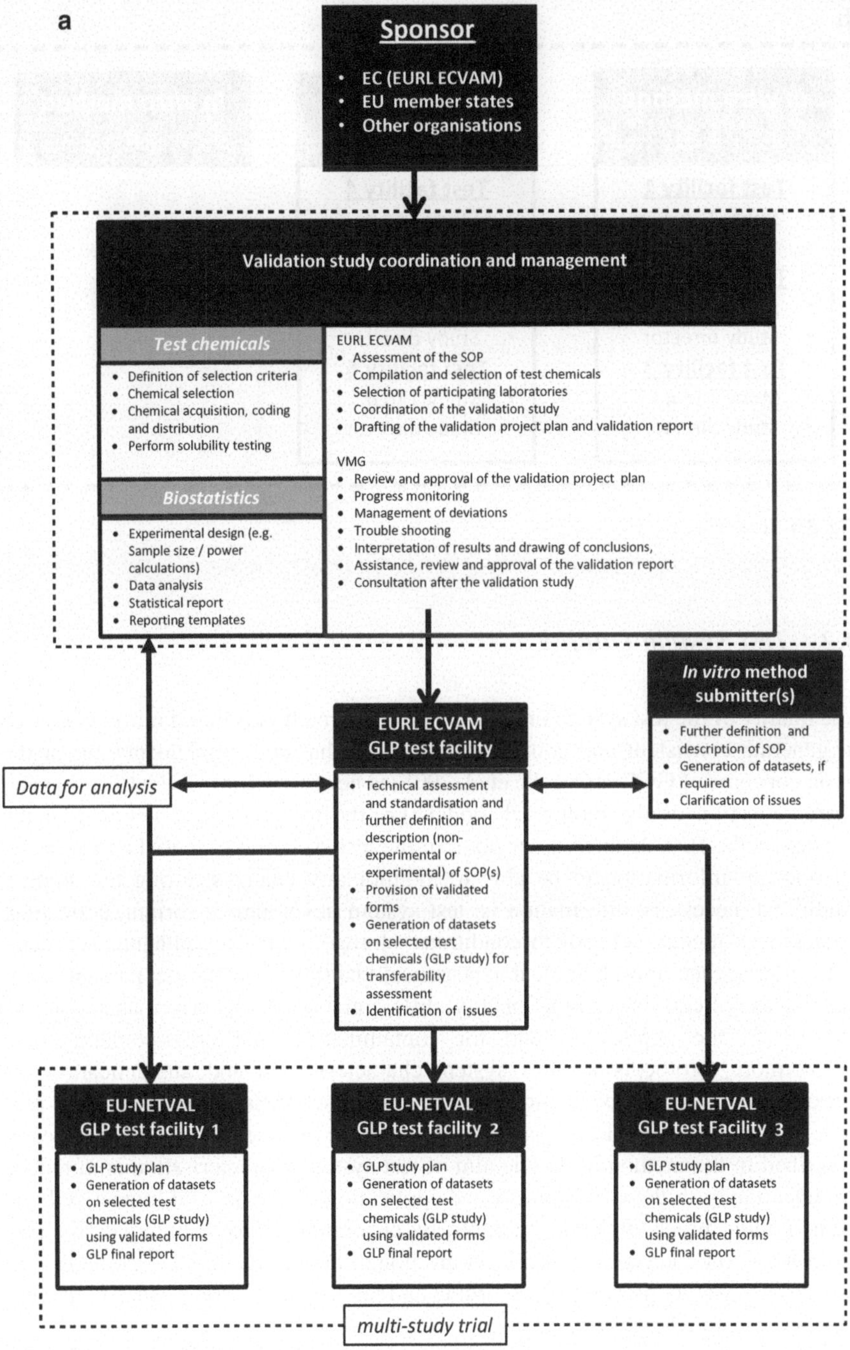

Fig. 5.4 Schematic representation of the organisation of a validation study at EURL ECVAM. Several quality assurance units might be involved in a multi-study validation trial. *Dashed lines* indicate quality assurance staff involvement

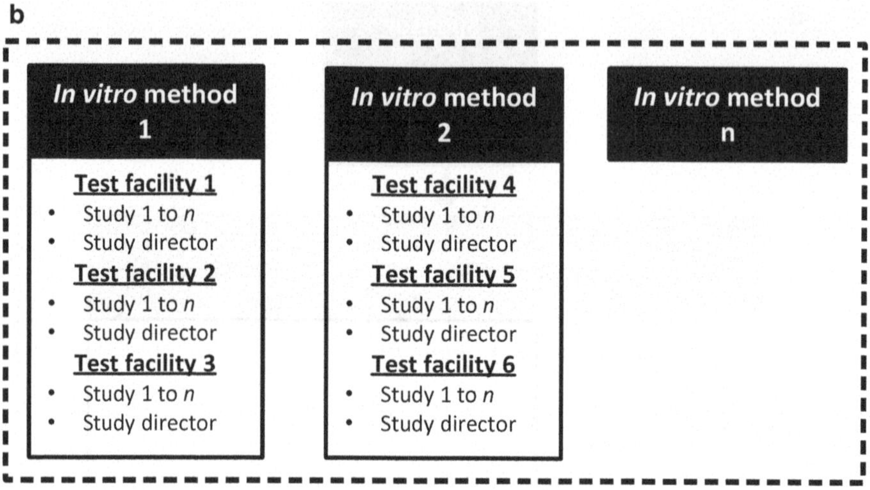

Fig. 5.4 (continued)

4.2 Test System Quality

The quality of the test system on which an *in vitro* method is based must be assured to generate consistent and comparable data. To that end, implementation of the basic concepts of GCCP (Coecke et al. 2005) is necessary for the identification and characterisation of the biological model (i.e. the test system component of the *in vitro* method). A detailed description of the test system is essential and includes all relevant information to be able to monitor any changes during the studies. Additional necessary information is: test system development (origin, collecting, processing); media and growth conditions; storage; recovery; authenticity; metabolic competence; morphological appearance; viability; growth rate; passage number (in case of cell lines); functionality; differentiation state; performance controls specific to the application and for contamination and cross-contamination. Furthermore, protocols for test system characterisation and authentication are important and also the evaluation criteria applied to assess if a test system is reliable. All of them should be based on scientific evidence and should be clearly described in the method SOP. The aim of test system characterisation is to reduce the uncertainty in the development and application of animal and human cell and tissue culture procedures and products, by encouraging greater international harmonisation, rationalisation and standardisation of laboratory practices, quality control systems, safety procedures, recording and reporting, and compliance with laws, regulations and ethical principles.

If test systems used in validation studies are genetically modified the Directive 2009/41/EC is applicable. This Directive lays down common measures for the contained use of genetically modified micro-organisms (GMMs), aimed at protecting human health and the environment. A notification has to be sent to the competent authorities before any use commences in the premises. A risk assessment of the GMMs used has to be performed. The Annexes to the Directive detail the criteria for assessing the risks of GMMs to health and the environment, as well as the protective measures for each of the four levels of containment. The Directive lays down the minimal standards applicable to the contained use of GMMs. Member States are permitted to take more stringent measures.

It is critical that quality control of the test system plus materials used (cell lines, media and other reagents) is adequately described. The main items to be described concern the quality of the test system, the quality of reagents and materials and the performance of the test system.

Quality controls for the integrity of the test system, e.g. microbial contamination, mycoplasma testing, should be described. The test system should also be fully characterised and authenticated in terms of DNA profile and species of origin and provided with a detailed data sheet. Contamination may also arise from the selection of reagents and materials. Good quality reagents and materials are available from numerous manufacturers who already perform a range of quality control tests and provide a Certificate of Analysis with their products. The performance of the test system should be evaluated with appropriate reference items, including positive, negative, and untreated and/or vehicle controls, as required, and performance acceptance criteria defined in the SOP. The performance should be continuously monitored against the acceptance criteria.

5 Role of ESAC in Evaluating the Design and Conduct of a Validation Study

At the end of the validation study, a validation report is produced. This report undergoes a scientific review by EURL ECVAM's Scientific Advisory Committee (ESAC), whose main role is to conduct independent peer-review of a validation study, assessing its technical and scientific validity for a given purpose. ESAC reviews the appropriateness of study design and management, the quality of the results obtained and the plausibility of the conclusions drawn. ESAC peer reviews are prepared by specialised ESAC Working groups composed of ESAC members, experts nominated by the ESAC and/or EURL ECVAM as well as scientists proposed by ICATM partner organisations. ESAC's advice is delivered to EURL ECVAM as formal "ESAC opinions" and "work group reports".

Building on ESAC's advice, EURL ECVAM prepares in close dialogue with regulators (PARERE), stakeholders (ESTAF) and international partners (ICATM), an "EURL ECVAM Recommendation" summarising EURL ECVAM's view on the validity of an *in vitro* method, and advising on its possible regulatory applicability, limitations and proper scientific use in a given regulatory context. It also identifies knowledge gaps and defines follow up actions. Finally, EURL ECVAM supports the international recognition and regulatory acceptance of the successful methods as well as their application by end users.

6 GLP Requirements for *In Vitro* Studies in the EU

6.1 Background

The quality and integrity of the data are crucial when *in vitro* studies are conducted for regulatory purposes, such as in the context of a marketing authorisation application for a pharmaceutical product, an application for approval of an active substance of a pesticide or the registration dossier of a chemical substance. The principles of Good Laboratory Practice (GLP) form an internationally recognised quality system, aimed at promoting the quality and validity of non-clinical safety data for regulatory purposes by allowing the reproducibility of the data and the reconstruction of the study from the paper records. As compliance with the principles of GLP is required by law for safety studies on chemical products around the world, it is important that newly developed *in vitro* methods can be performed in a GLP environment.

The United States (U.S.) Food and Drug Administration (FDA) developed good laboratory practice regulations in the 1970s after investigations at a number of test facilities uncovered widespread scientific misconduct, poor quality control and a lack of industry standards governing the recording and reporting of data (Baldeshwiler 2003). As other countries soon followed suit by establishing their own good laboratory practice standards, it became increasingly important to harmonise these standards (Seiler 2005). Divergent standards could lead to the duplication of tests, thereby increasing costs, resources and the use of experimental animals. Consequently, the OECD started its work on harmonised quality standards through its expert group on good laboratory practice in 1978. This led to the establishment of a system of mutual acceptance of test data (MAD) between countries, relying on both harmonised quality standards (GLP) and harmonised test guidelines. In this context, the OECD published its principles of GLP in 1981 together with a set of OECD test guidelines as part of the OECD Council Decision on MAD in the Assessment of Chemicals (OECD 1981). Together, GLP and OECD test guidelines would ensure that data can be accepted across borders and across regulatory systems:

> *data generated in the testing of chemicals in an OECD Member country in accordance with OECD Test Guidelines and OECD Principles of Good Laboratory Practice shall be accepted in other Member countries for purposes of assessment and other uses relating to the protection of man and the environment.* (OECD 1981)

The establishment of a harmonised set of GLP principles was only the first step in the development of the MAD system (Turnheim 2008). At the time, while the principles that test facilities needed to follow were harmonised, there was no harmonisation on how governments verified that test facilities actually complied with these principles. Therefore, countries were obliged to conduct inspections abroad to verify the compliance of foreign test facilities, from which they received data. This became infeasible given the swiftly increasing number of GLP test facilities around the world. Consequently, the OECD established harmonised procedures for monitoring GLP compliance through inspections and audits and for international liaison among authorities as part of an OECD Council Act in 1989. Based on these harmonised procedures, countries were able to recognise of the assurance of other countries that test data have been generated in accordance with the principles of GLP. The current MAD system applies to all 34 member countries of the OECD, as well as six non-member countries that have become full adherents to MAD. All these countries have incorporated the principles of GLP in national legislation. Many countries use the OECD principles of GLP, while some have adapted the principles: for instance, U.S. FDA and EPA (Environmental Protection Agency) have specified some further requirements in their GLP Regulations, which are applicable when studies are performed in the USA.

6.2 EU Legal Requirements

In the European Union, the OECD's principles of GLP and compliance monitoring practices were first incorporated in two Directives in 1987 and 1988, respectively (Council 1986, 1988). Following a revision in 2004, the currently applicable legislation is Directive 2004/10/EC on the harmonisation of laws, regulations and administrative provisions relating to the application of the principles of GLP and the verification of their applications for tests on chemical substances and Directive 2004/9/EC on the inspection and verification of GLP (European Parliament and Council 2004a, b). The former contains the OECD principles of GLP in its Annex, while the Annex to the latter includes the OECD guides for compliance monitoring procedures and guidance for the conduct of test facility inspections and study audits.

Directive 2004/10/EC stipulates in Article 1 that "*Member States shall take all measures to ensure that laboratories carrying out tests on chemical products, in accordance with Directive 67/548/EEC comply with the principles of good laboratory practice*" or "*where other Community provisions provide for the application of the principles of GLP*". These other provisions are numerous. A wide range of sector-specific legislation either requires or recommends the application of GLP for certain studies. This includes legislation on chemicals, pharmaceuticals, veterinary medicinal products, detergents, feed additives, food additives, genetically modified food or feed, pesticides, biocides and cosmetics (see Table 5.1). For instance, the REACH Regulation requires toxicological and ecotoxicological tests to be carried out in compliance with GLP or other international standards (European Parliament and Council 2006). The European Chemicals Agency has clarified in its guidance that no other international standard has so far been recognised as being equivalent (ECHA 2014). In some legislation, certain studies may also be carried out by laboratories accredited under the relevant ISO standard.

6.3 EU Authorities

In the European Union, there are two main players involved in the implementation of GLP: monitoring authorities and receiving authorities. Monitoring authorities are designated by the EU Member States and manage GLP compliance monitoring

Table 5.1 EU legislation with GLP provisions

Relevant legislation	
• *Chemicals*:	• *Feed additives*
– *Directive 2004/10/EC*	– *Regulation (EC) No 429/2008*
– *Regulation (EC) No 1907/2006*	• *Food additives*
– *Regulation (EC) No 1272/2008*	– *Regulation (EU) No 234/2011*
– *Directive 1999/45/EC*	• *Genetically modified food or feed*
• *Human medicinal products*	– *Regulation (EU) No 503/2013*
– *Directive 2003/63/EC*	• *Pesticides*
– *Regulation (EU) No 536/2014*	– *Regulation (EC) No 1107/2009*
• *Veterinary products*	• *Biocides*
– *Directive 2009/9/EC*	– *Regulation (EU) No 528/2012*
• *Detergents*	• *Cosmetics*
– *Regulation (EU) No 648/2004*	– *Regulation (EU) No 1223/2009*

programmes, inspect laboratories on a regular basis and conduct audits on studies carried out by these laboratories. While some countries have a single monitoring authority covering all GLP test facilities, others have multiple authorities, each covering different product areas. In total, these authorities monitor the GLP compliance of more than 700 laboratories. As long as they are part of a GLP monitoring programme, these laboratories are inspected on a regular basis. Routine study audits are conducted as part of such regular inspections. In addition, monitoring authorities can be requested to conduct triggered study audits in specific cases as a result of a regulatory submission. EU monitoring authorities share information through the EU GLP working group, an expert group managed by the European Commission.

Receiving authorities receive non-clinical safety data as part of regulatory submissions and must ensure that the aforementioned legal GLP requirements are met. They may verify whether the responsible test facility has been found in compliance by a national monitoring authority or request a study audit in case of doubt. Receiving authorities in Europe include the European Chemicals Agency (ECHA), European Medicines Agency (EMA), European Food Safety Authority (EFSA), as well as various national agencies that are responsible for assessing safety data, for instance as part of clinical trial applications or marketing authorisation applications for nationally approved pharmaceuticals.

6.4 Principles of GLP

The principles of GLP stipulate how a study should be organised, planned, performed, reported, reviewed and archived. Amongst other things, they cover the roles and responsibilities of laboratory staff, the quality assurance programme, the test facility, the facility's equipment, materials and reagents, the test systems, test items and reference items, the performance of the study in accordance with the study plan, the reporting of results in the final report, and the storage and retention of both records and materials.

Given the broad scope of GLP, the principles can apply both to *in vivo* and *in vitro* studies. Nevertheless, their application to *in vitro* studies may require special considerations. The OECD working group on GLP has described these considerations in a dedicated advisory document (OECD 2004). For instance, test facility management and personnel may require specific training and proper conditions of laboratory equipment need to be assured. The justification and characterisation of the test system is particularly important for *in vitro* studies. The study director needs to document that the *in vitro* method has been validated and that it provides the required performance. A well-documented validation of the *in vitro* method will support a claim that it is fit for purpose (Coecke et al. 2014a, b).

With a view to their regulatory acceptance, it is of great importance that *in vitro* studies are designed for use in according with the principles of GLP. Therefore these studies need to be carefully validated. This ensures that procedures and results are accurate, reliable, traceable, and reproducible and, where appropriate, comply with the relevant regulatory authorities' legislation. The OECD Guidance Document on the Validation and International Acceptance of new or updated test methods for Hazard Assessment No. 34 (OECD 2005) stipulates that data supporting the assessment of the validity of the test methods preferably should have been obtained in accordance with the OECD Principles of GLP.

Acknowledgements The authors would like to thank Enzo Genco, Johannes De Lange, Sotiris Moustakidis, Kamila Rzewucka, Athina Mitsiara, Priscilla Vaes (European Commission, Joint Research Centre, Ispra, Italy) for their assistance and advice during the preparation of this chapter.

References

AGIT (2007) Arbeitsgruppe Informationstechnologie: guidelines for the validation of computerised systems in GLP 2007; Verified December 2014. http://www.therqa.com/committees-working-parties/good-laboratory-practice/regulations-guidelines/agit-switzerland/

Baldeshwiler AM (2003) History of FDA good laboratory practices. Qual Assur J 7:157–161

Balls M (1995) Defining the role of ECVAM in the development, validation and acceptance of alternative tests and testing strategies. Toxicol *In Vitro* 9(6):863–869

Bouvier d'Yvoire M, Bremer S, Casati S, Ceridono M, Coecke S, Corvi R, Eskes C, Gribaldo L, Griesinger C, Knaut H, Linge JP, Roi A, Zuang V (2012) ECVAM and new technologies for toxicity testing. Adv Exp Med Biol 745:154–180

Coecke S, Balls M, Bowe G, Davis J, Gstraunthaler G, Hartung T, Hay R, Merten OW, Price A, Schechtman L, Stacey G, Stokes W (2005) Guidance on good cell culture practice. A report of the second ECVAM task force on good cell culture practice. Altern Lab Anim 33(3):261–287

Coecke S, Bowe G, Milcamps A, Bernasconi C, Bostroem A-C, Bories G, Fortaner TS, Gineste J-M, Gouliarmou V, Langezaal I, Liska R, Mendoza E, Morath S, Reina V, Wilk-Zasadna I, Whelan M (2014a) Considerations in the development of *in vitro* toxicity testing methods intended for regulatory use. *In vitro* toxicology systems. Springer, New York

Coecke S, Bowe G, Millcamps A, Bernasconi C, Bostroem AC, Bories G, Fortaner S, Gineste Jm, Gouliarmou V, Langezaal I, Liska R, Mendoza E, Morath S, Reina V, Wilk-Zasadna I, Whelan M (2014b) Considerations in the development of *in vitro* toxicity testing methods intended for regulatory use, Coecke S, et al (2014) In: Jennings P, Price A (eds) *In vitro* toxicology systems series: methods in pharmacology and toxicology. Springer, New York, pp 551–569

Council (1986) Directive 87/18/EEC of 18 December 1986 on the harmonisation of laws, regulations and administrative provisions relating to the application of principles of good laboratory practice and the verification of their applications for tests on chemical substances. OJ L 15, 17.1.1987, pp 29–30

Council (1988) Directive 88/320/EEC of 9 June 1988 on the inspection and verification of Good Laboratory Practice (GLP). OJ L 145, 11.6.1988, pp 35–37

European Parliament and Council (2004a) Directive 2004/10/EC of 11 February 2004 on the harmonisation of laws, regulations and administrative provisions relating to the application of the principles of good laboratory practice and the verification of their applications for tests on chemical substances. OJ L 50, 20.2.2004, pp 44–59

European Parliament and Council (2004b) Directive 2004/9/EC of 11 February 2004 on the inspection and verification of good laboratory practice (GLP). OJ L 50, 20.2.2004, pp 28–43

European Parliament and Council (2006) Regulation No. 1907/2006 concerning the Registration, Evaluation, Authorisation and Restriction of Chemicals (REACH), establishing a European Chemicals Agency

European Parliament and Council (2009) Directive 2009/41/EC on the contained use of genetically modified micro-organisms (GMMs). OJ L 125/75, 21.5.2009, pp 75–97

European Parliament and Council (2010) Directive 2010/63/EU of 22 September 2010 on the protection of animals used for scientific purposes. OJ L 276, 20.10.2010, pp 33–79

ECHA—European Chemical Agency (2014) Questions and Answers, REACH, Information requirements, test methods and quality of data, Version 1. http://echa.europa.eu/qa-display/-/qadisplay/5s1R/view/reach/Informationrequirementstestmethodsandqualityofdata

Gupta K, Rispin A, Stitzel K, Coecke S, Harbell J (2005) Ensuring quality of *in vitro* alternative test methods: issues and answers. Regul Toxicol Pharmacol 43:219–224

OECD (1981) Decision of the Council concerning the Mutual Acceptance of Data in the Assessment of Chemicals, C(81)30/FINAL

OECD (1989) Decision-Recommendation of the Council on Compliance with Principles of Good Laboratory Practice, C(89)87/FINAL

OECD (1995) OECD series on GLP and Compliance monitoring. Number 10. GLP consensus document on "The application of the principles of GLP to computerised systems"

OECD (2004) OECD series on GLP and Compliance monitoring. Number 14. GLP Advisory Document on "The application of the principles of GLP to computerised systems"

OECD (2005) OECD series on testing and assessment. Number 34. Guidance document on the validation and international acceptance of new or updated test methods for hazard assessment. ENV/JM/MONO(2005)14

OECD (2012) OECD QSAR toolbox. http://qsartoolbox.org

PIC/S (2007) Good Practices for the Computerised systems in regulated "GXP" environments. Verified December 2014. http://www.picscheme.org/pdf/27_pi-011-3-recommendation-on--computerised-systems.pdf

Rispin A, Harbell JW, Klausner M, Jordan FT, Coecke S, Gupta K, Stitzek K (2004) Quality assurance for *in vitro* alternative test methods: quality control issues in test kit production. Altern Lab Anim Suppl 1:725–729

Seiler JP (2005) Good laboratory practice. Springer, Berlin

Tice RR, Austin CP, Kavlock RJ, Bucher JR (2013) Improving the human hazard characterisation of chemicals: a Tox21 update. Environ Health Perspect 121:756–765

Turnheim D (2008) Current state of the implementation of the OECD GLP principles in the OECD member countries and non-member economies in light of the outcome of the 1998–2002 pilot project of mutual joint visits. Ann Ist Super Sanità 44(4):327–330

US EPA (Environmental Protection Agency) (2010a) ToxCastTM data sets and published research. http://epa.gov/ncct/toxcast

US EPA (Environmental Protection Agency) (2010b) Tox21. http://epa.gov/ncct/Tox21

Vedani A, Smiesko M, Spreafico M, Peristera O, Dobler M (2009) Virtual ToxLab—*in silico* prediction of the toxic (endocrine disrupting) potential of drugs, chemicals and natural products: two years and 2000 compounds of experience: a progress report. ALTEX 26:167–176

Chapter 6
Validation of Computational Methods

Grace Patlewicz, Andrew P. Worth and Nicholas Ball

Abstract In this chapter, we provide an overview of how (Quantitative) Structure Activity Relationships, (Q)SARs, are validated and applied for regulatory purposes. We outline how chemical categories are derived to facilitate endpoint specific read-across using tools such as the OECD QSAR Toolbox and discuss some of the current difficulties in addressing the residual uncertainties of read-across. Finally we put forward a perspective of how non-testing approaches may evolve in light of the advances in new and emerging technologies and how these fit within the Adverse Outcome Pathway (AOP) framework.

Keywords (Quantitative) Structure Activity Relationship [(Q)SAR] • OECD Validation Principles • Read-across • Adverse Outcome Pathways (AOPs) • Integrated Approaches to Testing and Assessment (IATA)

1 Introduction

The global regulatory landscape has changed significantly over the last decade as the volume and diversity of industrial chemicals manufactured has increased. Whilst each region may adapt its chemical management regulations to meet their own specific needs, all are comparable in terms of the general steps applied. These consist

---

G. Patlewicz (✉)
Dupont Haskell Global Centers for Health and Environmental Sciences,
Newark, DE 19711, USA

National Center for Computational Toxicology (NCCT), US Environmental Protection Agency (EPA), Research Triangle Park, NC 27711, USA
e-mail: patlewig@hotmail.com

A.P. Worth
European Commission, Joint Research Centre (JRC), Ispra, Italy

N. Ball
Toxicology and Environmental Research and Consulting (TERC), Environment, Health and Safety (EH&S), The Dow Chemical Company, Horgen, Zurich 8810, Switzerland

C. Eskes, M. Whelan (eds.), *Validation of Alternative Methods for Toxicity Testing*, Advances in Experimental Medicine and Biology 856,
DOI 10.1007/978-3-319-33826-2_6

of hazard identification/characterisation, an assessment of exposure and a risk assessment. The hazard identification step involves identifying all the hazards of potential concern and assigning a hazard classification irrespective of the exposures. The hazard characterisation step is usually dominated by *in vivo* toxicity test outcomes generated by standardised guidelines or protocols. Some regulations entail a tiered approach to satisfying hazard information requirements depending on manufacturing/import volumes for a given substance.

The potential time, cost and animal use to generate such hazard data can be significant and practically unrealistic to achieve given the number of chemicals under consideration. Furthermore, given that the on-going legislative mandates globally are growing, the number of chemical assessments will also increase significantly.

At the same time there has been a strong societal pressure to minimise the use of animals used in such chemical assessments. The EU's 7th Amendment to the Cosmetics Directive (EC 2003), now superseded by the Cosmetics Regulation (EC 2009), called for a ban on animal testing with certain deadlines for specific endpoints. The EU's REACH regulation (EC 2006) stipulates that vertebrate testing be carried out only as a last report and to consider all other options before performing or requiring testing as described by Articles 13(1) and 25(1).

These different drivers have motivated many efforts in the scientific community to investigate the feasibility of developing and applying alternative approaches to evaluate different hazard endpoints. Specifically, the types of alternative approaches that are considered within the scope of this chapter comprise non-testing approaches such as (Quantitative) Structure Activity Relationships ((Q)SARs), chemical categories and their associated read-across.

SAR and QSAR models, collectively referred to as (Q)SARs, are theoretical models that can be used to predict in a quantitative or qualitative manner the physicochemical, biological (e.g. an (eco)toxicological endpoint) and environmental fate properties of compounds from the knowledge of their chemical structure. A SAR is a (qualitative) association between a chemical substructure and the potential of a chemical containing the substructure to exhibit a certain biological effect. In contrast, a QSAR is a statistically established correlation relating (a) quantitative parameter(s) derived from chemical structure or determined by experimental chemistry to a quantitative measure of biological activity. In addition to (Q)SARs, a number of so-named expert systems have also been developed, generally as commercial products. The term "expert system" refers to a heterogeneous collection of computer-based estimation methods, which are based on the integrated use of databases (containing experimental data) and/or rulebases (containing either SARs, QSARs or both). Expert systems can be categorised as statistical in nature if they comprise QSARs, knowledge-based if they are based on SARs and hybrid if they are based on a mix of SARs and QSARs.

In terms of chemical assessment, the information on a chemical as provided by non-testing approaches can be used on its own or in conjunction with information from experimental test methods in the context of integrated approaches to testing and assessment (IATA) (Chap. 13). This chapter provides an overview of how (Q)SARs are validated and applied for regulatory purposes. It also outlines how chemical categories are derived to facilitate endpoint specific read-across and discusses the current difficulties in addressing residual uncertainties of read-across. Finally it puts forward a perspective of how non-testing approaches may evolve in light of the advances in new and emerging technologies and how these fit within the Adverse Outcome Pathway (AOP) framework.

2 OECD Validation Principles for QSARs

Consideration of the regulatory utility of (Q)SARs gathered significant momentum in the run-up to the REACH regulation coming into force. A number of activities were initiated; most notable of these was the International Council of Chemical Associations-Long Range Research Initiative (ICCA-LRI) Setubal workshop held in 2002 which brought together a wide international stakeholder group to formulate guiding principles for the development and application of (Q)SARs for regulatory purposes (Cronin et al. 2003; Cefic-LRI 2002). The principles became known as the "Setubal principles" and were subsequently taken up by the OECD and formally adopted as the OECD validation principles for (Q)SAR (OECD 2004) by the 37th Joint Meeting of the Chemicals Committee and Working Party on Chemicals, Pesticides and Biotechnology in November 2004. The five OECD principles are outlined as follows:

"To facilitate the consideration of a (Q)SAR model for regulatory purposes, it should be associated with the following information:

1. a defined endpoint
2. an unambiguous algorithm
3. a defined domain of applicability
4. appropriate measures of goodness-of-fit, robustness and predictivity
5. a mechanistic interpretation, if possible."

Preliminary guidance was drafted by the European Commission's Joint Research Centre (JRC) which provided context to help interpret each of these principles in turn (Worth et al. 2005). The JRC guidance was then modified and expanded by the OECD (Q)SAR Management group, and published as an OECD Guidance document in 2007 (OECD 2007a).

The formation of these Validation Principles marked a step change in how (Q) SARs should be evaluated and applied for regulatory purposes such as the REACH regulation. Rather than a formalised or institutionalised validation and acceptance process as is the case for *in vitro* or *in vivo* test methods, (Q)SARs were categorised as "existing" information. As such this categorisation permitted a (Q)SAR and its prediction to be evaluated on a case by case basis with respect to a chemical of interest and for a specific regulatory decision. In effect, the OECD principles provide a framework by which information is collected to characterise each principle in turn and to evaluate the extent to which the principles are satisfied for a given (Q)SAR of interest. Being able to satisfy all the principles or only a subset does not make a

(Q)SAR model become more or less acceptable though it may limit its practical utility in a regulatory context in terms of filling or substantiating a data gap. The OECD Principles provide a means of demonstrating the scientific validity of a (Q)SAR model. Of course this in itself only addresses one element of establishing whether the information derived from a (Q)SAR model meets the regulatory purpose of interest. The language under REACH, and specifically in Annex XI provides some helpful context for (Q)SAR use namely:

"Results obtained from valid qualitative or quantitative structure-activity relationship models (Q)SARs may indicate the presence or absence of a certain dangerous properties. Results of (Q)SARs may be used instead of testing when the following conditions are met:

- Results are derived from a (Q)SAR model whose scientific validity has been established
- The substance falls within the applicability domain of the (Q)SAR model
- Results are adequate for the purpose of classification and labelling (C&L) and/or risk assessment and
- Adequate and reliable documentation of the applied method is provided."

Here scientific validity makes reference to the OECD principles. The applicability domain describes the chemical space and scope of the (Q)SAR model as defined by its underlying training set. There are a number of different approaches that can be used to represent an applicability domain of a (Q)SAR which are described in more detail in Sect. 6. Results adequate for risk assessment or C&L refer to the regulatory purpose—is the endpoint of regulatory concern and whether the (Q)SAR meets the regulatory need. Adequate and reliable documentation stipulates how both the (Q)SAR model and its prediction need to be well documented. In the REACH guidance this is conveniently represented by a Venn diagram of three interrelated elements—a scientifically valid (Q)SAR model, applicable to the query chemical for an endpoint of regulatory interest. The intersection of these elements results in an "adequate" (Q)SAR result (see ECHA 2008).Whilst these conditions are specifically outlined for REACH, they could be applicable for other regulatory frameworks.

3 (Q)SAR Model Reporting Format (QMRF)

As stated above, under REACH one of the conditions for using (Q)SARs instead of experimental test data is that adequate and reliable documentation of the applied method is provided. To that end, the JRC in consultation with the EU QSAR Working Group, a subgroup under the then EU's Technical Committee for New and Existing Chemicals (TCNES) proactively developed a standard template to summarise and structure key information about a (Q)SAR model and its associated prediction. The standard template was termed a reporting format (RF). The intent was that the reporting format would be sufficiently flexible to accommodate any type of

(Q)SAR approach yet structured in a manner to provide sufficient and relevant information to facilitate decision making. As described in the REACH guidance (ECHA 2008), the RFs were expected to serve as an efficient and transparent exchange of (Q)SAR information between Industry and Regulatory Authorities by ensuring transparency, consistency, and acceptability:

Transparency: Models and prediction information would ideally be clearly documented and to enable a correct interpretation of the conclusions inferred.

Consistency: Information related to different approaches should be reported in a common format to allow different models and their predictions to be readily compared.

Acceptability: The reports should include all relevant information needed to evaluate the adequacy and completeness of the (Q)SAR and its prediction for a chemical of interest and the regulatory endpoint under consideration.

Two specific RFs were devised, a RF for (Q)SAR models termed the (Q)SAR Model Reporting Format (QMRF) and a RF for a prediction derived from an associated (Q)SAR called a (Q)SAR Prediction Reporting Format (QPRF). The general form of the QMRF and QPRF are described in brief. Comparable reporting formats also exist for analogue and chemical category approaches to facilitate their documentation and are described in Sect. 7.

The QMRF provides the framework for compiling robust summaries of (Q)SAR models and their corresponding validation studies, akin to the robust study summaries for a guideline toxicity test. The structure of the format was designed to reflect as far as possible the OECD principles. The QMRF contains information on the source, type, development, validation, and possible applications of the model. In the QMRF, each of the OECD principles is associated with a set of fields:

- Section 3: Identity of the endpoint
- Section 4: Description of the algorithm
- Section 5: Description of the applicability domain
- Sections 6 and 7: Description of the training and test sets respectively

Information about the identity of the chemicals contained in both training and test sets can also be included in the QMRF (where possible): (a) Chemical Name (IUPAC); (b) Chemical Name (Not IUPAC); (c) CAS Number; (d) SMILES (Simplified Molecular Input Line Entry System); (e) InChI (IUPAC International Chemical Identifier); (f) Mol file; (g) Structural formula; (h) Values for the dependent variable; (i) Values for the descriptors.

The above-described framework for documenting and assessing the validity of (Q)SAR models, while originally developed for REACH, should be sufficiently flexible to be applied in the hazard assessment of all types of chemicals and products, irrespective of the legislation under which they are regulated (industrial chemicals, biocides, etc.). However, other guidance would need to be developed to describe how models should be evaluated and their predictions interpreted in specific regulatory contexts.

The QSAR model reporting formats (QMRF and QPRF) were developed on the basis of consensus with the main stakeholders (industry and authorities), and they capture a level of detail that is a compromise between scientific rigour and practicality. As a consequence, it is not always clear how much detail should be included under the different headings, and what kind of information is pertinent for models developed by different methodological approaches, for example, QSAR models utilising Support Vector Machines, artificial neural networks, instance-based learning and consensus modelling. If models based on such methods are to gain acceptance, they need to be understandable to the assessor, and be described with a sufficient level of transparency to form the basis for regulatory decision making.

In some decision-making contexts, QSAR information is used to provide supplementary information and support weight-of-evidence (WoE) assessments, rather than directly filling an information requirement. An example could be in the assessment of toxicological profile of metabolites and degradates. In such cases, it could be argued that the amount of documentation could be relaxed, while focusing on the critical considerations to assess the adequacy of prediction. With this perspective in mind, a simplified checklist of 10 questions, drawing on both the QMRF and QPRF, has been explored, and illustrated in relation to the prediction of genotoxicity of pesticide metabolites (Worth et al. 2011).

4 JRC QSAR Model Database

In an effort to promote the re-use of commonly available (Q)SAR models and provide a means of readily identifying valid (Q)SAR models, a freely-accessible inventory of evaluated (Q)SARs was developed by the JRC known as the JRC QSAR Model Database. The database, accessible through a web-based interface (http://ihcp.jrc.ec.europa.eu/our_labs/predictive_toxicology/qsar_tools/QRF) allows for the retrieval of QMRFs in a suitable readable format and for the submission of a QMRF in different formats such as excel. A QMRF editor was also developed in parallel to facilitate completion of a QMRF and its submission into the Model Database. The QMRF Editor exists in two forms, as a Java web start application (v0.05) and as a standalone desktop application v2.0.0.

Developers and users of (Q)SAR models can submit a QMRF to the Model Database using one of these two editors. The JRC then performs a quality check of the QMRF submitted, to ensure that only properly documented summaries of (Q) SARs are included in the JRC QSAR Model Database. This quality check is a completeness check to determine whether a QMRF has been reasonably completed and to suggest revisions if specific information is lacking. It is not a scientific review of the model itself and inclusion of the model in the QSAR Model Database does not confer any acceptance or endorsement status by the JRC, the European Commission or the European Chemicals Agency (ECHA).

A QMRF can be searched and retrieved from the database on the basis of keywords or specific endpoint names. If a QMRF has been uploaded with associated training or test sets (as per sections 6 and 7 of the QMRF), a search can be performed to see whether a substance or structurally related analogue (on the basis of a Tanimoto

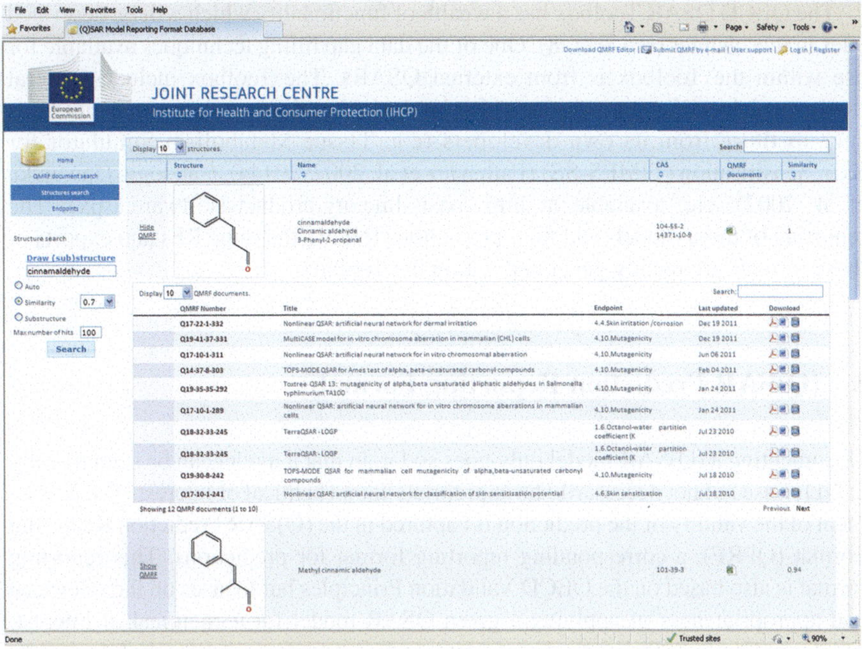

Fig. 6.1 Interface of the JRC Model Database

similarity score) is included in one of these data sets. Thus the experimental data underlying a given (Q)SAR model can be probed on the basis of a chemical identifier such as a CAS registry number or on the basis of a 2D chemical structure. Figure 6.1 provides an illustration of the JRC Model Database interface.

Each QMRF registered in the Model database is assigned a unique identifier; for example the QMRF for the Derek Skin sensitisation rule base is listed with record number Q13-34-36-315. Referencing this identifier in regulatory dossiers such as a REACH dossier could be an alternative to uploading a copy of the QMRF itself. At the time of writing (April 2015) there were over 80 QMRFs available in the Model database.

The documentation of (Q)SARs in this manner has applicability in the wider international arena. The OECD QSAR Toolbox is a case in point. The OECD QSAR Toolbox was first launched as a proof of concept in 2008. It was originally developed to mimic the category workflow that is described in the current Chapter R.6 of the REACH technical guidance (ECHA 2008, 2012a) and the OECD grouping document (OECD 2007b, 2014a). The intention was that this tool would facilitate the practical development, evaluation, justification and documentation of chemical categories and read-across. The most recent version of the Toolbox now available is QSAR Toolbox 3.3.5. This is available for free download from the OECD website (http://www.qsartoolbox.org/download.html).

The OECD QSAR Toolbox has a wealth of functionality which will be described later in this chapter (see Sect. 8). One of the data gap filling techniques available for use within the Toolbox is from external QSARs. The Toolbox includes several freely available QSARs and it can also be docked to third party commercial tools such as those from its own developers (e.g. TIssue Metabolism Simulator for Skin Sensitisation (TIMES-SS) (Patlewicz et al. 2007, 2014a), Catalogic (Jaworska et al. 2002), etc. available at http://oasis-lmc.org/products/software.aspx). The reporting of these models and their predictions follows the same RFs and exports of these formats are readily generated within the Toolbox.

5 (Q)SAR Prediction Reporting Format

Documenting a (Q)SAR model only goes so far, in that a model can be scientifically valid for use but not necessarily be appropriate for a chemical of interest. The assessment of the validity of the prediction is captured in the (Q)SAR Prediction Reporting Format (QPRF), a corresponding reporting format for predictions. This reporting format is also based on the OECD Validation Principles but focuses on the relevance and appropriateness of applying a given QSAR model for a chemical of interest. The format outlines the pertinent information to justify the scientific confidence of a prediction. In particular it gathers the information which assesses the relevance of a model for the chemical of interest in terms of whether that chemical falls within the applicability domain of the model as would have been described within the QMRF and explores the extent to which "similar" substances are correctly predicted. "Similar" in this context refers to substances that are similar with respect to the same driving/input parameters that were used to derive the (Q)SAR model itself. For example if a (Q)SAR was developed on the basis of parameters such as log Kow (log of the octanol-water partition coefficient) and MW (molecular weight), "similar" analogues would typically be those identified from the training set with comparable values for log Kow and MW. Determining how well the chemical of interest is predicted relative to other "similar" analogues is critical to establish whether related substances are reasonably predicted relative to their experimental results, and to provide the confidence that a robust and reliable prediction is feasible for the chemical of interest. Usually a search of appropriate analogues is made using the training set of the (Q)SAR model itself. Tools such as Toxmatch v1.07 (Patlewicz et al. 2008) allow a training set to be characterised on the basis of structural fragments or descriptors and a comparison to be made between the chemical of interest and the training set to help identify the most similar analogues. An assessment of the concordance between the estimated and experimental data for those analogues can then be made. If the training set is not available or limited, other tools such as the US EPA's Analog Identification Methodology (AIM) (http://www.epa.gov/oppt/sf/tools/aim.htm), Leadscope (www.leadscope.com), OECD Toolbox v3.3.5 (http://www.qsartoolbox.org) (discussed in ECETOC 2012; OECD 2014a) can be helpful in identifying relevant analogues with experimental data. An assessment of the agreement between experimental data and the (Q)SAR predictions can then be made.

The importance of the QPRF is to document why a prediction is considered relevant and hence how the information derived can be used in the context of decision making, for example filling a data gap as a replacement or used as part of a WoE or IATA. Examples of completed QPRFs may be found in the Appendices of ECETOC TR 116 (ECETOC 2012).

6 Applicability Domain

In order to obtain a reliable prediction, it is important to verify that the chemical of interest falls within the applicability domain of the model. The concept of applicability domain was introduced to assess the probability of a chemical being covered by the (Q)SAR training set and is hence related to the reliability of the prediction. However, even if a substance lies within the domain, this is not a guarantee of the validity and predictivity for that substance of interest.

Applicability domains were first considered as part of the background documents to the ICCA-LRI Setubal workshop on (Q)SARs. The background paper subsequently published by Eriksson et al. (2003) discussed the issue of how the applicability borders of a QSAR could be defined and how parameter and prediction uncertainty could be estimated. Following the Setubal workshop and as the discussion of the principles were taking shape at the OECD, the JRC held a workshop in 2004 on applicability domains. A definition was coined and specific considerations of how domains could be structured depending on the type of QSAR model were explored. The definition proposed was as follows: "The applicability domain of a (Q)SAR model is the response and chemical structure space in which the model makes predictions with a given reliability" (Netzeva et al. 2005).

There are many different approaches to characterise a domain, and whether a substance lies within the domain or not will ultimately depend on the approach applied. As a consequence, one methodology may categorise a substance of interest as being within the domain, whereas another might conclude that it is out of domain. The question of which domain approach to use therefore requires careful examination of the methodology used to characterise the domain and to what extent the model being used provides reasonable predictions for other related substances. The common means of characterising a domain for a QSAR model relies on its training set of chemicals, i.e. those substances that were used to derive the QSAR model. This type of information can be encoded in the form of structural keys, fragments or fingerprints—effectively these provide a means of characterising the domain in terms of its representative structural components. It also helps to address the question of whether the chemical of interest resembles any of the training set chemicals on the basis of structural similarity. Another means of characterising the domain can be in terms of the training set ranges of the descriptors used in the derivation of the model. This latter example illustrates one of the four approaches to defining interpolation regions in multivariate space that Jaworska et al. (2005) reviewed. The others they discussed included distance, geometrical and probability density distribution. An applicability domain need not be defined by only one approach. Nikolova-Jeliazkova and Jaworska (2005) presented a case study using the KOWWIN model and constructed an applicability domain on the basis of descriptor ranges coupled with a principal component rotation as a data preprocessing step. The methodologies described in Jaworska et al. (2005) were implemented into a tool called AMBIT Discovery v0.04 (released May 2006 by Ideaconsult Ltd, Bulgaria) freely available from SourceForge at http://ambit.sourceforge.net/download_ambitdiscovery.html.

Other researchers have considered how applicability domains could be constructed for expert systems. Dimitrov et al. (2005) formulated a stepwise domain for the TIMES expert system for skin sensitisation and mutagenicity which comprised a number of different domains from physicochemical property ranges to structural fragment based approaches through to interpolation and metabolic domains. A nearest neighbour atom centred fragment approach was incorporated in the TIMES system to characterise the structural domain. The methodology developed for TIMES was extracted as a standalone program called Domain Manager v1.0. It provides the methodology to codify domains on the basis of training set descriptor values and/or their structural fragments (e.g. by using atom centred fragments). Using the Domain Manager software, Patlewicz et al. (2011) exploited its functionality to construct a structural applicability domain on the basis of nearest neighbour atom centred fragments for the KOWWIN training set as a means to demonstrate "inclusion" of substances within the domain if their fragments overlapped 100 % with those of the training set itself. This approach was used to justify the validity of log Kow predictions relied upon in their subsequent REACH submissions.

Whilst many efforts have focused on developing approaches to characterise applicability domains of QSARs, much less attention has been paid to assessing the validity and scope of SARs or structural alerts. Structural alerts have enjoyed a renewed role in facilitating the development of chemical categories for read-across by providing a means to group chemicals on the basis of their presumed mechanistic similarity. Many such SARs are encoded as profilers within the OECD QSAR Toolbox.

Probably the earliest effort to consider the domain of structural alerts was that by Schultz et al. (2007). Here an experimental approach was taken to test out the scope of the Michael acceptor reaction domain that had been described for skin sensitisers by Aptula and Roberts (2006). A similar approach was undertaken by Enoch et al. (2012) for the SNAr reaction domain. SNAr is one of the six mechanistic domains described by Aptula and Roberts (2006) that were shown to be important in toxicological endpoints in which the ability to bind covalently to a protein is a key molecular initiating event. In Enoch et al. (2012) experimental data (2 h RC50 values from the glutathione assay (Schultz et al. 2005)) were generated for commercially available substituted benzenes. In Enoch et al. (2013) a similar exercise was undertaken for pyridines and pyrimidines which reside within the SNAr domain. The in-ring nitrogen(s) act as activating groups in the SNAr reaction. The position(s) of the in-ring nitrogen(s) as well as other activating groups, especially in relation to the leaving group, affect reactive potency. In both studies, the experimentally defined applicability domains resulted in a series of new structural alerts.

Investigating means of characterising applicability domain of alerts within expert systems have also been undertaken. One study attempted to define the domain for the skin sensitisation structural alert rule-base as contained within the Derek for Windows expert system (Ellison et al. 2009). Fragments generated for a test compound were queried against a training set of compounds represented by fragments. The approach was able to highlight test chemicals which differed from those in the training set. The information was used to designate chemicals as being either inside or outside the domain of applicability for the structural alert on which that training set was based. Ellison et al. (2011) also investigated the feasibility of deriving different applicability domain approaches for the mutagenicity alerts contained within the Derek for Windows expert system.

Chakravarti et al. (2012) explored the extent to which an applicability domain could be defined for individual alerts for the expert system CASE Ultra model. The domain for each alert was constructed using a set of fragments that were found to be statistically related to the endpoint in question as opposed to using overall structural similarity or physicochemical properties. Use of the applicability domains was found to reduce the number of false positive predictions. It is now possible to obtain ROC (receiver operating characteristic) profiles of CASE Ultra models by applying domain adherence cut-offs on the alerts identified in test chemicals.

The TIMES expert system relies on structural alerts to provide transparent mechanistic reasoning for predicting effects such as mutagenicity and skin sensitisation. As part of the on-going work within the TIMES-SS consortium (Patlewicz et al. 2014a), structural alerts underpinning the chemical reactivity to proteins were evaluated with a view to characterising their reliability and in so doing providing additional confidence in the robustness of a given prediction. Three key attributes for evaluating alert reliability were defined namely: Alert Performance, Number of chemicals and Mechanistic justification.

- Alert Performance was defined as the ratio between the number of correct (positive and negative) predictions and the total number of chemicals within the local training set that triggered the alert.
- Number of chemicals was defined as the absolute number of chemicals (n) that support an alert.
- Mechanistic justification made reference to the plausible rationale that related the chemistry with the sensitisation outcome.

In addition to attributes for alert reliability, four categories of reliability were defined. Alerts that were categorised as "High reliability" were those where the Alert Performance was equal or greater than 60 %; where the absolute number of chemicals underpinning an alert was equal or greater than 5 and where a mechanistic justification was available. Alerts that were categorised as "Low reliability" had a performance ratio of less than 60 % but where the number of chemicals was still 5 or greater and a mechanistic justification was available. An "Undetermined" reliability indicated that the number of chemicals was between 1 and 5 and mechanistic justification was available. Finally, an "Undetermined theoretical" reliability denoted that there were no chemicals underpinning an alert and only a mechanistic

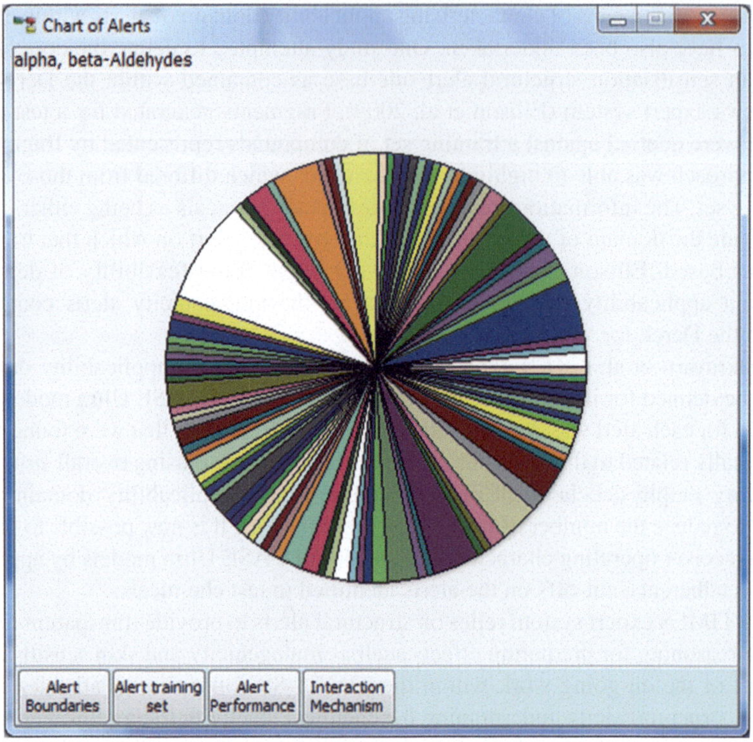

Fig. 6.2 Pie chart of all alerts within TIMES-SS

justification was being proposed. TIMES-SS comprises 98 Alerts, where each was evaluated in light of these reliability criteria. 26 out of 98 alerts (27 %) were found to be of "High reliability"; 39 alerts (40 %) were denoted as "undetermined" and 33 alerts (34 %) were "undetermined—theoretical". The 72 undetermined alerts were re-assessed in light of any new information and expert insight. Thirty four alerts were subsequently found to be "reliable" following the review. The alerts were adjusted and improvements were captured as refinements within TIMES-SS itself. Information that documented the reliability—in terms of annotating an alert, describing the mechanistic justification and documenting the substances underpinning the alert together with their experimental and predictive sensitisation outcomes are all now provided within TIMES-SS. Such refinements provide greater transparency and confidence in the predictions generated from TIMES-SS and demonstrates a step change in terms of substantiating the scientific confidence associated with a prediction. Figures 6.2 and 6.3 illustrate the type of information provided for one such alert. The pie chart in Fig. 6.2 represents all alerts, each of which is colour coded based on the depth of supporting information available. Each pie slice represents an alert, and clicking on the different buttons opens up information on the boundaries associated with the alert, the local training set, the performance of the alert and the mechanistic basis. Figure 6.3 shows a snapshot of the local training set

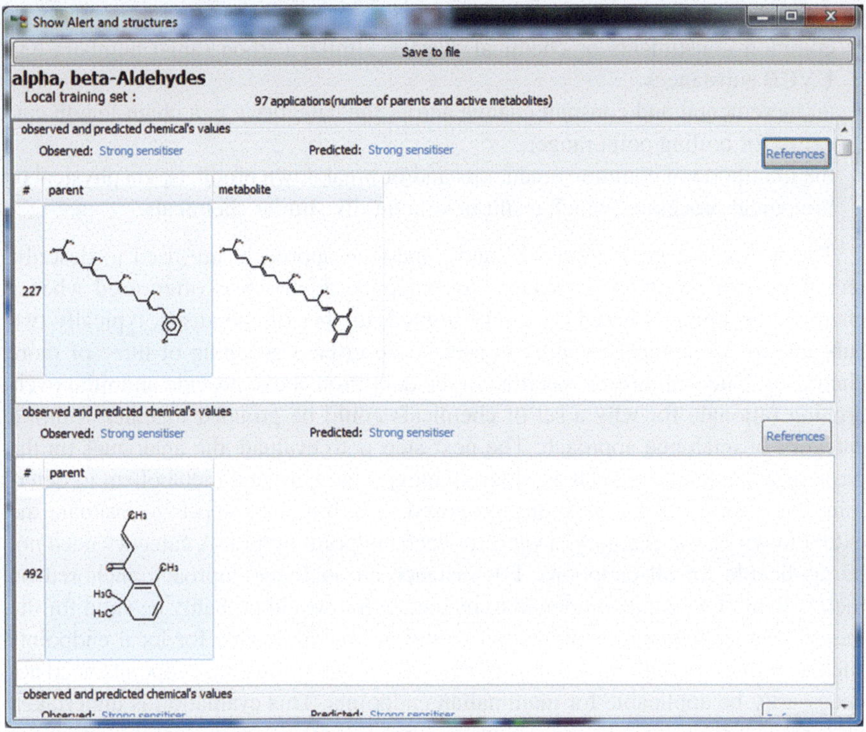

Fig. 6.3 Training set for the alpha, beta aldehyde alert

for the alpha, beta aldehydes alert in terms of the chemical structure and the outcomes from TIMES-SS and experimentally.

7 Chemical Categories and Read-Across

Chemical categories and their associated read-across form part of a continuum of non-testing approaches. Whilst (Q)SARs and chemical categories are underpinned by a relationship between structure and activity, (Q)SARs are more formalised in their construct.

Chemical categories for regulatory purposes have been conveniently defined within the REACH technical guidance (ECHA 2008) and the OECD grouping guidance (OECD 2007b, 2014a) as follows:

"A chemical category is a group of chemicals whose physico-chemical and human health and/or environmental toxicological properties and/or environmental fate properties are likely to be similar or follow a regular pattern as a result of structural similarity. The similarities may be based on the following:

- common functional group(s) e.g. aldehyde
- common constituents or chemical classes, similar carbon range numbers e.g. UVCB substances
- an incremental and constant change across the category e.g. a chain-length category for boiling point range;
- the likelihood of common precursors and/or breakdown products, via physical or biological processes, which result in structurally similar chemicals."

The terms 'category approach' and 'analogue approach' are used to describe techniques for grouping chemicals. An analogue approach is often used when a chemical grouping is based on a very limited number of chemicals, typically two substances. A chemical category is used to describe a grouping of three or more chemicals. The similarities specified in the definition above provide an initial overarching rationale for why a set of chemicals could be grouped together within a category or analogue approach. The next step is to evaluate the analogues on the basis of considerations such as bioavailability, reactivity and metabolism to determine the validity of the grouping proposed. A subsequent step is to evaluate the extent to which the category is valid for each endpoint in turn. A category need not be applicable for all endpoints. For instance an analogue approach anchored on metabolism of a parent compound to its metabolite would probably be valid for the majority of systematic endpoints but would not be applicable for local endpoints such as skin/eye irritation. A category derived for environmental endpoints need not necessarily be applicable for mammalian endpoints. This evaluation is undertaken with respect to what is known about the chemical determining features driving a given endpoint and how these concur with the experimental data and predicted information that are available for the analogues in turn.

Although the development of chemical categories and (Q)SARs are underpinned by the same principles of chemical similarity, there has been no specific requirement to validate a category per se. Most likely this is because *ad hoc* categories have been routinely used under the High Production Volume (HPV) programmes within the US and under the OECD. Within the EU under REACH, the adequacy and reliability of the category approach must be nonetheless substantiated on an endpoint basis. Specifically the following wording is described: "In all cases results should:

- be adequate for the purpose of classification and labelling and/or risk assessment,
- have adequate and reliable coverage of the key parameters addressed in the corresponding test method referred to in Article 13(3),
- cover an exposure duration comparable to or longer than the corresponding test method referred to in Article 13(3) if exposure duration is a relevant parameter, and
- adequate and reliable documentation of the applied method shall be provided."

Whilst a formal validation of a category is not required, a robust justification to demonstrate the validity of the endpoint data gap filling approach within a category or analogue approach needs to be presented. This is most readily documented in a format known as the Category (Analogue) Reporting Format (CRF/ARF). An ARF is used when a read-across is carried out from one substance (source substance) to the substance of interest (target substance). A CRF is used when the group comprises three or more members. These formats serve a similar purpose of assuring transparency, consistency and adequacy of the data gap filling prediction as already discussed for (Q)SARs.

At this point it is probably worthwhile noting some of the ancillary terms related to category and analogue approaches. Read-across is a data gap filling mechanism. Endpoint information for one chemical (target) is used to predict the same endpoint for another chemical (source), which is considered to be similar in some way (usually on the basis of structural similarity or same mode of action or other properties). Read-across can be used to fill data gaps in the context of both the analogue approach and the wider category approach. The approach can be qualitative or quantitative depending on the type of data available. For some endpoint specific categories, members may be related by a trend (e.g. increasing or decreasing molecular mass, carbon chain length or some other physicochemical property). Here trend analysis can be undertaken which essentially means that a local QSAR is derived using the category members themselves. An example could be that for a group of substances whose acute aquatic toxicity in fish were related to log Kow.

A final data gap filling technique is to use external QSARs. Each of these data gap filling approaches has also been implemented in the OECD Toolbox (see Sect. 8). There have been a several reviews that describe the state of the art of (Q)SARs for a number of different endpoints and these are well documented in the endpoint specific guidance for REACH (ECHA 2012b). Moreover, the availability of different (Q) SARs for each of the endpoints in turn have also been well described in ECETOC Technical Report (TR) 116 (ECETOC 2012) as well as the recently revised OECD grouping guidance (OECD 2014a). Whilst (Q)SARs may be used to directly fill a data gap for specific endpoints and in specific cases they are more typically used to provide supporting information in a WoE approach or simply to provide a means of rationalising the mechanistic similarity between category members.

In addition to the available regulatory guidance, practical guiding principles and considerations for developing analogue and category approaches are described in Wu et al. (2010), ECETOC (2012) and Patlewicz et al. (2013a). Wu et al. (2010) outlined a stepped process for analogue evaluation. ECETOC TR 116 builds on this with a regulatory focus and is structured along the lines of the ARF/CRF in terms of the types of information that should be provided (ECETOC 2012). In contrast, there has been little focus on deriving principles to assist in evaluating the scientific validity of an analogue/category approach. Under REACH, there is a strong preference for the use of interpolation within grouping approaches, presumably because this gives rise to less uncertainty than extrapolation. Extrapolation is therefore considered as less reliable due to this higher level of uncertainty associated with predictions. The exception to this is where an extrapolation from one substance to another leads to an equally severe or more severe hazard assessment for the target substance. Although it may seem logical to assume that interpolation is subject to less uncertainty than extrapolation, in reality the degree of uncertainty is not due to the interpolation or extrapolation of data, but rather the strength of the relationship forming the basis of the category/analogue approach itself. This in turn is dependent on the size of the category and the amount and quality of the experimental data for the category members themselves. If the relationship underpinning the category is poorly defined then interpolation or extrapolation can result in significant uncertainty. Building scientific confidence in an analogue or category approach relies upon WoE encompassing QSAR, *in vitro* assays or mechanistic/bridging studies to substantiate the validity of the relationship underpinning the grouping.

There will always be some degree of uncertainty associated with predictions of hazards and toxicity, indeed, there are inherent uncertainties with all test data not just when applying read-across approaches. Addressing these uncertainties during hazard characterisation depends on the type of endpoint. The uncertainties associated with endpoints where there is a 'presence/absence' of effect, will differ from those where there is a 'no effect level' as the result and uncertainties will also depend on the nature of the apical endpoint itself. The degree of scientific confidence required for a skin irritation outcome will arguably differ than that for a developmental toxicity outcome.

8 The OECD QSAR Toolbox

The OECD Toolbox was developed as a means to assist in the systematic development, evaluation, justification and documentation of endpoint specific categories. The workflow mimics that described in the OECD grouping guidance and the REACH guidance (ECHA 2008; OECD 2007b, 2014a). The Toolbox comprises regulatory inventories, experimental data from a number of sources including the ECHA REACH dissemination database, as well as a number of QSARs such as those from the US EPA's EpiSuite™. The Toolbox was developed in phases with the release of the first version (a technological proof-of-concept) in March 2008. Since then the Toolbox has been through several updates, with the current version being 3.3.5 and funded by ECHA.

A target substance is first introduced into the Toolbox and a search for existing experimental data is performed. If no or limited data are available, the substance is then profiled on the basis of a number of different profilers. There are different types of profilers within the Toolbox—predefined, general mechanistic, endpoint specific, empiric and toxicological. Predefined profilers include rulebases to help identify whether the target substance is a member or a potential member of an existing regulatory category such as those already established by the OECD or EPA as HPV categories. General mechanistic profilers comprise a set of different SAR rulebases that encode either general organic chemistry reaction principles which will be relevant for endpoints where covalent binding is a first step or other rulebases such as the Cramer structural classes (Cramer et al. 1978). Endpoint specific profilers include SARs developed for different endpoints such as the Verhaar alerts (Verhaar et al. 1992) for assigning MOA for aquatic toxicity, the ECOSAR classes, protein binding alerts and DNA binding alerts for skin sensitisation and genotoxicity endpoints as well as alerts for skin and eye irritation amongst others. Empiric profilers entail tools to categorise chemicals on the basis of their structural similarity, chemical elements or functional groups. In the current version of the Toolbox, only one toxicological profiler exists which is called Repeated Dose (HESS). This profiler was developed by National Institute of Technology and Evaluation (NITE) in Japan. It contains categories for substances expected to induce similar toxicological effects in repeated dose oral toxicity. Profilers also exist to simulate abiotic and biotic degradation as well as metabolism. After profiling, a search for related analogues based on the profiling outcomes can be performed to formulate a preliminary grouping. At this point, all or selected endpoint experimental data are extracted for all the analogues identified. Data gap filling techniques can then be performed for specific endpoints of interest. A series of subcategorisations are typically undertaken to filter the preliminary grouping into a smaller category of mechanistically similar analogues. The type of endpoint and the quantity of data will dictate whether a read-across or trend analysis data gap filling approach can be performed. Within the Toolbox, chemical descriptors such as log Kow are estimated to help evaluate the relationship between the analogue substances and the endpoint of concern. A read-across needs a minimum of two source analogues whereas at least three source analogues are needed to perform a trend analysis. Available QSARs within the Toolbox or QSARs docked to the Toolbox can also be used to fill specific data gaps. The profiling tools that have been best developed are those customised for aquatic toxicity, genotoxicity and skin sensitisation. This mirrors the maturity of available QSARs for these endpoints. The results of endpoint specific categories formed within the Toolbox can be documented in modified QMRF and QPRF reporting formats or exported as IUCLID files which can be readily included into REACH dossiers.

More information about the Toolbox including guidance and tutorials can be found at the OECD website at http://www.oecd.org/chemicalsafety/risk-assessment/theoecdqsartoolbox.htm#what_does_the_toolbox_do. A public discussion forum for users of the QSAR Toolbox is also available. Users can exchange experience with using the software, exchange databases or custom profilers, make improvement suggestions or seek technical help from the Toolbox developers.

9 Establishing Scientific Confidence (Validity) of Read-Across Approaches

Read-across approaches are conceptually accepted as a means to address data gaps within chemical category and analogue approaches, however many difficulties still remain in applying them consistently in practice for regulatory purposes. Efforts have been undertaken by Industry and ECHA to identify some of the barriers to broader acceptance of read-across approaches for at least the REACH regulation. Industry, through ECETOC, summarised the current practices for read-across (ECETOC 2012; Patlewicz et al. 2013a). An ECHA-Cefic LRI workshop provided a forum to exchange experiences in developing and evaluating read-across (Patlewicz et al. 2013b).

Whilst there has undoubtedly been success in the read-across of "simpler" endpoints such as aquatic toxicity and Ames mutagenicity, the key endpoints where progress is still limited but which represent the greatest need are reproductive, developmental and repeated dose toxicity. As part of the ongoing discussions within Cefic LRI and with ECHA, it has been recognised that assuring the scientific confidence of a read-across is dependent on identifying and addressing residual uncertainties. Thus establishing a systematic framework for characterising a read-across justification and its uncertainties would be a critical step. ECHA has been developing a read-across assessment framework (RAAF) to facilitate the evaluation of read-across justifications submitted as part of REACH dossiers (de Raat 2014). Elements of this RAAF have become public in terms of conceptually outlining the framework (see ECHA website: http://echa.europa.eu/en/view-article/-/journal_content/c6dd5b17-7079-433a-b57f-75da9bcb1de2). The RAAF has since been published, see http://echa.europa.eu/documents/10162/13628/raaf_en.pdf (ECHA 2015). Researchers from P&G who were part of a Cefic LRI read-across team have shared their own systematic framework for DART endpoints (Blackburn and Stuard 2014). The latter builds upon from their framework (Wu et al. 2010) for evaluating analogues for read-across which they tested out in 2011 with 14 different case studies (Blackburn et al. 2011). Whilst these initial constructs go some way towards identifying uncertainties and proposing assessment factors to address these, they fall short of proposing how uncertainty can be minimised in a scientifically robust manner. The dialogue of how to build scientific confidence in read-across has evolved with complementary frameworks proposed on how to identify and address uncertainties (Patlewicz et al. 2014c, 2015a; Schultz et al. 2015)

Advances in novel technologies such high throughput (HTS) and high content (HCS) screening methods present new opportunities for enhancing read-across approaches and reducing uncertainty particularly if these tools are anchored in an Adverse Outcome Pathway (AOP) framework to provide the biological context to aid interpretation (ECETOC 2009, 2012). An AOP is defined as a construct for representing existing knowledge concerning the causal linkages between initial molecular events and an adverse outcome at the individual and population level (Ankley et al. 2010).

The OECD has initiated a work programme to develop AOPs that can be applied to different regulatory purposes from test guideline development and refinement, development of IATA as discussed in Chap. 13 as well as chemical categories. The revised grouping guidance (OECD 2014a) discusses how information from AOPs can help in the development and application of toxicologically meaningful categories and read-across. Initial considerations were proposed during an OECD workshop in 2010 on "Using Mechanistic Information in Forming Chemical Categories" (OECD 2011). In addition the OECD has implemented the first published AOP for skin sensitisation (OECD 2012) into the OECD Toolbox to illustrate how information from key events can be used to substantiate a read-across. More information can be found in the tutorial developed (OECD 2014b) which is available at http://www.oecd.org/chemicalsafety/risk-assessment/theoecdqsartoolbox.htm#Guidance_Documents_and_Training_Materials_for_Using_the_Toolbox and this specific AOP application is also illustrated in Chap. 13. The scientific confidence considerations for AOPs for different applications is also discussed in Patlewicz et al. (2015b).

Applying novel technologies to support read-across remains a newly evolving field with notable efforts such as BASF's metabolomics approach (van Ravenzwaay et al. 2013). Other researchers from Industry are working on using transcriptomics data to elucidate modes of action (Naciff et al. 2013). As part of a Cefic LRI project researchers at TNO, University of Masstricht and elsewhere have developed a multi-disciplinary data infrastructure known as 'DIAMONDS' with statistical and computational tools on one side and 'Biological Verification' (BV) models on the other (see https://www.tno.nl/downloads/diamonds_leaflet.pdf). DIAMONDS aims to integrate data and knowledge from chemoinformatics, omics, (HTS-)assay and *in vivo* toxicity studies. DIAMONDS mines and visualises relevant (public) information to support mechanistic understanding and prediction of toxicological profiles. As part of a recent Cefic LRI project, the information and knowledge within DIAMONDS will be exploited to investigate the extent to which this can facilitate enhancement of read-across for particular endpoints (see http://www.cefic-lri.org/projects/50/172/AIMT4-UM-DECO2-Moving-from-DECO-towards-OECD/) (Jennen et al. 2014). Most recent efforts in the area of read-across have been described in Ball et al (2016).

As part of a European Commission-Cosmetics Europe FP7 Research Initiative, SEURAT-1 has also investigated the development of AOPs for different types of liver toxicity and their applicability in read-across. More information on SEURAT-1 can be found at the website http://www.seurat-1.eu/.

10 Future of QSARs

A major trend in the field of alternatives will be focused on the AOP framework as a means to construct and collate relevant mechanistic information for endpoints of regulatory concern. To that end, the types of QSARs developed and the sort of approaches used to support read-across within chemical categories will undoubtedly change. As has been described above with respect to the OECD Toolbox, the structural alert information encoded in the profilers provides a means to group chemicals on the basis of their chemical mechanistic similarity. This could be likened as a means to represent qualitative molecular initiating event (MIE) information within an AOP. Quantifying such information, would result in the development of new QSARs that predict the MIE or other key events (KEs) in the AOP. Certainly this represents a shift in the types of QSARs that will be developed in the future—ones that encode meaningful mechanistic information for molecular initiating events.

The AOP for skin sensitisation that has been implemented in the OECD Toolbox represents a step change to how non-testing approaches can be developed and utilised. This AOP could be likened to an IATA in terms of how structural information for different KEs and the underlying experimental data are organised. For each KE, profilers have been developed each with their own applicability domain that has been derived experimentally (akin to that described by Schultz et al. 2007 and Enoch et al. 2012, 2013). Outside of the OECD Toolbox infrastructure one could envisage similar IATA comprising different components that characterise structural alert information for different KEs as well as other relevant information from other endpoints that would impact an assessment for the endpoint of interest. Such IATA have already been constructed for respiratory sensitisation building upon the insights of the AOP for skin sensitisation (Mekenyan et al. 2014) as well as for skin sensitisation itself (Patlewicz et al. 2014b). Bringing several endpoint specific IATA together in one framework could also be envisaged to facilitate screening and prioritization on the basis of an overall hazard profile (Patlewicz et al. 2014d; Patlewicz and Fitzpatrick 2016).

References

Ankley GT, Bennett RS, Erickson RJ, Hoff DJ, Hornung MW, Johnson RD, Mount DR, Nichols JW, Russom CL, Schmieder PK, Serrrano JA, Tietge JE, Villeneuve DL (2010) Adverse outcome pathways: a conceptual framework to support ecotoxicology research and risk assessment. Environ Toxicol Chem 29:730–741

Aptula AO, Roberts DW (2006) Mechanistic applicability domains for nonanimal-based prediction of toxicological endpoints: general principles and application to reactive toxicity. Chem Res Toxicol 19(8):1097–1105

Ball N, Cronin MT, Shen J, Blackburn K, Booth ED, Bouhifd M, Donley E, Egnash L, Hastings C, Juberg DR, Kleensang A, Kleinstreuer N, Kroese ED, Lee AC, Luechtefeld T, Maertens A, Marty S, Naciff JM, Palmer J, Pamies D, Penman M, Richarz AN, Russo DP, Stuard SB, Patlewicz G, van Ravenzwaay B, Wu S, Zhu H, Hartung T (2016) Toward Good Read-Across Practice (GRAP) guidance. ALTEX 33:149–166

Blackburn K, Stuard SB (2014) A framework to facilitate consistent characterization of read across uncertainty. Regul Toxicol Pharmacol 68(3):353–362

Blackburn K, Bjerke D, Daston G, Felter S, Mahony C, Naciff J, Robison S, Wu S (2011) Case studies to test: a framework for using structural, reactivity, metabolic and physicochemical similarity to evaluate the suitability of analogs for SAR-based toxicological assessments. Regul Toxicol Pharmacol 60:120–135

Cefic-LRI (2002) (Q)SARs for human health and the environment. In: Workshop on regulatory acceptance, Setubal, Portugal, 4–6 March. Full report

Chakravarti SK, Saiakhov RD, Klopman G (2012) Optimizing predictive performance of CASE ultra expert system models using the applicability domains of individual toxicity alerts. J Chem Inf Model 52(10):2609–2618

Cramer GM, Ford RA, Hall RL (1978) Estimation of toxic hazard—a decision tree approach. Fd Cosmet Toxicol 16:255–276

Cronin MTD, Jaworska JS, Walker JD, Comber MHI, Watts CD, Worth AP (2003) Use of quantitative structure activity relationships in international decision-making frameworks to predict health effects of chemical substances. Environ Health Perspect 111:1391–1401

de Raat K (2014) Assessment of read-across: an ECHA perspective. Presented at the World Congress for Animal Alternatives, 24–28th August 2014, Prague

Dimitrov S, Dimitrova G, Pavlov T, Dimitrova N, Patlewicz G, Niemala J, Mekenyan O (2005) A stepwise approach for defining the applicability domain of SAR and QSAR models. J Chem Inf Comput Sci 45(4):839–849

EC (2003) Directive 2003/15/EC of the European parliament and the council of 27 February 2003 amending council directive 76/768/ EEC on the approximation of the laws of the members states relating to cosmetic products. Official J Eur Union L66:26–35

EC (2006) Regulation (EC) No 1907/2006 of the European Parliament and of the Council of 18 December 2006 concerning the Registration, Evaluation, Authorisation and Restriction of Chemicals (REACH), establishing a European Chemicals Agency, amending Directive 1999/45/EC and repealing Council Regulation (EEC) No 793/93 and Commission Regulation (EC) No 1488/94 as well as Council Directive 76/769/EEC and Commission Directives 91/155/EEC, 93/67/EEC, 93/105/EC and 2000/21/EC. Official J Eur Union, L396/1 of 30.12.2006. Commission of the European Communities
EC (2009) Regulation (EC) No 1223/2009 of the European Parliament and the Council of 30 November 2009 on cosmetic products. Official J Eur Union L342:59–209
ECETOC (2009) Advanced technologies in read-across for chemical risk assessment. Technical Report No. 109. European Centre for Ecotoxicology and Toxicology of Chemicals, Brussels, Belgium
ECETOC (2012) ECETOC Technical Report No. 116: Category approaches, Read-across, (Q)SAR
ECHA (2008) Guidance on information requirements and chemical safety assessment. Chapter R.6. http://echa.europa.eu/documents/10162/13632/information_requirements_r6_en.pdf
ECHA (2012a) Practical Guide 5 How to report (Q)SARs Version 2. December 2012. http://echa.europa.eu/documents/10162/13655/pg_report_qsars_en.pdf
ECHA (2012b) Guidance on information requirements and chemical safety assessment. Chapter R.7a: Endpoint Specific Guidance. In: Guidance for the implementation of REACH. Version 2.0. November 2012. http://echa.europa.eu/documents/10162/13632/information_requirements_r7a_en.pdf
ECHA (2015) Read-Across Assessment Framework (RAAF). May 2015. http://echa.europa.eu/documents/10162/13628/raaf_en.pdf
Ellison CM, Enoch SJ, Cronin MTD, Madden JC, Judson P (2009) Definition of the applicability domains of knowledge-based predictive toxicology expert systems by using a structural fragment-based approach. Altern Lab Anim 37(5):533–545
Ellison CM, Sherhod R, Cronin MTD, Enoch SJ, Madden JC, Judson PN (2011) Assessment of methods to define the applicability domain of structural alert models. J Chem Inf Model 51(5):975–985
Enoch SJ, Schultz TW, Cronin MTD (2012) The definition of the applicability domain relevant to skin sensitization for the aromatic nucleophilic substitution mechanism. SAR QSAR Environ Res 23(7–8):649–663
Enoch SJ, Cronin MTD, Schultz TW (2013) The definition of the toxicologically relevant applicability domain for the SNAr reaction for substituted pyridines and pyrimidines. SAR QSAR Environ Res 24(5):385–392
Eriksson L, Jaworska J, Worth AP, Cronin MTD, McDowell RM, Gramatica P (2003) Methods for reliability and uncertainty assessment and for applicability evaluations of classification- and regression-based QSARs. Environ Health Perspect 111(10):1361–1375
Jaworska J, Dimitrov S, Nikolova N, Mekenyan O (2002) Probabilistic assessment of biodegradability based on metabolic pathways: catabol system. SAR QSAR Environ Res 13(2):307–323
Jaworska J, Nikolova-Jeliazkova N, Aldenberg T (2005) QSAR applicability domain estimation by projection of the training set in descriptor space: a review. Altern Lab Anim 33:445–459
Jennen D, Polman J, van Delft J, Kleinjans J, Montoya-Parra G, Kamp H, van Someren E, Stierum R, Kroese D, Patlewicz G (2014) Data-integration for endpoints, chemoinformatics and omics. Toxicol Lett 229:S4–S5
Mekenyan O, Patlewicz G, Kuseva C, Popova I, Mehmed A, Kotov S, Zhechev T, Pavlov T, Temelkov S, Roberts DW (2014) A mechanistic approach to modelling respiratory sensitization. Chem Res Toxicol 27(2):219–239
Naciff JM, DeAbrew N, Overmann G, Adams R, Carr G, Settivari R, Tiesman J, Edward C, Daston G (2013) Identification of mode-of-action specific toxicity transcript profiles *in vitro* using a connectivity mapping approach. The Toxicologist PS 549:117
Netzeva TI, Worth AP, Aldenberg T, Benigni R, Cronin MTD, Gramatica P, Jaworska JS, Kahn S, Klopman G, Marchant CA, Myatt G, Nikolova-Jeliazkova N, Patlewicz GY, Perkins R, Roberts DW, Schultz TW, Stanton DT, van de Sandt JJM, Tong W, Veith G, Yang C (2005) Current status of methods for defining the applicability domain of (quantitative) structure-activity rela-

tionships. The report and recommendations of ECVAM workshop 52. Altern Lab Anim 33(2):155–173
Nikolova-Jeliazkova N, Jaworska J (2005) An approach to determining applicability domains for QSAR group contribution models: an analysis of SRC KOWWIN. Altern Lab Anim 33(5):461–470
OECD (2004) ENV/JM/MONO/(2004)24. http://www.oecd.org/env/ehs/risk-assessment/37849783.pdf
OECD (2007a) Guidance document on the validation of (quantitative) structure-activity relationships [(Q)SAR] models. OECD Environment Health and Safety Publications. Series on Testing and Assessment No. 69. Organisation for Economic Co-operation and Development, Paris, France
OECD (2007b) Guidance on grouping of chemicals. OECD Environment Health and Safety Publications, Series on Testing and Assessment No. 80. Organisation for Economic Co-operation and Development, Paris, France
OECD (2011) OECD Series on Testing and Assessment Number 138. Report of the workshop on using mechanistic information in forming chemical categories. ENV/JM/MONO(2011) 8. Organisation for Economic Co-operation and Development. Paris, France
OECD (2012) The adverse outcome pathway for skin sensitisation initiated by covalent binding to proteins Part 1: Scientific evidence. Series on Testing and Assessment No. 168 ENV/JM/MONO(2012)10/PART1
OECD (2014a) Guidance on grouping of chemicals, 2nd edn. OECD Environment Health and Safety Publications, Series on Testing and Assessment No. 194. Organisation for Economic Co-operation and Development, Paris, France
OECD (2014b) How to use the Toolbox AOP workflow for skin sensitization. http://www.oecd.org/env/ehs/risk-assessment/Tutorial_1_How%20to%20use%20AOP%20for%20Skin%20sensitization_F_28012014.pdf
Patlewicz G, Dimitrov S, Low LK, Kern PS, Dimitrova GD, Comber MI, Aptula AO, Phillips RD, Niemelä J, Madsen C, Wedebye EB, Roberts DW, Bailey PT, Mekenyan OG (2007) TIMES--SS—a promising tool for the assessment of skin sensitization hazard. A characterization with respect to the OECD validation principles for (Q)SARs and an external evaluation for predictivity. Regul Toxicol Pharmacol 48:225–239
Patlewicz G, Jeliazkova N, Gallegos Saliner A, Worth AP (2008) Toxmatch—a new software tool to aid in the development and evaluation of chemically similar groups. SAR QSAR Environ Res 19(3–4):397–412
Patlewicz G, Chen MW, Bellin CA (2011) Non-testing approaches under REACH—help or hindrance? Perspectives from a practitioner within industry. SAR QSAR Environ Res 22(1–2):67–88
Patlewicz G, Ball N, Booth ED, Hulzebos E, Zvinavashee E, Hennes C (2013a) Use of category approaches, read-across and (Q)SAR: general considerations. Regul Toxicol Pharmacol 67(1):1–12
Patlewicz G, Roberts DW, Aptula A, Blackburn K, Hubesch B (2013b) Workshop: use of 'read-across' for chemical safety assessment under REACH. Regul Toxicol Pharmacol 65(2):226–228
Patlewicz G, Kuseva C, Mehmed A, Popova Y, Dimitrova G, Ellis G, Hunziker R, Kerne P, Low L, Ringeissen S, Robert DW, Mekenyan O (2014a) TIMES-SS—recent refinements as a result of an Industrial skin sensitisation consortium. SAR QSAR Environ Res 25(5):367–391
Patlewicz G, Kuseva C, Kesova A, Popova I, Zhechev T, Pavlov T, Roberts DW, Mekenyan OM (2014b) Towards AOP application—implementation of an integrated approach to testing and assessment (IATA) into a pipeline tool for skin sensitization. Regul Toxicol Pharmacol 69(3):529–545
Patlewicz G, Ball N, Becker RA, Booth ED, Cronin MT, Kroese D, Steup D, van Ravenzwaay B, Hartung T (2014c) Read-across approaches—misconceptions, promises and challenges ahead. ALTEX 31(4):387–396
Patlewicz G, Becker RA, Rowlands JC, Mekenyan OM (2014d) Enhancing non-testing approaches using the AOP framework: a case study in building scientific confidence. Presented at the Workshop on Mitochondrial Toxicity and Pathway-Based Chemical Safety Assessment, An

Inaugural Symposium of the Society of Toxicological Alternatives and Translational Toxicology, CSOT, 13–14 October 2014, Beijing, China

Patlewicz G, Ball N, Boogaard PJ, Becker RA, Hubesch B (2015a) Building scientific confidence in the development and evaluation of read-across. Regul Toxicol Pharmacol 72(1):117–133. doi:10.1016/j.yrtph.2015.03.015

Patlewicz G, Simon TW, Rowlands JC, Budinsky RA, Becker RA (2015b) Proposing a scientific confidence framework to help support the application of adverse outcome pathways for regulatory purposes. Regul Toxicol Pharmacol 71(3):463–477

Patlewicz G, Fitzpatrick JM (2016) Current and Future Perspectives on the Development, Evaluation and Application of *in Silico* Approaches for Predicting Toxicity. Chem Res Toxicol 29(4):438–451

Schultz TW, Yarbrough JW, Johnson EL (2005) Structure-activity relationships for reactivity of carbonyl compounds with glutathione. SAR QSAR Environ Res 16:313–322

Schultz TW, Yarbrough JW, Hunter RS, Aptula AO (2007) Verification of the structural alerts for Michael acceptors. Chem Res Toxicol 20(9):1359–1363

Schultz TW, Amcoff P, Berggren E, Gautier F, Klaric M, Knight DJ, Mahony C, Schwarz M, White A, Cronin MT (2015) A strategy for structuring and reporting a read-across prediction of toxicity. Regul Toxicol Pharmacol 72(3):586–601

van Ravenzwaay B, Herold M, Kamp H, Montoya G, Fabian E, Looser R, Krennrich G, Mellert W, Prokoudin A, Strauss V, Walk T, Wiemer J (2013) Metabolomics and REACH: quantitative biological activity relationships. Toxicol Lett 221:S27–S28

Verhaar HJM, van Leeuwen CJ, Hermens JLM (1992) Classifying environmental pollutants. 1. Structure-activity relationships for prediction of aquatic toxicity. Chemosphere 25:471–491

Worth AP, Bassan A, Gallegos A, Netzeva TI, Patlewicz G, Pavan M, Tsakovska I, Vracko M (2005) The characterisation of (quantitative) structure-activity relationships: preliminary guidance. JRC report EUR 21866 EN. http://ihcp.jrc.ec.europa.eu/our_labs/predictive_toxicology/doc/QSAR_characterisation_EUR_21866_EN.pdf

Worth A, Lapenna S, Lo Piparo E, Mostrag-Szlichtyng A, Serafimova R (2011) A framework for assessing *in silico* toxicity predictions: case studies with selected pesticides. JRC report EUR 24705 EN. http://ihcp.jrc.ec.europa.eu/our_labs/predictive_toxicology/doc/EUR_24705_EN.pdf

Wu S, Blackburn K, Amburgey J, Jaworska J, Federle T (2010) A framework for using structural, reactivity, metabolic and physicochemical similarity to evaluate the suitability of analogs for SAR-based toxicological assessments. Regul Toxicol Pharmacol 56:67–81

Chapter 7
Implementation of New Test Methods into Practical Testing

Rodger D. Curren, Albrecht Poth and Hans A. Raabe

Abstract New toxicology test methods, especially those using *in vitro* methods, are continually being developed. Some are used by industry for screening purposes; others are eventually validated for regulatory use. However, for a new test method to be firmly adopted by industry it must be readily available, generally through an in-house industry laboratory, an academic laboratory, or a contract research organization. Regardless of the type of laboratory which intends to implement the test method, certain steps must be taken to ascertain that the method that is put into place is reproducible and performs identically to the test method that was published or has undergone validation. This involves developing protocols and standard operating procedures, training staff, developing historic positive and negative control data, establishing acceptable performance with proficiency chemicals, and addressing all the safety concerns that may accompany the assay. From experience within a contract research laboratory, we provide guidance on how to most efficiently accomplish these tasks.

Keywords *In vitro* • Test methods • Toxicology testing • GLPs • OECD Test Guidelines

1 Introduction

New toxicological test methods are constantly becoming available to scientists through word of mouth, poster presentations during scientific meetings, peer reviewed publications, regulatory agency announcements, national "VAMs" (e.g.

R.D. Curren (✉) • H.A. Raabe
Institute for *In Vitro* Sciences, Inc., 30 W. Watkins Mill Rd., Suite 100, Gaithersburg, MD 20878, USA
e-mail: rcurren@iivs.org

A. Poth
Eurofins BioPharma Product Testing, Munich, Germany

C. Eskes, M. Whelan (eds.), *Validation of Alternative Methods for Toxicity Testing*, Advances in Experimental Medicine and Biology 856,
DOI 10.1007/978-3-319-33826-2_7

the European Union Reference Laboratory for Alternatives to Animal Testing [EURL-ECVAM], the United States Interagency Coordinating Committee on the Validation of Alternative Methods [ICCVAM], the Japanese Center for the Validation of Alternative Methods [JaCVAM] , etc.), international authorities, e.g. Organization for Economic Cooperation and Development [OECD], and even YouTube (e.g. An *in vitro* cell-based test for skin sensitization prediction—https://www.youtube.com/watch?v=9eXgHwUtjpU). However, like any new product, some will prove to be very useful—and hence be quickly adopted by industry and academia, e.g. the Ames test (for review see (Mortelmans and Zeiger 2000))—while others, for various reasons, will languish, e.g. the cultured bovine lens assay for eye irritation (Sivak et al. 1994).

Several factors influence the acceptance of a new test method: its stage of development, what bodies endorse it, its purpose (screening versus regulatory submission), its performance (predictive capacity and reproducibility), its cost compared with similar methods, and its ease of performance (or general availability), among others. Clearly if the assay has been requested by a regulatory agency, it will quickly be adopted by those sections of industry that interact with that regulatory branch. On the other hand if the assay is just to be used as a screening assay (providing a yes/no answer as to some measure of toxicity) during product development its uptake by industry will be subject to a number of the other factors (not the least of which is performance) described above.

However a major part of any decision of when—or even if—to use a new toxicology test method is whether or not the test method is readily available. For initial adoption a company may be willing to conduct the test in its own laboratories, but for the longer term most companies prefer to have their testing conducted by an independent, third-party organization often referred to as a contract research organization (CRO). CRO's provide an important testing infrastructure that may be missing, or expensive to maintain, within an industrial setting where the primary activity is manufacturing a product. This infrastructure, as it relates to toxicology testing, consists of: (1) maintaining an extensive database of past control (positive and vehicle) values for the test, (2) technicians trained and routinely familiar with the test, (3) a standard protocol (often GLP-compliant), (4) specialized equipment necessary for the test, (5) experienced Study Directors competent to take responsibility for all phases of the study, and (6) a quality assurance staff prepared to audit, when required, selected phases of the study. Finally the CRO is generally seen as an independent party by the regulatory agencies and industry itself which strengthens their trust that the results of the study are unbiased. Since CROs generally stand at the forefront of implementing an often broadly defined new test method by developing it into a method that can be used confidently and economically on a routine basis, many of the following examples of implementation will be taken from the authors' own experiences in the CRO industry (for examples see (Norman 2014)).

2 Timeline for Implementation

There is generally considerable cost involved in implementing a new test method—even within an existing laboratory setting. This is a consequence of the time (and resulting labor costs) involved to address all of the necessary requirements of implementation. This often makes the decision on whether to implement a new assay somewhat difficult since the process could take anywhere from 2 to 6 months or even more in the case of more complicated assays! One of the first questions asked before the practical implementation of a new toxicology assay is whether there is actually a commercial or technical need for the assay. Numerous assays are developed and publicized each year that may have value to the academic research community, but have little interest for industry. Safety and efficacy studies are of paramount importance to industry, but even then the new test has to offer information that is either superior to an original test, or has been previously unavailable. Without a "pull" from industry that indicates a need, it would not be a sound business decision for a CRO to spend the significant amount of money normally required to develop a test method to the point where it is ready for commercial sale. Of course, if the test is already being considered for regulatory use—such as an OECD Guideline (Organisation for Economic Co-operation and Development 2015a) test method—then it could be advantageous to bring the test to a commercial offering in anticipation of a growing industry demand. Of importance for any commercial supplier of a test method is participation in prevalidation (Curren et al. 1995) or validation studies for that test method. This provides not only early experience with the test method but also an understanding of the level of interest from the regulatory community.

The potential regulatory use of any new assay is clearly an important aspect of any test method implementation decision; however lack of a foreseeable regulatory application should not automatically doom its implementation. Many new test methods are developed to measure efficacy endpoints rather than safety endpoints. Such methods, depending on their relevance and reproducibility, can be extremely useful in product development, and thus may be worthy of the expense of appropriate implementation even though their results may never be submitted to a regulatory agency.

3 Technical Implementation of a New Test Method

With any new toxicology method, it is rare—if ever—that the currently available information about the technical conduct of the test is sufficient to allow it to be used in a routine commercial or even an industrial setting. Personnel must be trained, reagents obtained, the target tissue or cells sourced, standard protocols and SOPs developed, historic control values (positive and negative [vehicle]) established, and evidence of proficiency with standard chemicals or products developed. This list (described in more detail later) is far from comprehensive. Even if the test method has gone through review at a high level, such as in the OECD Test Guidelines program (Organisation for Economic Co-operation and Development 2015a), there is generally not sufficient information available to begin conducting the test and immediately have confidence that the results obtained are actually predictive of a toxicological endpoint. OECD Test Guidelines really just give—as their name implies—guidance on the overall conduct of the test with certain critical test parameters described. OECD Test Guidelines are not working protocols—and are likely not compliant by themselves with Good Laboratory Practices (GLP) guidelines, although efforts have been initiated to improve harmonization of the OECD Test Guideline and GLP Programs. They provide a good starting point, and they are generally more useful to a laboratory wanting to implement a new assay than what is provided in the materials and methods section of most published new methods! Some additional help in utilizing a new test method can be found in OECD Guidance Documents (Organisation for Economic Co-operation and Development 2015b) which may provide additional points to consider, supporting documentation, additional approaches using the basic test method, thoughts on data interpretation, etc., about a new test method or related test methods. Other useful sources of test method information are: the EURL-ECVAM DataBase on ALternative Methods (DB-ALM) (https://eurl-ecvam.jrc.ec.europa.eu/databases/database-on-alternative-methods-db-alm), the Current Protocols series by John Wiley & Sons, Inc. (http://www.currentprotocols.com), and instructional videos which are often found on manufacturers' or test suppliers' sites.

(a) Training of personnel

No matter how well a test developer believes that they have described a new test method, experience has shown that without face-to-face, hands-on instruction, it is rare that new personnel can routinely obtain the same results as the test developer. There are always small, but important, steps that seem to be left out of written descriptions (generally inadvertently, sometimes intentionally), or if described are not correctly understood because of language and translation difficulties. Although the use of videos and teleconferencing over the last few years have made test method transfer more efficient than it was in the past, face-to-face instruction still seems to be preferable, especially if commercial use of the assay is being considered. We have found that most test developers are quite willing to find a suitable time to host visitors desiring to learn all the fine details of an assay. To make the process most efficient, personnel from the naïve laboratory should already be proficient in the general technical skills required for the assay, should have thoroughly read the published literature, and, for the best transfer efficiency, should have already conducted trial runs with the assay in their own laboratories. That way they will already have developed questions about unclear sections of the protocol, and will be able to describe how they interpreted and performed those unclear sections. It will be even better if they have conducted trials on several standard materials which can then be incorporated into the training program. This allows a more direct comparison of techniques and subsequent results between developer and routine user.

Technical personnel must also become familiar with how the raw data from various parts of the assay will be used in the final interpretation of the assay. In other words they must learn during the training about the specific "prediction model" (Bruner et al. 1996) used; exactly what the mathematical algorithm or written decision tree describes. The reason for this is that the level of precision (and documentation) for the conduct of various steps within the protocol may vary significantly according the construction of the prediction model. If the purpose of the method is just for identifying one end of a toxicity scale, e.g. a test material is non-toxic if a color change occurs within 1 min of the addition of a certain reagent, then it is extremely important to continually observe and make time records at close intervals at either side of the 1 min point, but less important to make recordings at close intervals as the distance increases past the critical time point. Similarly if the prediction model is based on subtle color changes in a reaction mixture, then technical staff must be trained to understand what ambient lighting conditions are acceptable for reading of the endpoint, e.g. readings might not be reliable if made inside a hood outfitted with the yellow lights commonly used to protect against the formation of toxic photoproducts during tissue culture operations.

(b) Specialized reagents

A common problem associated with moving a new testing method into commonplace, international usage, is that critical reagents may differ slightly—or even not be available—in different areas of the world. This adoption hurdle can result from many conditions, including shipping difficulties, cultural preferences, legal restrictions, and climate differences among many others. Specific examples might include: the requirement for serum factors that may be extremely labile, the need for animal organs discarded during food production which are not locally obtainable because local dietary customs do not include food from that type of animal, laws prohibiting the importation of human tissue constructs into certain countries, and the costly insulated packing necessary to protect against temperature extremes during shipping may make the purchase of certain reagents economically difficult.

New test method developers should try to consider scenarios such as those described above as they finalize the "design" of their new method. Of course, it is impossible to foresee all the difficulties that may arise as a test method begins to spread out internationally, but some foresight should certainly be given if routine use of the method around the world is expected. This is another area where CRO's are often able to assist, having surmounted similar hurdles in the past. They can likely make suggestions, such as: (a) can an isolation procedure be described to allow laboratories to isolate labile serum factors on-site (or can more stable factors be bioengineered and supplied directly as a reagent), (b) can new organ culture chambers be designed that would accommodate water buffalo corneas rather than corneas from beef cattle, (c) if national laws won't allow the importation of human tissue constructs, can a more general protocol be developed which would allow tissues manufactured within country to be used in the new test method, and (d) is it necessary to always use the temperature sensitive reagent with which the method was developed, or could a similar reagent without such concerns be substituted if other areas of the protocol were modified?

(c) Dealing with commercial sources of 3D tissue (when required by the test method)

A major improvement in test methodology over the last 25 years has been the validation, regulatory acceptance and increase in the use of 3D reconstructed human tissue to study chemical toxicity (see also Chap. 4, this volume). This is perhaps the ultimate example of a "specialized reagent", and the manufacture, shipping, and maintenance of the tissue must be standardized in order for test results to be reproducible. This becomes a challenge whose solution must be divided between the tissue manufacturer and the user. The tissue manufacturer must not only develop excellent manufacturing techniques and quality control procedures so that each batch of tissue is morphologically and functionally similar, but they must develop shipping procedures that maintain its reproducibility. Furthermore the establishment of a claim of reproducibility must be based on observations by the testing facility as well as the manufacturer. Quantitative performance standards should be established by the test manufacturer, and the resulting data from each batch of tissue should be available to the testing facility. The optimum situation is to have a very similar set of standards, e.g. identical chemical tested under identical conditions, which should be used by both the testing facility and the tissue manufacturer to facilitate data comparison between the two facilities. This will also facilitate troubleshooting the cause of failed experiments, since the two facilities can work together to determine if the tissue was manufactured correctly or if the shipping conditions were inadequate. See additional information about potential audits of tissue manufacturing facilities under "Developing GLP versions of the assay" below.

(d) Establishing historic control data

Before the data from any new method can be properly interpreted, the reproducibility of the method must be clearly established. The operator must know how the system responds normally (without the presence of a test chemical); this is generally called the assay negative control value. The effect of various solvents on the system should also be determined to create an historic assay solvent control value. At the same time the performance of the system with a chemical known to perturb the system must be ascertained; this is generally termed the assay positive control. Generally multiple runs (20 is often considered an adequate number (Hayashi et al. 2011)) are conducted and the mean and standard deviation calculated. Control charts, long used in monitoring performance in clinical and analytical laboratories (see, for example (Karkalousos and Evangelopoulos 2011) can be generated to monitor trends in the assay control values, and decision criteria can be established, e.g. the assay positive control must induce a response within two standard deviations of the historic mean, to determine the acceptability of any single assay run. Although establishing historic assay control data has been routine for many *in vitro* testing facilities, regrettably it has been ignored in others. In fact it is only recently that this has been addressed in OECD Test Guidelines. For example, TG431: *In Vitro* Skin Corrosion: Reconstructed Human Epidermis (Rhe) Test Method states in paragraph 20 that "Test method users should demonstrate reproducibility of the test methods over time with the positive and negative controls" and in paragraph 23 that "Concurrent negative and positive controls (PC) should be used in each run to demonstrate that viability (with negative controls), barrier function and resulting tissue sensitivity (with the PC) of the tissues are within a defined historical acceptance range" (Organisation for Economic Co-operation and Development 2014).

(e) Testing of proficiency chemicals

Data from numerous test chemicals are always generated by the test developer(s) during the development of any new assay. As the test method is transferred to new laboratories, a subset of these test chemicals should be identified which can serve as proficiency chemicals for any new laboratory adopting the method. The number of such proficiency chemicals may vary (recent OECD Test Guidelines have specified between 8 and 15), but the set should contain "negative" chemicals (those that cause little change in the test system or which cause changes below some cutoff value) and "positive" chemicals (those which cause changes above some predetermined cutoff value). The chemical set should also have representatives, as far as possible, of the chemical classes which make up the applicability domain of the method. It is preferable to have quantitative data for each of the proficiency chemicals rather than just "positive" or "negative" labels.

Each laboratory adopting the test method should then be able to show that they are able to reproduce the data for the proficiency chemicals before they conduct the test with any unknown materials. If they are not able to reproduce the values, then they should contact the test developer or another laboratory competent in conducting the assay to trouble shoot their methodology. If a CRO claims to have established the method, they should be willing to provide the results of their tests with the proficiency chemicals to any prospective client.

(f) Approval process for first commercial use of the assay

If the new test method is to be offered commercially, a CRO should have a standard approval process that precedes the commercial offering of the assay. This should include, at a minimum, many of the steps described above, such as: (1) developing a comprehension set of control data, both negative and positive, for the method, (2) showing proficiency with the assay by replicating data for a set of defined proficiency chemicals, or developing data for a set of representative chemicals when a formal chemical list is absent, (3) having sufficient number of trained laboratory staff, and (4) having established protocols and Standard Operating Procedures (SOPs) (see below).

(g) Investments needed for test set-up

When a potential user estimates the cost of setting up a new test method they should make sure that costs associated with all of the previously mentioned requirements are included in the calculation. That includes not only the cost of new equipment, reagents, and supplies, but also the often considerable cost of personnel training (which could include expenses for international travel) and the labor costs involved in establishing proficiency and historic control values. As was stated previously the time required to properly implement the test method should be carefully considered since it can take anywhere from 2 to 6 months (or in some cases even longer).

(h) Safety considerations

Before establishing any new test method in the laboratory, the potential safety concerns associated with the assay should be considered. Although there may be adequate safety standards in place to cover existing hazards in the laboratory, the new assay may require reagents or techniques which are new to the current staff. For example, in the case where radioisotopes are needed, the new test method may require different isotopes, or quantities of isotopes, than are currently covered by the laboratory's license. Similarly there may be new chemical disposal procedures that need to be arranged or additional personal protection devices (e.g. impervious gloves or full face respirators) that will need to be procured. In addition there might be changes in the techniques for treating the test system that put the technician at higher risk for unwanted exposures. For example, additional containment procedures may be required when switching from simple direct application of dosing solutions to the application of light powders or sprays to the test system. Of course the costs associated with implementing these new standards will need to be included in the analysis of the startup costs, as well as the continuing costs, of performing the assay.

In addition to the potential new chemical hazards addressed above, new biological safety hazards may accompany a new test method, especially since many new *in vitro* methods use reagents and biologicals derived from animals, as well as human cells or tissues as the reporter system. With the use of human cells and tissues comes the risk of exposure to a number of virus including Hepatitis B, Hepatitis C, Human Immunodeficiency virus, Epstein-Barr virus and several others. Therefore human cell lines should be handled using BSL-2 safety precautions and the U.S. Centers for Disease Control's "universal precautions" (Centers for Disease Control 1988, June 24) should be followed. Primary human cells obtained directly from donors should be screened for pathogens before use. Purchasers of commercially available human 3D reconstructed tissue models should request a screening report for human pathogens (and other common adventitious agents) directly from the manufacturer of the tissues. Further information on safety precautions for the use of human cells and tissues can be found in many textbooks on tissue culture methods, e.g. (Freshney 2016). In addition, it is prudent to consider providing an immunization program for laboratory staff, especially against Hepatitis B.

Whereas the handling of bovine tissues from slaughterhouse operations might be considered a relatively safe activity, we simply need to recall the impact of the outbreak of bovine spongiform encephalopathy in cattle in England in the 1990s (European Food Safety Authority 2012) that restricted the use of most bovine-derived biologicals or tissues within the European community. This event adversely affected the marketing and sales of European derived bovine serum medium supplements, as well as delayed the adoption of a new test method—the Bovine Corneal Opacity and Permeability (BCOP) assay—within the European Community. The event also changed the way that these bovine tissues are routinely handled with the laboratory; often requiring the handling of these tissues using BSL-2 safety precautions.

However, just putting adequate health and safety policies into place does not entirely assure the safety of the laboratory staff. Thorough training is mandatory, and many laboratories require a yearly refresher training course covering the hazards that currently exist in the laboratory. It is all too easy to become complacent with safety procedures, especially when running a "routine" test method.

4 Developing GLP Versions of the Test Method

Throughout the process of assay development, optimization and validation, considerable efforts are applied to ensuring that the assay is technically robust, relevant for purpose, and readily portable or transferable, especially to laboratories that may be participating in a formal prevalidation or validation study. These efforts are targeted at optimizing the test methods to maximize their predictive power, and to making specifications clear and unambiguous so that there is less likelihood for variations in technical interpretations which could adversely affect the reproducibility of the assay. The refinements generally address the technical execution of procedures and processes, and controlling test system conditions (equipment parameters and environment). Furthermore, anticipated downstream regulatory requirements and GLP compliance expectations influence the refinement of assay SOPs and protocols towards clearer definition of specifications and more thorough and transparent capture and documentation of test data. Indeed a major goal of the formal validation of assays for regulatory purposes should be the refinement of the protocols and the documentation tools to support real-world regulatory compliance requirements. To this end, validation studies that were conducted in full GLP compliance are more likely to have been optimized to meet these requirements. However, many multi-site validation studies are not conducted in full GLP compliance, and many of the participating laboratories may not be experienced in the real-world GLP compliance issues such as those routinely encountered by industry laboratories and CROs. Thus, significant GLP compliance gaps may readily persist in validated test method SOPs and protocols. Therefore, each end user must take on the responsibility to develop GLP-compliant versions of the validated test method protocols to fit their organization's regulatory framework and compliance programs.

(a) Refining protocols to meet international GLP standards

It is evident that much of the refinement of assay protocols towards GLP compliance involves clarification of technical specifications. For example, when defining the delivery of a dose volume to the test system, specifying the acceptable range for the precision of the delivered dose provides more explicit information than simply presenting the target dose volume (e.g., specifying a dose of 100±1 μL is more likely to improve reproducibility than simply specifying a dose of 100 μL). Hopefully details such as those given in the example have already been implemented in the robust test method SOP. However, the challenge in implementing a truly GLP compliant program lies in the approaches taken to assure that the 100±1 μL dose delivery was achieved. There may be several ways to achieve this goal, and depending upon budgets, criticality of the specification, and available practical solutions, each organization must find "acceptable" means to this end. Whereas one laboratory may consider the annual calibration certificate for the 100 μL micropipette to be sufficient, the next laboratory may require more direct evidence such as verifying each delivered dose gravimetrically. What is ultimately required for GLP compliance is a defined approach to ensuring the stated specifications are met. The specific approaches may be added to the test method protocol, or may be presented in referenced equipment and procedural SOPs.

In addition to the technical refinements, protocols must also address the relevant requirements of the specific regulatory agencies intended for submission. Since there is not universal agreement in all of the GLP elements among all of the GLP compliance systems (i.e., US FDA, US EPA, OECD, Japanese Ministry of Health, etc.), the protocol must be modified to meet the specific regulatory requirements. There may be country or region-specific requirements for protocol format and approval processes. For example, whereas the Testing Facility Study Director is the only one required to ratify a pre-clinical study protocol for US FDA GLP compliance, the Japan Ministry of Health requires approval by Test Facility management personnel as well. There may also be differences in the specific data analyses or interpretation of test results that must be defined. Although a test method may be scientifically validated for predicting a toxicological outcome, each regulatory agency may differ in the acceptance of the test results as "standalone" results, and some agencies may only recognize positive predictions. How the data are to be analyzed, and how the resultant predictions are to be presented for classification and labeling purposes or hazard communication purposes should be clearly defined in the GLP compliant protocol.

Lastly, most regulatory agencies have specific requirements for the characterization of the test chemical or formulation, as well as the evaluation of test chemical stability under the test conditions to support the submission for regulatory review and approval. These regulatory requirements and the methods envisioned to meeting them must be added to the validated test method protocols to comply with most GLP preclinical testing requirements since these elements would rarely be included as part of the test method validation exercise. In fact, protocols prepared for test method validation are prepared using blind-coded chemicals, and thus would not be expected to include requirements for test chemical characterization, or evaluation of test chemical stability under the test method conditions. Even if test chemical characterizations were conducted on the set of chemicals used in the validation, neither the characterization requirements nor the characterization results would have been included in the test method protocol.

(b) Interactions with Quality Assurance personnel.

It is extremely important to engage the organization's quality assurance and regulatory compliance personnel early in the adoption of any technology. There are many aspects of implementing test methods where the support and approval of the quality assurance and regulatory compliance personnel would benefit the process. First of all, the quality assurance unit (QAU) can provide considerable help in ensuring that the protocol and supporting SOPs and documentation address the myriad of GLP compliance requirements. They can also provide an objective evaluation of the methods proposed for ensuring technical and regulatory compliance long before any laboratory activities are initiated. The QAU can also provide guidance on the installation and operation of any equipment needed to support the test method, as well as ensuring that GLP-compliant SOPs and supporting use, maintenance, and calibration records are developed. Their independent evaluations provide a broad perspective on the range of compliance requirements often overlooked when attempting to address solely the test method compliance requirements.

Similarly personnel from each company's regulatory assurance and corporate compliance program should be engaged during the establishment of the test method to assure that the test methods applied, and the data analyses conducted, fit within the requirements of the relevant regulatory authorities. These same personnel can provide guidance on the preparation and submission of study reports and executive summaries for the specific authorities.

In a robust GLP compliance program, the QAU routinely audits study related documents at the initiation of a study, during the in-life phase, and through to completion of the final report. Typically, upon initiation of the study, the protocol and study authorization and placement letters are audited, test material receipt and disposition documents are audited and the master schedule is updated to reflect the specific regulatory requirements. In-life phases are selected for random auditing during the execution of the study, and the raw data and analyses, batch records, notebooks, and test material usage and disposition documents are audited after the in-life phase is completed. Finally, the study reports are audited to assure that they reflect the content and findings of the raw data files. Early in the adoption of any test method, the QAU must work closely with the Study Director and other laboratory operations personnel to identify those procedures and activities that have the greatest impact upon the execution and outcome of the study. Candid discussions of how variances in procedures or specifications can affect the outcome of the study results will help identify key steps for auditing, as well as provide a consensus for evaluating the impacts when inevitable deviations from the study protocol occur.

As mentioned earlier, each laboratory should validate its test method implementation by testing of a set of proficiency chemicals and establishing criteria for evaluating the implementation. The authors recommend that the testing of the proficiency chemicals be conducted either in full GLP compliance or at least under non-GLP conditions that still allow for the QAU audit functions to be exercised. This may indeed be the first opportunity to evaluate the integration of the QAU into auditing the novel technologies and correct any variances that adversely affect GLP compliance. Once the test method has been fully adopted the Study Director and QAU work closely together to develop a systems approach to support the efficient execution and auditing of the validated studies. This includes developing and monitoring the performance of the test method historically by collecting and analyzing assay control data, developing "normal" historical assay control ranges, and establishing test method acceptance criteria.

(c) GLP training of technicians

Basic technician GLP training programs focus upon ensuring that the laboratory staff have received training in, and have demonstrated proficiency at, documenting their conduct of the test method and properly recording data. A training file documenting the subject areas covered should be kept for each technician participating in work with the test method. Periodic (generally yearly) retraining should be conducted and recorded in the training files.

(d) Interactions with, and auditing of, suppliers

In a GLP-compliant study the Study Director is responsible for assuring the suitability of reagents used in the study. With standard, commonly used chemicals produced by major suppliers this assurance is usually not difficult. Current Certificates of Analysis and other technical literature dealing with the reagent are generally sufficient to support their use. However many new test methods now utilize commercially available human tissue constructs, or very unique chemicals, whose properties can be quite complex and difficult to completely standardize. In such cases it may be appropriate to ascertain vendor qualification by conducting periodic GLP audits of the supplier. This will allow first hand observations of their Quality Control procedures, technician training, internal documentation of manufacturing processes, and certain critical stages of the product production. It should not be expected, however, that the manufacturer will reveal all aspects of their manufacturing process. Certain stages of production, or specific ingredients, may be proprietary and hence not revealed during an audit. This should not invalidate the audit as long as it can be shown that the process is controlled, that proper records are kept, and that the results of lot release testing are available to the user.

5 Communicating Assay Results

(a) Designing report format

The design of the report format is dependent on whether the study is performed in line with the GLP requirements or not. If the study has been conducted only for screening purpose and carried out without the GLP requirements, a letter report consisting of one or two pages is sufficient. This will include a short description of the method, the presentation of the results in tabular form and the conclusion(s) of the study.

In case the studies are carried out under GLP conditions a standard report format is used, including the necessary GLP requirements for a study report. Most of the CRO's have their standard report formats, which are used for new *in vitro* methods as a template. The standard report format can vary to a certain extent from one CRO to another; however the content is nearly the same. A GLP standard report format includes a front page indicating the Study Director, the test facility, and the sponsor including the study monitor. It also includes the assigned study number and the sponsor's study number if available. The second page includes the table of contents. The next section is reserved for the Study Director's statement of GLP compliance and the quality assurance statement. This is followed by a summary page, where the results are described in brief and a conclusion on the study results is given. The next section can include general information on the schedule of the study, any deviation from the study plan and the archiving information according to the GLP requirements. An introduction section will follow including the aims of the study and any guidelines and regulations relevant for the test system. Then a material and methods section will include the description of the test system, the test item preparation, the controls (positive and negative controls where relevant) and the experimental performance together with the endpoint measurement and determination. Included in the material and methods is the recording of the data and the evaluation of the results together with the acceptance criteria and historical data. The next section is related to the results, their discussion and the final conclusions. References related to the test system and cited in the report are listed next. The individual results can be included in the main part of the final report, but generally these data are included in an appendix to the final report. Appendices can also include other relevant information on the test system (e.g. Certificates of Analysis). It is recommended to use International System of Units (SI) in the final report. The final report has to be signed and dated by the Study Director and sometimes also the signature of the management is included.

It should be noted that some regulatory agencies require that study reports are submitted to them in a very specific format, and that reports not complying may be rejected. An example is the US Environmental Protection Agency's Office of Chemical Safety and Pollution Prevention which prescribes the format for submission of results from studies addressing its Health Effects Test Guidelines (2011, November 30).

(b) Understanding needs of individual clients

Before starting testing with a new *in vitro* method it is important to know for what purpose this method is used. Therefore, close interaction with the client—be that an internal laboratory down the hall or an industrial firm halfway around the globe—is recommended from the beginning. It should be clarified whether the new method is to be used for screening purposes, for weight-of-evidence evaluation, for read-across, or for waiving arguments. It should also include clarification of whether the new test method is used for classification and labelling, or even for regulatory submission. Such a new *in vitro* test once validated can be a stand-alone method regarded as a full replacement of an *in vivo* test method or it can be considered to be used as part of an integrated test strategy. Based on the needs, the interaction with the client will be more or less intensive. Discussions with the client may include whether benchmark chemicals should be used as part of the testing. If the new *in vitro* test is used within a weight-of-evidence assessment, the discussion will be more intensive as an hypothesis for the occurrence of a certain effect has to be elaborated, for which the new *in vitro* test will provide information on whether or not that hypothesis can be confirmed.

(c) Determining appropriate language for Conclusions

In the discussion section the Study Director interprets the data in terms of any patterns that were observed, any relationships among experimental variables that are important and any correlations between variables that are discernible. In this section all relevant and important assay results should be described and discussed. Major differences or trends are important and need to be explained. Statistics, when appropriate and historical control data are included in the data evaluation. The acceptance criteria of the assay should be included and discussed, as well as the negative and positive control data. The conclusions section should use some elements from the introduction and their structure should be similar. The conclusions should be written in a clear and straightforward language. The conclusion section should simply state what the Study Director believes the data mean and, as such, should relate directly back to the problem/question stated in the introduction section.

6 Conclusions

Although it might seem that when a new test method has successfully passed a "validation" stage, or has been accepted for use by a regulatory agency, that there are no further barriers to the general use and acceptance of the assay by industry or academia. This is clearly far from the truth. Considerable work is still ahead to make the assay easily available to all potential users, from very small businesses to major international corporations. Decisions have to be made by potential conductors of the test method concerning its economic feasibility, what new hazards it might represent, what new techniques might have to be learned, how to demonstrate competence, whether to implement with full GLP compliance, how to train staff, and how to communicate the results. However there is some standard guidance that has been presented in the previous pages that can make this formidable process manageable, and will result in further steps forward for the new toxicology.

Acknowledgments The authors would like to thank Dr. Gertrude-Emilia Costin of IIVS for her careful review of the manuscript and her constructive comments.

References

Bruner L, Carr G, Chamberlain M, Curren R (1996) No prediction model, no validation study. Altern Lab Anim 24:139–142

Centers for Disease Control (1988) Perspectives in disease prevention and health promotion update: Universal precautions for prevention of transmission of human immunodeficiency virus, Hepatitis B virus, and other bloodborne pathogens in health-care settings. In: MMWR: Morb Mortal Wkly Rep. http://www.cdc.gov/mmwr/preview/mmwrhtml/00000039.htm. Accessed 7 Jan 2016

Curren R, Southee J, Spielmann H, Liebsch M, Fentem J, Balls M (1995) The role of prevalidation in the development, validation and acceptance of alternative methods. Altern Lab Anim 23:211–217

European Food Safety Authority (2012) Successful EU response to BSE. http://www.efsa.europa.eu/en/press/news/120130f.htm. Accessed 7 Jan 2016

Freshney RI (2016) Culture of animal cells: a manual of basic technique and specialized applications. Wiley, New York

Hayashi M, Dearfield K, Kasper F, Lovell D, Martus H, Thybauld V (2011) Compilation and use of genetic toxicology historical control data. Mutat Res 723:87–90

Karkalousos P, Evangelopoulos A (2011) Quality control in clinical laboratories. In: Ivanov O (ed) Applications and experiences in quality control. Intech, Rijeka

Mortelmans K, Zeiger E (2000) The Ames Salmonella/microsome mutagenicity assay. Mutat Res 455(1–2):29–60

Norman K (2014) Cosmetic safety assessments in the 21st century. Specialty Chemicals Magazine. p 22–23

Organisation for Economic Co-operation and Development (2014) *In vitro* skin corrosion: reconstructed human epidermis (Rhe) test method. http://www.oecd-ilibrary.org/environment/test-no-431-*in-vitro*-skin-corrosion-reconstructed-human-epidermis-rhe-test-method_9789264224193-en. Accessed 7 Jan 2016

Organisation for Economic Co-operation and Development (2015a) OECD guidelines for the testing of chemicals. http://www.oecd.org/env/ehs/testing/oecdguidelinesforthetestingofchemicals.htm. Accessed 7 Jan 2016

Organisation for Economic Co-operation and Development (2015b) Series on testing and assessment: testing for human health. http://www.oecd.org/env/ehs/testing/seriesontestingandassessmentadoptedguidanceandreviewdocuments.htm. Accessed 7 Jan 2016

Sivak JG, Herbert KL, Segal L (1994) Ocular lens organ culture as a measure of ocular toxicity: the effect of surfactants. Toxicol Meth 4:56–65

U.S. Environmental Protection Agency (2011) Pesticide Registration (PR) Notice 2011-3, Standard Format for Data Submitted Under the Federal Insecticide, Fungicide, and Rodenticide Act (FIFRA) and Certain Provisions of the Federal Food, Drug, and Cosmetic Act (FFDCA). http://www.epa.gov/sites/production/files/2014-04/documents/pr2011-3.pdf. Accessed 7 Jan 2016

Chapter 8
Pathway Based Toxicology and Fit-for-Purpose Assays

Rebecca A. Clewell, Patrick D. McMullen, Yeyejide Adeleye, Paul L. Carmichael and Melvin E. Andersen

Abstract The field of toxicity testing for non-pharmaceutical chemicals is in flux with multiple initiatives in North America and the EU to move away from animal testing to mode-of-action based *in vitro* assays. In this arena, there are still obstacles to overcome, such as developing appropriate cellular assays, creating pathway-based dose-response models and refining *in vitro-in vivo* extrapolation (IVIVE) tools. Overall, it is necessary to provide assurances that these new approaches are adequately protective of human and ecological health. Another major challenge for individual scientists and regulatory agencies is developing a cultural willingness to shed old biases developed around animal tests and become more comfortable with mode-of-action based assays in human cells. At present, most initiatives focus on developing *in vitro* alternatives and assessing how well these alternative methods reproduce past results related to predicting organism level toxicity in intact animals. The path forward requires looking beyond benchmarking against high dose animal studies. We need to develop targeted cellular assays, new cell biology-based extrapolation models for assessing regions of safety for chemical exposures in human populations, and mode-of-action-based approaches which are constructed on an understanding of human biology. Furthermore, it is essential that assay developers have the flexibility to 'validate' against the most appropriate mode-of-action data rather than against apical endpoints in high dose animal studies. This chapter demonstrates the principles of fit-for-purpose assay development using pathway-targeted case studies. The projects include p53-mdm2-mediated DNA-repair, estrogen receptor-mediated cell proliferation and PPARα receptor-mediated liver responses.

R.A. Clewell (✉) • P.D. McMullen • M.E. Andersen
ScitoVation, 6 Davis Drive, PO Box 110566, Research Triangle Park, NC 27709, USA
e-mail: rclewell@scitovation.com

Y. Adeleye • P.L. Carmichael
Unilever Safety and Environmental Assurance Centre,
Colworth Science Park, Sharnbrook, Bedfordshire, UK

C. Eskes, M. Whelan (eds.), *Validation of Alternative Methods for Toxicity Testing*,
Advances in Experimental Medicine and Biology 856,
DOI 10.1007/978-3-319-33826-2_8

Keywords Toxicity pathways • Case study approach • *In vitro* toxicity testing • Fit-for-purpose safety assessment • DNA damage • Nuclear receptor activation • Estrogen signaling

1 Background

In the 1930s consumer poisonings occurred from ethylene glycol containing sulfanilamide preparations (Wax 1995). This episode led to efforts by the USFDA to rely on toxicity testing in animals and to develop safety factor approaches for using this information for establishing safe human exposures (Lehman and Fitzhugh 1954). Over the ensuing years, there were calls for more and more testing for multiple endpoints with individual compounds and the numbers of compounds in commerce requiring testing grew. Animal testing became increasingly expensive, exorbitant in the use of animals, and took a long time for completion for any specific chemical. And yet, the relevance of these animal studies to human biology remains in question. In recent years, there has been a dawning realization that the system of in life animal testing requires change.

In the USA, the EPA's ToxCast program, in collaboration with the Tox21 initiative and other research partners developed approaches to screen chemicals through a diverse suite of repurposed, commercially available assays using quantitative high throughput screening (q-HTS). Phase I of the ToxCast program included more than 600 assays for over 300 compounds (Judson et al. 2010). The express goal of the program was to develop bioactivity signatures that would assist in prioritizing compounds for further testing. The q-HTS results across the multiple assays could rank compounds in terms of hazard potential and support a tiered approach to reserve animal testing for those chemicals more likely to have specific forms of toxicity, high potency or high levels of human exposure. Many papers describing the ToxCast research are now available (Judson et al. 2010; Martin et al. 2010; Kleinstreuer et al. 2011; Martin et al. 2011; Sipes et al. 2011). Even though the predictive potential of the assays for in-life rodent toxicity from Phase I appears low (Thomas et al. 2012), there are still proposals for using ToxCast results to estimate "Toxicity-Related Biological Pathway Altering Doses for High-Throughput Chemical Risk Assessment" (Judson et al. 2011). This process of risk assessment utilizes both the *in vitro* concentrations causing responses in various assays, conservative estimates of human exposures and high throughput dosimetry. However, despite the momentum pushing incorporation of these efforts into risk assessment strategies, there is really little evidence supporting the relevance of these assays for assessing particular cellular endpoints and confirming similarities of potency in assays using human cells or tissues.

Many of the European Union-based initiatives focus more on animal alternatives, especially in light of the restrictions on using animals to test safety of cosmetics. Russell and Burch set the stage for increasing considerations of the humane use of animals in their 1959 book (Russell and Burch 1959). This effort instigated widespread calls to end unnecessary animal testing. Programs such as ECVAM in the EU and ICCVAM in the US focused on developing alternative *in vitro* assays to replace specific animal tests. In general, the validation process in these organizations was to show that new, alternative methods provide results equivalent to historical toxicity tests using live animals. It has become increasingly apparent that this definition—accurately predicting results of an animal test—is not appropriate as the toxicity testing field moves to embrace new *in vitro*, cell-based assays using human cells/tissues for assessing risks and safety of chemicals in the human population.

The National Research Council (NRC) report from the US National Academy of Sciences, "Toxicity Testing in the twenty-first Century: A Vision and A Strategy", proposed a shift from toxicity testing using animal studies to evaluation of perturbation of so-called toxicity pathways in mode-of-action-based *in vitro* assays using human cells or human cell lines (NRC 2007; Krewski et al. 2010). Toxicity pathways are simply normal signaling pathways in cells that can cause toxic responses if they are sufficiently perturbed by chemical treatments. The NRC report was fundamentally different from other approaches in stressing that these *in vitro* cell-based methods were the *preferred* approach for toxicity testing of environmental compounds in the twenty-first century because they are based in human biology. Newer *in vitro* methods would allow broad evaluation of chemical dose-response, including concentrations equivalent to those arising from ambient human exposures. The read-out of the assays would include both measures of adverse responses *in vitro* and the dose response for the pathway, to support pathway-based dose-response modeling (Boekelheide and Andersen 2010; Andersen et al. 2011).

Adverse Outcome Pathways (AOPs). The strategies described in this chapter focus on developing fit-for-purpose *in vitro* assays for several toxicity pathways. In this usage, fit-for-purpose means that the results from the *in vitro* assays will suffice for safety assessments without resorting to animals studies. As such, the goal is to develop assays that can be used to define chemical risk/adversity without requiring measures of adversity from animal studies. In contrast, in-life apical responses play a central role in defining Adverse Outcome Pathways (AOPs). Nonetheless, the structure of the AOP framework dovetails with the process of predicting molecular initiating events, identifying key events, developing assays for biomarkers of these key events *in vitro* and comparing these *in vitro* responses to the suite of short-term *in vivo* responses that map with key events. Figure 8.1 captures the relationship between *in vitro* assays measuring biomarkers of key events with the AOP framework describing the linkage from these key events onto the more conventional *in vivo* responses used in the past as the benchmark responses for risk assessment.

In vitro based safety assessments. While most current q-HTS efforts focus on repurposing available assays as screening tools and prioritizing compounds for further *in vivo* testing (Collins et al. 2008; Kavlock et al. 2012), the goal described in the 2007 NRC report is to transition to *in vitro* based risk assessments using well-designed assays that can account for key biological processes responsible for toxicological outcomes. Our strategy for *in vitro* based safety assessments is shown in Fig. 8.2. The process begins with the development of "fit-for-purpose" *in vitro* assays to examine cellular pathway responses. These fit-for-purpose assays should

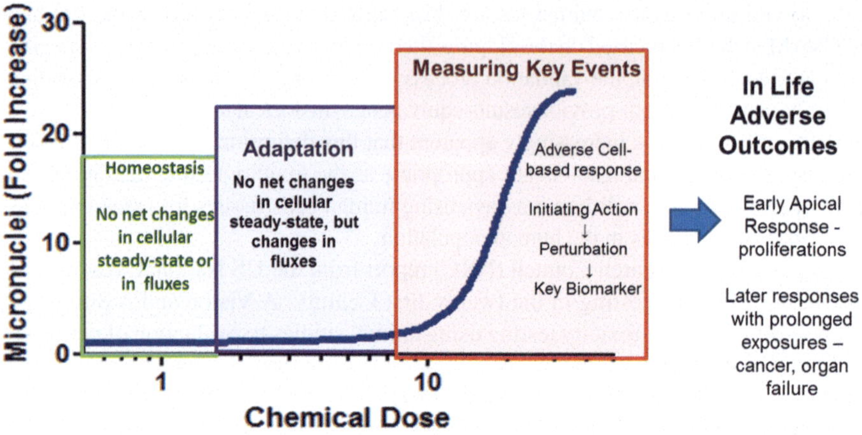

Fig. 8.1 Placing *in vitro* cell-based assays for key events into the Adverse Outcome Pathway context

Fig. 8.2 A roadmap for developing prototype risk assessments based on results from fit-for-purpose *in vitro* assays for specific toxicity pathways. We describe on-going efforts to develop assays for three pathways—p53-DNA damage-repair in human cell lines, PPARα nuclear receptor pathway in liver cells, and estrogen-pathway signaling in uterine cells tissues

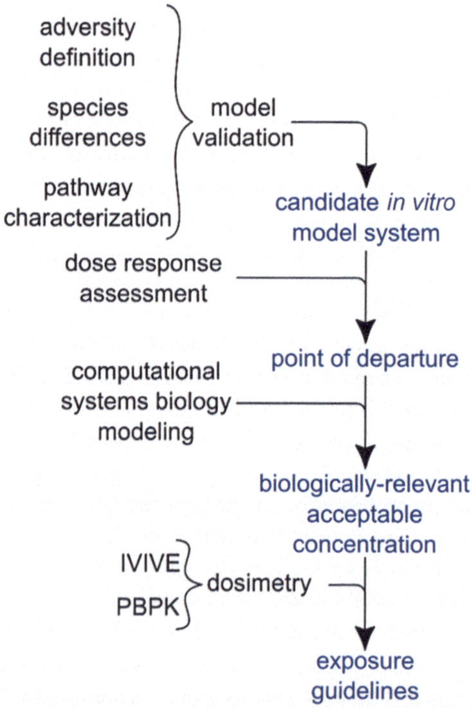

have read-outs that represent key signaling processes for a particular cellular outcome given a chemical's mode of action and clear measures of biomarkers of adversity. Dose-response data, together with computational pathway models, provide mechanistic understanding of the shape of dose-response curves to support low dose extrapolation from *in vitro* test results (Bhattacharya et al. 2011) and predict regions of safety for human exposure without resorting to in-life animal assays (Andersen and Krewski 2010). Quantitative *in vitro-in vivo* extrapolation (QIVIVE), a form of reverse dosimetry, can then be used to estimate safe human exposures from the active concentrations in the fit-for-purpose pathway assays (Clewell et al. 2008; Rotroff et al. 2010; Wetmore et al. 2012). In this chapter we focus on three examples of creating, characterizing and validating *in vitro*, fit-for-purpose assays.

2 Case Study 1: Identifying Appropriate *In Vitro* Models for PPARα Mediated Responses in Human Hepatocytes

A challenge in moving to new test methods with human cells, cell lines, organotypic cultures, etc. is developing comfort that these assays produce safety assessments protective of human health. The extant information on chemical toxicity derives largely from in life studies with laboratory animals, primarily rats and mice. The first efforts with ToxCast were motivated by the extensive animal data required for registration of pesticides. While program has produced a wealth of data across chemicals and assay endpoints, one difficult issue was that the *in vitro* assays used human cells and cell constituents whereas the existing data were from studies conducted primarily in rodents. Another challenge was that none of the ToxCast *in vitro* assays were structured in such a way as to look at corresponding doses required for an equivalent response *in vitro* and *in vivo*. *In vitro* to *in vivo* comparisons are difficult to make with chronic responses since the existing cell culture systems do not allow long-term exposures. Nonetheless, other screening programs include multiple shorter-term *in vivo* assays that provide insight about molecular initiating events and key events. In our case studies the anchoring occurs on the basis of doses and exposures required to produce similar outcomes on a cellular or tissue level for key events. One key aspect of validation, as we begin the change to new *in vitro* tests, is to develop equivalent anchors for *in vitro* and short-term *in vivo* studies to show similarities in cellular responses and in the dose at which responses occur. A challenge arises for creating assays for pathway in which there appears to be significant difference in biology between species and we continue to rely on these older studies to infer human risks. Several examples are available from extensive work with mode-of-action and the human relevance framework for liver nuclear receptors such as CAR and PPARα (Andersen et al. 2014; Corton et al. 2014; Elcombe et al. 2014). In rodents, persistent activation of these receptors leads to liver cancer through non-genotoxic processes associated with increased cell replication in livers. Proliferation does not occur in human hepatocytes. How do we take these differences into account in creating an assay for liver cell responses?

Understanding PPARα pathway biology. This pathway represents a prototype of a nuclear receptor mediated toxicity pathway with important species differences (Klaunig et al. 2003). Activation of the peroxisome proliferator-activated receptor alpha (PPARα) nuclear receptor in liver parenchymal cells from humans results in a series of coordinated events leading to downstream alterations in gene expression with alterations in lipid and fatty acid metabolism (McMullen et al. 2014). We have used a combination of microarray-based gene expression data, regulatory interactions inferred from protein-DNA transcription factor arrays and published chromatin immunoprecipitation (ChIP) data (van der Meer et al. 2010) to develop a picture of PPARα-mediated transcriptional regulation after treatment with a PPARα-selective ligand (GW7647) in rats and in humans. This agonist altered expression of about 200 genes in human primary hepatocytes. Only a limited number of genes that were differentially regulated by GW7647 treatment bind PPARα either directly at a PPARα response element (PPRE) or indirectly where PPARα binds genes in the absence of a PPRE. Approximately half of the differentially regulated genes showed no PPARα binding at all. Because they are not regulated by transcription factor-DNA interactions, these genes are considered nongenomic targets of PPARα. We then inferred the transcription factors involved in gene regulation, leading to a clearer picture of the hierarchical organization of the PPARα response network and the concentration- and time-dependent structure of the network (McMullen et al. 2014). This human network and the human responses should form the basis of a safety assessment for PPARα agonist compounds (Fig. 8.3). However, the responses in rodents, studied extensively *in vivo* using xenobiotics that target PPARα, include changes other than alteration in fatty acid metabolism genes (Rosen et al. 2008a, b, 2010). These alternative responses include hypertrophy and cell proliferation on short-term exposures and liver cancer in life-time exposures, which are apparently associated with persistent proliferation as a key event when maintained over a longer time. We have to carefully consider how *in vitro* responses in human cells, a species that does not show a similar proliferative response (Klaunig et al. 2003), replace long-term animal studies that have a proliferation-dependent response.

Human and Rat differences in biology. Detailed systems biology mapping of the pathway during assay validation for rat and human (McMullen et al. 2015) also provided better understanding of the biological differences in the human and rat responses. Among the genes downregulated in rats were two transcription factors—Ets1 and Hnf6. The former was fairly densely connected in the human network developed from gene expression and ChIP data streams (McMullen et al. 2014). Hnf6 is a central transcription factor in differentiation of precursor states into final hepatic lineages (Odom et al. 2004). In addition, the downregulated genes in rats that contained a PPRE had a different nucleotide binding sequence in the flanking regions of the core PPARα-RXR binding regions. These adjacent regions lack conservation of nucleotide binding motifs that were present with upregulated genes containing PPRE sites. The differential liver responses to nuclear receptors that form heterodimers with RXR may be related to down regulation of several key transcription factors in rodents that cause a form of "dedifferentiation" to earlier phenotypes in at least some regions of the rat liver acinus. The downregulation of these pathways then primes cells to proliferate with arrival of signals from other cells in the liver. Our experience with PPARα shows the value of integrating transcriptomics, ChIP, and network analysis to understand the key differences between the rodent and human response. It also highlights the importance of developing *in vitro*, systems biology tools that evaluate the mode of action in the relevant species of interest, i.e., humans.

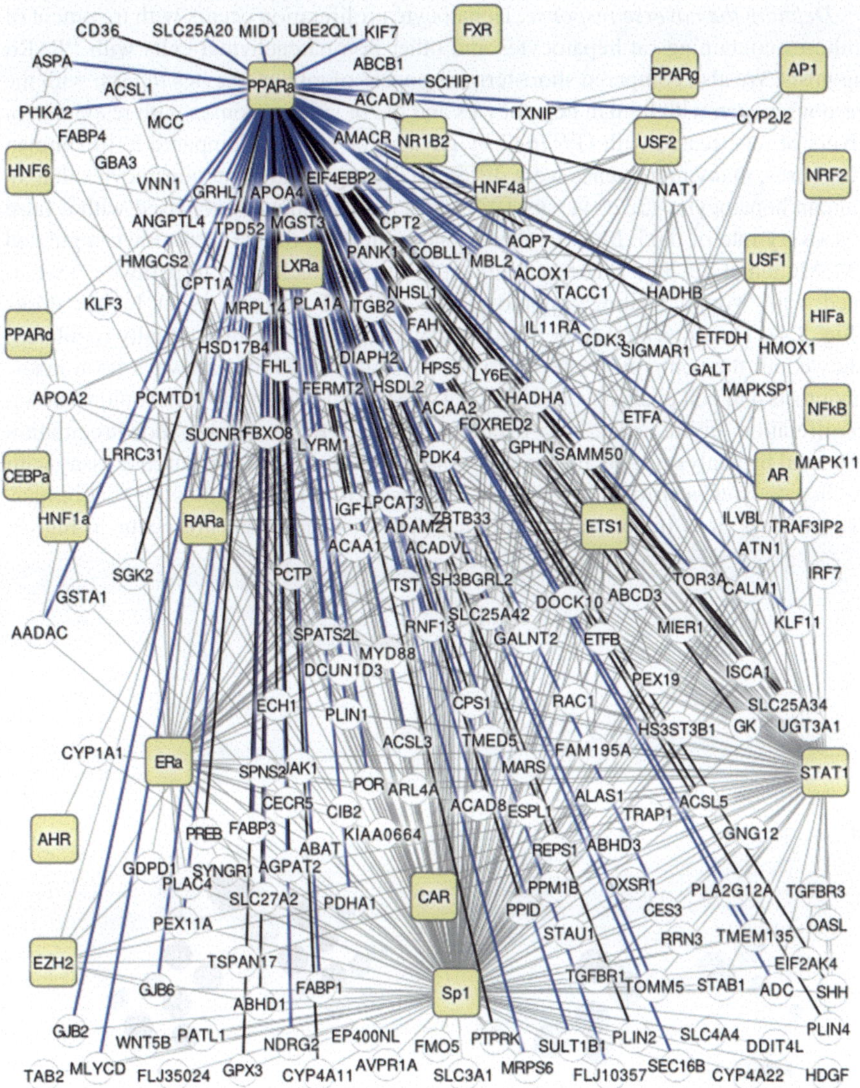

Fig. 8.3 Developing a Transcription regulatory network for PPARα activation in human hepatocytes. *Yellow nodes* are transcription factors and *white nodes* are target genes that are regulated by GW7647—a selective PPARα agonist. *Blue lines* reflect direct genomic regulation by PPARα (*blue*), *black lines* are indirect genomic regulation by PPARα (*black*), and *grey lines* reflect non-genomic regulation (no PPARa binding with the gene). The general scheme is reprinted with permission (McMullen et al. 2014)

Defining the adverse response. Hepatocyte proliferation occurs with treatment of cultures containing rat hepatocytes and other non-parenchymal cells with PPARα agonists. We also compared short-term responses of rat hepatocytes *in vitro* with the responses seen with human hepatocytes and the *in vitro* responses with responses in livers of rats treated with GW7647 by gavage for 3-days. Compared to the human responses, many more genes were differentially regulated in rat hepatocytes than in human hepatocytes (2320 versus 192) for genes statistically increased with a false discovery rate of 0.05. Most of these genes unique to the rat were down-regulated (McMullen et al. 2015). The enrichment in these rat cells was in pathways associated with wound healing, apoptosis and cell-cell interactions at higher doses (Fig. 8.4). The *in vivo* rat liver responses (with 3830 genes differentially regulated) showed all the components seen *in vitro* plus enrichment of pathways for mitosis, apoptosis and cell cycle checkpoints (Fig. 8.5). While *in vitro* assays could focus on proliferation in rat cells, key events associated with pathway downregulation assessed by analyzing specific genes would also serve as a fit-for-purpose assay with rat hepatocytes *in vitro* to look at dose response with PPARα activating xenobiotics. Nonetheless, it is important to note that these responses did not occur in human

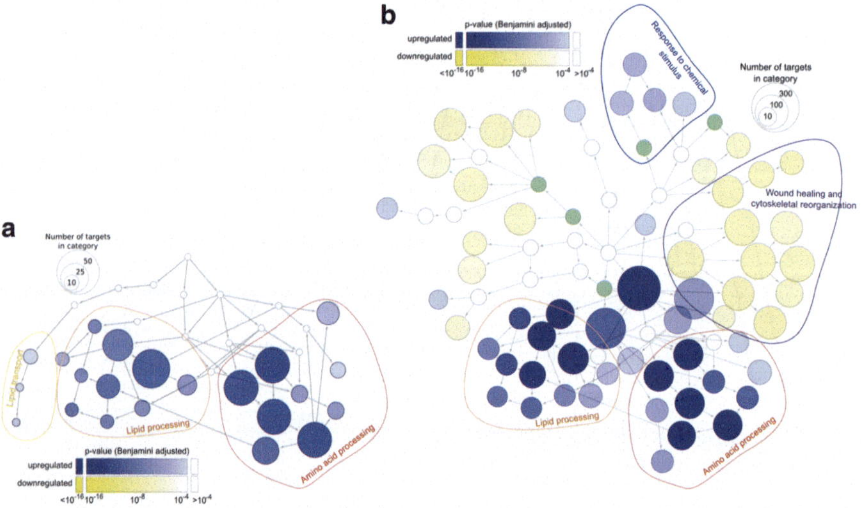

Fig. 8.4 Functional gene expression response to PPARα activation in rats and humans. (**a**) Gene ontology enrichment analysis confirms that the transcriptional program upregulated in human hepatocytes consists of lipid and amino acid metabolic machinery. Here, significant categories are connected according to the Gene Ontology hierarchy, highlighting the relationships between categories. Circle size denotes the number of genes represented within the category and the color saturation denotes the significance of enrichment. (**b**) In rat hepatocytes, this lipid/amino acid metabolic machinery is conserved, but it is accompanied by changes in cytoskeletal remodeling and response to external stimuli. The genes in these latter pathways represent downregulated responses absent in human hepatocytes. Details of this analysis appeared in earlier work (McMullen et al. 2015)

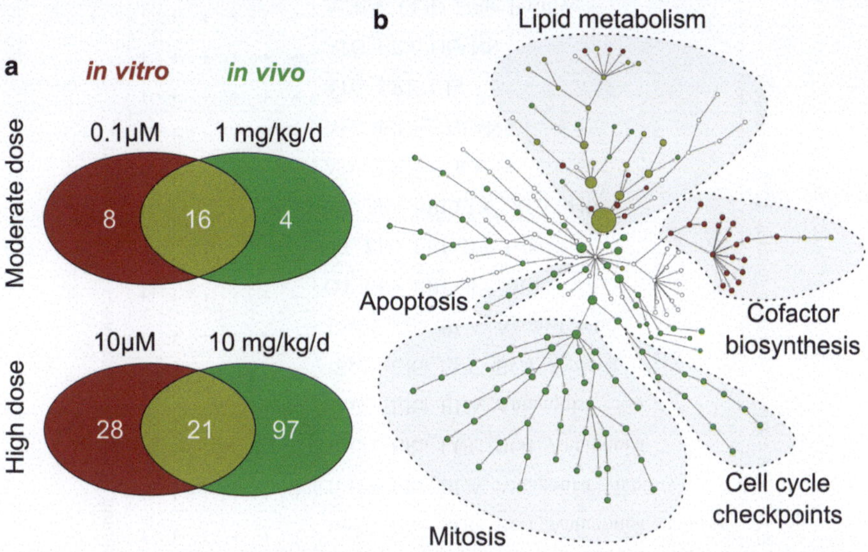

Fig. 8.5 Low/moderate levels of PPARα activation drive similar gene expression signatures *in vitro* and *in vivo*; high levels activation cause differences in gene expression profiles. (**a**) Venn diagrams of Reactome pathways enriched by genes upregulated by GW7647. Most activated pathways are common to both *in vitro* and *in vivo* contexts at moderate doses (*red* and *orange* pathways). (**b**) At higher relative doses, there are multiple pathway signals only noted *in vivo* (*green nodes*) that are consistent with cell proliferation and hypertrophy. See an earlier paper for more details (McMullen et al. 2015)

cells—the species of interest for risk assessment purposes. Thus, evaluations with humans need to assess dose response for components of the upregulation of genes associated with fatty acid/lipid metabolism, not proliferation.

Current status of fit-for-purpose hepatocyte assays for risk assessments of PPARα agonists. By assuring ourselves that the biological differences relating to the rat proliferative responses are not relevant to humans, the focus for a safety assessment becomes using human primary hepatocytes. This source is not always available, is costly and depends on access to surgical samples or accident victims. New tools available for creating stem cell derived hepatocytes (e.g., iCell-Hepatocytes from Cellular Dynamics International; http://www.cellulardynamics.com/products/hepatocytes.html) promise to provide a well-defined steady source of cells and the availability of cells derived from a more diverse human population. As the phenotype of these stem cell-derived hepatocytes become better understood they may become the gold standard for looking at multiple adverse cellular responses of the liver. Finally, we note the clear advantage of starting with a highly selective ligand (GW7647) to map the pathway. Further work with xenobiotics can then see to what extent another compound overlaps with a relatively pure agonist and to what extent other pathways contribute.

Table 8.1 Comparing results for selected compounds with Uterotrophic assay validation (OECD 2007) and estrogen receptor-specific ToxCast assays

Name	CAS	Uterotrophic	Tox21_Aromatase_Inhibition_viability	Tox21_Aromatase_Inhibition	NVS_NR_hER	NVS_NR_mERa	NVS_NR_bER	OT_ER_ERaERa_0480	OT_ER_ERaERa_1440	OT_ER_ERaERb_0480	OT_ER_ERaERb_1440	OT_ER_ERbERb_0480	OT_ER_ERbERb_1440	Tox21_ERa_LUC_BG1_Antagonist_viability	Tox21_ERa_BLA_Antagonist_viability	Tox21_ERa_LUC_BG1_Antagonist	Tox21_ERa_BLA_Antagonist_ratio	Tox21_ERa_LUC_BG1_Agonist	Tox21_ERa_BLA_Agonist_ratio	OT_ERa_EREGFP_0120	OT_ERa_EREGFP_0480	OT_ERa_ERELUC_AG_1440	ATG_ERRa_TRANS	ATG_ERRg_TRANS	ATG_ERE_CIS	ATG_ERa_TRANS	ACEA_T47D_80hr_Positive	ACEA_T47D_80hr_Negative
4-(2-Methylbutan-2-yl)phenol	80-46-6	+	−	+	−	−	−	+	+	+	+	+	+	−	−	−	−	+	+	+	+	−	−	−	+	+	+	−
4-tert-Butylphenol	98-54-4	+	+	−	−	−	−	+	+	+	+	+	+	−	−	−	−	−	+	+	+	−	−	−	+	+	−	−
17-Methyltestosterone	58-18-4	+	−	−	+	−	−	−	−	+	+	+	+	+	−	−	−	+	+	−	−	−	−	−	+	+	+	−
Ethylparaben	120-47-8	−	−	−	−	−	−	+	+	+	+	+	+	−	−	−	−	+	+	−	−	−	−	−	+	+	+	−
2-Naphthalenol	135-19-3	−	−	−	+	−	+	+	+	+	+	+	+	−	−	−	−	−	+	−	−	−	−	−	+	+	−	−
Dipentyl phthalate	131-18-0	−	−	−	−	−	−	+	−	+	+	+	−	−	−	−	−	−	−	+	+	−	−	−	−	+	−	−

Despite similar patterns of responses in various *in vitro* assays, there are discrepancies when examining uterotrophic responses in the intact rat

3 Case Study 2: Ensuring Biological Relevance of *In Vitro* Assays: An Example with Estrogenic Activity in the Uterus

There are two main approaches to identifying data gaps in the battery of *in vitro* assays for particular toxicity pathways: (1) comparing *in vitro* predictions to "gold standard" *in vivo* assays and (2) identifying key events in the toxicity pathway based on the sum of the literature on mode of action and chemical toxicity, and then building conceptual frameworks to which the current set of assays can be compared. The previously described work in rodent hepatocyte response to PPARα agonists demonstrates the first approach, which may be called "ground truthing". As discussed, care must be taken when using this approach to introduce steps that ensure that the methods developed are relevant to human and not merely rodent biology. The second approach has been demonstrated in recent publications for the DNA damage response pathway (Adeleye et al. 2015; Clewell et al. 2014) and thyroid hormone disruption (Murk et al. 2013). This approach, which can be considered a "ground-up" approach, focuses on mapping the key events in the signaling pathway and building fit-for-purpose assays based on pathway knowledge. Both approaches—ground-truthing and ground-up pathway building—can help to identify data gaps, design appropriate assays, and, ideally, can be used to ensure more accurate predictions of regions of safety for chemical exposures based on *in vitro* assays.

Current strategies for identifying estrogenic chemicals. The *in vivo* rat uterotrophic assay (OCSPP 890.16) is a gold standard for testing estrogen disruption. Various *in vitro* assays have been incorporated into the current high throughput screening programs to replace the uterotrophic assay for identifying estrogenic compounds. These assays primarily focus on binding and activation of the estrogen receptors, ERα and ERβ (Table 8.1). Despite relative success of the EDSP and ToxCast assays in predicting estrogenic chemicals from the OECD guideline E assays, i.e., 91 % sensitivity, 65 % specificity (Rotroff et al. 2013) or EPA estrogen receptor reference library (Rotroff et al. 2014), there are some chemicals that are not correctly identified as estrogenic with the current *in vitro* methods. A survey of compounds that are difficult to classify (i.e., compounds that are positive in approximately half of the relevant ToxCast™ assays, but test negative in the uterotrophic assay) highlights this difficulty (Table 8.1). It would be difficult to predict whether these compounds are uterotrophic in rats based on the ToxCast results alone.

Identifying a model for estrogen receptor mediated uterine proliferation. To identify the appropriate *in vitro* assays to predict uterine responses to estrogenic compounds, we first developed an Adverse Outcome Pathway (AOP)-like framework for the estrogen-mediated proliferation pathway. The effort is similar to development of an AOP, in that we develop a framework that begins with a molecular initiating event (i.e., ligand-receptor binding) and outlines the ensuing signaling events that lead to a particular outcome (i.e., proliferation). However, this process differs from the AOP, in that events beyond cellular outcome (i.e., tissue, body, population) are not formally considered. Instead, we focus on defining key events governing cellular outcome in order to identify the most important characteristics of the required *in vitro* assays (Fig. 8.1).

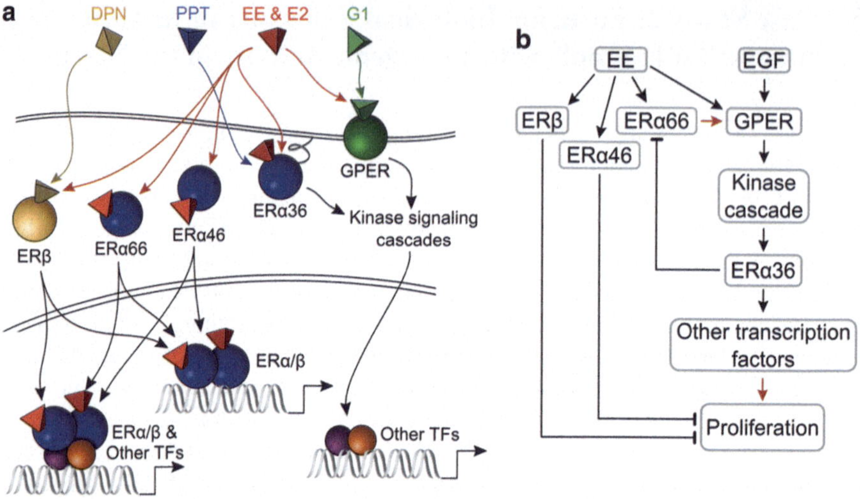

Fig. 8.6 (**a**) Overview of the estrogen receptor signaling pathway and (**b**) a proposed network regulating estrogen-mediated proliferative response. PPT, DPN and G1 are selective agonists for ERα, ERβ, and GPER, respectively. EE (ethinyl estradiol) and E2 (17β-estradiol) bind all of the receptors. ERα66 (also known as ESR1) is full-length ERα. ERα46 and ERα36 are shorter isoforms of ERα with ligand binding and signaling properties distinct from ERα66. GPER is a G-coupled protein receptor that binds estrogen and promotes proliferation

Several estrogen receptors (ERs) play key roles in proliferative signaling (Fig. 8.6a). In addition to ERβ and full-length ERα (ERα66), two shorter isoforms of ERα (ERα46, ERα36) and an estrogen binding G-coupled protein receptor (GPER) mediate estrogenic compound-induced proliferation (Filardo et al. 2000; Penot et al. 2005; Wang et al. 2006). All of these receptors bind estrogen. However, differences in the shape and structure of the binding domains lead to differential binding affinity for both native ligands (estrogens) and exogenous chemicals (Lin et al. 2013). ERα66, ERα46 and ERβ exert their effects through ligand binding, dimerizing, translocating to the nucleus, and binding estrogen response elements (EREs) in the promoter regions of specific ER-regulated genes. ERα66 generally promotes proliferation, while ERβ and ERα46 have inhibitory effects on proliferation (Hall and McDonnell 1999; Penot et al. 2005; Klinge et al. 2010; Abot et al. 2013). GPER and ERα36 are localized to the cell membrane and appear to exert their effect through activation of kinase cascades (Filardo et al. 2000; Wang et al. 2006), facilitating growth factor pathways. This type of signaling, leading to gene transcription without direct binding of the estrogen receptor to EREs, is generally referred to as "non-genomic". GPER and ERα36 appear to have a pro-proliferative effect. Despite clear contributions from these multiple receptors to estrogenic response, the current suite of HTS assays focuses on agonism and antagonism of only two forms of ER—ERα66 and ERβ (see Table 8.1). This limitation highlights the need for a fit-for-purpose assay conducted in intact cells and recapitulating phenotypic response (uterine proliferation). Currently, the EDSP/ToxCast program includes one such assay: the breast cell line (T47D) proliferation assay.

The T47D cell line is derived from human breast. Yet, there are clear instances in which chemicals that selectively bind certain ER isoforms or recruit tissue-specific cofactors (i.e., selective estrogen receptor modulators; SERMs) cause differential responses in the breast and uterus. Tamoxifen inhibits breast cell proliferation but induces uterine cell proliferation, ultimately increasing risks of uterine cancer in women. Tamoxifen is an ERα antagonist, but it is also a GPER agonist. In many human endometrial cancers high expression of GPER associates with high-grade uterine carcinoma (He et al. 2009). Additionally, endometrial GPER protein expression correlates with tamoxifen-induced uterine bleeding and abnormal endometrial thickening in women receiving tamoxifen for treatment of breast cancer (Ignatov et al. 2010). Tamoxifen likely inhibits breast cell proliferation through inhibition of ERα, while inducing uterine proliferation through activation of the membrane bound GPER. Tamoxifen acts as a SERM, and the different cellular context of the uterus and breast causes opposite phenotypic responses. Clearly, assays for estrogenic activity in cells derived from different tissues might need to be tailored based on differential content of E2 binding receptors in the tissues.

Ensuring that the in vitro model represents the appropriate biology. A human uterine cell line that proliferates in response to estrogen and estrogen-like compounds is the Ishikawa human uterine adenocarcinoma cell line. These cells are available through the European Collection of Cell Cultures (ECACC) and have been genotyped for confirmation of cell type. To show that Ishikawa cells have the appropriate biology to recapitulate *in vivo* uterine effects, we undertook preliminary studies to ensure that these Ishikawa cells (1) contain all relevant ER isoforms, (2) recapitulate ER-mediated cellular responses and (3) are a reliable, robust *in vitro* model.

ER protein and mRNA expression were examined in Ishikawa cells under a variety of conditions—growth media, charcoal-stripped serum, and estrogen treated media (Clewell et al. 2015a). All of the major estrogen receptor isoforms (ERα66, ERα46, ERα36, ERβ and GPER) were expressed in Ishikawa cells and receptor expression was stable out to passage 11. To determine whether these cells maintained responsiveness to estrogens, we evaluated their ability to recapitulate phenotypic responses to ethinyl estradiol (EE), including transcriptional activation of ER-mediated genes (PGR, GREB1) and enzyme activity induction (alkaline phosphatase; ALP). qPCR and whole genome arrays evaluated EE-induced expression across doses that caused no effect, minimal effect, or maximal effect on proliferation, looking at seven doses from 10^{-13} to 10^{-9} M. Transcriptional response was observed as early as 6 h and at low picomolar concentrations of EE (Fig. 8.7a). Induction of ALP enzyme activity, a response associated with uterine but not breast cells, was also recapitulated with the Ishikawa cells (Fig. 8.7b).

Cellular proliferation is arguably the most important endpoint as it represents the phenotypic response of interest in a risk assessment context (e.g., proliferation as a preliminary event for carcinogenesis). Further, a proliferative response to estrogen treatment confirms functionality of the estrogen receptors and their downstream signaling networks. Proliferation was measured in the Ishikawa cells using a variety of markers (BrdU incorporation, DNA content, enzymatic viability, impedance measurements). The various methods showed similar results: maximal induction of

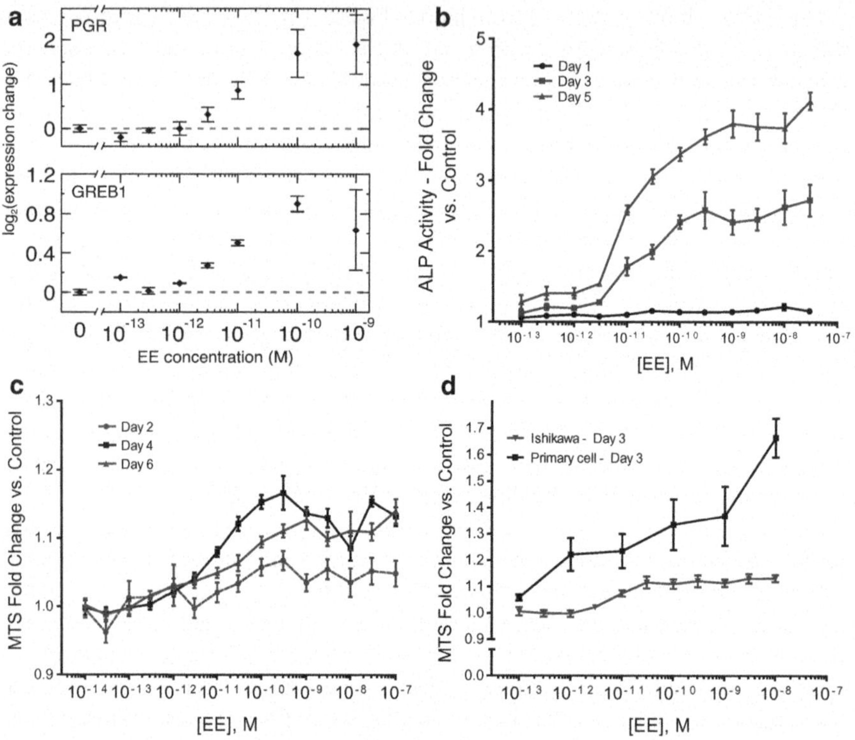

Fig. 8.7 Showing that Ishikawa Cells have representative phenotypic responses following treatment with ethinyl estradiol. These panels show: (**a**) transcription of ER regulated *ALPP* and *GREB1* genes in Ishikawa cells, (**b**) alkaline phosphatase (ALP) enzyme activity in Ishikawa cells, (**c**) EE-mediated proliferation in Ishikawa cells, and (**d**) a comparison of proliferation response in Ishikawa and primary uterine cells. Details appeared previously (Clewell et al. 2015a)

1.2-fold over control at 3 days of treatment. In-depth dose-response studies (Fig. 8.7c) were completed with enzymatic viability (MTS) assays (18 doses, 10^{-14} to 10^{-6} M EE); proliferation occurred at low picomolar concentrations of EE (3×10^{-12} M).

The ability of the Ishikawa cells to express the appropriate estrogen receptors and to recapitulate the phenotypic responses to estrogen treatment indicates its value as a fit-for-purpose assay for estrogenic activity. However, the more important goal is to ensure that the model is quantitatively predictive of human response. To this end, we compared proliferative response of the Ishikawa cells to those of primary human endometrial epithelial cells (Fig. 8.7d). While the primary cells showed a stronger proliferative response to EE (1.8 vs. 1.2-fold increase), the doses required to induce proliferation were similar (~10 pM). In addition, 17β-estradiol (E2) induces transcriptional and proliferative responses in the Ishikawa cells at concentration consistent with serum E2 levels associated with estrogenic responses in women *in vivo* (low picomolar). Finally, we know that Ishikawa cells proliferate in response to the uterine-specific SERM tamoxifen (Johnson et al. 2007). From these preliminary studies, we drew the following conclusions:

1. Ishikawa cells are an adequate platform for evaluating ER-mediated signaling networks and identifying key signaling processes responsible for epithelial cell proliferation in uterine tissues.
2. This cell model provides a reasonable *in vitro* platform for dose-response evaluations and predicting regions of safety for human exposures.

Moving towards an in vitro risk assessment strategy for estrogen receptor mediated uterotrophic response. The next stage in assay development will be the use of Ishikawa and primary human endometrial epithelial cells to infer the ER signaling network, perform dose-response evaluations for prototype ER agonists, and underpin development of a computational systems biology model of ER-mediated proliferation (Fig. 8.6b). Identifying the ER isoforms responsible for a proliferative response is essential to determining whether the fit-for-purpose assays can adequately account the processes that define the shapes of the dose-response curves for ER ligands. This type of evaluation employs two complementary approaches to identify the key ERs responsible for mediating proliferation in the uterine epithelial cell: (1) evaluation of proliferative response in Ishikawa cells following treatment with selective ER agonists and (2) evaluation of proliferative response following EE treatment in cells with overexpression of the ER isotypes. These studies and published mechanistic data support computational models of the ER signaling networks and evaluate the feasibility of these proposed models for generating observed dose-response curves. The understanding of the pathway structure and dynamics of signaling through the suite of receptors will provide a more biologically realistic assay for assessing estrogenic activity in uterine cells and the necessary pathway information to moving to allow *in vitro* only assessment of risks of estrogenic compounds.

4 Case Study 3: Defining Adversity Using *In Vitro* Assays for a DNA Damage Response Pathway

A significant challenge facing the movement towards total replacement of animal testing in toxicological risk assessments is finding a way to define adversity *in vitro*. Clearly, without the use of animal testing, conventional hazard identification is not possible. Instead, we must begin to think in terms of cellular integrity and maintenance of cellular function. For this example, we describe current efforts in defining safety through *in vitro* assays for the DNA damage stress response pathway (Fig. 8.8).

The p53 signaling network is activated in response to various types of DNA damage and functions in multiple ways to conserve genome stability and prevent heritable mutation. Activated p53 is both a recruitment factor for nuclear DNA repair enzymes and a transcription factor (Fig. 8.8). It transcriptionally regulates key DNA damage response proteins that induce cell cycle arrest, DNA repair, and apoptosis in mammalian cells (Barak et al. 1993; el-Deiry et al. 1993; Harper et al. 1993; Miyashita and

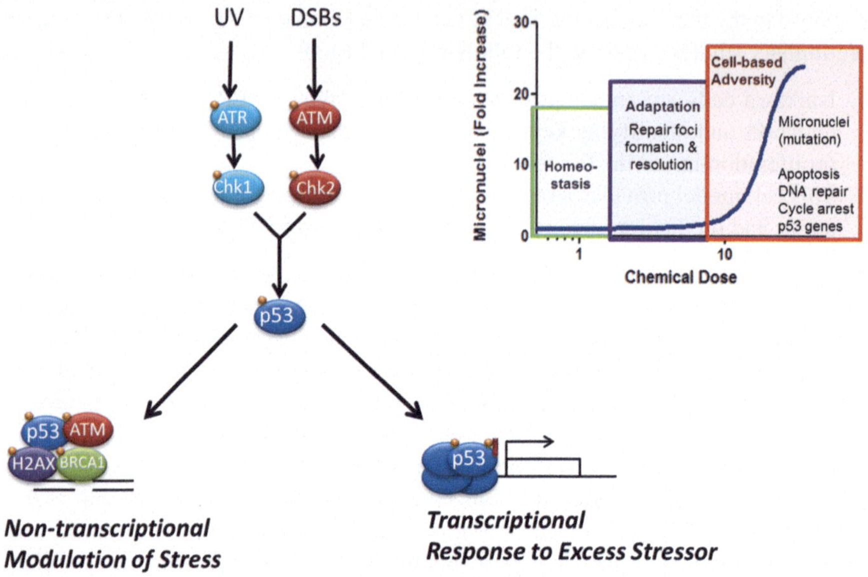

Fig. 8.8 Lower levels of damage lead to formation of DNA repair foci. Lower stress levels cause p53 to bind at sites of DNA-damage and recruit other proteins to a repair complex. Higher stress levels saturate the ability for repair and lead to transcriptional activation of stress response genes by a p53 tetramer. The combination of these processes may result in threshold shaped dose-response curves for permanent DNA damage. Portions of the figure on the left are published from earlier work (Clewell et al. 2014) with permission

Reed 1995; Oda et al. 2000; Thornborrow et al. 2002; Chen and Sadowski 2005; Arias-Lopez et al. 2006). Levels of DNA-damage that exceed repair capacities will produce adverse consequences, including mutation, cancer and cytotoxicity.

Defining assay endpoints and conducting dose-response evaluations. Initially, the p53 pathway project focused on defining the key readouts for the p53-mediated DNA damage response, developing "fit-for-purpose" assays for these readouts, and conducting dose-response assessments using prototype DNA damaging chemicals (Adeleye et al. 2015; Clewell et al. 2014). Initial studies were performed in a human fibrosarcoma cell line (HT1080), which expresses native p53. Key regulators of DNA damage have been described based on studies with ultraviolet (UV) and gamma irradiation (Lahav et al. 2004; Batchelor et al. 2011; Purvis et al. 2012). We developed high-throughput assays to study the dose-response for these key aspects of the DNA damage response, including: DNA damage (p-H2AX), p53 activation (p53, p-p53 (ser15)), cellular response to p53 (cell cycle arrest, apoptosis), and a marker of fixed chromosomal mutation (micronuclei formation). Micronuclei (small pieces of DNA or whole chromosomes lost during mitosis) are a measure of permanent (unrepaired) DNA damage. Additionally, we used whole genome transcriptomic dose-response studies for each of the three prototype chemicals. Three chemicals with different modes of action were used to probe the cellular response to different types of DNA damage: etoposide (ETP; topoisomerase II inhibitor and double strand break inducer); methyl methanesulfonate (MMS; methylating agent and single strand break inducer) and quercetin (QUE; oxidative DNA damage). The resulting data were then used to evaluate the shape of the dose-response curves for each of the measured biomarkers in an effort to define a region of safety (or point of departure) for these chemicals (Clewell et al. 2014).

When the dose-response trends were compared across endpoints, it was clear that many of the endpoints conventionally recognized as protective against DNA-damage—transcriptional regulation of cell cycle arrest and apoptosis by p53—occurred at higher doses than induction of micronuclei formation (Fig. 8.9). In fact, with all of the chemicals micronuclei induction occurred at doses equal to, or lower than, doses required to activate p53-mediated gene transcription. Thus, any protective effect of p53 against MN formation is unlikely to result from changes in transcriptional programs in the cells. Instead, it appears that the ability of p53 to prevent changes in the net level of permanent DNA damage at low chemical doses is likely due to post-translational processes, i.e., p53 serving as recruitment factor for repair proteins at the site of DNA damage. This is consistent with studies that looked at the time course and dose response for formation and resolution of DNA repair centers (DRCs) with γ-irradiation (Neumaier et al. 2012). Formation of DRCs and DNA repair were considerably more efficient at lower doses than at high exposures.

Identifying key cellular processes for maintaining homeostasis in the presence of low chemical doses (e.g., region of adaptive cellular response). Transcriptional up-regulation of stress response genes constitutes a major cellular defense program against variety of chemical induced cellular stresses. Cellular activation of transcriptional programs, however, requires significant time for RNA and protein synthesis and consumes considerable cellular energy stores. For many types of cellular stress (oxidative damage, DNA damage, heat and osmotic shock, etc.), post-translational processes also work to protect cells (Zhang et al. 2015). Post-translational responses are rapid and do not require transcriptional activation of genes. The rapid response by post-translational control can maintain cellular homeostasis in the presence of transient, low-level damage. With sustained or higher levels of damage, these post-translational processes become overwhelmed, forcing a transition to the slower responding, less efficient transcriptional controls.

Defining regions of adaptive response and a point of departure for chemical safety assessment. The p53 DNA damage response pathway employs both transcriptional and post-translational processes to prevent heritable changes to the DNA. In addition to its activity as a transcription factor, p53 also serves as recruitment factor for repair proteins at the site of double strand breaks. p53, together with several other scaffold proteins, kinases and repair proteins, localizes at the sites of double strand breaks and forms DRCs (Al Rashid et al. 2005). These DRCs repair DNA damage without requiring activation of transcription. As a proof of concept, we conducted detailed studies of DRC formation at doses below those causing micronucleus-formation (Clewell et al. 2016). Neocarzinostatin (NCS) was used to confirm the role of post-translational repair in preventing genotoxic outcomes. NCS causes a short burst of oxidative damage that forms double strand breaks. NCS is destroyed during this process, however, and the resulting double strand breaks are susceptible to normal repair processes.

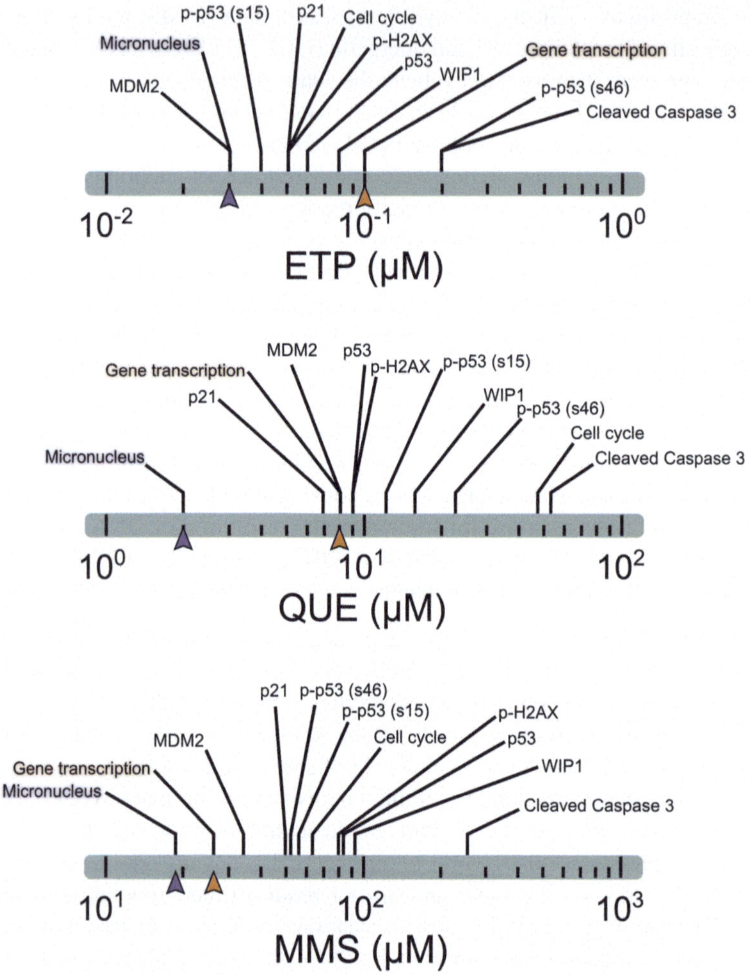

Fig. 8.9 Comparison of responses across endpoints for DNA damage response. Values shown for BMDL (lower 95 % confidence limit for the Benchmark Dose). For each chemical, micronucleus induction (*purple*) occurred at lower doses than gene transcription (*orange*)

Using high-content imaging with confocal microscopy, HT1080 cells were exposed to varying doses of NCS and the number of DRCs was counted across doses and times. With this technique the absolute number foci (DNA repair centers) can be counted in each cell (Fig. 8.10). NCS-mediated double strand breaks were more rapidly resolved (measured as foci dissolution) at low levels of exposure (Fig. 8.10; lower left). More importantly, the dose- and time-response data for NCS indicates that post-translational repair processes can prevent long-term damage at low doses (Fig. 8.10; lower right), and that higher doses that saturate these post-translation processes activate transcriptional processes and are the same as those that cause permanent chromosomal changes (i.e., micronuclei induction). Thus, this transition from post-translational

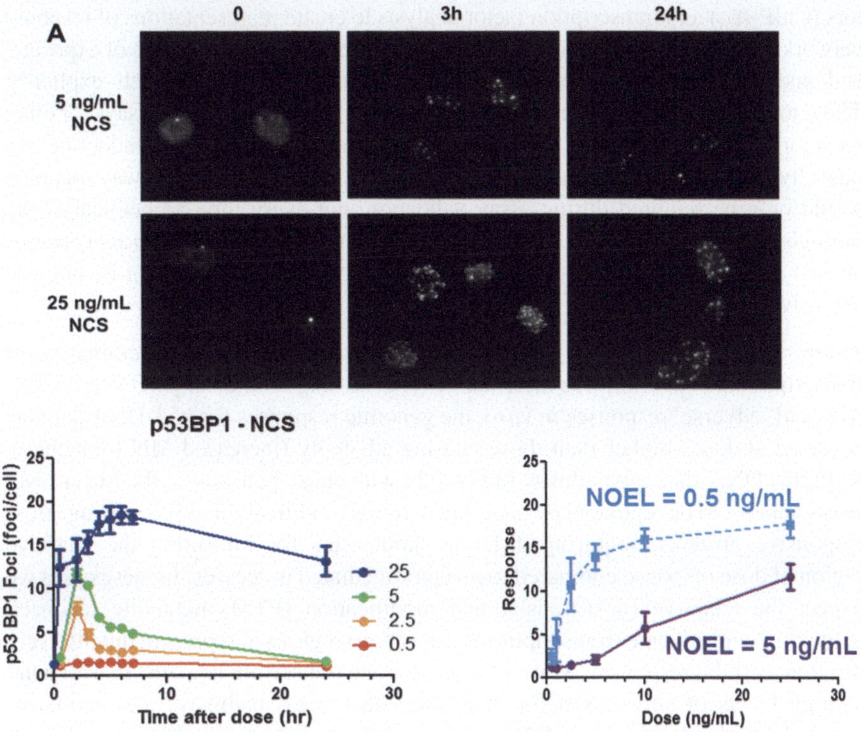

Fig. 8.10 Time-Course Behaviors of DNA-Repair Centers (DRCs) serve to differentiate regions of adaptation from those showing overt adversity responses. (**a**) Images of DRC foci following treatment with neocarzinostatin (NCS). *Lower left*: Dose- and time-response for DRC foci (as foci per nucleus) following treatment with NCS. At lower exposures, foci resolve quickly. At higher, doses they persist out beyond 24 h. *Lower right*: This panel shows plots of foci remaining at 24 h (in *purple*) and total number of foci produced (in *blue*). Representative of studies reported in an earlier paper (Clewell et al. 2016)

response (DNA repair center formation) to transcriptional response represents a "tipping point" between adaptation and adversity (Fig. 8.8).

Discussion: In this chapter we highlighted steps in developing mode-of-action based assays for three "toxicity pathways"—estrogen receptor (ER) signaling, PPARα activation and DNA-damage stress responses. In conventional parlance these fall into receptor-mediated (ER and PPAR) and stress response (DNA-damage) pathways. To some extent, the difference in these two broad categories is whether the responses arise through structural recognition by a receptor or through chemical reactivity with various cellular macromolecules. Despite these differences, we pushed forward initially with all three pathways looking at multiple endpoints for adversity, including the comparison of patterns of altered gene expression in terms of enrichment of genes in specific Gene Ontology or Reactome (www.reactome.org) categories. These studies required development of improved methods to visualize networks and compare responses across dose, time and treatment conditions. In addition, mapping and modeling pathway function led us to develop platforms to assess DNA-binding of receptors (ChIP-seq) and transcription factor analysis to create representations of response networks and to examine visually differential responses across dose, time of exposure, and species. The PPARα example here demonstrates this approach explicitly. These tools for pathway mapping will also be important for validating assays for other receptor-mediated pathways, including estrogen and eventually other endocrine and other liver nuclear receptor pathways. Importantly, this process of pathway mapping would only be required during assay validation, not every time a chemical is run through the assay. In addition, the efforts described here to anchor *in vitro* assay results against short-term measures of key events in animal studies would not be pursued extensively beyond the assay validation activities.

Transcriptional versus Post Translational Regulation. While the examination of transcriptional activation was informative for assessing relationship between 'adaptive' and 'adverse' responses *in vitro*, the genomic responses for p53-DNA damage occurred at doses higher than those causing adversity (increased MN frequency). With this DNA-damage pathway (and likely with stress pathways), the lower level stress appears to be regulated by post-translational modification of preexisting stress responsive proteins, requiring different biomarkers for informing the adaptive region of dose response compared to receptor mediated processes. In these cases we expect the adaptive (post-translational modification (PTM)/metabolic feedback activation) and adverse (transcriptional activation) regions to represent qualitatively different cellular responses. Molecular sensors and transcription factors that respond to high levels of stress exist for other canonical stress pathways, i.e., oxidative stress, heat shock response, DNA damage response, hypoxia, ER stress, metal stress, inflammation and osmotic stress (Simmons et al. 2009). For many of these pathways, there is strong evidence for PTM or metabolic feedback pathways that regulate cell stress before getting to doses that activate entirely new transcriptional programs (Zhang et al. 2015). Assays for stress response pathways will need to measure changes in pathway activity in the absence of altered gene expression—such as the transient formation of DRCs with p53.

Pathway dynamics differ across modes-of-action. The measure of adversity with PPARα in rat liver cells or with estrogens in human uterine cells/tissues is increased proliferation (or a marker of proliferative capacity) secondary to the expression of new programs of gene expression in the cells. These responses move the cell from one state with a particular pattern of gene expression to a different state with alterations in many genes and many cellular response components. These broad changes in cell state do not occur gradually, but tend to be dichotomous—abruptly turning on after reaching a sufficiently high dose/perturbation (Bars et al. 1989; Andersen et al. 1997; Louis and Becskei 2002; Sarangapani et al. 2002). At the Hamner, we refer to these dichotomous behaviors as the cells 'going someplace else'. Pathways regulating cellular homeostasis are referred to as cells 'staying put'.

Fit-for-What-Purpose. The purpose here is to have assays whose outputs are sufficient to complete a risk assessment without resorting to in life animal studies. Thus, the validated, fit-for-purpose assay needs to provide an understanding of pathway dose response to assist in 'high dose to low dose' extrapolation and to provide a point-of-departure concentration that becomes the starting point for *in vitro-in vivo* extrapolation. One significant advantage of moving to mode-of-action based cellular assays is that the processes of understanding pathway dynamics and dose response are now fairly well-developed areas of investigation. The biomedical engineering community has contributed new modeling tools and methodologies for understanding the structure and dynamic behaviors of cell signaling networks (Sipes et al. 2013; Attene-Ramos et al. 2015). They have largely focused on components and interactions of highly non-linear signaling processes with less interest on dose response. Pathway models examining dose response become important tools for interpreting these fit-for-purpose assays. Pathways leading to proliferation or leading to new gene expression programs within cells have highly non-linear, ultrasensitive signaling components (called motifs) that create switch-like behaviors (Zhang et al. 2013). A good example here is work on platelet-derived growth factor signaling in 3T3 cells (Bhalla et al. 2002). Homeostasis (staying put) requires negative feedback or feedforward motifs to ensure the cell maintains function within relatively narrow bounds despite altered stressor input. The dose-response of many of these feedback networks show thresholds—regions of increasing stressors that do not lead to any increase in the cellular stress level (Zhang et al. 2014). Work with Hog (high osmolarity glycerol) signaling in yeast provides excellent reading to understand the properties of a well-studied stress pathway (Mettetal et al. 2008). We envision that the validation of mode-of-action based cellular response assays provides (1) appropriate endpoints for measurement, (2) understanding of the signaling structure (a pathway map) and (3) computational models that predict the expected dose response below regions accessible to direct experimental measurement. To a very large extent, we see that the work with validating each of these assays brings a systems toxicology/network biology focus to understanding dose-response behaviors of toxicity pathways.

Quantitative high-throughput screens (q-HTS) and mode-of-action based assays. In this chapter, we focused on developing assays that fulfill the vision of the 2007 NRC report—assays that would examine perturbations of toxicity pathways and apply results from these mode-of-action based assays for human health risk assessment. The NRC report showed the overlap of these new methods (Fig. 8.11) with the original risk assessment process outlined in another National Academy report on risk assessment in the federal government (NRC 1983). This representation of the risk assessment based on perturbations of toxicity pathways includes q-HTS to assess modes-of-action/molecular initiating events with assays that may indicate possible hazard, but are themselves inadequate for risk assessment. Results from q-HTS and chemical characterization tools then direct testing toward specific cell-based, mode of action assays as discussed here. The risk assessments arising from these new approaches differ from those generated from animal test results. The risk assessment based on an apical observation attempts to provide an estimate of the incidence of the apical endpoint at lower exposures. Mode-of-action based assays provide a different focus for the risk/safety assessment. They predict safe doses, i.e., doses that do not appreciably perturb the pathway. This different emphasis will lead to a safety, rather than risk, assessment.

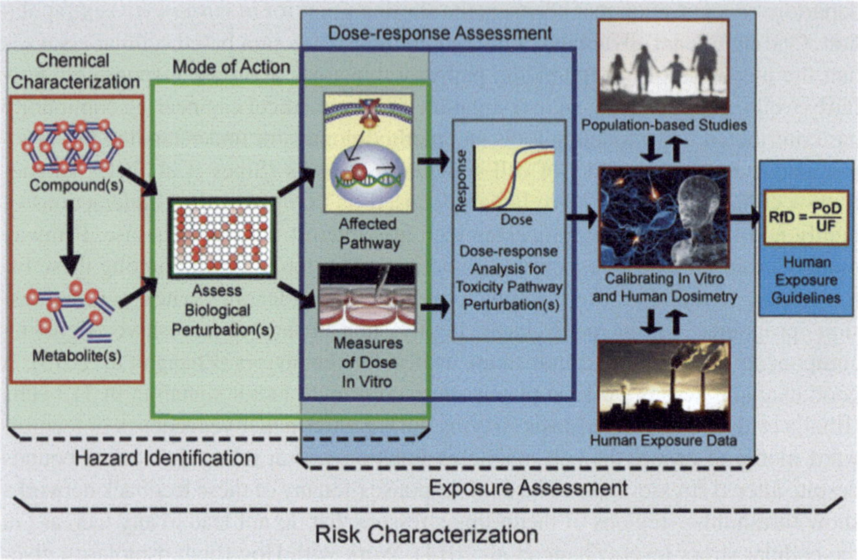

Fig. 8.11 The schematic of the risk assessment process using *in vitro* mode-of-action based assays with human cells. The hazard identification process, arising from chemical characterization and broad assessment of biological perturbations from q-HTS methods, identifies the most sensitive mode-of-action. The mode-of-action assays, as discussed in this chapter, and dose response models then contribute to more quantitative risk assessments by providing more relevant points-of-departure that can be taken to *in vitro-in vivo* extrapolation. Reprinted from an earlier publication with permission (Krewski et al. 2011)

Summary: Our experience with developing mode-of-action based assays and moving on to validate them has aided the development of a fairly standard workflow to design assays, map pathway responses to positive control compounds, and create dose response tools for the "toxicity pathway". None of the three pathways were entirely unique with respect to the methods used for validation. As we move forward to examine other "toxicity" pathways the process will become simpler (learning from past experiences) and allow development of assays for the next sets of pathways to be more rapid and more efficient.

Acknowledgments The ACC-LRI (Long Range Research Initiative of the American Chemistry Council), Dow Chemical, Unilever, the ExxonMobil Foundation, Dow Corning have supported method development and data generation needed in pursuing case study approaches to bringing the TT21C vision to life more quickly. Unilever has supported the p53-case study activities at The Hamner.

References

Abot A, Fontaine C, Raymond-Letron I, Flouriot G, Adlanmerini M, Buscato M, Otto C, Berges H, Laurell H, Gourdy P, Lenfant F, Arnal JF (2013) The AF-1 activation function of estrogen receptor alpha is necessary and sufficient for uterine epithelial cell proliferation *in vivo*. Endocrinology 154:2222–2233

Adeleye Y, Andersen M, Clewell R, Davies M, Dent M, Edwards S, Fowler P, Malcomber S, Nicol B, Scott A, Scott S, Sun B, Westmoreland C, White A, Zhang Q, Carmichael PL (2015) Implementing toxicity testing in the 21st century (TT21C): making safety decisions using toxicity pathways, and progress in a prototype risk assessment. Toxicology 332:102–111. doi:10.1016/j.tox.2014.02.007

Al Rashid ST, Dellaire G, Cuddihy A, Jalali F, Vaid M, Coackley C, Folkard M, Xu Y, Chen BP, Chen DJ, Lilge L, Prise KM, Bazett Jones DP, Bristow RG (2005) Evidence for the direct binding of phosphorylated p53 to sites of DNA breaks *in vivo*. Cancer Res 65:10810–10821

Andersen ME, Krewski D (2010) The vision of toxicity testing in the 21st century: moving from discussion to action. Toxicol Sci 117:17–24

Andersen ME, Birnbaum LS, Barton HA, Eklund CR (1997) Regional hepatic CYP1A1 and CYP1A2 induction with 2,3,7,8-tetrachlorodibenzo-p-dioxin evaluated with a multicompartment geometric model of hepatic zonation. Toxicol Appl Pharmacol 144:145–155

Andersen ME, Clewell HJ, Carmichael PL, Boekelheide K (2011) Can case study approaches speed implementation of the NRC report: "toxicity testing in the 21st century: a vision and a strategy?". ALTEX 28:175–182

Andersen ME, Preston RJ, Maier A, Willis AM, Patterson J (2014) Dose-response approaches for nuclear receptor-mediated modes of action for liver carcinogenicity: results of a workshop. Crit Rev Toxicol 44:50–63

Arias-Lopez C, Lazaro-Trueba I, Kerr P, Lord CJ, Dexter T, Iravani M, Ashworth A, Silva A (2006) p53 modulates homologous recombination by transcriptional regulation of the RAD51 gene. EMBO Rep 7:219–224

Attene-Ramos MS, Huang R, Michael S, Witt KL, Richard A, Tice RR, Simeonov A, Austin CP, Xia M (2015) Profiling of the Tox21 chemical collection for mitochondrial function to identify compounds that acutely decrease mitochondrial membrane potential. Environ Health Perspect 123:49–56

Barak Y, Juven T, Haffner R, Oren M (1993) mdm2 expression is induced by wild type p53 activity. EMBO J 12:461–468

Bars RG, Mitchell AM, Wolf CR, Elcombe CR (1989) Induction of cytochrome P-450 in cultured rat hepatocytes. The heterogeneous localization of specific isoenzymes using immunocytochemistry. Biochem J 262:151–158

Batchelor E, Loewer A, Mock C, Lahav G (2011) Stimulus-dependent dynamics of p53 in single cells. Mol Syst Biol 7:488

Bhalla US, Ram PT, Iyengar R (2002) MAP kinase phosphatase as a locus of flexibility in a mitogen-activated protein kinase signaling network. Science 297:1018–1023

Bhattacharya S, Zhang Q, Carmichael PL, Boekelheide K, Andersen ME (2011) Toxicity testing in the 21 century: defining new risk assessment approaches based on perturbation of intracellular toxicity pathways. PLoS One 6:e20887

Boekelheide K, Andersen ME (2010) A mechanistic redefinition of adverse effects—a key step in the toxicity testing paradigm shift. ALTEX 27:243–252

Chen J, Sadowski I (2005) Identification of the mismatch repair genes PMS2 and MLH1 as p53 target genes by using serial analysis of binding elements. Proc Natl Acad Sci U S A 102:4813–4818

Clewell HJ, Tan YM, Campbell JL, Andersen ME (2008) Quantitative interpretation of human biomonitoring data. Toxicol Appl Pharmacol 231:122–133

Clewell RA, Sun B, Adeleye Y, Carmichael P, Efremenko A, McMullen PD, Pendse S, Trask OJ, White A, Andersen ME (2014) Profiling dose-dependent activation of p53-mediated signaling pathways by chemicals with distinct mechanisms of DNA damage. Toxicol Sci 142:56–73

Clewell RA, Miller ME, Alyea R, Andersen ME (2015a) Developing fit-for-pupose assays for estrogenic responses in uterine cells. Toxicol. Sci

Clewell RA and Andersen ME (2016) Approaches for characterizing threshold dose-response relationships for DNA-damage pathways involved in carcinogenicity *in vivo* and micronuclei formation *in vitro*, Mutagenesis, 31, 333–340

Collins FS, Gray GM, Bucher JR (2008) Toxicology. Transforming environmental health protection. Science 319:906–907

Corton JC, Cunningham ML, Hummer BT, Lau C, Meek B, Peters JM, Popp JA, Rhomberg L, Seed J, Klaunig JE (2014) Mode of action framework analysis for receptor-mediated toxicity: the peroxisome proliferator-activated receptor alpha (PPARalpha) as a case study. Crit Rev Toxicol 44:1–49

Elcombe CR, Peffer RC, Wolf DC, Bailey J, Bars R, Bell D, Cattley RC, Ferguson SS, Geter D, Goetz A, Goodman JI, Hester S, Jacobs A, Omiecinski CJ, Schoeny R, Xie W, Lake BG (2014) Mode of action and human relevance analysis for nuclear receptor-mediated liver toxicity: a case study with phenobarbital as a model constitutive androstane receptor (CAR) activator. Crit Rev Toxicol 44:64–82

el-Deiry WS, Tokino T, Velculescu VE, Levy DB, Parsons R, Trent JM, Lin D, Mercer WE, Kinzler KW, Vogelstein B (1993) WAF1, a potential mediator of p53 tumor suppression. Cell 75:817–825

Filardo EJ, Quinn JA, Bland KI, Frackelton AR Jr (2000) Estrogen-induced activation of Erk-1 and Erk-2 requires the G protein-coupled receptor homolog, GPR30, and occurs via trans-activation of the epidermal growth factor receptor through release of HB-EGF. Mol Endocrinol 14:1649–1660

Hall JM, McDonnell DP (1999) The estrogen receptor beta-isoform (ERbeta) of the human estrogen receptor modulates ERalpha transcriptional activity and is a key regulator of the cellular response to estrogens and antiestrogens. Endocrinology 140:5566–5578

Harper JW, Adami GR, Wei N, Keyomarsi K, Elledge SJ (1993) The p21 Cdk-interacting protein Cip1 is a potent inhibitor of G1 cyclin-dependent kinases. Cell 75:805–816

He YY, Cai B, Yang YX, Liu XL, Wan XP (2009) Estrogenic G protein-coupled receptor 30 signaling is involved in regulation of endometrial carcinoma by promoting proliferation, invasion potential, and interleukin-6 secretion via the MEK/ERK mitogen-activated protein kinase pathway. Cancer Sci 100:1051–1061

Ignatov T, Eggemann H, Semczuk A, Smith B, Bischoff J, Roessner A, Costa SD, Kalinski T, Ignatov A (2010) Role of GPR30 in endometrial pathology after tamoxifen for breast cancer. Am J Obstet Gynecol 203(595):e599–516

Johnson SM, Maleki-Dizaji M, Styles JA, White IN (2007) Ishikawa cells exhibit differential gene expression profiles in response to oestradiol or 4-hydroxytamoxifen. Endocr Relat Cancer 14:337–350

Judson RS, Houck KA, Kavlock RJ, Knudsen TB, Martin MT, Mortensen HM, Reif DM, Rotroff DM, Shah I, Richard AM, Dix DJ (2010) *In vitro* screening of environmental chemicals for targeted testing prioritization: the ToxCast project. Environ Health Perspect 118:485–492

Judson RS, Kavlock RJ, Setzer RW, Hubal EA, Martin MT, Knudsen TB, Houck KA, Thomas RS, Wetmore BA, Dix DJ (2011) Estimating toxicity-related biological pathway altering doses for high-throughput chemical risk assessment. Chem Res Toxicol 24:451–462

Kavlock R, Chandler K, Houck K, Hunter S, Judson R, Kleinstreuer N, Knudsen T, Martin M, Padilla S, Reif D, Richard A, Rotroff D, Sipes N, Dix D (2012) Update on EPA's ToxCast program: providing high throughput decision support tools for chemical risk management. Chem Res Toxicol 25:1287–1302

Klaunig JE, Babich MA, Baetcke KP, Cook JC, Corton JC, David RM, DeLuca JG, Lai DY, McKee RH, Peters JM, Roberts RA, Fenner-Crisp PA (2003) PPARalpha agonist-induced rodent tumors: modes of action and human relevance. Crit Rev Toxicol 33:655–780

Kleinstreuer NC, Judson RS, Reif DM, Sipes NS, Singh AV, Chandler KJ, Dewoskin R, Dix DJ, Kavlock RJ, Knudsen TB (2011) Environmental impact on vascular development predicted by high-throughput screening. Environ Health Perspect 119:1596–1603

Klinge CM, Riggs KA, Wickramasinghe NS, Emberts CG, McConda DB, Barry PN, Magnusen JE (2010) Estrogen receptor alpha 46 is reduced in tamoxifen resistant breast cancer cells and re-expression inhibits cell proliferation and estrogen receptor alpha 66-regulated target gene transcription. Mol Cell Endocrinol 323:268–276

Krewski D, Acosta D Jr, Andersen M, Anderson H, Bailar JC III, Boekelheide K, Brent R, Charnley G, Cheung VG, Green S Jr, Kelsey KT, Kerkvliet NI, Li AA, McCray L, Meyer O, Patterson RD, Pennie W, Scala RA, Solomon GM, Stephens M, Yager J, Zeise L (2010) Toxicity testing in the 21st century: a vision and a strategy. J Toxicol Environ Health B Crit Rev 13:51–138

Krewski D, Westphal M, Al-Zoughool M, Croteau MC, Andersen ME (2011) New directions in toxicity testing. Annu Rev Public Health 32:161–178

Lahav G, Rosenfeld N, Sigal A, Geva-Zatorsky N, Levine AJ, Elowitz MB, Alon U (2004) Dynamics of the p53-Mdm2 feedback loop in individual cells. Nat Genet 36:147–150

Lehman AJ, Fitzhugh OG (1954) 100-fold margin of safety. Association Food Drug Office U.S.Q. Bulletin 18:33–35

Lin AH, Li RW, Ho EY, Leung GP, Leung SW, Vanhoutte PM, Man RY (2013) Differential ligand binding affinities of human estrogen receptor-alpha isoforms. PLoS One 8:e63199

Louis M, Becskei A (2002) Binary and graded responses in gene networks. Sci STKE 2002(143):pe33

Martin MT, Dix DJ, Judson RS, Kavlock RJ, Reif DM, Richard AM, Rotroff DM, Romanov S, Medvedev A, Poltoratskaya N, Gambarian M, Moeser M, Makarov SS, Houck KA (2010) Impact of environmental chemicals on key transcription regulators and correlation to toxicity end points within EPA's ToxCast program. Chem Res Toxicol 23:578–590

Martin MT, Knudsen TB, Reif DM, Houck KA, Judson RS, Kavlock RJ, Dix DJ (2011) Predictive model of rat reproductive toxicity from ToxCast high throughput screening. Biol Reprod 85:327–339

McMullen PD, Bhattacharya S, Woods CG, Sun B, Yarborough K, Ross SM, Miller ME, McBride MT, LeCluyse EL, Clewell RA, Andersen ME (2014) A map of the PPARalpha transcription regulatory network for primary human hepatocytes. Chem Biol Interact 209:14–24

McMullen PD, Andersen ME, Pendse S, Bhattacharya S, Clewell RA, LeCluyse EL (2015) Identifying qualitative differences in PPARa signaling networks in human and rat hepatocytes. Toxicol Sci Rev. Identifying qualitative differences in PPARa signaling networks in human and rat hepatocytes – significance for risk assessment

Mettetal JT, Muzzey D, Gomez-Uribe C, van Oudenaarden A (2008) The frequency dependence of osmo-adaptation in Saccharomyces cerevisiae. Science 319:482–484

Miyashita T, Reed JC (1995) Tumor suppressor p53 is a direct transcriptional activator of the human bax gene. Cell 80:293–299

Murk AJ, Rijntjes E, Blaauboer BJ, Clewell R, Crofton KM, Dingemans MM, Furlow JD, Kavlock R, Kohrle J, Opitz R, Traas T, Visser TJ, Xia M, Gutleb AC (2013) Mechanism-based testing strategy using *in vitro* approaches for identification of thyroid hormone disrupting chemicals. Toxicol *In Vitro* 27:1320–1346

Neumaier T, Swenson J, Pham C, Polyzos A, Lo AT, Yang P, Dyball J, Asaithamby A, Chen DJ, Bissell MJ, Thalhammer S, Costes SV (2012) Evidence for formation of DNA repair centers and dose-response nonlinearity in human cells. Proc Natl Acad Sci U S A 109:443–448

NRC (1983) Risk assessment in the federal government: managing the process. National Academy Press, Washington, DC

NRC (2007) Toxicity testing in the 21st century: a vision and a strategy. The National Academies Press, Washington, DC

Oda K, Arakawa H, Tanaka T, Matsuda K, Tanikawa C, Mori T, Nishimori H, Tamai K, Tokino T, Nakamura Y, Taya Y (2000) p53AIP1, a potential mediator of p53-dependent apoptosis, and its regulation by Ser-46-phosphorylated p53. Cell 102:849–862

Odom DT, Zizlsperger N, Gordon DB, Bell GW, Rinaldi NJ, Murray HL, Volkert TL, Schreiber J, Rolfe PA, Gifford DK, Fraenkel E, Bell GI, Young RA (2004) Control of pancreas and liver gene expression by HNF transcription factors. Science 303:1378–1381

OECD. (2007), Test No. 440: Uterotrophic Bioassay in Rodents: A short-term screening test for oestrogenic properties, OECD Guidelines for the Testing of Chemicals, Section 4, OECD Publishing, Paris. DOI: http://dx.doi.org/10.1787/9789264067417-en

Penot G, Le Peron C, Merot Y, Grimaud-Fanouillere E, Ferriere F, Boujrad N, Kah O, Saligaut C, Ducouret B, Metivier R, Flouriot G (2005) The human estrogen receptor-alpha isoform hERalpha46 antagonizes the proliferative influence of hERalpha66 in MCF7 breast cancer cells. Endocrinology 146:5474–5484

Purvis JE, Karhohs KW, Mock C, Batchelor E, Loewer A, Lahav G (2012) p53 dynamics control cell fate. Science 336:1440–1444

Rosen MB, Abbott BD, Wolf DC, Corton JC, Wood CR, Schmid JE, Das KP, Zehr RD, Blair ET, Lau C (2008a) Gene profiling in the livers of wild-type and PPARalpha-null mice exposed to perfluorooctanoic acid. Toxicol Pathol 36:592–607

Rosen MB, Lee JS, Ren H, Vallanat B, Liu J, Waalkes MP, Abbott BD, Lau C, Corton JC (2008b) Toxicogenomic dissection of the perfluorooctanoic acid transcript profile in mouse liver: evidence for the involvement of nuclear receptors PPAR alpha and CAR. Toxicol Sci 103:46–56

Rosen MB, Schmid JR, Corton JC, Zehr RD, Das KP, Abbott BD, Lau C (2010) Gene expression profiling in wildtype and PPARalpha-null mice exposed to perfluorooctane sulfonate reveals PPARalpha-independent effects. PPAR Res 794739:1–23. http://dx.doi.org/10.1155/2010/794739

Rotroff DM, Wetmore BA, Dix DJ, Ferguson SS, Clewell HJ, Houck KA, Lecluyse EL, Andersen ME, Judson RS, Smith CM, Sochaski MA, Kavlock RJ, Boellmann F, Martin MT, Reif DM, Wambaugh JF, Thomas RS (2010) Incorporating human dosimetry and exposure into high-throughput *in vitro* toxicity screening. Toxicol Sci 117:348–358

Rotroff DM, Dix DJ, Houck KA, Knudsen TB, Martin MT, McLaurin KW, Reif DM, Crofton KM, Singh AV, Xia M, Huang R, Judson RS (2013) Using *in vitro* high throughput screening assays to identify potential endocrine-disrupting chemicals. Environ Health Perspect 121:7–14

Rotroff DM, Martin MT, Dix DJ, Filer DL, Houck KA, Knudsen TB, Sipes NS, Reif DM, Xia M, Huang R, Judson RS (2014) Predictive endocrine testing in the 21st century using *in vitro* assays of estrogen receptor signaling responses. Environ Sci Technol 48:8706–8716

Russell WMS, Burch RL (1959) The principles of humane experimental technique. Methuen, London

Sarangapani R, Teeguarden J, Plotzke KP, McKim JM Jr, Andersen ME (2002) Dose-response modeling of cytochrome p450 induction in rats by octamethylcyclotetrasiloxane. Toxicol Sci 67:159–172

Simmons SO, Fan CY, Ramabhadran R (2009) Cellular stress response pathway system as a sentinel ensemble in toxicological screening. Toxicol Sci 111:202–225

Sipes NS, Martin MT, Reif DM, Kleinstreuer NC, Judson RS, Singh AV, Chandler KJ, Dix DJ, Kavlock RJ, Knudsen TB (2011) Predictive models of prenatal developmental toxicity from ToxCast high-throughput screening data. Toxicol Sci 124:109–127

Sipes NS, Martin MT, Kothiya P, Reif DM, Judson RS, Richard AM, Houck KA, Dix DJ, Kavlock RJ, Knudsen TB (2013) Profiling 976 ToxCast chemicals across 331 enzymatic and receptor signaling assays. Chem Res Toxicol 26:878–895

Thomas RS, Black M, Li L, Healy E, Chu TM, Bao W, Andersen M, Wolfinger R (2012) A comprehensive statistical analysis of predicting *in vivo* hazard using high-throughput *in vitro* screening. Toxicol Sci 128(2):398–417

Thornborrow EC, Patel S, Mastropietro AE, Schwartzfarb EM, Manfredi JJ (2002) A conserved intronic response element mediates direct p53-dependent transcriptional activation of both the human and murine bax genes. Oncogene 21:990–999

van der Meer DL, Degenhardt T, Vaisanen S, de Groot PJ, Heinaniemi M, de Vries SC, Muller M, Carlberg C, Kersten S (2010) Profiling of promoter occupancy by PPARalpha in human hepatoma cells via ChIP-chip analysis. Nucleic Acids Res 38:2839–2850

Wang Z, Zhang X, Shen P, Loggie BW, Chang Y, Deuel TF (2006) A variant of estrogen receptor-{alpha}, hER-{alpha}36: transduction of estrogen- and antiestrogen-dependent membrane-initiated mitogenic signaling. Proc Natl Acad Sci U S A 103:9063–9068

Wax PM (1995) Elixirs, diluents, and the passage of the 1938 Federal Food, Drug and Cosmetic Act. Ann Intern Med 122:456–461

Wetmore BA, Wambaugh JF, Ferguson SS, Sochaski MA, Rotroff DM, Freeman K, Clewell HJ III, Dix DJ, Andersen ME, Houck KA, Allen B, Judson RS, Singh R, Kavlock RJ, Richard AM, Thomas RS (2012) Integration of dosimetry, exposure, and high-throughput screening data in chemical toxicity assessment. Toxicol Sci 125:157–174

Zhang Q, Bhattacharya S, Andersen ME (2013) Ultrasensitive response motifs: basic amplifiers in molecular signalling networks. Open Biol 3:130031

Zhang Q, Bhattacharya S, Conolly RB, Clewell HJ III, Kaminski NE, Andersen ME (2014) Molecular signaling network motifs provide a mechanistic basis for cellular threshold responses. Environ Health Perspect 122(12):1261–1270

Zhang Q, Bhattacharya S, Pi J, Clewell RA, Carmichael PL, Andersen ME (2015) Adaptive posttranslational control in cellular stress response pathways and its relationship to toxicity testing and safety assessment. Toxicol Sci 147(2):302–316

Chapter 9
Evidence-Based Toxicology

Sebastian Hoffmann, Thomas Hartung and Martin Stephens

Abstract Evidence-based toxicology (EBT) was introduced independently by two groups in 2005, in the context of toxicological risk assessment and causation as well as based on parallels between the evaluation of test methods in toxicology and evidence-based assessment of diagnostics tests in medicine. The role model of evidence-based medicine (EBM) motivated both proposals and guided the evolution of EBT, whereas especially systematic reviews and evidence quality assessment attract considerable attention in toxicology.

Regarding test assessment, in the search of solutions for various problems related to validation, such as the imperfectness of the reference standard or the challenge to comprehensively evaluate tests, the field of Diagnostic Test Assessment (DTA) was identified as a potential resource. DTA being an EBM discipline, test method assessment/validation therefore became one of the main drivers spurring the development of EBT.

In the context of pathway-based toxicology, EBT approaches, given their objectivity, transparency and consistency, have been proposed to be used for carrying out a (retrospective) mechanistic validation.

In summary, implementation of more evidence-based approaches may provide the tools necessary to adapt the assessment/validation of toxicological test methods and testing strategies to face the challenges of toxicology in the twenty first century.

Keywords Evidence-based toxicology • Test method assessment • Validation • Mechanistic validation • Diagnostic test assessment

S. Hoffmann (✉)
seh consulting + services, Paderborn, Germany
e-mail: sebastian.hoffmann@seh-cs.com

T. Hartung
Johns Hopkins Bloomberg School of Public Health, Center for Alternatives to Animal Testing (CAAT), Baltimore, MD, USA

University of Konstanz, CAAT-Europe, Konstanz, Germany

M. Stephens
Johns Hopkins Bloomberg School of Public Health, Center for Alternatives to Animal Testing (CAAT), Baltimore, MD, USA

C. Eskes, M. Whelan (eds.), *Validation of Alternative Methods for Toxicity Testing*, Advances in Experimental Medicine and Biology 856,
DOI 10.1007/978-3-319-33826-2_9

1 Introduction

1.1 The Roots of Evidence-Based Toxicology

Evidence-based toxicology (EBT) was introduced independently by two groups in 2005. Guzelian et al. (2005) used EBT in the context of toxicological risk assessment and causation. Hoffmann and Hartung (2005) were motivated to coin the term EBT by parallels between the evaluation of test methods in toxicology and evidence-based assessment of diagnostics tests in medicine. The concept of adopting and adapting evidence-based approaches to toxicology was further elaborated (Hoffmann and Hartung 2006). In order to more fully implement the evidence-based principles of transparency, consistency and objectivity, steps such as application of systematic review techniques, critical review of toxicological test methods and improvement of practices for quality assurance of toxicological studies were proposed.

1.2 The Role Model of Evidence-Based Medicine

Evidence-based approaches have first been developed in medicine. Spurred in the 1970s by the challenge of comprehensively and adequately accounting for the growing amount of evidence becoming available and translating this into clinical practice, evidence-based medicine (EBM), which later was expanded to evidence-based health care (EBHC), started to emerge. EBM/EBHC, concisely described by Eddy (2005), has developed into a widely accepted approach for assessing clinical practices, especially in diagnosis, prognosis and treatment, on the basis of research evidence, clinical expertise and the expectations of patients (Sackett et al. 1996).

To transparently and objectively synthesize the ever-growing evidence while minimizing the impact of potential biases, systematic review was developed as the core tool of EBM (Horvath and Pewsner 2004). Systematic reviews comprise several steps. After framing the research question, a protocol is developed describing the literature search, inclusion/exclusion criteria, the quality appraisal of studies, and the synthesis of the evidence, which may include a meta-analysis. The protocol itself is often peer-reviewed, and then implemented, with the whole process well-documented. This ensures that a systematic review can be readily reproduced or updated, which is usually not the case for narrative reviews, the standard approach to literature review in toxicology. The synthesised research evidence is made available to the medical community via publication. For example, systematic reviews registered with the Cochrane Collaboration (Cochrane reviews) are published through the Cochrane Library.

Guidance on systematic reviews is readily available. The 'Cochrane Handbook for Systematic Review of Interventions' may be considered as the most comprehensive source of guidance (Higgins and Green 2008). In addition, the Cochrane Collaboration Diagnostic Test Accuracy Working Group is preparing a handbook for diagnostic test assessment (DTA) reviews (see www.srdta.cochrane.org).

For the literature search, information specialists should be consulted to identify the sources to search, to develop an appropriate search strategy and to help document the search process. Search results should be centrally compiled in reference management software.

The data appraisal step of systematic reviews requires a harmonised and widely accepted scheme to assess the data quality of included studies or their potential for bias. Evidence levels are used to structure the evidence according to its nature, ranging for example from expert opinion to studies of high quality, such as randomised clinical trials, and systematic reviews thereof. Evidence levels have been described for most, if not all, clinical fields. The Centre for Evidence Based Medicine (CEBM), for example, provides an overview on their website.

Once the individual pieces of evidence, such as studies or case reports, have been assigned to an evidence level, the quality of them needs to be evaluated in detail. A wealth of appraisal tools is available, covering all clinical fields and evidence levels. In the field of diagnosis, examples of such tools are QUADAS-2 (Whiting et al. 2011) and QAREL (Lucas et al. 2010). In general, these tools ask a number of questions, which are to be answered for each piece of evidence.

Finally, the selected and appraised evidence is systematically analysed, often, but not necessarily, by using meta-analysis (Egger and Smith 1997; Egger et al. 1997).

1.3 The Evolution of EBT

Based on parallels in the problems faced in medicine and toxicology, the usefulness of the principles and approaches of EBM for toxicology were explored (Guzelian et al. 2005; Hoffmann and Hartung 2006). At an international forum in 2007, the concepts of EBT were discussed and defining characteristics were proposed (Griesinger and Guzelian 2009).

After these preparatory steps, two evidence-based approaches have attracted considerable attention in toxicology. Systematic reviews have been taken-up and implemented by various governmental institutions to comprehensively assess the human health hazards associated with chemical substances of interest (EFSA 2010; Rooney et al. 2014; NRC 2014). Such reviews seem well suited to assessing toxicological evidence, where the stakes are often high with respect to public health, environment, industry interests, and animal use.

In addition, the need for an objective assessment of evidence quality—a pivotal step in a systematic review—has been acknowledged. Initial attempts, such as ToxRTool (Schneider et al. 2009), have been followed by a critical appraisal (Krauth et al. 2013) and further research on the topic (e.g. Maxim and van der Sluijs 2014).

Additionally, Bus and Becker (2009) and Hartung (2010) have proposed to apply the concepts of evidence-based toxicology as an objective means for evaluating the methods needed and developed for implementing the 'Toxicity Testing for the twenty first Century—a Vision and a Strategy' proposed by the US National Research Council (NRC 2007).

Scientists from academia, governmental agencies and industry from Europe and North America established an Evidence-Based Toxicology Collaboration (EBTC) in 2011 (see www.ebtox.com), which committed itself to translating evidence-based approaches from medicine to toxicology by furthering the conceptual development of EBT and by serving as a hub for the various developments.

The EBTC has especially developed interest in exploring evidence-based approaches to toxicological test assessment. For this, methodology and tools developed in the field of evidence-based evaluation of diagnostic tests were considered to bear considerable potential. For example, systematic review techniques developed for diagnostic test assessment apply in a similar manner to the retrospective assessment of toxicological test methods. In addition, the Collaboration sees potential of EBT concepts for test assessment to support the vision of a pathway-driven toxicology (NRC 2007; Andersen et al. 2010). By definition, pathway toxicology needs to be based on mechanisms applicable to the target species, i.e. usually humans, in order to minimise the uncertainty introduced by species differences. This opens up the possibility to reduce or eliminate the need to rely on animal toxicity data in general and, more specifically, potentially diminishes the need of animal toxicity data to serve as the default reference standard in new test assessments. In addition, pathway-driven toxicology generates large amounts of data. In order to make best use of these data and preclude discussions on potentially various interpretations, a homogenous and widely acceptable way to extract evidence from these data—as well as synthesize such data—are key. Evidence-based approaches would be highly qualified to achieve this by providing the means to systematically and transparently review the evidence as well as to provide an objective synthesis thereof (Hartung 2010).

2 The Evidence-Base of Validation

2.1 *Diagnostic Test Assessment (DTA)*

The principles and approaches for the validation of toxicological test methods were developed in the 1990s, when—primarily as a response to political pressure build-up by the animal protection movement—*in vitro* test methods emerged for the assessment of local toxic effects. In order to demonstrate to the scientific community (including regulators) the appropriateness of these *in vitro* methods, a rigorous assessment framework, named validation, was developed (Balls et al. 1990) and subsequently fine-tuned (Curren et al. 1995; Hartung et al. 2004). This resulted in validation as being a formal requirement for a new test method to be accepted as a test guideline by the Organization for Economic Cooperation and Development (OECD), as detailed in the 'Guidance document on the validation and international acceptance of new or updated test methods for hazard assessment' (OECD 2005).

In essence, validation is the process by which the relevance and reproducibility of a toxicological test method is assessed for a particular purpose (Leist et al. 2012). A test's relevance has three components (Fig. 9.1): predictivity (its ability to predict

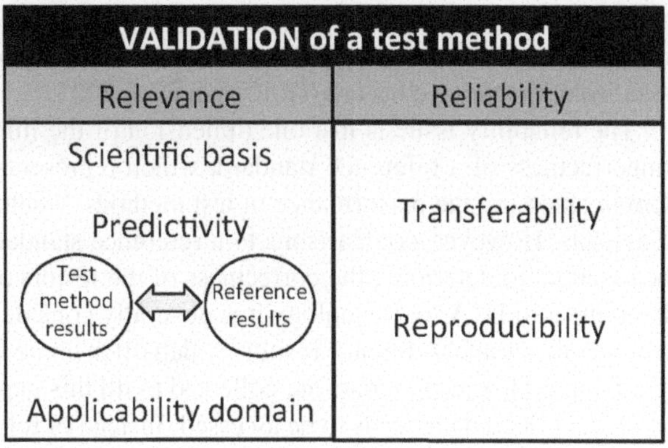

Fig. 9.1 Essential elements of test method validation

the results of the reference test it is intended to replace or supplement), scientific basis (how well it reflects the biology or mechanism of action underlying the end-point in question) and the applicability domain (a description of the classes of substances that can and that cannot be tested with the test method). The reliability of a test method is evaluated by assessing its transferability, i.e. how well a test method can be established at other sites and laboratory, and its reproducibility, focused on the reproducibility of results within laboratories and between laboratories.

In practice, predictivity has been emphasized in validation exercises. However, various problems related to this aspect of validation have been identified, mainly due to the comparative assessment approach, such as the imperfectness of the reference standard, assurance of sufficient quality of the retrospective reference data, and the need to incorporate mechanistic evidence. Others problems also arose from the emphasis on predictivity, such as ensuring an appropriate test substance selection, and translating the results of a validation study to the use scenarios of a test method in practice. Some of these have been described by Hoffmann et al. (2008a).

The field of DTA was identified as a potential resource for solutions to these problems because of the many similarities between diagnostic and toxicological test assessment (Hoffmann and Hartung 2005). DTA being an EBM discipline, test method assessment (validation) therefore became one of the main drivers spurring the development of EBT.

Before addressing some of the above-mentioned issues in more detail, it has to be acknowledged that validation features various characteristics that are in essence evidence-based. Objectivity and reduction of potential biases were considered of utmost importance to validation and have led to several formal requirements, such as independence in the management of a validation study including the substance selection and the statistical data analysis, specification of the statistics to be applied before initiating the study, and a thorough review process. It may be argued that the reliability aspect of a validation also contributes to the evidence-base as this component provides helpful information for assessing the confidence in results. Reliability assessments may ultimately inform the predictivity assessment, as inherent variability of the results of the test method under review as well as the reference standard, can be taken into account by better defining a realistic optimal predictive performance. This has, for example, been done for the *in vivo* reference test methods for skin irritation and eye hazard (Hoffmann et al. 2005; Adriaens et al. 2014).

The reliability issue is just one dimension of the limitations contributing to the imperfectness of a reference standard, which represents an inherent general problem in assessing the performance of test methods, whether in toxicology or clinical diagnosis. However, comparisons to a reference standard have been considered to be essential to determine the correctness of the results of the test method assessed. In contrast to DTA, in toxicology it is generally considered not ethical and feasible to generate parallel reference results, which often come from *in vivo* reference tests. Therefore, retrospective data are collected to fill this gap. In this process the quality of the identified data needs to be assessed. Instead of relying on the often subjective opinion of an expert panel, specific quality appraisal tools could be applied for this purpose. Another factor adding to the imperfectness of the *in vivo* reference test is that it represents only an imperfect model for the human target population. DTA has developed several approaches to at least reduce the impact of this imperfect source, including the correction of predictive parameter estimates and the construction of a composite reference standard that combines multiple test results through a predefined rule (Reitsma et al. 2009).

Even more fundamentally, the predictive performance of a toxicological test method or testing strategy can often be described in more detail. For example, validation studies are focused on the estimation of the predictive parameters of sensitivity (a test's ability to detect positives in a population of positives) and specificity (a test's ability to detect negatives in a population of negatives). The mutual dependence of these two parameters is frequently disregarded, but could be better described by application of receiver operation characteristics (ROC) curves and by calculation of additional predictive parameters, such as predictive values or odd ratios. Such approaches would permit (1) a more comprehensive test method/strategy assessment and (2) the translation of validation study results to other contexts and settings, for example, to populations of chemicals with prevalences different from the one used in the validation study (Hoffmann and Hartung 2005). Validation studies focused on estimation of specificity and sensitivity adopt more or less balanced designs, i.e. they test similar numbers of negative and positive substances, as determined by the reference test (i.e. a prevalence of 50 %). In this way the confidence in both estimates is comparable. In practice, however, numbers of negatives and positives may not be balanced, but may be substantially different. For example, the proportion of substance being carcinogenic has been estimated to be 5–10 % (Fung et al. 1995).

The Bayesian approach of incorporating a priori information, such as prevalence, into the interpretation of test results can be applied not only to populations, but also be used on the level of individual test substances. Consider, for example, two structurally similar substances. For one of them a positive result from a specific toxicological test is available. Submitting also the other substance to the same toxicological test, a positive result would confirm our a-priori expectation, while a negative result may be considered as insufficient to come to a conclusion. This intuitive approach of incorporating prior biological knowledge can be made transparent and more objective, by quantifying the pre-test likelihood of the substance being positive.

2.2 Mechanistic Validation

Like predictivity, the scientific basis of a test is also an important component of a test's relevance, and therefore its validation status. However, in contrast to predictivity, the scientific basis of a test—though a critical consideration during test development—has largely been in the background of validation exercises.

For human health assessment, animal data have served as the default standard of comparison when assessing the predictivity of new methods, owing to ethical constraints on human testing and the paucity of existing high-quality human data for most toxicological endpoints. This introduces the confounding variable of interspecies extrapolation into validation efforts.

Using predictivity as the indicator of relevance has additional limitations in the context of twenty first century toxicology (Tox21C), with its emphasis on pathway- or mechanism-based methods, which are typically run with human cells, often in high-throughout platforms. First, Tox21C methods are not intended to be one-to-one replacements of animal tests, so one-to-one comparison cannot be applied. Second, the new methods often emphasize human biology in their choice of cells and pathways, so optimizing them to predict animal biology is not ideal.

The limitations of using animal data as the default standard in predictivity comparisons constitute one of several factors leading to calls for a reinvention of validation (Birnbaum 2013; Judson et al. 2013). Another factor has been the mismatch between the multi-year duration of typical validation exercises and the rapid pace of evolution with Tox21C methods and approaches (Hartung 2010).

Hartung et al. (2013) proposed to address many of the validation shortcomings, at least for pathway-based methods, by assessing relevance with respect to the assay's scientific basis, and not its correlation to animal data. Building on an earlier suggestion of Frazier (1994), they called for 'mechanistic validation,' that is demonstrating that an assay is a reliable indicator of whether a chemical adversely perturbs a given biological pathway. One can also take this a step further by first demonstrating that a putative pathway accurately reflects the (human) biology in question, and then move on to assess the fidelity of an assay to reflect perturbations in that pathway.

Simply put, the following steps would be part of mechanistic validation:

- Condense the knowledge of biological/mechanistic circuitry (in the absence of xenobiotic challenge) underlying the hazard in question
- Compile evidence that reference chemicals leading to the hazard in question perturb the biology in question, i.e., mainly pathway identification by using reference substances in valid(ated) models and experimental proof of their role
- Develop a test that purports to reflect this biology
- Verify that toxicants shown to employ this mechanism also do so in the model
- Verify that interference with this mechanism hinders positive test results

Thus mechanistic validation could be central to building scientific confidence in assays intended to implement the vision of the National Research Council report on "Toxicity Testing in the twenty first Century" (NRC 2007). The key questions for this implementation are 'what are the key toxicologically relevant pathways?' and 'are assays that accurately reflect chemically induced perturbations to these pathways available?' These are just the sorts of questions that mechanistic validation is intended to address.

Hartung et al. (2013) proposed that evidence-based approaches be used to carry out a (retrospective) mechanistic validation, given their objectivity, transparency and consistency. These defining characteristics would aid any assessment—whether based on mechanism or correlation, in surviving peer scrutiny—but they are especially suited to guide solving the challenges of assessing the validation status of pathway-based assays. In this context, systematic review could and should be used to guide framing the questions at hand, identifying and selecting relevant existing studies, extracting data from included studies, appraising their quality (internal validity or risk of bias), analysing their data, and writing-up the results. Consequently, systematic review, in contrast to narrative reviews, could help limit disagreements among stakeholders over the design, conduct, and reporting of reviews (Stephens et al. 2013).

Of course, one has to know the right questions to ask in any type of review. The Bradford-Hill criteria (Hill 1965) for causality have been proposed as a starting point for assessing the validity of putative pathways and pathway-based assays (Hartung et al. 2013). These criteria can be summarized as relating to strength (of association), consistency, specificity, temporality, biological gradient, plausibility, coherence, and experiment.

A downside of the mechanistic approach to validation is that one needs to know the key toxicologically-relevant pathways. This is a knowledge gap that is being filled incrementally. Fortunately, this is a key aim of the Human Toxome Project (Hartung and McBride 2011; Bouhifd et al. 2015). The end result will mean that confidence in a test will not or not only be based on correlation but also on the accumulated knowledge of how a particular exposure leads to particular effects.

3 Summary

Validation, or more generally, toxicological test assessment, was one of the drivers of developing the concepts of an evidence-based toxicology. While several aspects of validation can be considered evidence-based, solutions for problems faced when validating toxicological test methods were found in diagnostic test assessment (DTA), a discipline of evidence-based medicine (EBM).

Systematic review techniques developed under DTA are applicable for various purposes. Retrospective validation could in essence be conducted as a systematic review. But also prospective validation studies feature a retrospective component: reference test results could be obtained by applying at least some steps of a systematic review, such as the search and the quality assessment. In addition, systematic reviews have been proposed as a tool to implement the pathway-driven toxicology of the twenty first century, through a process of 'mechanistic validation.' These reviews could provide the means to synthesise the huge amount of data generated in an objective and transparent manner.

Other DTA approaches that pertain to toxicological test assessment/validation are methods to describe the predictive performance in greater details as compared to what is usually done. ROC curves describing the dependence of a test's specificity and sensitivity have occasionally been used (e.g., Eskes et al. 2014). Also the concept of evaluating test methods in various prevalence settings has been demonstrated, for example in the context of a testing strategy assessment (Hoffmann et al. 2008b).

In summary, implementation of more evidence-based approaches may provide the tools necessary to adapt the assessment/validation of toxicological test methods and testing strategies to face the challenges of toxicology in the twenty first century.

References

Adriaens E, Barroso J, Eskes C, Hoffmann S, McNamee P, Alépée N, Bessou-Touya S, De Smedt A, De Wever B, Pfannenbecker U, Tailhardat M, Zuang V (2014) Retrospective analysis of the Draize test for serious eye damage/eye irritation: importance of understanding the *in vivo* endpoints under UN GHS/EU CLP for the development and evaluation of *in vitro* test methods. Arch Toxicol 88:701–723

Andersen ME, Al-Zoughool M, Croteau M, Westphal M, Krewski D (2010) The future of toxicity testing. J Toxicol Environ Health B Crit Rev 13:163–196

Balls M, Blaauboer B, Brusick D, Frazier J, Lamb D, Pemberton M, Reinhardt C, Roberfroid M, Rosenkranz H, Schmid B, Spielmann H, Stammati AL, Walum E (1990) Report and recommendations of the CAAT/ERGATT workshop on the validation of toxicity test procedures. Altern Lab Anim 18:313–337

Birnbaum LS (2013) 15 years out: reinventing ICCVAM. Environ Health Perspect 121:40

Bouhifd M, Andersen ME, Baghdikian C, Boekelheide K, Crofton KM, Fornace AJ Jr, Kleensang A, Li H, Livi CB, Maertens A, McMullen PD, Rosenberg M, Thomas R, Vantangoli M, Yager JD, Zhao L, Hartung T (2015) The human toxome project. ALTEX 32:112–124

Bus JS, Becker RA (2009) Toxicity testing in the 21st century: a view from the chemical industry. Toxicol Sci 112:297–302

Curren RD, Southee JA, Spielmann H, Liebsch M, Fentem JH, Balls M (1995) The role of prevalidation in the development, validation and acceptance of alternative methods. Altern Lab Anim 23:211–217

Eddy DM (2005) Evidence-based medicine: a unified approach. Health Aff 24:9–17

EFSA (European Food Safety Authority) (2010) Application of systematic review methodology to food and feed safety assessments to support decision making. EFSA J 8:1–90

Egger M, Smith GD (1997) Meta-analysis: potentials and promise. BMJ 315:1371–1374

Egger M, Smith GD, Phillips AN (1997) Meta-analysis: principles and procedures. BMJ 315:1533–1537

Eskes C, Hoffmann S, Facchini D, Ulmer R, Wang A, Flego M, Vassallo M, Bufo M, van Vliet E, d'Abrosca F, Wilt N (2014) Validation study on the Ocular Irritection(®) assay for eye irritation testing. Toxicol *In Vitro* 28:1046–1065

Frazier JM (1994) The role of mechanistic toxicology in test method validation. Toxicol *In Vitro* 8:787–791
Fung VA, Barrett JC, Huff J (1995) The carcinogenesis bioassay in perspective: application in identifying human cancer hazards. Environ Health Perspect 103:680–683
Griesinger C, Guzelian P (2009) Proceedings of the 1st international forum towards evidence-based toxicology. Hum Exp Toxicol 28:71–163
Guzelian PS, Victoroff MS, Halmes NC, Janes RC, Guzelian CP (2005) Evidence-based toxicology: a comprehensive framework for causation. Hum Exp Toxicol 24:161–201
Hartung T (2010) Evidence-based toxicology - the toolbox of validation for the 21st century? ALTEX 27:253–263
Hartung T, McBride M (2011) Food for thought … on mapping the human toxome. ALTEX 28:83–93
Hartung T, Bremer S, Casati S, Coecke S, Fortaner S, Gribaldo L, Halder M, Hoffmann S, Janusch-Roi A, Prieto P, Sabbioni E, Scott L, Worth A, Zuang V (2004) A modular approach to the ECVAM principles on test validity. Altern Lab Anim 32:467–472
Hartung T, Hoffmann S, Stephens M (2013) Mechanistic validation. ALTEX 30:119–130
Higgins JPT, Green S (eds) (2008) Cochrane handbook for systematic reviews of interventions. Wiley, Chichester (UK)
Hill AB (1965) The environment and disease: association or causation? Proc R Soc Med 58:295–300
Hoffmann S, Hartung T (2005) Diagnosis: toxic! – trying to apply approaches of clinical diagnostics and prevalence in toxicology considerations. Toxicol Sci 85:422–428
Hoffmann S, Hartung T (2006) Toward an evidence-based toxicology. Hum Exp Toxicol 25:497–513
Hoffmann S, Cole T, Hartung T (2005) Skin irritation: prevalence, variability, and regulatory classification of existing *in vivo* data from industrial chemicals. Regul Toxicol Pharmacol 41:159–166
Hoffmann S, Edler L, Gardner I, Gribaldo L, Hartung T, Klein C, Liebsch M, Sauerland S, Schechtman L, Stammati A, Nikolaidis E (2008a) Points of reference in the validation process. Altern Lab Anim 36:343–352
Hoffmann S, Saliner AG, Patlewicz G, Eskes C, Zuang V, Worth AP (2008b) A feasibility study developing an integrated testing strategy assessing skin irritation potential of chemicals. Toxicol Lett 180:9–20
Horvath AR, Pewsner D (2004) Systematic reviews in laboratory medicine: principles, processes and practical considerations. Clin Chim Acta 342:23–39
Judson R, Kavlock R, Martin M, Reif D, Houck K, Knudsen T, Richard A, Tice RR, Whelan M, Xia M, Huang R, Austin C, Daston G, Hartung T, Fowle JR III, Wooge W, Tong W, Dix D (2013) Perspectives on validation of high-throughput assays supporting 21st century toxicity testing. ALTEX 30:51–56
Krauth D, Woodruff TJ, Bero L (2013) Instruments for assessing risk of bias and other methodological criteria of published animal studies: a systematic review. Environ Health Perspect 121:985–992
Leist M, Hasiwa M, Daneshian M, Hartung T (2012) Validation and quality control of replacement alternatives – current status and future challenges. Toxicol Res 1:8–22
Lucas NP, Macaskill P, Irwig L, Bogduk N (2010) The development of a quality appraisal tool for studies of diagnostic reliability (QAREL). J Clin Epidemiol 63:854–861
Maxim L, van der Sluijs JP (2014) Qualichem *in vivo*: a tool for assessing the quality of *in vivo* studies and its application for bisphenol A. PLoS One 9:e87738
NRC (National Research Council) (2007) Toxicity testing in the 21st century: a vision and a strategy. National Academy Press, Washington, DC
NRC (National Research Council) (2014) Review of EPA's Integrated Risk Information System (IRIS) process. National Academy Press, Washington, DC

OECD (2005) Guidance document on the validation and international acceptance of new or updated test methods for hazard identification. 1-96. Organisation for Economic Cooperation and Development. Environmental Health and Safety Monograph Series on Testing and Assessment No. 34

Reitsma JB, Rutjes AWS, Khan KS, Coomarasamy A, Bossuyt PM (2009) A review of solutions for diagnostic accuracy studies with an imperfect or missing reference standard. J Clin Epidemiol 62:797–806

Rooney AA, Boyles AL, Wolfe MS, Bucher JR, Thayer KA (2014) Systematic review and evidence integration for literature-based environmental health science assessments. Environ Health Perspect 122:711–718

Sackett DL, Rosenberg WMC, Gray JAM, Haynes RB, Richardson WS (1996) Evidence based medicine: what it is and what it isn't. BMJ 312:71–72

Schneider K, Schwarz M, Burkholder I, Kopp-Schneider A, Edler L, Kinsner-Ovaskainen A, Hartung T, Hoffmann S (2009) "ToxRTool", a new tool to assess the reliability of toxicological data. Toxicol Lett 189:138–144

Stephens ML, Andersen M, Becker RA, Betts K, Boekelheide K, Carney E, Chapin R, Devlin D, Fitzpatrick S, Fowle JR III, Harlow P, Hartung T, Hoffmann S, Holsapple M, Jacobs A, Judson R, Naidenko O, Pastoor T, Patlewicz G, Rowan A, Scherer R, Shaikh R, Simon T, Wolf D, Zurlo J (2013) Evidence-based toxicology for the 21st century: opportunities and challenges. ALTEX 30:74–103

Whiting PF, Rutjes AW, Westwood ME, Mallett S, Deeks JJ, Reitsma JB, Leeflang MM, Sterne JA, Bossuyt PM, QUADAS-2 Group (2011) QUADAS-2: a revised tool for the quality assessment of diagnostic accuracy studies. Ann Intern Med 155:529–536

Chapter 10
Validation of Transcriptomics-Based *In Vitro* Methods

Raffaella Corvi, Mireia Vilardell, Jiri Aubrecht and Aldert Piersma

Abstract The field of transcriptomics has expanded rapidly during the last decades. This methodology provides an exceptional framework to study not only molecular changes underlying the adverse effects of a given compound, but also to understand its Mode of Action (MoA). However, the implementation of transcriptomics-based tests within the regulatory arena is not a straightforward process. One of the major obstacles in their regulatory implementation is still the interpretation of this new class of data and the judgment of the level of confidence of these tests. A key element in the regulatory acceptance of transcriptomics-based tests is validation, which still represents a major challenge. Although important advances have been made in the development and standardisation of such tests, to date there is limited experience with their validation. Taking into account the experience acquired so far, this chapter describes those aspects that were identified as important in the validation process of transcriptomics-based tests, including the assessment of standardisation, reliability and relevance. It also critically discusses the challenges posed to validation in relation to the specific characteristics of these approaches and their application in the wider context of testing strategies.

Keywords Transcriptomics • Validation • Toxicogenomics • *In vitro* tests • Bioinformatics workflow

---

R. Corvi (✉) • M. Vilardell
European Commission, Joint Research Centre (JRC), Ispra, Italy
e-mail: raffaella.corvi@ec.europa.eu

J. Aubrecht
Pfizer Global Research and Development, Groton, CT, USA

A. Piersma
Center for Health Protection, National Institute for Public Health and the Environment RIVM, Bilthoven, The Netherlands

Institute for Risk Assessment Sciences, Utrecht University, Utrecht, The Netherlands

C. Eskes, M. Whelan (eds.), *Validation of Alternative Methods for Toxicity Testing*, Advances in Experimental Medicine and Biology 856,
DOI 10.1007/978-3-319-33826-2_10

1 Introduction

Transcriptomics, in particular the gene-array technology, allows the detection of mRNA levels of multiple genes in parallel. This provides genome-wide information on biological processes and associated molecular pathways and can be used as a powerful application to screen compounds for toxicological potential. Importantly, such information may also facilitate biomarker identification for compound and tissue-specific human toxicity assessment allowing better classification and earlier prediction of toxicity.

When initial studies (Hamadeh et al. 2001, 2002a, b) indicated that "gene signatures" could be used for discriminating toxic versus non-toxic agents, toxicogenomics stepped in as a scientific sub-discipline of toxicology. During the last two decades this field, and in particular transcriptomics, has expanded rapidly. The advantage of this approach is the ability of a single investigation to query and quantify the levels of hundreds to tens of thousands transcriptional gene products in a single assay. These methodologies provide an exceptional framework to study not only molecular changes underlying the adverse effects of a given compound, but also to understand its Mode of Action (MoA) through developing hypotheses and setting up the appropriate experimental designs. They are currently used in basic research and as screening tests, but researchers are also investigating their application in the regulatory context for hazard and risk assessment of compounds (Bourdon-Lacombe et al. 2015). The implementation of transcriptomics within the regulatory arena is not a straightforward process. One of the major obstacles in their regulatory implementation is still the interpretation of this new class of data and the judgment of the level of confidence of these tests (Pettit et al. 2010; Goetz et al. 2011).

A key element in the regulatory acceptance of transcriptomics-based tests is indeed validation, which represents a major challenge (Corvi et al. 2006). Although important advances have been made in the development and standardisation of such tests, to date there is limited experience with their validation, i.e. the way to assess the validation status is not clearly defined, yet.

Generally, validation studies focus on the assessment of test reliability and test relevance for a well-defined purpose. *Reliability* is linked to test reproducibility and provides information about the level of standardisation of the technology through controlling the major sources of variation. *Relevance* is the extent to which a test method correctly predicts the biological effect of interest. Consequently, relevance is a concept that is directly related to the mechanism of toxicity inducing the adverse effect *in vivo*.

In order to assess the scientific validity of a test method, the EU Reference Laboratory for Alternatives to Animal Testing (EURL ECVAM) follows the modular approach to validation (Hartung et al. 2004). This consists of six modules: (1) test definition, (2) within-laboratory reproducibility, (3) transferability, (4) between-laboratory reproducibility, (5) predictive capacity, (6) applicability domain and (7) performance standards. The question here is whether this approach is also applicable to the evaluation of transcriptomics-based tests.

Transcriptomics-based tests can be divided into two main applications based on the specific use of the data generated: (1) to reveal molecular mechanisms of action, based on defined sets of genes or pathways related to known Modes of Action (MoA); and (2) to find gene signatures using the whole transcriptome to make predictions of different toxicity classes. Although different levels of validation might be needed for the different approaches, a high level of standardisation and reproducibility of each individual test remains essential whatever regulatory use is envisaged.

Taking into account the experience acquired so far, this chapter aims at describing and discussing those aspects that were identified as critical in the validation process of transcriptomics-based tests, including the assessment of standardisation, reliability and relevance.

2 Test Definition and Standardisation

2.1 Experimental Design and Test Development

The definition of the test method is essential and it is the first step that needs to be considered during its development and the assessment of its scientific validity. The toxicogenomics methodology is based on the assumption that toxicity is accompanied by changes in the gene expression profiles that are causally linked or represent a response to a toxic insult (Steiner et al. 2004). This implies that toxicity-induced changes need to be differentiated from the experimental noise and from adaptive physiological changes through the selection of an appropriate experimental design (Goetz et al. 2011). Several factors should be taken into consideration in order to set up the appropriate experimental design when developing a transcriptomics-based test: (1) the objective of the study and the purpose of the test; (2) the appropriate representation of the chemical space; (3) the biological model; (4) the test item concentrations and exposure time to observe effects; (5) the controls (two classes of matched controls can be considered: a vehicle control and a compound of similar structure that does not produce the adverse outcome); (6) the number of experiments that can be performed simultaneously; (7) the presence of sources of variation which encompass, among others, batch effects, operators, sample variability, vehicles used; (8) possible pooling of samples; (9) the complex bioinformatics analysis; and (10) other aspects if known. In particular, the high dimensionality of data and the large number of parameters that need to be considered, including the bioinformatics analysis, pose a huge challenge in the experimental design of a transcriptomics-based test method and in the following validation as compared with that of a conventional test method.

The experimental design of a transcriptomics-based test is critical and must reflect the question that is being asked. As for any approach that undergoes validation an important consideration in both the development and the validation of a transcriptomics-based test is whether the different classes of compounds are sufficiently represented and what is the ideal number of chemicals in a training set and in a validation set. The training set serves to identify informative biomarkers during the development of the test in order to establish a classifier (or prediction model); the validation set is aimed at testing the appropriateness of the classifier. There is no clear definition on how to calculate the size of these classes of compounds. However, the number of compounds representing each class may be lower than that proposed in the early days of microarray development where testing thousands of parameters at the same time (high-dimensionality) was considered as a limitation of the technique. Some rules have nowadays been proposed by different authors. For example, Allison et al. (2006), proposed to use at least four to six compounds per class for a well-defined class comparison. On the other hand, Dobbin and Simon (2007) claimed that this was an over-simplification of the problem and proposed at least 20–30 compounds per class in order to build an appropriate predictor. Despite these estimates, considerations about resources and costs, as well as the availability of sufficiently well characterised reference chemicals often limit the number of compounds used and the associated experiments that can be conducted. In a study by Doktorova and colleagues (2014) the number of chemicals per class was reduced to 15, as it was considered to represent a well defined chemical space characterised by chemicals that are not too diverse.

In summary, a good transcriptomics-based test submitted for validation should be transparently described in a standard template (a kind of Standard Operating Procedure [SOP] that includes the bioinformatics workflow). The minimal sample size for the development and the validation of a test (i.e. necessary for getting trustable results) depends on how diverse the dataset is and how precise the question and purpose of the test are.

2.2 Standardisation

Huge progress has been made in the last decade in the standardisation of transcriptomics methodology and the development of bioinformatics approaches applied to it (ECETOC report 2013). Several consortia were established and projects undertaken in order to tackle these issues. Among these is the Micro Array Quality Control (MAQC) project, led by the US FDA, the main goal of which was to assess microarray variability and to develop standards and quality measures for transcriptomics data (Shi et al. 2010; Wen et al. 2010; Luo et al. 2010).

Several reports have suggested that adhering to standard laboratory practices and careful analysis of data can lead to high quality, reproducible results that reflect the biology of the system (Bammler et al. 2005; Dobbin et al. 2005; Irizarry et al. 2005; Larkin et al. 2005). While standardisation is a necessary condition for a test that should undergo validation, it may also limit the fast pace of discoveries that characterise an emerging field. There is thus a need to find the optimal balance of standardisation necessary, as too little standardisation will generate low quality data, while too much standardisation may inhibit progress.

Different steps of standardisation can be identified in the experimental and bioinformatics workflow (see Fig. 10.1):

1. Laboratory experimentation
2. Pre-processing, array quality control, normalization procedures and data filtering
3. Collapsing of data (i.e. reducing the high dimensionality of the data matrix)
4. Data analysis and interpretation

2.3 The Bioinformatics Workflow

To date some robust laboratory protocols or semi-automatic tools are available that can support standardisation of some steps related to laboratory procedures, array quality control measures, data normalisation and data filtering (i.e. bullet points 1 and 2 above). However, consensus is still required about the level of complexity that should be taken into account in the analysis of the data, which may vary with the specific research question considered. Statistical analysis may be conducted at different levels (i.e. probe, gene or pathway) and many different approaches have been proposed (Allison et al. 2006), which could be combined or interrelated to achieve better results. For example, pathway analysis, usually using hypergeometric tests, Gene Set Enrichment Analysis or by assigning scores (Kamburov et al. 2013; Subramanian et al. 2005; Yildirimann et al. 2011) seems to be more robust than analysis based on single gene changes. This can be explained either by the reduction of the number of tests necessary for a pathway analysis when compared with that conducted when all genes are considered, or by the fact that different compounds can alter different genes from the same pathways to produce a similar outcome. For instance, hepatotoxicity of metapyrilene (MP) has been repeatedly studied using various techniques including the transcriptomics approach (Hamadeh et al. 2002b). This compound induces marked and reproducible hepatic injury in rodents, and was used to assess the validity of toxicogenomics analyses among a multicentre platform (Waring et al. 2004; Chu et al. 2004). The study concluded that the microarray experiments did not supply reliable results because of a high variability between research facilities, explained by the low number of genes that appeared commonly changed across facilities. However, latter studies revealed that the robustness of the results regarding the change of certain biological pathways was deemed sufficient to consider the method reliable, although the fitness of the individual genes was somewhat questionable.

There are many resources related to biological pathway definitions (e.g. Ingenuity Pathways Analysis [http://www.ingenuity.com]; KEGG [http://www.genome.jp/kegg/]; Reactome [http://www.reactome.org/]; and agglomerative pathways resources like ConsensusPathDB [http://cpdb.molgen.mpg.de/] Croft et al. 2013; Kamburov et al. 2013), however none of them is complete. Moreover, the gene and pathway relationships among species are also not well defined, making it sometimes

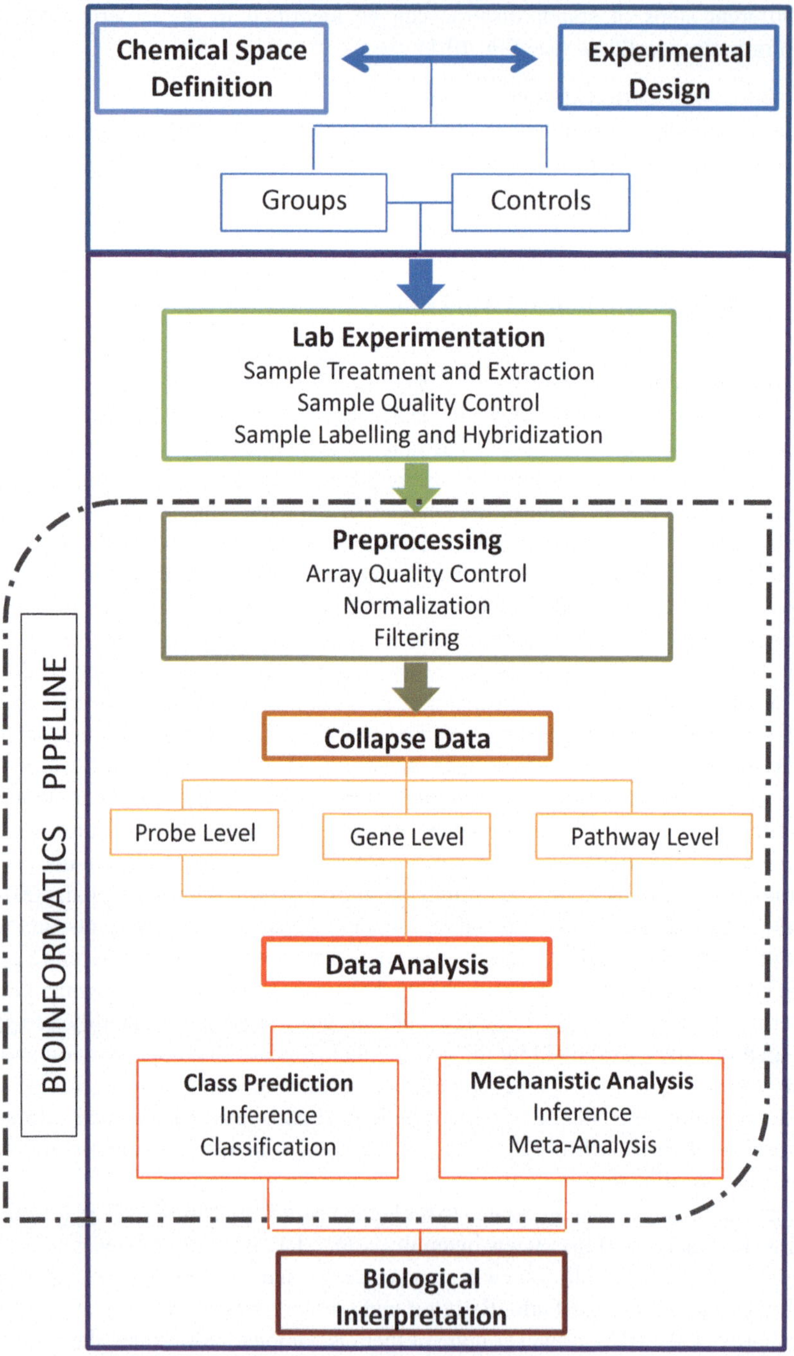

Fig. 10.1 The test method workflow shows different steps that should be taken into consideration during the validation process. First (*dark blue square*), it is necessary to define the experimental design taking into account the objectives of the study and the chemical space of interest. In a second step (*lilac square*), standardised comprehensive procedures for laboratory experimentation

necessary to look for species ontologies using other tools, e.g. Biomart (http://www.biomart.org/, Kasprzyk 2011). To avoid these difficulties some authors advocate the importance of the use of Gene Ontology that addresses the need for consistent descriptions of gene products across databases, in terms of their associated biological processes, cellular components and molecular functions in a species-independent manner (www.geneontology.org).

Finally, in relation to the most appropriate statistical analysis, the "IMPROVER" project (Industrial Methodology for Process Verification of Research) concluded that different strategies based on different statistical methodologies can result optimal, if there is a good selection of the methods at each stage of the analytical strategy (Tarca et al. 2013). Thus, the statistical analysis should be seen as a flexible approach during the development of a test, which integrates different sources of data and/or different levels of information during the analysis. This can improve both data interpretation and avoid spurious statistical associations found in some studies (Venet et al. 2011). Network representations of the results, as understanding gene/pathways relationships or looking for driving genes, can also contribute to data interpretation. In addition, the use of tested and open source softwares to implement the different steps of the analysis can facilitate the validation of the test method and eventually its subsequent regulatory use, due to the possibility to transparently describe the analysis pipelines (R project, http://www.r-project.org/).

One of the main issues regarding standardisation of the transcriptomics analysis is the continuous update of biological databases (probe annotation, gene annotation, ontologies, pathways, etc.) and their related software packages. To ensure that the proposed methodology is reproducible, it will be necessary to provide a description of the versions of either databases or softwares used.

In summary, for laboratory experimentation and data pre-processing some robust protocols or semi-automatic tools are available, that have improved the standardisation of the test method. Although some recommendations on the appropriate strategy for microarray data analysis (Allison et al. 2006; Hahne et al. 2008) have been made, none of them has clearly shown better performance over the others. Consequently, flexibility should be allowed in the choice of the strategy to be used in the data analysis. However, once a test method is deemed ready for standard use and validation, each step of the analysis should be embedded in the bioinformatics pipeline (computational code, Fig. 10.1) and should not be modified anymore. This should ensure transparency of the methods when open source software tools are used (e.g. R project, http://www.r-project.org/), allowing the repeatability of the analysis across experiments, which is a key factor in the validation of a method and its use in support of regulatory decisions.

Fig. 10.1 (continued) (*light green square*), pre-processing of the data (*dark green square*), collapsing of data which can be executed using different levels of complexity (*orange square*), data analysis (*light red square*) and biological interpretation (*dark red square*). Moreover, open source code or pipelines should be provided for data pre-processing, collapsing and analysis enhancing the transparency of the bioinformatics workflow, which is critical for a method undergoing validation

3 Reliability Aspects: Case Studies

Reliability accounts for the test method reproducibility within- and between-laboratories and over time. This is an essential aspect in the validation of all tests that will be used for regulatory assessment. On the other hand, for tests applied as screening tools in priority setting or on a case by case basis to answer specific mechanistic questions, in-house validation might suffice. Reproducibility between-laboratories might not always be necessary in these cases, or when there is good data reproducibility within-laboratory or with another technique (e.g. RT-PCR) or when other sources of data exist allowing data comparison.

A few studies have been published so far that investigated the reproducibility of transcriptomics-based tests. Among these, in the EU 6th Framework Programme project "PREDICTOMICS" an inter-laboratory comparison was conducted to test the reproducibility of transcriptomics changes induced by the immunosuppressive agent, Cyclosporine A (CsA) on the human renal proximal tubular HK-2 cell line (Jennings et al. 2009). Four European laboratories took part in this study. Analysis of the transcription profiles demonstrated that one laboratory clustered away from the other laboratories, potentially due to an inclusion of a cell trypsinisation step by this laboratory. Once the genes responsible for this separate clustering were removed, all laboratories showed similar expression profiles. The authors concluded that under standardised conditions, whole genome expression analysis can be reasonably reproducible between different laboratories. However, confounding factors such as medium exhaustion must also be considered in such analyses, showing the importance of having a Standard Operating Procedure (SOP) that is as detailed as possible.

Another prospective collaborative study was designed to determine the level of intra- and inter- laboratory reproducibility between three independent test facilities (Scott et al. 2011). As in the previous study all laboratories adopted the same protocols for all aspects of the toxicogenomic experiment including cell culture, chemical exposure, RNA extraction, microarray data generation and analysis. The genotoxic carcinogen benzo[a]pyrene (B[a]P) and the human hepatoma cell line HepG2 were used to generate three comparable toxicogenomic data sets. High levels of reproducibility were demonstrated using a widely employed gene expression microarray platform. While differences at the global transcriptome level were observed between the laboratories, a common subset of B[a]P responsive genes ($n=400$ gene probes) was identified at all laboratories which included many genes previously reported in the literature as B[a]P responsive. These data showed promise that the current generation of microarray technology, in combination with a standard *in vitro* experimental design, can produce robust data that can be generated reproducibly in independent laboratories.

CarcinoGENOMICS, another project of the European Union, offered an excellent platform for the investigation of reproducibility of the omics-based tests in general and for the assessment of various bioinformatics approaches employed to deal with this issue (http://www.carcinogenomics.eu/). Its major goal was to develop and select appropriate omics-based *in vitro* methods for assessing the carcinogenic potential of compounds. The idea was to design a battery of mechanism-based *in vitro* tests covering major target organs for carcinogenic action e.g. liver, lung and kidney. "Omics" responses (genome-wide transcriptomics as well as metabolomics) were generated following exposure to a well-defined set of model compounds, namely genotoxic carcinogens, non-genotoxic carcinogens and non-carcinogens. Among others, one of the objectives of the study was to make a preliminary assessment of test method transferability and between-laboratory reproducibility in a blinded inter-laboratory study. Three coded chemicals were tested in three laboratories for each test system (cell model) using the same agreed SOPs and controlled conditions. The two systems employed were human-based and organ- specific, namely, HepaRG for the liver and RPTEC/TERT1 for the kidney. Several bioinformatics approaches were identified to judge data reproducibility and were used independently by different bioinformaticians. These approaches ranged from evaluation of response gene lists, correlation analyses to multivariate statistical methods such as support vector machine classification and analysis of variance (Herwig et al. 2015). Independently from the bioinformatics approaches applied, the liver model generated reproducible transcriptomics results, with the exception of a single experiment in one laboratory (Doktorova et al. 2014). Regarding the RPTEC/TERT1 model, two laboratories showed highly reproducible results, while one laboratory generated results which did not appear to be reproducible. This outcome was in line with experimental observations due to problems related to the culturing of cells in one of the laboratories (much slower cell growth in comparison to the other laboratories). It was subsequently identified that the outlier laboratory had a significant mycoplasma contamination, which is known to lead to gene expression alteration and to interfere with growth rates (Miller et al. 2003). Interestingly, despite these results the three coded chemicals were classified in the correct classes by all laboratories, suggesting that the prediction model was quite robust. In addition, it was very reassuring in view of regulatory use of transcriptomics data that the different bioinformatics tools used in parallel assessments of reproducibility all led to consistent results.

Overall, the results described above represent a proof of concept that *in vitro* test systems may be considered suitable to be used for standardised transcriptomics analysis, as long as SOPs for cell culture preparation, chemical exposure and data processing are strictly followed in the laboratory. Moreover, the carcinoGENOMICS study demonstrated that different bioinformatics approaches to the analysis of reproducibility can be considered appropriate. It would be beneficial to develop a general guidance document on the validation process, which describes in detail the different approaches that can be used in the assessment of a transcriptomics-based test.

4 Relevance Aspects

As mentioned above, the purpose of a test needs to be well defined to allow an assessment of its relevance. Moreover, both its predictive capacity and applicability domain should be considered, including the definition of chemical classes and/or ranges of test method endpoints for which the test method makes reliable predictions. In drug development applications the assessment of assay relevance is called "Context of Use" (COU) and it is required for assays/biomarkers that are being proposed for qualification via FDA's Center for Drug Evaluation and Research (CDER) Biomarker Qualification Program (BQP) (FDA 2014). It details the manner of use, interpretation, and purpose of an assay/biomarker in drug development.

As mentioned above, transcriptomics-based tests can be divided in two groups:

1. **Hypothesis-driven approaches**, which are based on defined sets of genes or pathways associated with known Mode of Action (MoA) that have been previously published and accepted by the scientific community. These approaches are used to predict specific toxic effects and elucidate the underlying mechanisms. They might be preferable to the methods described below due to the defined biological domain of such assays which is crucial for understanding the usefulness of the assay. One could stipulate that, not just effect sizes, but the relevance of the genes affected in terms of their function in the toxicological process investigated may provide a measure of validity of the prediction.
2. **Approaches based on the use of gene expression signatures to predict different toxicity classes**. These approaches make no assumptions about a chemical's specific MoA and utilise the full transcriptome to identify and characterise relevant transcriptional changes that discriminate between chemical toxicity classes. These discriminatory patterns of gene expression are sometimes not fully elucidated from the mechanistic point of view. Despite the lack of full mechanistic understanding of the relationship between genes/pathways and the adverse effect, there still may be a coherent link between them. However, these predictive approaches will require a high level of validation and documentation to gain confidence in the results. It is encouraged that further mechanistic studies are conducted *a posteriori* to investigate the mechanistic plausibility of the significantly differentially expressed genes and pathways of the test method. These test methods should be prioritised based on the extent of mechanistic understanding available. Ideally, one would like to fully understand mechanisms of all genes differentially expressed and the pathways to which they are linked, however this might be impeded sometimes by a lack of scientific knowledge. Moreover, these pathways might be different among species and the relevance to human cells should be established.

Recently, the Genotoxicity Working Group of the HESI Genomics Committee spearheaded an effort to develop and qualify an *in vitro* genomic biomarker assay to facilitate the risk assessment of frequently occurring positive findings in the *in vitro* chromosome damage assays. First, the group analyzed gene expression profiles of agents with known mechanisms of genotoxicity to identify a transcriptomic signature, a gene set also called "genomic biomarker", which is indicative of DNA damage in human cells *in vitro*. The data from initial multi-laboratory studies provided a foundation for the development of a standardised protocol that was evaluated via a multi-laboratory study coordinated by the HESI genomics committee (reviewed in Ellinger-Ziegalbauer et al. 2009). Although the resulting genomic biomarker was developed via statistical evaluation of a complex data set, the gene set consisted of biologically relevant transcripts and pathways known to be involved in genotoxic stress response making the biological relevance of the assay obvious (Li et al. 2015; Buick et al. 2015). Furthermore, these results were shared with the FDA via the Voluntary Exploratory Data Submission (VXDS) process (Goodsaid et al. 2010) to gain direct feedback from the regulatory agency on the potential utility of this approach (Context of Use) to provide mechanistic context in follow-up to positive findings in *in vitro* chromosome damage assays. Following on from the VXGDS submission, a standardised protocol consisting of a combined RT-PCR and microarray-based approach was established and currently is being validated by the HESI genomics committee by generating expression signatures for defined classes of compounds in a multi-laboratory fashion. The validation and potential qualification by the FDA of the genomic biomarker approach for regulatory use is anticipated in 2015. To our knowledge, this is the most advanced biomarker validation/qualification project and learning from this endeavour will be helpful to other projects to come.

Overall, the nature and extent of validation that is needed will depend both on the type of test and the context of application (Fig. 10.2). Transcriptomics assays analysing global gene expression as a general inventory of possible compound MoAs may need a different approach from assays focusing on effects of compounds on a particular gene pathway. These different approaches will also be reflected in different positions of assays within testing strategies. In general, it is most probable that, as is true for most alternative assays in general, transcriptomics-based tests won't stand alone in the same way as animal studies are used in traditional toxicology (Adler et al. 2011; Worth et al. 2014). Rather, the combination of complementary assays in test batteries as part of integrated testing strategies seems to be most promising (Piersma et al. 2014). Therefore, the relevance of individual tests will need to be evaluated in the context of the positioning of the test within an integrated testing strategy or IATA (see Chap. 13), or when applied to support grouping and read-across (Patlewicz et al. 2014). This means that false negative or false positive results may be considered worthwhile findings as they may indicate that the toxic mechanism of the compound is not covered within the biological applicability domain of an individual assay. However, the same compound may trigger its toxicity pathway in another test within the battery, and thus be correctly scored in the battery as a whole. In light of the above, the approach to assess the relevance or predictivity remains a challenging task for which specific rules still need to be devised, and might for the time being be left open to expert judgement.

5 Conclusions and Future Perspectives

A high level of standardisation of transcriptomics experiments has been acquired so far, especially in the experimental steps and the treatment of data. Therefore, *in vitro* models may be considered suitable to be used for transcriptomics analysis, and generation of reproducible results is possible. On the other hand, a general guidance

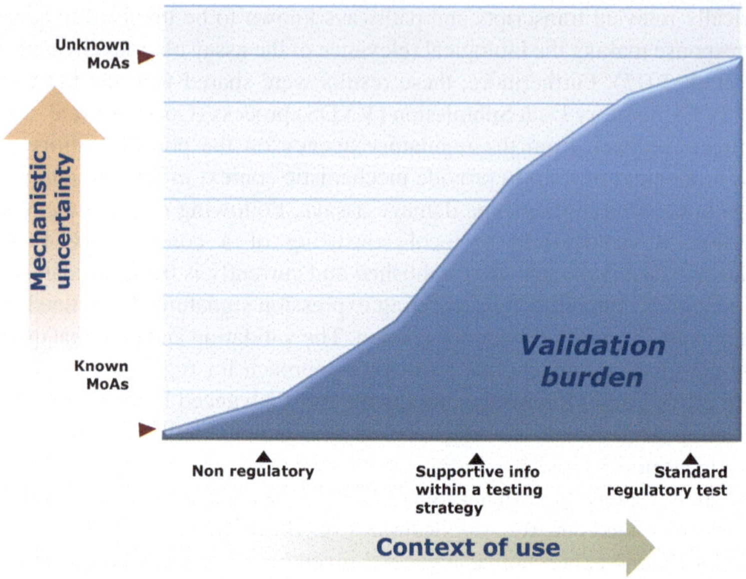

Fig. 10.2 The approach to validation needed to assess a transcriptomics-based test might depend both on the context of use of the test under consideration and the level of mechanistic understanding and knowledge of the differentially expressed genes or pathways. This means that most probably there is not one rule that fits all. For example, for tests used as screening tools for priority setting, in-house validation might suffice, while a test method that will be used as a standard regulatory test and for which MoA of the signature is not completely understood will require the highest level of validation for regulators to acquire sufficient confidence

and consensus for the assessment of the predictive capacity of transcriptomics-based tests depends on the application of the tests in the wider context of testing strategies. For the time being, we might need to rely on a flexible approach involving expert judgement.

Due to the rapid pace with which transcriptomics technologies and data analysis are developing, it is foreseeable that validity considerations given to a test method or a testing strategy need to be regularly reviewed to reflect scientific progress. The need to consider these assays in the context of an integrated testing strategy complicates the assessment of relevance and acceptability for each individual assay and this is perhaps the most important bottleneck for regulatory implementation (Kinsner-Ovaskainen et al. 2012; Römer et al. 2014). The development of adverse outcome pathways (AOPs) for modes of chemical toxicant action leading to adverse toxicological outcomes may also provide context of use for new molecular biology assays and represents a framework to interpret toxicogenomics *in vitro* data, and provide mechanistic information for key toxicity pathways (Landesmann et al. 2013). This opens the way for validation of assays in the context of the AOP.

An issue that will need further attention in the future is the threshold of adversity. This represents an essential aspect in the interpretation of data and is related to the concentration dependent testing that defines adversity *in vitro*. There is some concern that enhanced sensitivity of the assays will lead to lower specificity. Therefore, better definition and understanding of how to specify and interpret thresholds of adversity at the molecular level will improve the interpretation of toxicogenomics tests and testing strategies (Piersma et al. 2011; Boverhof and Gollapudi 2011).

To facilitate the development, validation and finally successful application of transcriptomic-based assays and biomarkers in risk assessment, collaborative efforts and a dialog among all stakeholders i.e. assay developers and scientists from academia, industry and regulatory agencies, are needed (Paules et al. 2011). For instance, large scientific projects coordinated by large consortia of stakeholders such as carcinoGENOMICS and HESI Genomics Committee are essential to share ideas and resources. In addition, the biomarker qualification process (reviewed in Goodsaid et al. 2010) that was developed at the US FDA provides an excellent opportunity for essential feedback from regulators.

A flexible, case by case approach to validation is likely be required for transcriptomics-based assays. The definition of performance standards might partially circumvent this problem for the aspect of assay reproducibility. As to predictivity, classical validation typically compares alternative test results with a 'gold standard', usually an animal study. With the advent of the principles of AOPs and integrated testing strategies the realization has been strengthened that one-to-one replacement and associated one-to-one validation practice are not fit for purpose. Rather, combinations of tests (test batteries) are needed to cover complex endpoint areas such as reproductive toxicology. Thus, validation should focus on batteries as a whole rather than on individual tests (Piersma et al. 2013). This shifts the practice of validation, whilst basic principles of reliability and relevance remain intact. Transcriptomics assays are typical in this realm, as they may represent different biological domains based on the biological system employed and the extent of gene expression analysis applied. This poses challenges to validation, but opens the way for novel approaches dedicated to reliable alternative testing strategies towards the implementation of alternative approaches in hazard and risk assessment.

References

Adler S, Basketter D, Creton S et al (2011) Alternative (non-animal) methods for cosmetics testing: current status and future prospects-2010. Arch Toxicol 85(5):367–485

Allison DB, Cui X, Page GP, Sabripour M (2006) Microarray data analysis: from disarray to consolidation and consensus. Nat Rev Genet 7(1):55–65, Erratum in: Nat Rev Genet 7(5):406

Bammler T, Beyer RP, Bhattacharya S et al (2005) Standardizing global gene expression analysis between laboratories and across platforms. Nat Methods 2(5):351–356, Erratum in: Nat Methods 2(6):477

Bourdon-Lacombe JA, Moffat ID, Deveau M et al (2015) Technical guide for applications of gene expression profiling in human health risk assessment of environmental chemicals. Toxicol Appl Pharmacol 289(3):573–588

Boverhof D, Gollapudi BB (eds) (2011) Applications of toxicogenomics in safety evaluation and risk assessment. Wiley, New York

Buick JK, Moffat I, Williams A et al (2015) Integration of metabolic activation with a predictive toxicogenomics signature to classify genotoxic versus nongenotoxic chemicals in human TK6 cells. Environ Mol Mutagen 56:520–534. doi:10.1002/em.21940

Chu TM, Deng S, Wolfinger R et al (2004) Cross-site comparison of gene expression data reveals high similarity. Environ Health Perspect 112(4):449–455

Corvi R, Ahr HJ, Albertini S et al (2006) Meeting report: validation of toxicogenomics-based test systems: ECVAM-ICCVAM/NICEATM considerations for regulatory use. Environ Health Perspect 114(3):420–429

Croft D, Mundo AF, Haw R et al (2013) The Reactome pathway knowledgebase. Nucleic Acids Res 42(Database issue):D472–D477. doi:10.1093/nar/gkt1102

Dobbin K, Simon R (2005) Sample size determination in microarray experiments for class comparison and prognostic classification. Biostatistics 6(1):27–38, Erratum in: Biostatistics (2005) 6(2):348

Dobbin KK, Simon RM (2007) Sample size planning for developing classifiers using high-dimensional DNA microarray data. Biostatistics 8(1):101–117

Doktorova TY, Yildirimman R, Ceelen L et al (2014) Testing chemical carcinogenicity by using a transcriptomics HEPARG-based model? EXCLI J 13:623–637

ECETOC Workshop report No 25 (2013) Omics and risk assessment. http://www.ecetoc.org/publication

Ellinger-Ziegelbauer H, Fostel JM, Aruga C et al (2009) Characterization and interlaboratory comparison of a gene expression signature for differentiating genotoxic mechanisms. Toxicol Sci 110(2):341–352

Food and Drug Administration Center for Drug Evaluation and Research (CDER) (2014) Guidance for industry and FDA staff—qualification process for drug development tools. http://www.fda.gov/downloads/Drugs/GuidanceComplianceRegulatoryInformation/Guidances/UCM230597.pdf

Goetz AK, Singh BP, Battalora M et al (2011) Current and future use of genomics data in toxicology: opportunities and challenges for regulatory applications. Regul Toxicol Pharmacol 61:141–153

Goodsaid FM, Amur S, Aubrecht J et al (2010) Voluntary exploratory data submissions to the US FDA and the EMA: experience and impact. Nat Rev Drug Discov 9(6):435–445

Hahne F, Mehrle A, Arlt D et al (2008) Extending pathways based on gene lists using InterPro domain signatures. BMC Bioinformatics 9:3

Hamadeh HK, Bushel P, Paules R, Afshari CA (2001) Discovery in toxicology: mediation by gene expression array technology. J Biochem Mol Toxicol 15(5):231–242

Hamadeh HK, Bushel PR, Jayadev S et al (2002a) Gene expression analysis reveals chemical-specific profiles. Toxicol Sci 67(2):219–231

Hamadeh HK, Knight BL, Haugen AC et al (2002b) Methapyrilene toxicity: anchorage of pathologic observations to gene expression alterations. Toxicol Pathol 30(4):470–482

Hartung T, Bremer S, Casati S et al (2004) A modular approach to the ECVAM principles on test validity. Altern Lab Anim 32(5):467–472

Herwig R, Gmuender H, Corvi R et al (2015) Inter-laboratory study of human *in vitro* toxicogenomics-based tests as alternative methods for evaluating chemical carcinogenicity: a bioinformatics perspective. Archiv Toxicol, doi:10.1007/s00204-015-1617-3

Irizarry RA, Warren D, Spencer F et al (2005) Multiple-laboratory comparison of microarray platforms. Nat Methods 2(5):345–350

Jennings P, Aydin S, Bennett J et al (2009) Inter-laboratory comparison of human renal proximal tubule (HK-2) transcriptome alterations due to Cyclosporine A exposure and medium exhaustion. Toxicol *In Vitro* 23:486–499

Kamburov A, Stelzl U, Lehrach H, Herwig R (2013) The ConsensusPathDB interaction database: 2013 update. Nucleic Acids Res 41(Database issue):D793–D800

Kasprzyk A (2011) BioMart: driving a paradigm change in biological data management. Database (Oxford) 2011:bar049

Kinsner-Ovaskainen A, Maxwell G, Kreysa J et al (2012) Report of the EPAA-ECVAM workshop on the validation of Integrated Testing Strategies (ITS). Altern Lab Anim 40(3):175–181

Landesmann B, Mennecozzi M, Berggren E, Whelan M (2013) Adverse outcome pathway-based screening strategies for an animal-free safety assessment of chemicals. Altern Lab Anim 41(6):461–471

Larkin JE, Frank BC, Gavras H et al (2005) Independence and reproducibility across microarray platforms. Nat Methods 2(5):337–344

Li HH, Hyduke DR, Chen R et al (2015) Development of a toxicogenomics signature for genotoxicity using a dose-optimization and informatics strategy in human cells. Environ Mol Mutagen 56:505–519. doi:10.1002/em.21941

Luo J, Schumacher M, Scherer A et al (2010) A comparison of batch effect removal methods for enhancement of prediction performance using MAQC-II microarray gene expression data. Pharmacogenomics J 10(4):278–291

Miller CJ, Kassem HS, Pepper SD et al (2003) Mycoplasma infection significantly alters microarray gene expression profiles. Biotechniques 35(4):812–814

Patlewicz G, Ball N, Becker RA et al (2014) Read-across approaches - misconceptions, promises and challenges ahead. ALTEX 31(4):387–396

Paules RS, Aubrecht J, Corvi R et al (2011) Moving forward in human cancer risk assessment. Environ Health Perspect 119(6):739–743

Pettit S, des Etages SA, Mylecraine L et al (2010) Current and future applications of toxicogenomics: results summary of a survey from the HESI Genomics State of Science Subcommittee. Environ Health Perspect 118(7):992–997

Piersma AH, Hernandez LG, van Benthem J et al (2011) Reproductive toxicants have a threshold of adversity. Crit Rev Toxicol 41(6):545–554

Piersma AH, Bosgra S, van Duursen MB et al (2013) Evaluation of an alternative *in vitro* test battery for detecting reproductive toxicants. Reprod Toxicol 38:53–64

Piersma AH, Ezendam J, Luijten M et al (2014) A critical appraisal of the process of regulatory implementation of novel *in vivo* and *in vitro* methods for chemical hazard and risk assessment. Crit Rev Toxicol 24:1–19

Römer M, Eichmer J, Metzger U et al (2014) Cross-platform toxicogenomics for the prediction of non-genotoxic hepatocarcinogenesis in rat. PLoS One 9(5):e97640. doi:10.1371/journal.pone.0097640

Scott DJ, Devonshire AS, Adeleye YA et al (2011) Inter- and intra-laboratory study to determine the reproducibility of toxicogenomics datasets. Toxicology 290:50–58

Shi L, Campbell G, Jones WD et al (2010) The Microarray Quality Control (MAQC)—II study of common practices for the development and validation of microarray/based predictive models. Nat Biotechnol 28(8):827–838

Steiner G, Suter L, Boess F et al (2004) Discriminating different classes of toxicants by transcript profiling. Environ Health Perspect 112(12):1236–1248

Subramanian A, Tamayo P, Mootha VK et al (2005) Gene set enrichment analysis: a knowledge-based approach for interpreting genome-wide expression profiles. Proc Natl Acad Sci 102(43): 15545–15550

Tarca AL, Lauria M, Unger M et al (2013) Strengths and limitations of microarray-based phenotype prediction: lessons learned from the IMPROVER Diagnostic Signature Challenge. Bioinformatics 29(22):2892–2899

Venet D, Dumont JE, Detours V (2011) Most random gene expression signatures are significantly associated with breast cancer outcome. PLoS Comput Biol 7(10):e1002240

Waring JF, Ulrich RG, Flint N et al (2004) Interlaboratory evaluation of rat hepatic gene expression changes induced by methapyrilene. Environ Health Perspect 112(4):439–448

Wen Z, Wang C, Shi Q et al (2010) Evaluation of gene expression data generated from expired Affymetrix GeneChip® microarrays using MAQC reference RNA samples. BMC Bioinformatics 11(Suppl 6):S10

Worth A, Barroso J, Bremer S et al (2014) Alternative methods for regulatory toxicology – a state-of-the-art review. EUR 26797. http://publications.jrc.ec.europa.eu/repository/handle/JRC91361

Yildirimman R, Brolén G, Vilardell M et al (2011) Human embryonic stem cell derived epatocyte-like cells as a tool for *in vitro* hazard assessment of chemical arcinogenicity. Toxicol Sci 124(2):278–290

Chapter 11
Ensuring the Quality of Stem Cell-Derived *In Vitro* Models for Toxicity Testing

Glyn N. Stacey, Sandra Coecke, Anna-Bal Price, Lyn Healy, Paul Jennings, Anja Wilmes, Christian Pinset, Magnus Ingelman-Sundberg, Jochem Louisse, Simone Haupt, Darren Kidd, Andrea Robitski, Heinz-Georg Jahnke, Gilles Lemaitre and Glenn Myatt

Abstract Quality control of cell cultures used in new *in vitro* toxicology assays is crucial to the provision of reliable, reproducible and accurate toxicity data on new drugs or constituents of new consumer products. This chapter explores the key scientific and ethical criteria that must be addressed at the earliest stages of developing toxicology assays based on human pluripotent stem cell (hPSC) lines. It also identifies key considerations for such assays to be acceptable for regulatory,

---

G.N. Stacey (✉)
UK Stem Cell Bank, Advanced Therapies Division, NIBSC-MHRA, London, UK

SEURAT-1 Stem Cell Group, Paris, France
e-mail: glyn.stacey@nibsc.org

S. Coecke • A.-B. Price
European Commission, Joint Research Centre (JRC), Ispra, Italy

SEURAT-1 Stem Cell Group, Paris, France

L. Healy
Haematopoietic Stem Cell Laboratory, The Francis Crick Institute,
44 Lincoln's Inn Fields, London WC2A 3LY, UK

SEURAT-1 Stem Cell Group, Paris, France

P. Jennings • A. Wilmes
Division of Physiology, Medical University of Innsbruck, Innsbruck, Austria

SEURAT-1 Stem Cell Group, Paris, France

C. Pinset • G. Lemaitre
I-Stem, INSERM/UEVE U861, Evry, France

SEURAT-1 Stem Cell Group, Paris, France

M. Ingelman-Sundberg
Karolinska Institutet, Solna, Sweden

SEURAT-1 Stem Cell Group, Paris, France

C. Eskes, M. Whelan (eds.), *Validation of Alternative Methods for Toxicity Testing*,
Advances in Experimental Medicine and Biology 856,
DOI 10.1007/978-3-319-33826-2_11

laboratory safety and commercial purposes. Also addressed is the development of hPSC-based assays for the tissue and cell types of greatest interest in drug toxicology. The chapter draws on a range of expert opinion within the European Commission/Cosmetics Europe-funded alternative testing cluster SEURAT-1 and consensus from international groups delivering this guidance such as the International Stem Cell Banking Initiative. Accordingly, the chapter summarizes the most up-date best practices in the use and quality control of human Pluripotent Stem Cell lines in the development of *in vitro* toxicity assays from leading experts in the field.

Keywords Toxicology • Stem cell differentiation • Quality control • Stem cell characterisation

1 Introduction

As more types of *in vitro* human cell and tissue models become available, toxicologists have an ever-broadening range of alternative methods that those which are already established and validated at the regulatory level. Human pluripotent stem cells provide an exciting and potentially powerful source of *in vitro* cell and tissue models which could generate *in vitro* methods for testing toxicokinetic or toxicodynamic effects aiming to more closely mimic the *in vivo* response of cells in the human body. Furthermore, the ability to generate induced pluripotent stem cell lines from any patient raises significant hopes for new innovative tools with bespoke genotypes to study certain diseases and specific adverse outcome pathways (AOPs) and

J. Louisse
Wageningen University and Research Centre, Wageningen, Netherlands

SEURAT-1 Stem Cell Group, Paris, France

S. Haupt
Life and Brain, Bonn, Germany

SEURAT-1 Stem Cell Group, Paris, France

D. Kidd
Covance Laboratories Limited, Harrogate, UK

SEURAT-1 Stem Cell Group, Paris, France

A. Robitski • H.-G. Jahnke
University of Leipzig, Leipzig, Germany

SEURAT-1 Stem Cell Group, Paris, France

G. Myatt
Leadscope, Columbus, OH, USA

SEURAT-1 Stem Cell Group, Paris, France

accelerate our understanding of deregulation of normal human biological processes resulting in adverse outcome. Furthermore, besides their use in the development of *in vitro* methods for toxicological applications, human pluripotent stem cells offer a wide array of applications for therapy, drug discovery and efficacy testing.

However, creating an *in vitro* cell or tissue culture system which represents dividing or differentiated cell types *in vivo* is just a starting point. The next step is to qualify such *in vitro* cell and tissue models for specific toxicological applications. This will require validation to comply with the additional regulatory requirements for toxicological methods and involves producing data to demonstrate the reliability and relevance of the *in vitro* toxicological methods for human safety assessment. Furthermore, it is important to remember that *in vitro* methods should be established for a particular purpose, and a clearly defined application should be identified, such as hazard identification, capturing key events in AOPs, or information about the risk assessment for targeting either toxicokinetic or toxicodynamic processes.

Careful ongoing evaluation is needed to identify how stem cell and tissue-based *in vitro* methods can fill critical gaps where information is needed, that currently cannot be obtained with the available non-stem cell based systems. The first stem cell-based *in vitro* method was established in the field of reproductive toxicity i.e. the embryotoxicology test using mouse embryonic stem cells in their undifferentiated state (Genschow et al. 2004). It is anticipated that a range of *in vitro* toxicity methods derived from human stem cell lines will soon be available since over the last decennia much progress has been made in generating diverse cell types. However, these will need to meet a range of technical and regulatory criteria which are considered here as well as reviewing important aspects that will be required to demonstrate the reliability and relevance of the *in vitro* toxicological methods based on differentiation of human pluripotent stem cells.

It has been 15 years since toxicologists formerly announced the need for guidance on the principles of best practice for cell cultures used in laboratory testing (Hartung et al. 2000), and principles of good cell culture practice were established in 2005 (Coecke et al. 2005). The advent of human pluripotent stem cell lines in 1998 (Thomson et al. 1998; Takahashi et al. 2007) has offered exciting new possibilities for *in vitro* toxicology assays (Scott et al. 2013) but also new challenges for these complex and potentially variable cultures.

However, the specific challenges in the maintenance, differentiation and quality control of such cultures also prescribe that we revisit GCCP with them in mind. Here we seek to present a consensus on the best practice in the development of stem cell and tissue-based *in vitro* methods, drawing on the experience of the SEURAT-1 cluster of consortia (www.SEURAT-1.eu/) which focused on the development of alternative product safety testing paradigms based on human stem cell lines. Within SEURAT-1, the ToxBank and Scr&Tox consortia have collaborated on aspects of best practice relating to sourcing and quality control of pluripotent stem cell lines. These efforts have been utilised and are referenced here together with other international collaborations on consensus in the banking, testing and supply of human pluripotent stem cell (hPSC) lines including the hESCreg database of hPSC lines (www.hescreg.eu/) and the International Stem Cell Banking Initiative (www.stem_cell_forum.net/).

2 Selection of Cell Lines

Choosing a cell line for early research without appropriate attention to key ethical, regulatory and scientific issues can result in a significant waste of time and resources. Failure to carry out such an assessment could mean that, after commitment of much research effort, they will nevertheless prove unsuitable for the establishment of an appropriate *in vitro* toxicological method, which can be utilised by industry for development, screening and regulatory purposes. A number of specific questions need to be considered in making an initial selection of a cell line and these include (Stacey et al. 2012a):

- Does the cell line meet key scientific criteria for a hPSC line and is it of an appropriate genotype e.g. ethnic background, to carry the required normal or disease associated genetic state?
- Has the provider of the cell line carried out appropriate cell characterisation and testing?
- Was the original tissue consented appropriately?
- Are intellectual property rights clear and would they permit use in industry for testing new drugs/compounds?

In the following sections we now consider approaches to meet each of these criteria in turn.

2.1 Preliminary Scientific Cell Line Selection Criteria

The key scientific criteria for initial cell line selection were established for the SEURAT-1 project and are summarised in Table 11.1. These also provide a useful aide memoir on core hPSC line characteristics that should be included as a minimum when reporting a new cell line and furthermore, should be considered when reviewing cell lines which the reader may wish to use in their research. Clearly, a broad range of biological markers may be reported for individual cell lines and Pistolatto et al. (2012) have reviewed potential biological markers of relevance to quality control of human embryonic stem cell (hESC) lines and human induced pluripotent stem cell (hiPSC) lines.

The European Commission has funded a searchable database of hESC and hiPSC lines (www.hescreg.eu/) which provides a facility for uploading a large amount of regulatory and scientific data on specific cell lines from the originators. It holds information on more than 700 international hPSC lines and is closely linked to the SEURAT-1 programme via the ToxBank data warehouse and a newly-funded iPSC bank for Europe called EBiSC (www.ebisc.eu/). The hESCreg database provides a platform where all registered lines are evaluated and presented in a consistent way which facilitates direct comparison of data on different lines and selection of lines suitable for a end-user's specific needs

Table 11.1 Key questions to address when selecting cell lines for development of stem cell based assays (developed from a SEURAT-1 collaboration between the Scr&Tox and ToxBank consortia by Stacey et al. (2014a) and also Luong et al. (2011)

Criterion	Description
Is the source of the original somatic cells described?	Cell type, tissue, donor age, commentary on suitability of donor consent and for iPSC lines only the passage number of the parental cell line
Is the derivation method reported in detail	*hESC* (including blastocyst quality and preparation, cell line isolation method, passage/seeding and culture conditions) *hiPSC*—Including the reprogramming method (e.g. vector system, small molecules, protein, mRNA, or miRNA transduction), cell line isolation method, passage/seeding and culture conditions
Is sufficient characterisation reported?	See Sect. 3
Is an adequate assessment of pluripotency potential reported?	See Sect. 3
Is microbial contamination screening reported	As a minimum this should include mycoplasma testing. See Sect. 3

(including the development/optimization of *in vitro* toxicological methods for regulatory applications) and for compliance with requirements for inclusion in European Commission funded research.

2.2 Selecting Suppliers of Cell Lines

Unfortunately, obtaining cells from convenient sources such as colleagues and academic collaborators has been associated with dissemination of significant numbers of misidentified or microbially contaminated cell lines, even when these have been obtained from the originator (Stacey et al. 2000; MacLeod et al. 1999; Rispin et al. 2004; Gupta et al. 2005).

A variety of suppliers of human pluripotent stem cell lines exist and have been reviewed (Luong et al. 2012) and also more recently a directory of suppliers has been produced by the ToxBank project (Stacey et al. 2014a) which includes specific criteria for selection of cell line supply source.

2.3 Donor Consent

Obviously, local and national rules, regulations, and laws must be considered when obtaining human tissue or cell lines. The ToxBank consortium has published a document 'Points to consider in gaining access to human tissue and cell lines' which has a special focus on the transfer of cell lines within European countries (Stacey et al. 2012b). Records of donor consent should confirm that the donor understood certain specific issues, including that the tissue would be used to generate a cell line that may be used in a wide range of research, including genetic testing and development for commercial exploitation. In addition, in the interests of smooth translation to industrial use, it is helpful if it is clearly documented that the donor agreed they would not retain any financial interest in the use of the cells or the research and products arising from it. Appendix 1 shows an exemplar from the SEURAT-1 project of a number of key questions to be addressed when sourcing cell lines for development as tools for acute toxicity testing in industry and the selection process is also available in (Stacey et al. 2012a). For acceptability of hESC derived cells in some countries it may be necessary to establish that the original embryonic cells were not established for the specific purpose of research and were produced from supernumery embryos produced for reproductive treatments.

Where the consent lacks such information it may be necessary to seek additional consent from the donor or consider an alternative cell line. The former approach may require significant effort and time and is not to be undertaken lightly.

2.4 Commercial Exploitation Issues

Failure to address key commercial issues relating to the use of any materials key to a toxicity assay may result in restrictions on the ability to use the assay (i.e. for research only) or inability to agree terms on access to intellectual property. Such issues could delay or even prohibit their use in industry and therefore it is important that users apply suitable due diligence when obtaining hPSCs and other materials that may be critical in assays at a later stage.

Key criteria for selection of hPSCs identified for the SEURAT-1 project (Stacey et al. 2012a) included:

- The owner of the cell line is clearly identifiable (NB numerous cell lines have shared ownership)
- Permission has been granted by the owner/s or their agents for the intended use or is the line released for general research without constraint (see also donor consent).
- Intellectual property rights relating to the cell line or any components used to derive the cell line (e.g. DNA constructs, reprogramming technique) are clear and would not influence their use for commercial application.

If a potential adverse impact on ultimate use of materials for commercial purposes is discovered this should be discussed with industry partners or sponsors who will ultimately be using the *in vitro* toxicity method. Any limitations on the use of the materials can then be understood before making further significant investment of resources.

3 Standardisation and Control of Seed Stocks of Undifferentiated Cells

3.1 General Considerations

Principles of Good Cell Culture Practice (GCCP) were established by an international Task Force led by the European Union Reference Laboratory for Alternatives to Animal Testing (EURL ECVAM), then the European Centre for the Validation of Alternative methods (Coecke et al. 2005), and more specific guidance on banking hESC lines has been established by the International Stem Cell Banking Initiative (2009). These established the core criteria for reliable preparation of quality controlled stock cultures for laboratory use including toxicological studies and testing in regulatory protocols. These generic criteria are applicable to all hPSC lines and here we describe additional controls for use with hiPSCs.

GCCP prescribes the preparation of a low passage master cell bank from which cells are expanded to create working cell stocks for use in routine operations. Such a system will permit long term use of each cell line whilst enabling cultures in use to be maintained within acceptably low passage levels, to avoid the risk of culture switching, contamination or loss due to laboratory accidents. Banking procedures and their subsequent release testing should be carefully documented and traceable to ensure that each batch of cells used can be subject to troubleshooting. This is an important quality assurance tool that will be required if particular batches of reagents are discovered to be contaminated or if users of certain cell lines report problems with cell performance or contamination (for an example of basic cell bank documentation see Stacey and Masters (2008)). Such procedures and documentation would be expected for any cell line intended for commercial exploitation. Each bank of cells should be subjected to a consistent and documented regime of quality control and safety testing to ensure that the quality of cells supplied from different cell lines is standardised at a minimum level. Table 11.2 shows the key types of characterisation

Table 11.2 Characterisation of stocks of undifferentiated hPSC

Characteristic	Exemplar of acceptable criteria
Analysis of colony and cell morphology	Round shape cells, large nucleolus, not abundant cytoplasm in flat and tightly-packed colonies
Analysis of proliferation	mitotically active, self-renewal
Analysis of gene expression	Positivity for a panel of markers selected from but not exclusively, Oct-4, SSEA3, SSEA-4, Tra-1-60, Tra-1-81, Sox 2, CD30, CD9
	Negativity for: SSEA-1 and lineage specific markers (e.g. nestin)
Karyological status	Giemsa-banded karyotypes (for cell banks and ideally every 10 passages)
	NB analysis of nucleotide polymorphism (SNP) and array comparative genome hybridization (aCGH) may yield additional data.
Other genetic status tests	See Sect. 3.7

NB Other quality control and safety tests are documented elsewhere in the text and Table 11.3

Table 11.3 Summary of QC criteria for release of seed stocks of undifferentiated hPSCs (adapted from ISCBI (2009) and Pistollato et al. (2012))

Analytical technique	Required characteristic reported for each cell line
Identity e.g. DNA profile	Matches parent cell line or tissues (if available)
Bacteria/fungi	Contamination not Detected
Mycoplasma	Contamination not Detected
Karyotype	Report karyotype from 20 metaphase analyses (see ISCBI 2009)
Post-Thaw Recovery	Viable colonies recovered (quantified efficiency of recovery of each bank/lot should be given) *NB viable colonies should also be predominantly free of differentiated cells.*
Pluripotency	Report data available or traceable to stocks tested for pluripotency[a]
Growth Characteristics	Report value
Cell antigen expression	High proportion of cells (approx. 70 %) positive for each marker[a]
Cell gene expression	Report data available[a]
Genetic stability	Report data available[a]

[a]Precise requirements for hESC lines are also discussed in (ISCBI 2009) and a detailed review of potential additional quality control markers of phenotype and epigenetic stability are discussed by Pistollato et al. (2012)

that need to be performed on undifferentiated hPSCs and Table 11.3 shows a typical testing regime and release criteria for a stock of cryopreserved hPSCs which should be completed before cells are used or released to other laboratories.

Cryopreservation of master and working cell banks should be carefully controlled and the following issues should be born in mind:

- Culture selection: cultures used for cryopreservation should exhibit high viability and low levels of differentiation, to maximise the numbers of undifferentiated stem cells submitted to preservation. Cells should ideally be in exponential phase of growth when cells tend to have a low ratio of nuclear to cytoplasmic volume which promotes penetration of cryoprotectants such as dimethylsuplhoxide.
- Selection of an appropriate grade of cryoprotectant such as the pharmaceutical specification for dimethyl-sulphoxide (European Pharmacopoeia, vol. 5.0, pp 1445–1446).
- Avoid suspending cells to be preserved in media containing chemicals such as antibiotics, which may be highly toxic at the concentrations that will be achieved in residual liquid remaining around cells in the final stages of preservation and early thawing.
- Use a validated cooling process which achieves the cooling rates intended. Ideally the temperature profile of a reference sample should be recorded for each preservation run so that satisfactory cooling can be confirmed.
- Remove cryoprotectants by centrifugation prior to cell culture, as they may affect cell biology as in the case of dimethyl-sulphoxide which can cause a range of effects on cell biology including cell differentiation.

- Recover a sample vial from each cell bank immediately after freezing is complete to confirm successful preservation and adequate viability of cells.
- Ensure that storage conditions are carefully monitored.

The most common and reliable method for preservation of cell lines is cryopreservation of aqueous suspensions containing approximately 10 % dimethylsulphoxide subjected to controlled rate cooling (typically −1 °C min^{-1}). However, some stem cell laboratories may use vitrification methods involving staged additions of increasingly concentrated cryoprotectant and ultra-rapid cooling which may be difficult to establish in routine use. For further information on cryopreservation and vitrification see USP (2013) and Stacey and Day (2007).

Also implicit in GCCP is the requirement to use good aseptic techniques to avoid contamination (typically bacteria, fungi and mycoplasma). Cultures should be monitored for evidence of contamination by daily microscopic examination and cell banks should also be subject to 'sterility tests' by recovery of frozen cells and inoculation into micro biological growth media. It is important to realise that so-called "sterility tests" will not isolate all contaminants, but are designed to alert lab workers that there has been a breakdown in aseptic procedures. Broader spectrum tests for contamination may be used as discussed below. All incoming stem cell lines should be treated as potentially contaminated and quarantined accordingly, until viable cryopreserved stocks have been established and subjected to sterility and mycoplasma testing (see below).

3.2 *Viability and Measurement of Growth*

A viability test will give a "snap shot" of a certain aspect of cellular function or state, and does not necessarily give any information on the competency of a stem cell culture to survive and replicate, yield more stem cells or sustain its pluripotency. However, cell viability and altered cell growth are important indices of toxicity. They can be measured by several parameters including cell numbers, cell membrane function, DNA and protein content, redox potential, the ATP/ADP levels, or enzyme activity. Therefore, common biomarkers that can be quantified include ATP, NADH, Caspases, LDH, live- and dead-cell proteases, and membrane integrity, organelle function. A common form of viability assay is the trypan blue dye exclusion test, which enables an evaluation of membrane integrity. Use of such tests immediately after thawing cryopreserved cells may overestimate the number of cells that will survive as many cells will already be committed to the apoptotic cell death pathways. Enzyme-based methods (e.g. MTT, Alamar Blue, ATP assay), which employ a colorimetric or fluorometric assay, are often considered superior due to their ease-of-use, reproducibility and scalability. In general, cell viability assays enable either the analyses of whole populations or of individual cells, and multiplexing of cell viability and functional assays will be necessary to gain mechanistic insights into biological, toxicological and pathological processes.

In order to establish a cell viability assay the potential limitations of assay chemistries have to be determined, in particular with respect to manual or automated high-throughput applications. Depending on the assay chemistry and read out, some assays may be more suited to certain cell lines and culture models (2D vs. 3D, proliferative vs. non-proliferative cells, etc.). Assay responsiveness is influenced by a number of factors including e.g.: (1) culture medium, supplements and matrix proteins (2) dosage and exposure time to solvents and compounds, (3) buffering capacity, (4) pH, (5) cell density, (6) evaporation/edge-effects of the microtiter plate, (7) incubation time and temperature, (8) linearity and chemical interactions between media, test compound and assay chemistry, (9) stability of assay reagents for time course analysis, (10) intra- and inter-assay variability, (11) and costs. Nevertheless, it is crucially important that the data obtained *in vitro* must reliably predict *in vivo* biological implications. Many key toxic events and pathways are common to most cell types. Hence, a particular toxicant can cause cell type specific pathologies, although the upstream event may be common to the modes-of-action triggered by the substance. Thus, it is fundamentally important to select a viability assay that is qualified for the cultures and methods in question and to use the assay in combination and/or correlation with functional read-outs in order to reflect cell type-specific toxicity.

Clearly, a crucial issue is to demonstrate cell growth, but the nature of growth measurements will depend on whether cells are passaged as single cell suspensions or colony fragments. However, a general criterion can be set for a thawed vial of cells to be able to expand and "regenerate a 'representative' culture within an acceptable period of time" (ISCBI 2009), for example between 2 and 5 days (NB in some case this may be as long as 24 days).

Growth rate is an important characteristic to monitor as it may reveal fundamental changes in the cell line such as mycoplasma contamination or transformation. For single cell passaging methods it can be measured by estimating the population doublings at each passage i.e. number of cells harvested compared to the number of cells seeded. More accurate doubling estimations can be made by taking account of plating efficiency i.e. (number of colonies/number of cells inoculated) × 100 %. For "cut and paste" passaging this is more difficult to estimate but an indication may be obtained from the rate of growth of colonies and if colony fragment seeding density at subculture and harvesting point is consistent, the time between passages may also be a helpful way to measure growth rate.

3.3 Identity Testing

It is an important part of Good Cell Culture Practice (Coecke et al. 2005) to ensure that cell lines used in laboratory testing and research work have been authenticated. Cell line authentication by DNA profiling is a critical step in the banking process, to give a unique "bar-code" for the cell line which can be used to identify and resolve cases of cell line cross-contamination and avoid dissemination of misidentified lines (Barallon et al. 2010a, b; Nims et al. 2010; Capes-Davis et al. 2013).

Methodologies for individual specific genetic identification using short tandem repeat (STR) DNA profiling have been standardised within the field of forensic science, and commercial services and kits are readily accessible. These kits typically do not comprise PCR primers for exactly the same set of alleles but typically produce results for five or more common STR alleles, which facilitates direct comparison of cell line profiles even when generated in different laboratories using different kits (ISCBI 2009). Useful guidance on performance of STR typing has been published in an ANSI standard (Kerrigan and Nims 2011). Reporting of DNA profile data should be considered carefully as donors could be identified (Knoppers et al. 2011; Isasi et al. 2014).

It should be born in mind that multiple cell lines isolated from the same embryo or donor tissue (or identical twins), are not likely to be discriminated using DNA fingerprinting. Such cells should be clearly identified in their naming (see Luong et al. 2011). Where there is no other scientific means of demonstrating their unique identity (such as detection of stable differences in microsatellite DNA), physical isolation of cell lines may be used to manage the risk of such lines becoming switched.

3.4 Microbiological Testing

Routine microbiological screening should include testing for mycoplasma which can cause permanent deleterious effects on cellular physiology and genetics. Pharmacopoeia tests for culture of these organisms have been established including broth/agar culture and DNA staining following culture in susceptible cells. Such industry standards are sensitive and the cell line inoculation method will also identify contamination with strains that will grow in cell culture only. However, such techniques require from several days (cell culture inoculation and DNA stain) up to 3 weeks (broth culture) incubation for a final result. In addition, a range of commercial kits (including RT-PCR) are available to rapidly monitor for mycoplasma. Novel and rapid test systems should be qualified for their sensitivity and specificity for detection of different strains so that their performance compared to standard methods is understood by users.

Bacterial and fungal contamination in cell culture can easily cause catastrophic loss of cultures. Routine use of antibiotics will not always prevent this and may protract contamination events by inhibiting but not eliminating contaminants or even inducing antibiotic resistance. Pharmacopoeia methods are established for detection of such contamination but, whilst still representing the industry standard, those used for cell culture samples rely on traditional culture media and conditions which will not enable all microorganisms to grow. A range of rapid detection techniques have also been developed including non-specific methods (e.g., ATP bioluminescence, laser particle detection), detection of microbial products (e.g. bacterial endotoxin, fungal glycans) and specific detection methods including RT-PCR amplification of ribosomal RNA gene sequences. Use of these rapid techniques is currently a subject of investigation and at this stage they have value when more than one method is used in combination. However, at this time established sterility testing methods remain the accepted test for vials from banks of cell lines.

Another reason for carrying out microbiological testing is to give some assurance that the cell lines will not represent an infectious hazard to laboratory workers. Whilst viral infection might be expected to be evident due to lysis of the cells, such infection can be persistent and non-cytopathic. One approach is to only use cell lines from donors for which screening for the most prevalent serious blood-borne viruses has been performed. However, it is probably more relevant to test seed or master stocks of cell lines directly using PCR techniques (see ISCBI 2009) and to assess the most likely risks associated with the tissue of origin in addition to considering blood-borne pathogens prevalent within the donor population. Of course it is important to use test methods that would not cross-react with vectors used to generate PSC lines, for example HIV primers and lentiviral vectors. Microarray systems are now available that cover a broader range of human pathogens and other potential cell line contaminants, but these will require validation particularly regarding their sensitivity. It is also important to recognise that even non-pathogenic viral infection will have an impact on stem cell biology and in the near future it may be feasible to use next generation sequencing to screen cell lines for any virus.

3.5 Key Phenotypic Markers Required for Quality Control of Undifferentiated hPSC Lines

Two types of marker are required for the control of undifferentiated hPSC lines: firstly those which identify the cell line as having a phenotype typical of hPSC and secondly, those which are capable of indicating changes in the expanding cell populations away from the original undifferentiated state. Key generic phenotypic marker profiles of undifferentiated cultures have been reported and an exemplar has been the consensus established by a multi-centre collaboration performed by the International Stem Cell Initiative (www.stem_cell_forum.net) for the typical antigenic and expression profiles for both hESCs and hiPSCs (Adewumi et al. 2007) (for a summary of hESC vs. hiPSC characteristics see Pistollato et al. (2012)). These studies identified that expression of a profile of certain markers was typical of all hESC and hiPSC lines, although it should be recognised that these markers may also be found individually in other cell types (Adewumi et al. 2007). Table 11.2, developed by the SEURAT-1 Stem Cell Group from Pistollato et al. (2012), shows typical quality control markers. It is also difficult to assign markers which will unequivocally identify change in the composition of an undifferentiated hPSC culture. Markers which have been used to indicate levels of differentiated cells in such cultures are SSEA-1 and SSEA-4 (Adewumi et al. 2007), which may increase and decrease respectively as the level of differentiated cells in a population increases in hPSC lines. Loss of markers of self-renewal such as Nanog and Oct-4, are also taken as an indication of significant changes in the stem cell population. However, the exact nature of culture effects that cause such changes in stem cell populations remain to be elucidated.

3.6 Functional Potential and Pluripotency in Stem Cell Lines

The ultimate assay of pluripotency (germline competency) would not be permitted in humans. However, there are various *in vitro* assays which can be used to measure potential pluripotency including generation of teratomas in immune compromised mice, *in vitro* growth of embryoid bodies and directed differentiation. Whilst, none of these has yet been able to prove that an hESC or hiPSC line can generate all cells of the human body i.e. truly pluripotent, *in vitro* differentiation methods can be used to show that the cell line is capable of generating cells representative of each of the three germ layers required to create all the cells of the human body. They also provide an indication that the cell line has not been altered by *in vitro* culture. If a cell line fails to demonstrate potential pluripotency using one of these *in vitro* differentiation protocols, it may indicate that the cells were not fully pluripotent or that they have undergone deleterious changes during their isolation and culture. However, cases have been identified (members of the International Stem Cell Banking Initiative, personal communications) where a cell line appearing to fail one pluripotency assay may reveal potential for full pluripotency in another assay. The establishment of reliable pluripotency assays is a challenge for the stem cell field. There is clearly a significant challenge in performing routine reliable pluripotency assays as part of quality control regime and it is not possible at this time to make firm conclusions about the most suitable methods for routine laboratory use and they may ultimately involve a combination of *in vitro* and molecular assays. Suppliers of stem cell lines need to consider what method is most appropriate in their hands to confirm the desired characteristics of the cells they release. It is recommended that researchers should perform pluripotency assays in early evaluation of cell lines, on cells recovered from seed stocks and repeated where cells are maintained for extended periods.

3.7 Stem Cell Type Specific Quality Control

The means by which hPSC lines are derived may require certain additional features in their quality control regime. In particular, supplementary criteria for quality control of hESC lines derived from pre-implantation genetic diagnosis (PGD) and those subject to recombinant DNA modification have already been described by Pistollato et al. (2012).

To assure the scientific quality of iPSC lines it is important to demonstrate that expression of exogenous reprogramming factors has been silenced and/or that the reprogramming vectors have been removed. In retroviral systems, incomplete silencing may be an indication of incomplete reprogramming which may influence their biological performance for *in vitro* assays. It is also important to remember that recombinant non-integrating virus can persist for a number of passages and testing should be performed some time after isolation of the iPSC line to give assurance that virus is no longer expressed.

3.8 Release of Seed Stocks

It is part of Good Cell Culture Practice to document that each bank of stem cells has passed key quality control procedures. It is therefore recommended that the various test procedures described above should be captured on a single QC/characterisation summary record sheet for each cell bank demonstrating that this bank is fit to release for experimental purposes. This will require setting acceptability criteria for each test. Such criteria are relatively straightforward for tests such as identity, mycoplasma and virus screens where the result is usually clear cut i.e. a positive signal is present or absent. However, such tests will need to be qualified with validation data (sensitivity and specificity in particular) for the method used and records of controls used in each test. Other tests such as viability, karyology, phenotype and pluripotency will generate quite variable data where it may be more challenging to set cut off limits for acceptability. In the case of viability it may be possible to establish such limits, but for others the range of data may vary between cell lines and culture media used. In such cases data may need to be recorded "for information" rather than as a release criterion (see Table 11.3). It is important that suppliers of cells can provide specific detailed certification for each cell bank or batch of differentiated cells and not just historical testing from early seed stocks.

4 Selecting and Developing Differentiation Protocols

4.1 General Considerations

As for any procedure that is going to form part of a formal regulated system, a prescriptive document (Standard Operating Procedure in a Good Laboratory Practice (GLP) environment) that incorporates all essential reagents, components and steps of the method description, should be developed. In particular, for stem cell-based *in vitro* methods, SOPs should be established to provide detailed descriptions of the propagation and selection criteria for undifferentiated cells (see Sect. 3), culture differentiation, the cell expansion and differentiation media and manipulation procedures. In addition, the criteria for successful and sufficient differentiation and assessment and measurement of its reproducibility, together with the methods that are used for characterisation purposes, should be documented to assure that users can replicate original work and understand the test system they are working with (Coecke et al. 2005, 2014).

It is also important to understand the temporal profile of changes that occur during the differentiation process and it may be necessary to establish certain levels of biomarker expression for intermediate stages of differentiation, to determine reproducibility of development in culture profile. This is especially relevant if a progenitor population is generated and cryopreserved to provide cells for the final stages of differentiation. Differentiated cells intended for use in an assay should be evaluated not only by the characterization of cell specific gene or protein expression but also (where possible) by cell specific functional assays.

4.2 Selection of Biomarkers for Key Cell Types Required in Toxicology

In order to confirm that a particular differentiation process has generated appropriate cells for use, it is evident that markers for each common hPSC-derived cell type are required, which identify a sufficient level of cell differentiation/maturation and the suitability of these cells for use in a particular *in vitro* toxicity method. Acceptance criteria will need to be defined as the list of essential phenotypic or molecular features that must be satisfied prior to the user accepting each batch of differentiated cells for use in the *in vitro* toxicity assay. Typically acceptability is based on level of expression of certain markers characteristic of the desired cell type. Table 11.4 shows typical markers and functional assays that may be used for the quality control of hPSC-derived *in vitro* models obtained by directed differentiation protocols to yield neural, hepatic, cardiac, keratinocyte and "mesenchymal" cell types as developed by the SEURAT-1 partners (SEURAT-1 Stem Cell Group). The specific metrics used to monitor these characteristics and provide tolerance limits and acceptability criteria for use, may vary depending on the specific differentiation protocol, cell line used and aim of the studies. Thus, it will be necessary to generate qualification data for the protocol with the proposed cell lines to understand the most informative biomarkers to use in culture quality control and acceptable levels of biomarker expression and variability.

4.2.1 Development of Human Neuronal Models Derived from Pluripotent Stem Cells: Neuronal/Glial-Like Cells for Toxicity Studies

Human embryonic stem cells (hESCs) and human induced pluripotent stem cells (hiPSCs) have ability to differentiate to various somatic-like cell types, including neuronal and glial cells. However, developing well-defined conditions to consistently generate a pure population of neural stem cells (NSCs) is critical to achieve this goal. There are several neural induction methods that enrich NSCs of neuronal cell population using spontaneous differentiation, chemical induction or mouse stromal feeder cells. NSCs can be manually isolated and propagated as monolayer cultures for many passages. In principle, these cells can be differentiated to neurons and glial cells, providing an endless supply of cells for *in vitro* assays. Unfortunately, the robustness of these methods is hampered by batch to batch variability of isolated NSCs. Moreover, differentiation of NSCs often results in variable and heterogeneous cultures of neurons, glial cells as well as undifferentiated cells which hampers many downstream applications such as *in vitro* assays. One promising approach is to identify cell surface markers (immune-phenotyping screens) specifically expressed on NSCs, glia and neurons to purify distinct cell type population using FACS (Yuan et al. 2011).

Table 11.4 Candidate biomarkers for quality control of different hPSC-derived cell models studied in the SEURAT-1 partner consortium Sr&Tox (http://www.scrtox.eu/)

Cell types	Cellular markers	Functional assays
Cardiac cells	Immunological analysis of cellular marker expression: Tropomyosin, Troponin I, Actinin, Atrial Natriuretic Peptide, Desmin	Microscopic evaluation of contracting cells/areas produced in an efficient differentiation protocol (Lian et al. 2012)
	Costaining of MLC-2a and MLC2v for determination of subtype cell fate (atrial/ventricle) and maturation (atrial occurs first in both subtypes, in ventricular CMs than coexpressed and later only MLC-2v)	When such 2D cultures are used for generation of 3D cultures the number of contracting clusters may be scored as a proportion of seeded undifferentiated cells (microscopic evaluation and field potential recoding)
	Analysis of gene expression: brachyury, Nkx2.5, alpha-cardiac actin, nppa	Generation of action potentials (for example as measured by micro-electrode array)
		Sensitivity to channel blockers (for example as measured by micro-electrode array)
		Functional hormonal regulation
Hepatic cells	Analysis of markers/genes expression: CYP3A4, CYP2B6, CYP1A1/2, CYP2C9, CYP2C19, CYP2D6, AFP, ALB, Sox17, CXCR4, HGF, HNF4, TAT, TTR	Urea synthesis
		Glycogen uptake
		Albumin secretion
		Fibrinogen secretion
		ATP levels
		Glutathione (GSH) levels
		Drug-metabolizing cytochrome P450 (CYP450) activities (in particular CYP3A)
		Analysis of Albumin synthesis
		Phase II activities: Measurement of activities of glutathione S-transferase (GST) isoenzymes; Measurement of activities of UDP-glucuronosyltransferase (UGT) isoenzymes
		Drug transporter capacity: analysis of ATP-binding cassette (ABC) transporter expression and activity; analysis of solute carrier (SLC) transporter expression

(continued)

Table 11.4 (continued)

Cell types	Cellular markers	Functional assays
Neural cells	**Immunological markers:**	Measurement of neurite and axon formation and extension
	Neural stem/progenitor cells:	
	Sox1, Sox2, Pax6, Nestin (neuroepithelial stem cells)	
	Sox2, Pax6, Nestin, 3CB2, BLBP (radial glia-like stem cells)	
	Neurons (generic):	
	β-III-tubulin, MAP2, NF200, Synapsin-I	
	Neuronal subtype-specific	
	Dopaminergic neurons: Tyrosine-hydroxylase (TH), FoxA2, En-1, Nurr-1	
	Cholinergic neurons: ChAT, VAChT, Acetylcholin	
	Serotonergic neurons: 5HT	
	GABAergic neurons GABA, GAD65/67	
	Glutamatergic interneurons: VGLUT1/2	
	Layer-specific cortical neurons: FOXP1, FOXP2, CTIP2, calbindin, DARPP-32, Tbr1, Tbr2, Satb2, Cux1/2	
	Sensory neurons: Peripherin, Brn3a	
	Motoneurons: HB9, SMI-32	
	Analysis of gene expression/ immunocytochemistry for neural progenitors:	Generation of action potentials (as generated using micro-electrode arrays)
	FoxG1, Otx1/2 (forebrain fate)	
	Pax2, En1/2, Lmx1Aa/b (midbrain fate)	
	Gbx2, HoxA2, HoxA4 (hindbrain fate) (*specific markers for ventral and dorsal identities should/could be implemented for detailed characterization*)	
		Presence of ion channel activity

(continued)

Table 11.4 (continued)

Cell types	Cellular markers	Functional assays
Keratinocytes	Analysis of immunological marker expression (FACS, IF): K5, K14, K10	Capacity to formed a stratified epidermis confirmed by histological H&E staining on the 3D tissue to identify the presence of Stratum Basal, Stratum Spinosum, Stratum Granulosum and Stratum Corneum. Presence of Stratum Corneum is critical and tolerances should be set for its thickness based on user experience. N.B. too thin giving poor barrier function and false positive toxicity and too thick yielding false negative results
	Pan-CK antibody (CK14, CK15, CK16 and CK19) performed on the terminally differentiated keratinocytes as a QC step immediately prior to setting up for 3D differentiation into 3D epidermis	
	Analysis of gene expression: K5, K14, DeltaNP63 (marker of proliferative keratinocytes) involucrin (marker of senescent keratinocytes)	
Mesenchymal progenitor cells	Analysis of immunological marker expression (FACS): CD29, CD44, CD73, CD105, CD166	Analysis of the proliferation capacity in presence of increasing serum concentration e.g. using Cell Titer Glo™
		Analysis of cells response to statin treatment and rescue by mevalonate

There are several diverse neural differentiation protocols for hESCs and hiPSCs, based on embryoid body formation (Carpenter et al. 2001; Zhang et al. 2001), direct differentiation into a neural lineage in a suspension cultures (Nat et al. 2007; Schulz et al. 2004) or in adherent cultures in coated well plates or in co-culture with mouse stromal cells or astrocytes (Baharvand et al. 2007). These differentiation protocols include several differentiation- and proliferation-inducing factors such as basic FGF, EGF, fetal bovine serum, inhibitory protein noggin, retinoic acid, BDNF, GDNF cAMP etc. as well as conditioned medium. The current neural differentiation protocols are not equally effective for different hESCs and hiPSCs lines, probably due to the influence of genetic background or derivation and culture methods.

One of the most important issues in the neuronal/glial differentiation is the detailed characterization of the terminal differentiated cell populations. It is important to show that markers for pluripotency disappear during neuronal differentiation and markers specific for neuronal and glial maturation are up regulated. It has also been shown that in neurosphere cultures the process of neuronal differentiation is more advanced than in monolayer cultures. Typically neurospheres do not express the pluripotency marker Oct-4 after 3–6 weeks of differentiation but express Musashi, Nestin and Pax-6 indicating their neural progenitor nature (Lappalainen et al. 2010).

In order to evaluate the suitability of the differentiated neuronal cell population for neurotoxicological studies the expression of specific differentiation-related markers should be characterized. In the case of neuronal differentiation the most obvious candidate proteins include neurofilament 200 (NF200), Synapsin-I and synaptophysin for synaptogenesis (Pistollato et al. 2012). Studies of neurotoxicity *in vitro* studies should be performed in the presence of glial cells (due to their *in vivo* supportive role), therefore it is important to identify and quantify the ratio of neurons to astrocytes/microglia/oligodendrocytes in the terminal cell population. Astrocytes are usually identified by the expression e.g. of GFAP, oligodendrocytes by the presence of a marker such as Olig1 and oligodendrocytes are stained by antibodies for OX-42 (Lappalainen et al. 2010). A list of commonly used markers for each of the neurodevelopmental stages is given in Table 11.5. After selection of a marker panel, specific quality control methods are needed to establish acceptability criteria (i.e. acceptable level of expression of cell specific markers at different stages of cells development and maturation) as well as evaluation of neuronal functionality. Mature neuronal cultures derived from PSC should be proven to be electro-physiologically active, generating action potentials. One of the commonly used techniques to characterize action potentials of mature neurons is the multi-electrode array (MEA) (Hogberg et al. 2011), which is used as an alternative to the more classical and challenging 'patch-clamping' technique. In this case, specific quality control metrics for the functional activity and threshold levels for positive controls need to be defined in order to properly judge the neuronal maturation of an individual cell preparation. In general, a well-defined set of quality control analyses should serve as basis for acceptance criteria supporting a reduction of intra- and inter-laboratory variability of the test system as has been shown in ring-trial neurotoxicity studies based on MEA measurements (Novellino et al. 2011).

Current neuronal differentiation protocols developed for hESCs and hiPSCs usually yield a high percentage of neural precursors (>80 %) (Zhou et al. 2010) and a significant number of target cells (up to 60 %) (Zeng et al. 2010). hiPSCs also give an opportunity to create a range of patient genotype-specific or disease-specific models for neurotoxicity testing. Recently, there have been a number of key developments in neurotoxicological assays, including tests developed that measure the effects of chemicals on dopaminergic neurons (Zeng et al. 2006). A further test system has been developed based on measurement of neurite outgrowth providing automated high-content image analysis and high-throughput screening (Harrill et al. 2010). Screening using neural progenitors and differentiated neural cells (Han et al. 2009) has also been established. These test systems could be adapted for the screening of compounds for neurotoxic effects at different stages of neuronal differentiation. However, in this field one of the main aims is to establish human relevant test systems which can predict the effect of a chemical on the function of neuronal networks measured by MEA techniques which can be applied to high throughput/content screening.

Table 11.5 Markers used to discriminate between the different stages of renal development and expected markers of differentiated target cells

	Target cell	Expression	Additional characteristics
Development	Pluripotent stem cells	Nanog 16 (Silva et al. 2009), Oct4 (POU5F1) (Pan et al. 2002)	Highly proliferative, can differentiate into all 3 germ layers
	Mesendoderm cells	Brachyury, Mixl1 (Lam et al. 2014)	Precursor of the intermediate mesoderm
	Intermediate mesoderm (IM)	Pax2[a], Osr1[b], Lhx1[b] (Dressler 2009; Xia et al. 2013)	Precursor of the metanephric mesenchyme and ureteric bud
	Metanephric mesenchyme (MM)	Six2, Sall1, Hox11, Eya1 (Dressler 2009), Cited1/Cited 2 (Boyle et al. 2007)	Glomerulus and tubular nephron progenitor
	Ureteric bud (UB)	HoxB7, Gfra1, c-Ret (Xia et al. 2013), Cited4	Collecting duct progenitor
Differentiated	Podocyte cells	Synaptopodin (Faul et al. 2007) podocin (Boute et al. 2000; Roselli et al. 2002; Schwarz et al. 2001), nephrin (Holthofer et al. 1999; Holzman et al. 1999; Kestila et al. 1998; Ruotsalainen et al. 1999), podocalyxin (Horvat et al. 1986; Kerjaschki et al. 1984), CD2AP (Li et al. 2000; Shih et al. 2001)	Large cell body, interdigitated foot processes, VEGF and prostaglandin secretion (Jennings et al. 2003)
	Proximal tubular cells	Aquaporin 1 (Nielsen et al. 2002), claudin 2 and 10 (Muto et al. 2010; Van Itallie et al. 2006; Wilmes et al. 2014), parathyroid hormone receptor 1 (Stacey et al. 2014b), organic cation transporter 2 (SLC22A2), MATE1 (SLC47A1) and MATE2-K (SLC47A2), organic anion transporter 1 (SLC22A6) and organic anion transporter 3 (SLC22A8) (Motohashi et al. 2013), glutamyl transferase (GGT) (Glenner and Folk 1961)	Cobble stone morphology. PTH dependent cAMP induction, low to medium transepithelial electrical resistance (TEER), dome formation and paracellular water transport (Jennings et al. 2003; Wilmes et al. 2014)
	Collecting duct cells	aquaporin 2, 3 and 4 (Nielsen et al. 2002), pendrin (SLC26A4) (Soleimani 2015)	AVP dependent cAMP induction, very high TEER (Jennings et al. 2003)

Please note these markers are not necessarily exclusive to the designated cell types, but they discriminate from the other cell types in the table

[a]Some authors show that Pax2 expression persists into MM and UB stages

[b]Osr1 and Lhx1 also expressed in the lateral plate mesoderm

4.2.2 Development of Human Hepatic Models Derived from Pluripotent Stem Cells

Hepatocytes derived from pluripotent stem cells are generally foetal in their phenotype. One major additional problem in order to obtain well differentiated hepatocytes is the fact that freshly isolated hepatocytes rapidly dedifferentiate in most *in vitro* systems used and lose their ability to perform basic hepatocyte functions (Richert et al. 2006; Schwartz et al. 2014) (for typical hepatocyte markers see Table 11.5). The key differences between stem cell derived hepatocytes and adult human hepatocytes have hitherto limited the use of stem cells as a source for *in vitro* modelling of liver responses.

Several different protocols have been used for the differentiation from stem cells to hepatocyte like cells. A key component is the differentiation of stem cells to DE-HEP cells using Activin A (Hay et al. 2008). However, most of these protocols do not provide cells with a useful phenotype. Some improved phenotype is obtained by using 3D systems for the differentiation process either in hollow fibre bioreactors (Sivertsson et al. 2013) or in spheroids (Subramanian et al. 2014), where the expression of key genes encoding e.g. albumin production and drug metabolism is improved.

Toxicity assays have been performed using known hepato-toxins and high content image analysis based determination of toxicity (Sirenko et al. 2014) or ATP based toxicity assays (Ulvestad et al. 2013; Szkolnicka et al. 2014). The results show that in some cases the sensitivity for drug toxicity in these stem cell derived hepatocytes are not too far from those produced by primary hepatocytes cultivated for 48 h, but still they do not yet provide a robust model of drug induced hepatotoxicity *in vivo* in man. To a certain extent the latter extrapolation cannot be possible unless systems are developed that can be used for assays of chronic drug toxicity, which *in vivo* often develops only after 4–12 weeks of development. Furthermore, it is important to include immunologically relevant cells which can mimic the drug induced idiosyncratic reactions, which in many cases are dependent on the action of specific HLA class II antigens. Usually stem cell derived hepatocytes are stable only for a maximum of 2 weeks, hampering the use of such cells for chronic drug induced hepatotoxicity. However, using a 3D collagen matrix culture (3D clump cultures) CYP3A4 expression at rather relevant levels have been achieved for 75 days (Gieseck et al. 2014). The introduction of immune cells and non-parenchymal cells together with the stem cell derived hepatocytes into 3D *in vitro* systems would of course be very challenging. In addition, it would be valuable to use cells derived from patients susceptible to drug induced liver toxicity in comparison with unaffected controls to highlight potential adverse effects in the liver *in vivo*. However, because of the current limited knowledge about differentiation of stem cells into non parenchymal cells and the lack of a useful hepatocyte phenotype in models derived from hESC or hiPSC,, it is anticipated that such integrated systems will not be achieved in the near future.

A key issue in the field of stem cell derived hepatocytes are conditions for better differentiation and 2D or 3D systems for cultivation of hepatocytes that prevent dedifferentiation. Interesting approaches have been taken where iPSC derived hepatocyte like cells have been co-cultured with endothelial cells and mesenchymal stem cells on a pre-solidified matrix forming 3D spheroids (liver buds) in 2 days which are transplanted after 4 days into immunodeficient mice. The resulting human liver tissue has been found to be highly vascularized with many hepatic functions, although bile ducts are lacking (Takebe et al. 2013, 2014).

Another interesting approach is the use of small molecules for differentiation of stem cells to hepatocytes and for proliferation of hepatocytes *in vitro* as discussed by Shan et al. (2013). This involves cultivating iPSCs on matrigel supported by conditioned media from primary mouse embryonic fibroblasts in the presence of Activin A and growth factors, where after the small molecules were added 21 days post cultivation for 9 days. The hepatocyte phenotype that these authors achieve is encouraging for hPSC-based hepatic models, with high expression of several true hepatic genes. It will be interesting to see if this or similar protocols can be successfully reproduced in other labs and further developed.

In conclusion, we currently have stem cell derived hepatocytes that are relatively fetal, undifferentiated and not yet of the phenotypic level that can replace human primary hepatocytes with respect to screening for effects such as drug induced hepatotoxicity. Two novel protocols do indeed generate promising liver functionality of such cells, but are complicated and labour intensive. We would also like to see these protocols reproduced in other labs. However, finding of novel key factors for the differentiation of immature cells into hepatocytes might play an important role for future development of the field. An interesting aspect in this respect is the simple overexpression of HNF4α for differentiation of HepaRG cells to generate highly functional differentiated cells (Chen et al. 2014). Further identification of key transcription factors and other gene products necessary to activate in the right window of cell differentiation might take this field into a new era.

4.2.3 Development of Human Cardiac Models Derived from Pluripotent Stem Cells

Recently, the use of hiPSC-derived cardiomyocytes (hiPSC-CM) has increased tremendously in the study of basic cardiac (disease) biology, to assess the effects of drugs on the heart (efficacy), and to assess possible toxic chemical effects. Recently, Acimovic and colleagues published a review on the available human pluripotent stem cell-derived cardiomyocytes as research and therapeutic tools (Acimovic et al. 2014). In the following section, the available methods as described in their review are shortly discussed.

One of the first protocols describing cardiomyocyte formation from pluripotent stem cells consists of a co-culture of hESC with mouse visceral-endoderm-like (END-2) cells (Mummery et al. 2003). END-2 cells secrete factors that have a direct effect on cardiomyocyte differentiation, such as bone morphogenetic factors (BMPs), nodal/Activin A, fibroblast growth factors (FGFs) and repressors of the canonical Wnt/β-catenin signalling pathway (Acimovic et al. 2014). Generally, the efficiency to produce cardiomyocytes using this original protocol was quite low, but modifications to serum- and insulin-free culture conditions, as well as addition of L-ascorbic acid increased the cardiomyocyte yield (Passier et al. 2005; Freund et al. 2010).

Cardiomyocytes can also be obtained by culturing stem cells as three dimensional cell aggregates called embryoid bodies (EBs). To increase the cardiomyocyte yield, specific growth factors and small molecules can be added. For example, short term BMP4 treatment can be added to promote mesoderm induction (Zhang et al. 2008) and also p38 mitogen activated protein kinase (MAPK) inhibition is used to increase the cardiomyocyte yield (Graichen et al. 2008). Further, the canonical Wnt/β-catenin signalling pathway has an important role in cardiomyocyte differentiation. For efficient cardiomyocyte differentiation, this pathway should be activated in the early phase, but inhibited in a later phase of differentiation (Lian et al. 2012). Wnt signalling inhibition increases the efficiency of BMP4-directed cardiac differentiation (Ren et al. 2011). Also, application of the histone deacetylase inhibitor trichostatin A has been found to enhance EB-mediated cardiac differentiation (Lim et al. 2013).

Growth factors and small molecules are also applied in monolayer-based cardiac differentiation protocols. Like with EB-mediated cardiac differentiation, efficient differentiation results can be obtained if with BMP4-directed cardiac differentiation Wnt signalling is activated in an early phase and inhibited in a late phase of differentiation (Paige et al. 2010). Further improvements can be obtained when insulin is removed and FGF2 is added (Uosaki et al. 2011). However, it must be noted that different differentiation protocols result in different ratios of ventricular and atrial-like cardiomyocytes (Acimovic et al. 2014).

During the differentiation steps, signs of immaturity are clearly visible (e.g. expression of pluripotency and mesenchymal markers, such as stage-specific antigen 1 (SSEA1) and mesoderm posterior 1 (MESP1) (Blin et al. 2010) or cardiac progenitor markers LIM, homeodomain transcription factor Isl1 (Cai et al. 2003) and homeobox protein Nkx-2.5 (Stanley et al. 2002). When maintained in culture for a long time, phenotypic features similar as those of adult cardiomyocytes can be found (Acimovic et al. 2014; Ivashchenko et al. 2013; Lee et al. 2011; Lundy et al. 2013). It also has been suggested that stimuli like electrical stimulation and mechanical stretching improve maturity and functionality of stem-cell derived cardiomyocytes (Nunes et al. 2013).

For a correct interpretation and reproducibility of experiments, it is of utmost importance that the stem-cell derived cardiomyocytes used are thoroughly characterized. There are no published consensus guidelines for this, but Mordwinkin and colleagues recently published a list of criteria that could be used for stem-cell derived cardiomyocyte characterization (Mordwinkin et al. 2013). These are listed in Table 11.5.

4.2.4 Development of Human Keratinocyte Models Derived from Pluripotent Stem Cells

Human keratinocytes and human skin models are key to predicting skin irritation and may be valuable in predicting other toxic effects in skin. Currently, there are a number of *in vitro* systems that have regulatory acceptance but these are all derived from primary tissue. The provision of cell lines that can expand indefinitely and are capable of robustly and reliably producing stratified epidermis *in vitro* would be of value to this area of testing. Pluripotent stem cells have been shown to be capable of generating dermis and express a range of biomarkers characteristic of the skin (Green et al. 2003; Guenou et al. 2009; Itoh et al. 2011). These markers include the keratins 5, 10, and 14, as well as involucrin and filaggrin, ITGA6, ITGB4, integrins α6 and β4, collagen VII and laminin 5 (Guenou et al. 2009; Laustriat et al. 2010; Dinella et al. 2014). The generation of 3D models of skin have also been reported (Petrova et al. 2014). Systems are also in development to construct micro-physiological skin models allowing the interactions between the skin and other organs to be studied (Guo et al. 2013). These systems are very promising areas of development but as with other cell types derived from PSCs, to date, they have yet to be developed into reproducible differentiation protocols that can be used across a range of PSC lines.

4.2.5 Development of Human "Mesenchymal" Models Derived from Pluripotent Stem Cells

Mesenchymal stromal cells (MSCs) have the ability to differentiate into a number of cell types including adipocytes, chondrocytes, osteocytes and muscle. These cells have a great therapeutic potential due to their regenerative and immune-regulatory properties (Augello et al. 2010; Glenn and Whartenby 2014; Sutton and Bonfield 2014), they can also be isolated from a number of tissue sources and minimal criteria to characterise bone marrow derived cell types has been established (Dominici et al. 2006). However, there is considerable confusion in the literature arising from poor reporting of MSC cultures and it is important to recognise that this is not one cell type but probably represents at least three fundamentally different groups. Primary human MSCs have also been investigated for use *in vitro*, in acute toxicity testing and results from these studies indicate that MSCs could be a potential candidate cell type for use in basal cytotoxicity assays (Scanu et al. 2011). However, as with many primary cells, MSCs suffer from a number of issues including; source of starting tissue, availability of cells and batch to batch variability. These could potentially be circumvented by using hPSCs to derive MSCs. Indeed there are a number of reports of the successful generation of MSCs from PSCs both in 2D and 3D systems (Hematti 2011; Chen et al. 2012; Li et al. 2013; TheinHan et al. 2013; Tang et al. 2014). Human pluripotent stem cell derived MSCs have been shown to be comparable to their bone marrow derived counterparts in radiosensitivity assays (Islam et al. 2015). The similarity between fibroblasts and mesenchymal cells is often debated and indeed there is a school of thought that suggests that these two cell types are identical (Haniffa et al. 2009; Hematti 2012). The use of fibroblasts, derived from human embryonic stem cells, as models for genotoxity has recently been explored (Vinoth et al. 2014) and the study revealed that these hPSC-derived fibroblasts were as sensitive to genotoxic challenge as other somatic cell types used in this assay, Again, as with other cell type specific models being established using PSCs, the robustness and reproducibility of this system is still a challenge but with time and effort these issues should be resolved.

4.3 Development of New In Vitro Stem Cell-Based Cell Models for Toxicology

Clearly, a range of *in vitro* tissue cell types are achievable with increasingly better defined hPSC directed differentiation protocols under development. However, the stem cell community will need to clearly define the use-cases where stem-cell based *in vitro* methods have an added value to other test systems or identify use-cases where none of the *in vitro* methods based on non-stem cell based systems can be used. An area of strong relevance to toxicology which is now advancing in terms of stem cell-derived models is neuronal 2D and 3D cultures and kidney-derived models. The following section draws on the toxicologically relevant exemplar of kidney to identify the kinds of scientific considerations needed to progress the early stage development of new stem cell based models.

The kidneys are vital organs which control the constituents of the blood and thereby regulate whole body homeostasis. Blood is continuously filtered in the glomerulus, passes into the renal tubule where essential substances such as sodium, glucose and amino acids are reabsorbed and waste products and excess substances are secreted. Due to the multitude of transporting and metabolising systems required to perform these tasks, cells of the kidney interact with a wide variety of chemicals entities. This, coupled with its ability to concentrate and metabolise compounds, makes it susceptible to injury by a wide variety of xenobiotics. The cells of the nephron exhibit a high degree of physiological, morphological and biochemical heterogeneity (Anonymous 1988; Kriz and Kaissling 2000). These properties determine site-specific sensitivities to xenobiotics. The cells of the glomerulus and the proximal tubule are the most frequently studied in the context of renal disease and toxicity due to their critical roles in filtration and reabsorption, respectively. Cultured renal cells, either primary cultures or immortalised cell, have been extensively employed in physiological and toxicological studies (Wilmes and Jennings 2014; Jennings et al. 2008, 2014; Dressler 2006). However, there is now a growing interest in the utilisation of stem cells to derive renal target cells, mostly for tissue engineering purposes. The use of stem cells is also very attractive for the field of toxicology, not least due to the fact that cells can be derived from target populations.

The use of pluripotent stem cells to derive renal phenotypes, either from embryonic or somatic sources, brings new challenges. And the major challenge currently faced is the ability to acquire target cells with the desired phenotype. One strategy being pursued is to drive pluripotent stem cells through the critical stages in renal development. The kidney and gonads arise from the intermediate mesoderm which progress into the primary nephric duct consisting of the pronephros, mesonephros and metanephros (Sariola 2002). The permanent adult kidney derives from the latter. Within the metanephros the ureteric bud invades the surrounding metanephric mesenchyme and two-way signalling between these two tissues induces branched morphogenesis, leading to mature nephron development (Davies 2002). The ureteric bud will finally develop into the collecting duct, whereas the metanephric mesenchyme gives rise to both glomerulus and the cells of the renal tubule (Dressler

2009). As pluripotent stem cells develop into multipotent lineages they lose the expression of critical pluripotent genes such as Nanog and Oct4 (or POU5F1 in humans). The loss of the expression of these genes is often used to demonstrate successful progression to multipotent lineages. The intermediate mesoderm expresses factors such as Pax2, Osr1 and Lhx1, which may or may not be lost as differentiation continues to metanephric mesenchyme and ureteric bud lineages (Xia et al. 2013; Takasato et al. 2014). Different protocols have been used to generate reasonably pure populations of intermediate mesoderm, including sequential addition of BMP4/FGF2, retinoic acid/activin A/BMP2 (Lam et al. 2014) or by activation of Wnt signaling with the small molecule agonist CHIR99021 (CHIR) to create brachyury and Mixl1 positive mesendoderm cells (Araoka et al. 2014; Narayanan et al. 2013). Several different mixes of developmental growth factors have been used to differentiate the intermediate mesoderm further into metanephric mesenchyme and ureteric bud cells and even to podocyte and proximal-like phenotypes (Song et al. 2012; Silva et al. 2009). A list of commonly used markers for each of the developmental stages and also markers and characteristics of the mature phenotypes, which are present *in vivo* and maintained in primary culture and some cell lines are given in Table 11.5.

While there has been great success in the derivation of target renal cells from pluripotent stem cells, there is still a great deal of work that needs to be done. For example, most of the protocols developed to-date give mixed populations of cells, in different differentiation states. Traditional strategies for *in vitro* toxicity studies rely on relatively pure cultures of the target cell types. However, probably more troublesome is the lack of temporal phenotypic stability of the derived cells, which would be problematic for reproducibility and interpretation of chemical exposures. However, the field is in its infancy and it is hoped that many of these challenges will be overcome in the near future. Table 11.5 gives examples of some to the key markers which may be useful in the development and control of stem cell-derived models of kidney tissue.

5 General Acceptability Criteria of Stem Cell-Derived *In Vitro* Toxicology Assays

In order that *in vitro* toxicity methods based on stem cells-derived cell or tissue model (test system) can be considered reliable and relevant with global applicability, they must be reliable and robust showing technical reproducibility between different experimental runs, operators, laboratories and source of equipment and reagents. A key component in assuring such reliability and standardisation of data outputs is the use of suitable positive and negative controls. The generation of clear and unambiguous data, and a clear defined concept on how to use the results in the context of hazard and risk assessment contexts are of high importance. For new *in vitro* toxicity methods based on stem cells as a test system, enough historical data should be generated in the development phase to define specific acceptance criteria for all elements of the *in vitro* method and a defined level of expected performance of the method or the specific measurements of interest.

In order to define such stem-cell based phenotypic criteria the stem cells must have a specific functionality, which can be measured, whilst for stem cell-based molecular features a very well defined characteristics need to be identified to establish specific acceptance criteria. Acceptance criteria related to the biological function of stem cell-derived models are used to qualify the stem cells for use in a specific *in vitro* toxicity method and have been based largely on marker phenotype as described in Sect. 3 above. These can help to assure consistent performance of the stem cell-derived cell model based on monitoring activities carried out at specific points in preparation and use of the stem cell derived culture. However, the association between marker phenotype and predictability of toxicity with different compounds clearly requires validation and control of appropriate functional features of the cellular model. Exemplars of these are also discussed above in Sect. 3. Such acceptance criteria may need to be specific to the stem cell line used or its genotype and also the intended application of the assay. However, for *in vitro* methods where different stem cell models may need to be used for the same application (e.g. read-out, multiple genotypes tested against the same compound), it is important to define performance standards that the different methods should comply with to enable the stem cell user community to compare results from different stem cell-based *in vitro* methods.

Regulatory acceptance and validation of new *in vitro* assays can be a time-consuming process and given the variety of new alternative *in vitro* methods now becoming available, a system of capturing early stage protocols and qualification data is now being developed in a collaboration between the ToxBank consortium and the Joint Research Centre in Ispra (Stacey et al. 2014b). Developing *in vitro* methods to achieve regulatory acceptance usually follows a series of steps:

(1) method development (e.g. carried out in academic, industrial or regulatory environments)
(2) method optimisation (e.g. carried out by the original developer, new users of a particular method, or by validation bodies)
(3) method validation (with or without the involvement of validation bodies)
(4) method acceptance (e.g. facilitated by the involvement of the Organisation of Economic Cooperation and Development, OECD, through the development of test guidelines)

Acceptance and performance criteria, covering all aspects of the test system and test method should be developed and optimised as early as possible to expedite the overall validation process. Validation is a pivotal step towards the regulatory acceptance and the international recognition of *in vitro* methods for a range of scientific purposes by a variety of end-users, as described in internationally accepted guidance (OECD 2005), and throughout this book.

6 Establishment of Control Materials

The process of regulatory acceptance must include evaluation of the performance of novel *in vitro* toxicity methods and their comparison with established *in vitro* and *in vivo* toxicological methods. However, it is also important to control the essential components of the *in vitro* method including the exposure and purity of the test chemicals (test items), the *in vitro* biological models (test systems including any stem cell based test system), the analytical techniques used and the experimental design. Endpoint controls are increasingly PCR-based and it is likely that there will be a role for DNA or RNA-based reference materials to provide quantitative controls for assuring suitability of cultures for use in toxicology assays.

Control compounds (positive controls) whose mode of action and *in vitro* response is well characterised are clearly important to demonstrate consistent functionality of cell-based models and a number have been established in the ToxBank project and by EURL ECVAM at the European Commission Joint Research Centre. Such reference materials will be vital for international standardisation in development of these assays and reference materials for analytical techniques will probably be an important influence in standardisation of stem cell–based toxicological assays in the long term. However, such control materials are highly specific to the cell type and induced mechanism of toxicity. Another useful approach for developers of *in vitro* stem cell or tissue-based methods for specific toxicological applications is to consult lists of commercially available chemicals that can be used to assess the performance of their new developed methods. A range of sets of test compounds being established for different purposes can be found at http://chelist.jrc.ec.europa.eu/.

Control materials will clearly also be important for the control of safety testing for viral contaminants and numerous international reference materials for virus detection in certain products have already been established. For more information see http://nibsc.org/.

7 Conclusions and Future Perspectives

Future toxicological applications in routine testing using stem cell or tissue test systems will be dependent on the ability to deliver consistency in their molecular and phenotypic functions. Another critical factor in their successful uptake in industry will be the availability of cell preparations which can be used directly in *in vitro* methods. This may involve more efficient differentiation protocols and also new cryopreservation methods for differentiated cells and progenitor cultures. Careful attention to good cell culture practice in coordination with attention to regulatory and industry requirements will be critical. This will probably require the qualification of initial cultures (prior to differentiation), using new phenotypic and/or epigenetic screens to establish batch to batch consistency and maintenance of pluripotency.

Complex systems comprising multiple cell types that generate more sophisticated and predictive data sets could be extremely valuable and are just beginning to be developed. Approaches have also been developed within the SEURAT-1 programme using 3D culture to establish stable systems to study repeat dose and chronic drug effects and increased comparability with *in vivo* function.

The recent successes in direct differentiation to certain cell types outlined in this chapter provide future potential *in vitro* models with more efficient differentiation but these are still at a very early stage of research development. Major challenges remain regarding our ability to achieve sufficient numbers of cells reproducibly for large scale assays and our ability to assure that the responses of culture models produced by artificial cell differentiation replicate those of cells created via natural pathways of cellular development and differentiation. However, new strategies and developments including bead-based combinatorial differentiation (Tarunina et al. 2014; Efthymiou et al. 2014) are providing promising potential solutions.

The importance of technological developments and systems biology approaches for stem cell models of the future will also require progress in the following areas:

- stem cell "omics" technologies and more readily accessible bioinformatics systems
- stem cell culture automation as well as high-throughput techniques to promote reproducibility and capacity for industry use
- development of stem cell cultures in systems biological approaches to generate information regarding toxicological pathways or the mode of action of test items of various kinds.
- establishment of the potential toxicological applications using iPSC and the donor concept (e.g. DILI project)
- use of stem cell-based *in vitro* methods for integrated testing strategies
- validated stem cell-based mechanistic tools targeting key events in adverse outcome pathway, especially specific for human cells.
- the adverse outcome pathways (AOPs) concept that has been designed to be used for human risk assessment. Therefore to be useful for regulatory purposes it has to demonstrated that the key events described in the AOP are relevant to human cells, and vice versa. For this reason the critical molecular mechanisms of toxicity that are unique for human cells have to be studied using human models derived from hPSCs since the available human cells originated from cancer tissue do not represent the physiological, normal human situation and have very limited application
- role of stem cell-based toxicological methods for future regulatory applications (mixtures, grouping of substances) in combination with profiling methods such as omics etc.
- enhancing communication on progress in development of qualified stem cell based assays

It is hoped that stem cell and tissue-based *in vitro* models will increasingly help to elucidate critical toxicological questions and hopefully more elusive chronic toxic effects. Going forward, it will be important to continue identifying gaps and opportunities regarding the role of stem cell and tissue-based *in vitro* toxicological methods in assuring complementarity with other *in vitro* methods by exploiting the unique features of stem cells (especially of human origin) as a toxicological test systems. Much progress has been made in the development of stem cell-based neural cell models. However, key challenges remain for the development of accurate *in vitro* human cell-based models of *in vivo* tissue types including, muscle (especially cardiac tissue), liver and kidney. We have highlighted some advances in these areas but further work is required to provide accurate models of adult human tissue which can be generated reliably and efficiently for use in the setting of routine industrial scale screening.

8 Appendix 1: Ethics Criteria for Cell Lines Selection (hiPSCs and hESCs)

In order to establish that all cell lines were obtained from tissue that has been ethically sourced the researchers must be able to provide evidence for the following:

- That fully informed consent was obtained and recorded for the donor tissue
- That consent permits the intended uses of the hPSC lines derived from the donor's tissue
- That the donor's identity was anonymised
- A validated copy of the original consent form (with donor details redacted) is available and/or a statement is available from a person authorised by the owner or derivation centre on the ethical provenance of the cell line including a contact that would facilitate confirmation of the original consent without breaking donor anonymity.
- There should be a clear statement on any constraints applied by the donor on the use of derivatives from their cells/tissues.
- Cell lines are registered within the hESCreg database
- Copies of blank consent form (or an English translation) and any information provided to the donor are available.
- Evidence from the donation process that the donor was aware that:
- Derived lines may be exploited commercially but that donors would not receive personal financial benefit.
- The donors decision to donate tissue would not influence their personal treatment an there would be no feedback on data from the cell line derived from their tissue. Derived hPSCs could be used for a wide range of purposes in different laboratories and may be tested for genetic characteristics, microbiological contamination and other features of the cells.

References

Anonymous (1988) A standard nomenclature for structures of the kidney. The Renal Commission of the International Union of Physiological Sciences (IUPS). Pflugers Arch 411(1):113–120

Acimovic I, Vilotic A, Pesl M, Lacampagne A, Dvorak P, Rotrekl V, Meli AC (2014) Human pluripotent stem cell-derived cardiomyocytes as research and therapeutic tools. Biomed Res Int 2014:512831

Adewumi O, Aflatoonian B, Ahrlund-Richter L, Amit M, Andrews PW, Beighton G, Bello PA, Benvenisty N, Berry LS, Bevan S, Blum B, Brooking J, Chen KG, Choo AB, Churchill GA, Corbel M, Damjanov I, Draper JS, Dvorak P, Emanuelsson K, Fleck RA, Ford A, Gertow K, Gertsenstein M, Gokhale PJ, Hamilton RS, Hampl A, Healy LE, Hovatta O, Hyllner J, Imreh MP, Itskovitz-Eldor J, Jackson J, Johnson JL, Jones M, Kee K, King BL, Knowles BB, Lako M, Lebrin F, Mallon BS, Manning D, Mayshar Y, McKay RD, Michalska AE, Mikkola M, Mileikovsky M, Minger SL, Moore HD, Mummery CL, Nagy A, Nakatsuji N, O'Brien CM, Oh SK, Olsson C, Otonkoski T, Park KY, Passier R, Patel H, Patel M, Pedersen R, Pera MF, Piekarczyk MS, Pera RA, Reubinoff BE, Robins AJ, Rossant J, Rugg-Gunn P, Schulz TC, Semb H, Sherrer ES, Siemen H, Stacey GN, Stojkovic M, Suemori H, Szatkiewicz J, Turetsky T, Tuuri T, van den Brink S, Vintersten K, Vuoristo S, Ward D, Weaver TA, Young LA, Zhang W, ISCI, International Stem Cell Initiative (2007) Characterization of human embryonic stem cell lines by the International Stem Cell Initiative. Nat Biotechnol 25:803–816

Araoka T, Mae S, Kurose Y, Uesugi M, Ohta A, Yamanaka S, Osafune K (2014) Efficient and rapid induction of human iPSCs/ESCs into nephrogenic intermediate mesoderm using small molecule-based differentiation methods. PLoS One 9(1), e84881. doi:10.1371/journal.pone.0084881

Augello A, Kurth T, De Bari C (2010) Mesenchymal stem cells: a perspective from *in vitro* cultures to *in vivo* migration and niches. Eur Cell Mater 20:121–133

Baharvand H, Mehrjardi NZ, Hatami M, Kiani S, Rao M, Haghighi MM (2007) Neural differentiation from human embryonic stem cells in a defined adherent culture condition. Int J Dev Biol 51(5):371–378

Barallon R, Steven R, Bauer SR, Butler J, Capes-Davis A, Dirks WG, Elmore E, Furtado M, Kline MC, Kohara A, Los GV, MacLeod RAF, Masters JRW, Nardone M, Nardone RM, Nims RW, Price PJ, Reid YA, Shewale J, Sykes G, Steuer AF, Storts DR, Thomson J, Taraporewala Z, Alston-Roberts C, Kerrigan L (2010a) Recommendation of short tandem repeat profiling for authenticating human cell lines, stem cells, and tissues. *In Vitro* Cell Dev Biol Anim 46:727–732

Barallon R, Steven R, Bauer SR, Butler J, Capes-Davis A, Dirks WG, Elmore E, Furtado M, Kline MC, Kohara A, Los GV, MacLeod RAF, Masters JRW, Nardone M, Nardone RM, Nims RW, Price PJ, Reid YA, Shewale J, Sykes G, Steuer AF, Storts DR, Thomson J, Taraporewala Z, Alston-Roberts C, Kerrigan L (2010b) on behalf of ATCC® SDO Workgroup ASN-0002. Cell line misidentification: the beginning of the end. Nature Rev. Cancer 10:441–448

Blin G, Nury D, Stefanovic S, Neri T, Guillevic O, Brinon B, Bellamy V, Rücker-Martin C, Barbry P, Bel A, Bruneval P, Cowan C, Pouly J, Mitalipov S, Gouadon E, Binder P, Hagège A, Desnos M, Renaud JF, Menasché P, Pucéat M (2010) A purified population of multipotent cardiovascular progenitors derived from primate pluripotent stem cells engrafts in postmyocardial infarcted nonhuman primates. J Clin Invest 120(4):1125–1139

Boute N, Gribouval O, Roselli S, Benessy F, Lee H, Fuchshuber A, Dahan K, Gubler MC, Niaudet P, Antignac C (2000) NPHS2, encoding the glomerular protein podocin, is mutated in autosomal recessive steroid-resistant nephrotic syndrome. Nat Genet 24(4):349–354

Boyle S, Shioda T, Perantoni AO, de Caestecker M (2007) Cited1 and Cited2 are differentially expressed in the developing kidney but are not required for nephrogenesis. Dev Dyn 236(8):2321–2330

Cai CL, Liang X, Shi Y, Chu PH, Pfaff SL, Chen J, Evans S (2003) Isl1 identifies a cardiac progenitor population that proliferates prior to differentiation and contributes a majority of cells to the heart. Dev Cell 5(6):877–889

Capes-Davis A, Reid YA, Kline MC, Storts DR, Strauss E, Dirks WG, Drexler HG, MacLeod RAF, Sykes G, Kohara A, Nakamura Y, Elmore E, Nims RW, Alston-Roberts C, Barallon R, Los GV, Nardone RM, Price PJ, Steuer A, Thomson J, Masters JRW, Kerrigan L (2013) Match criteria for human cell line authentication: where do we draw the line? Intl J Cancer 132:2510–2519

Carpenter MK, Inokuma MS, Denham J, Mujtaba T, Chiu CP, Rao MS (2001) Enrichment of neurons and neural precursors from human embryonic stem cells. Exp Neurol 172(2):383–397

Chen Y, Pelekanos R, Ellis R, Horne R, Wolvetang E, Fisk N (2012) Small molecule mesengenic induction of human induced pluripotent stem cells to generate mesenchymal stem/stromal cells. Stem Cells Transl Med 1(2):83–95

Chen KT, Pernelle K, Tsai YH, Wu YH, Hsieh JY, Liao KH, Guguen-Guillouzo C, Wang HW (2014) Liver X receptor α (LXRα/NR1H3) regulates differentiation of hepatocyte-like cells via reciprocal regulation of HNF4α. J Hepatol 61(6):1276–1286

Coecke S, Balls M, Bowe G, Davis J, Gstraunthaler G, Hartung T, Hay R, Merten OW, Price A, Schechtman L, Stacey G, Stokes W (2005) Guidance on good cell culture practice. Altern Lab Anim 33:261–287

Coecke S, Bowe G, Millcamps A, Bernasconi C, Bostroem AC, Bories G, Fortaner S, Gineste JM, Gouliarmou V, Langezaal I, Liska R, Mendoza E, Morath S, Reina V, Wilk-Zasadna I, Whelan M (2014) Considerations in the development of *in vitro* toxicity testing methods intended for regulatory use. In: Paul J, Anna P (eds) *In vitro* toxicology systems, Methods in pharmacology and toxicology. Springer, New York, pp 551–569

Davies JA (2002) Morphogenesis of the metanephric kidney. ScientificWorldJournal 2:1937–1950. doi:10.1100/tsw.2002.854

Dinella J, Koster M, Koch P (2014) Use of induced pluripotent stem cells in dermatological research. J Invest Dermatol 134, e23. doi:10.1038/jid.2014.238

Dominici M, Le Blanc K, Mueller I, Slaper-Cortenbach I, Marini F, Krause D et al (2006) Minimal criteria for defining multipotent mesenchymal stromal cells. The International Society for Cellular Therapy position statement. Cytotherapy 8(4):315–317

Dressler GR (2006) The cellular basis of kidney development. Annu Rev Cell Dev Biol 22:509–529. doi:10.1146/annurev.cellbio.22.010305.104340

Dressler GR (2009) Advances in early kidney specification, development and patterning. Development 136(23):3863–3874. doi:10.1242/dev.034876

Efthymiou A, Chen G, Rao M, Chen G, Boehm M (2014) Self-renewal and cell lineage differentiation strategies in human embryonic stem cells and induced pluripotent stem cells. Expert Opin Biol Ther 14(9):1333–1344

Faul C, Asanuma K, Yanagida-Asanuma E, Kim K, Mundel P (2007) Actin up: regulation of podocyte structure and function by components of the actin cytoskeleton. Trends Cell Biol 17(9):428–437

Freund C, Davis RP, Gkatzis K, Ward-van Oostwaard D, Mummery CL (2010) The first reported generation of human induced pluripotent stem cells (iPS cells) and iPS cell-derived cardiomyocytes in the Netherlands. Neth Heart J 18(1):51–54

Genschow E, Spielmann H, Scholz G, Pohl I, Seiler A, Clemann N, Bremer S, Becker K (2004) Validation of the embryonic stem cell test in the international ECVAM validation study on three *in vitro* embryotoxicity tests. Altern Lab Anim 32:209–244

Gieseck IRL, Hannan NR, Bort R, Hanley NA, Drake RA, Cameron GW, Wynn TA, Vallier L (2014) Maturation of induced pluripotent stem cell derived hepatocytes by 3D-culture. PLoS One 9(1), e86372

Glenn J, Whartenby K (2014) Mesenchymal stem cells: Emerging mechanisms of immunomodulation and therapy. World J Stem Cells 6(5):526–539

Glenner GG, Folk JE (1961) Glutamyl peptidases in rat and guinea pig kidney slices. Nature 192:338–340

Graichen R, Xu X, Braam SR, Balakrishnan T, Norfiza S, Sieh S, Soo SY, Tham SC, Mummery C, Colman A, Zweigerdt R, Davidson BP (2008) Enhanced cardiomyogenesis of human embryonic stem cells by a small molecular inhibitor of p38 MAPK. Differentiation 76(4):357–370

Green H, Easley K, Iuchi S (2003) Marker succession during the development of keratinocytes from cultured human embryonic stem cells. Proc Natl Acad Sci U S A 100:15625–15630

Guenou H, Nissan X, Larcher F, Feteira J, Lemaitre G, Saidani M, Del Rio M, Barrault CC, Bernard FX, Peschanski M et al (2009) Human embryonic stem-cell derivatives for full reconstruction of the pluristratified epidermis, a preclinical study. Lancet 374:1745–1753

Guo Z, Higgins C, Gillette B, Itoh M, Umegaki N, Gledhill K, Sia S, Christiano A (2013) Building a microphysiological skin model from induced pluripotent stem cells. Stem Cell Res Ther 4(Suppl 1):S2. doi:10.1186/scrt363

Gupta K, Rispin A, Stitzel K, Coecke S, Harbell J (2005) Ensuring quality of *in vitro* alternative test methods: issues and answers. Regul Toxicol Pharmacol 43:219–224

Han Y, Miller A, Mangada J, Liu Y, Swistowski A, Zhan M, Rao MS, Zeng X (2009) Identification by automated screening of a small molecule that selectively eliminates neural stem cells derived from hESCs but not dopamine neurons. PLoS One 4(9), e7155. doi:10.1371/journal.pone.0007155

Haniffa M, Collin M, Buckley C, Dazzi F (2009) Mesenchymal stem cells: the fibroblasts' new clothes? Haematologica 94(2):258–263

Harrill JA, Freudenrich TM, Machacek DW, Stice SL, Mundy WR (2010) Quantitative assessment of neurite outgrowth in human embryonic stem cell-derived hN2 cells using automated high-content image analysis. Neurotoxicology 31(3):277–290

Hartung T, Gstraunthaler G, Balls M (2000) Bologna statement on good cell culture practice (GCCP). ALTEX 17:38–39

Hay DC, Fletcher J, Payne C, Terrace JD, Gallagher RC, Snoeys J, Black JR, Wojtacha D, Samuel K, Hannoun Z, Pryde A, Filippi C, Currie IS, Forbes SJ, Ross JA, Newsome PN, Iredale JP (2008) Highly efficient differentiation of hESCs to functional hepatic endoderm requires Activin A and Wnt3a signaling. Proc Natl Acad Sci U S A 105(34):12301–12306

Hematti P (2011) Human embryonic stem cell-derived mesenchymal progenitors: an overview. Methods Mol Biol 690:163–174

Hematti P (2012) Mesenchymal stromal cells and fibroblasts: a case of mistaken identity? Cytotherapy 14(5):516–521

Hogberg HT, Sobanski T, Novellino A, Whelan M, Weiss DG, Bal-Price AK (2011) Application of micro-electrode arrays (MEAs) as an emerging technology for developmental neurotoxicity: evaluation of domoic acid-induced effects in primary cultures of rat cortical neurons. Neurotoxicology 32(1):158–168

Holthofer H, Ahola H, Solin ML, Wang S, Palmen T, Luimula P, Miettinen A, Kerjaschki D (1999) Nephrin localizes at the podocyte filtration slit area and is characteristically spliced in the human kidney. Am J Pathol 155(5):1681–1687

Holzman LB, St John PL, Kovari IA, Verma R, Holthofer H, Abrahamson DR (1999) Nephrin localizes to the slit pore of the glomerular epithelial cell. Kidney Int 56(4):1481–1491

Horvat R, Hovorka A, Dekan G, Poczewski H, Kerjaschki D (1986) Endothelial cell membranes contain podocalyxin—the major sialoprotein of visceral glomerular epithelial cells. J Cell Biol 102(2):484–491

International Stem Cell Banking Initiative, Andrews PW, Arias-Diaz J, Auerbach J et al (2009) Consensus guidance for banking and supply of human embryonic stem cell lines for research purposes. Stem Cell Rev 5:301–314

Isasi R, Andrews PW, Baltz JM, Bredenoord AL, Burton P, Chiu IM, Hull SC, Jung JW, Kurtz A, Lomax G, Ludwig T, McDonald M, Morris C, Ng HH, Rooke H, Sharma A, Stacey GN, Williams C, Zeng F, Knoppers BM (2014) Identifiability and privacy in pluripotent stem cell research. Cell Stem Cell 14(4):427–430

Islam M, Stemig M, Takahashi Y, Hui S (2015) Radiation response of mesenchymal stem cells derived from bone marrow and human pluripotent stem cells. J Radiat Res 56(2):269–277
Itoh M, Kiuru M, Cairo MS, Christiano AM (2011) Generation of keratinocytes from normal and recessive dystrophic epidermolysis bullosa-induced pluripotent stem cells. Proc Natl Acad Sci U S A 108:8797–8802
Ivashchenko CY, Pipes GC, Lozinskaya IM, Lin Z, Xiaoping X, Needle S, Grygielko ET, Hu E, Toomey JR, Lepore JJ, Willette RN (2013) Human-induced pluripotent stem cell-derived cardiomyocytes exhibit temporal changes in phenotype. Am J Physiol Heart Circ Physiol 305(6):H913–H922
Jennings P, Koppelstaetter C, Helbert MJ, Pfaller W (2003) Renal culture models: Contribution to the understanding of nephrotoxic mechanisms. Clinical Nephrotoxins: Renal Injury from drugs and chemicals, 2nd edn, pp. 115–147
Jennings P, Koppelstaetter C, Lechner J, Pfaller W (2008) Renal culture models: contribution to the understanding of nephrotoxic mechanisms. In: Broe ME, Porter GA (eds) Clinical nephrotoxins: renal injury from drugs and chemicals, 3rd edn. Springer, New York, pp 223–250
Jennings P, Aschauer L, Wilmes A, Gstraunthaler G (2014) Renal cell culture. In: Jennings P, Bal-Price A (eds) *In vitro* toxicology systems. Methods in pharmacology and toxicology. Springer, New York, pp 79–101. doi:10.1007/978-1-4939-0521-8_4
Kerjaschki D, Sharkey DJ, Farquhar MG (1984) Identification and characterization of podocalyxin—the major sialoprotein of the renal glomerular epithelial cell. J Cell Biol 98(4):1591–1596
Kerrigan L, Nims RW (2011) Authentication of human cell-based products: the role of a new consensus standard. Regen Med 6:255–260, http://standards.atcc.org/kwspub/home/the_international_cell_line_authentication_committee-iclac_/)
Kestila M, Lenkkeri U, Mannikko M, Lamerdin J, McCready P, Putaala H, Ruotsalainen V, Morita T, Nissinen M, Herva R, Kashtan CE, Peltonen L, Holmberg C, Olsen A, Tryggvason K (1998) Positionally cloned gene for a novel glomerular protein—nephrin—is mutated in congenital nephrotic syndrome. Mol Cell 1(4):575–582
Knoppers BM, Isasi R, Benvenisty N, Kim OJ, Lomax G, Morris C, Murray TH, Lee EH, Perry M, Richardson G, Sipp D, Tanner K, Wahlström J, de Wert G, Zeng F (2011) Publishing SNP genotypes of human embryonic stem cell lines: policy statement of the International Stem Cell Forum Ethics Working Party. Stem Cell Rev 7(3):482–484
Kriz W, Kaissling B (2000) Structural organization of the mammalian kidney, vol 1, 3rd edn, The kidney, physiology and pathophysiology. Lippincott Williams & Wilkins, Philadelphia
Lam AQ, Freedman BS, Morizane R, Lerou PH, Valerius MT, Bonventre JV (2014) Rapid and efficient differentiation of human pluripotent stem cells into intermediate mesoderm that forms tubules expressing kidney proximal tubular markers. J Am Soc Nephrol 25(6):1211–1225
Lappalainen RS, Salomäki M, Ylä-Outinen L, Heikkilä TJ, Hyttinen JA, Pihlajamäki H, Suuronen R, Skottman H, Narkilahti S (2010) Similarly derived and cultured hESC lines show variation in their developmental potential towards neuronal cells in long-term culture. Regen Med 5(5):749–762
Laustriat D, Gide J, Peschanski M (2010) Human pluripotent stem cells in drug discovery and predictive toxicology. Biochem Soc Trans 38:1051–1057
Lee YK, Ng KM, Lai WH, Chan YC, Lau YM, Lian Q, Tse HF, Siu CW (2011) Calcium homeostasis in human induced pluripotent stem cell-derived cardiomyocytes. Stem Cell Rev 7(4):976–986
Li C, Ruotsalainen V, Tryggvason K, Shaw AS, Miner JH (2000) CD2AP is expressed with nephrin in developing podocytes and is found widely in mature kidney and elsewhere. Am J Physiol Renal Physiol 279(4):F785–F792
Li O, Tormin A, Sundberg B, Hyllner J, Le Blanc K, Scheding S (2013) Human embryonic stem cell-derived mesenchymal stroma cells (hES-MSCs) engraft *in vivo* and support hematopoiesis without suppressing immune function: implications for off-the shelf ES-MSC therapies. PLoS One 8(1), e55319. doi:10.1371/journal.pone.0055319

Lian X, Hsiao C, Wilson G, Zhu K, Hazeltine LB, Azarin SM, Raval KK, Zhang J, Kamp TJ, Palecek SP (2012) Robust cardiomyocyte differentiation from human pluripotent stem cells via temporal modulation of canonical Wnt signaling. Proc Natl Acad Sci U S A 109(27):E1848–E1857

Lim SY, Sivakumaran P, Crombie DE, Dusting GJ, Pébay A, Dilley RJ (2013) Trichostatin A enhances differentiation of human induced pluripotent stem cells to cardiogenic cells for cardiac tissue engineering. Stem Cells Transl Med 2(9):715–725

Lundy SD, Zhu WZ, Regnier M, Laflamme MA (2013) Structural and functional maturation of cardiomyocytes derived from human pluripotent stem cells. Stem Cells Dev 22(14):1991–2002

Luong MX, Auerbach J, Crook JM, Daheron L, Hei D, Lomax G, Loring JF, Ludwig T, Schlaeger TM, Smith KP, Stacey G, Xu RH, Zeng F (2011) A call for standardized naming and reporting of human ESC and iPSC lines. Cell Stem Cell 8(4):357–359

Luong MXL, Smith KP, Crook JM, Stacey GN (2012) Biobanks for pluripotent stem cells (Chapter 8). In: Loring JF, Petersen SE (eds) Human stem cell manual, 2nd edn. Elsevier, London, pp 105–125

MacLeod RAF, Dirks WG, Matsuo Y, Kaufman M, Milch H, Drexler HG (1999) Widespread intraspecies cross-contamination of human tumour cell lines arising at source. Int J Cancer 83:555–563

Mordwinkin NM, Burridge PW, Wu JC (2013) A review of human pluripotent stem cell-derived cardiomyocytes for high-throughput drug discovery, cardiotoxicity screening, and publication standards. J Cardiovasc Transl Res 6(1):22–30

Motohashi H, Nakao Y, Masuda S, Katsura T, Kamba T, Ogawa O, Inui K (2013) Precise comparison of protein localization among OCT, OAT, and MATE in human kidney. J Pharm Sci 102(9):3302–3308

Mummery C, Ward-van Oostwaard D, Doevendans P, Spijker R, van den Brink S, Hassink R, van der Heyden M, Opthof T, Pera M, de la Riviere AB, Passier R, Tertoolen L (2003) Differentiation of human embryonic stem cells to cardiomyocytes: role of coculture with visceral endoderm-like cells. Circulation 107(21):2733–2740

Muto S, Hata M, Taniguchi J, Tsuruoka S, Moriwaki K, Saitou M, Furuse K, Sasaki H, Fujimura A, Imai M, Kusano E, Tsukita S, Furuse M (2010) Claudin-2-deficient mice are defective in the leaky and cation-selective paracellular permeability properties of renal proximal tubules. Proc Natl Acad Sci U S A 107(17):8011–8016

Narayanan K, Schumacher KM, Tasnim F, Kandasamy K, Schumacher A, Ni M, Gao S, Gopalan B, Zink D, Ying JY (2013) Human embryonic stem cells differentiate into functional renal proximal tubular-like cells. Kidney Int 83(4):593–603

Nat R, Nilbratt M, Narkilahti S, Winblad B, Hovatta O, Nordberg A (2007) Neurogenic neuroepithelial and radial glial cells generated from six human embryonic stem cell lines in serum-free suspension and adherent cultures. Glia 55(4):385–399

Nielsen S, Frokiaer J, Marples D, Kwon TH, Agre P, Knepper MA (2002) Aquaporins in the kidney: from molecules to medicine. Physiol Rev 82(1):205–244

Nims RW, Sykes G, Cottrill K, Ikonomi P, Elmore E (2010) Short tandem repeat profiling: part of an overall strategy for reducing the frequency of cell misidentification. *In Vitro* Cell Dev Biol Anim 46:811–819

Novellino A, Scelfo B, Palosaari T, Price A, Sobanski T, Shafer TJ, Johnstone AF, Gross GW, Gramowski A, Schroeder O, Jügelt K, Chiappalone M, Benfenati F, Martinoia S, Tedesco MT, Defranchi E, D'Angelo P, Whelan M (2011) Development of micro-electrode array based tests for neurotoxicity: assessment of interlaboratory reproducibility with neuroactive chemicals. Front Neuroeng 4:4. doi:10.3389/fneng.2011.00004

Nunes SS, Miklas JW, Liu J, Aschar-Sobbi R, Xiao Y, Zhang B, Jiang J, Massé S, Gagliardi M, Hsieh A, Thavandiran N, Laflamme MA, Nanthakumar K, Gross GJ, Backx PH, Keller G, Radisic M (2013) Biowire: a platform for maturation of human pluripotent stem cell-derived cardiomyocytes. Nat Methods 10(8):781–787

OECD (2005) OECD series on testing and assessment. Number 34. Guidance document on the validation and international acceptance of new or updated test methods for hazard assessment. ENV/JM/MONO(2005)14

Paige SL, Osugi T, Afanasiev OK, Pabon L, Reinecke H, Murry CE (2010) Endogenous Wnt/beta-catenin signaling is required for cardiac differentiation in human embryonic stem cells. PLoS One 5(6), e11134

Pan GJ, Chang ZY, Scholer HR, Pei D (2002) Stem cell pluripotency and transcription factor Oct4. Cell Res 12(5–6):321–329

Passier R, Oostwaard DW, Snapper J, Kloots J, Hassink RJ, Kuijk E, Roelen B, de la Riviere AB, Mummery C (2005) Increased cardiomyocyte differentiation from human embryonic stem cells in serum-free cultures. Stem Cells 23(6):772–780

Petrova A, Celli A, Jacquet L, Dafou D, Crumrine D, Hupe M, Arno M, Hobbs C, Cvoro A, Karagiannis P, Devito L, Sun R, Adame L, Vaughan R, McGrath J, Mauro T, Ilic D (2014) 3D *in vitro* model of a functional epidermal permeability barrier from human embryonic stem cells and induced pluripotent stem cells. Stem Cell Reports 2(5):675–689

Pistollato F, Bremer-Hoffmann S, Healy L, Young L, Stacey G (2012) Standardization of pluripotent stem cell cultures for toxicity testing. Expert Opin Drug Metab Toxicol 8:239–257

Ren Y, Lee MY, Schliffke S, Paavola J, Amos PJ, Ge X, Ye M, Zhu S, Senyei G, Lum L, Ehrlich BE, Qyang Y (2011) Small molecule Wnt inhibitors enhance the efficiency of BMP-4-directed cardiac differentiation of human pluripotent stem cells. J Mol Cell Cardiol 51(3):280–287

Richert L, Liguori MJ, Abadie C, Heyd B, Mantion G, Halkic N, Waring JF (2006) Gene expression in human hepatocytes in suspension after isolation is similar to the liver of origin, is not affected by hepatocyte cold storage and cryopreservation, but is strongly changed after hepatocyte plating. Drug Metab Dispos 34(5):870–879

Rispin A, Harbell JW, Klausner M, Jordan FT, Coecke S, Gupta K, Stitzek K (2004) Quality assurance for *in vitro* alternative test methods: quality control issues in test kit production. Altern Lab Anim 1(Suppl):725–729

Roselli S, Gribouval O, Boute N, Sich M, Benessy F, Attie T, Gubler MC, Antignac C (2002) Podocin localizes in the kidney to the slit diaphragm area. Am J Pathol 160(1):131–139

Ruotsalainen V, Ljungberg P, Wartiovaara J, Lenkkeri U, Kestila M, Jalanko H, Holmberg C, Tryggvason K (1999) Nephrin is specifically located at the slit diaphragm of glomerular podocytes. Proc Natl Acad Sci U S A 96(14):7962–7967

Sariola H (2002) Nephron induction. Nephrol Dial Transplant 17(Suppl 9):88–90

Scanu M, Mancuso L, Cao G (2011) Evaluation of the use of human Mesenchymal Stem Cells for acute toxicity tests. Toxicol *In Vitro* 25(8):1989–1995

Schulz TC, Noggle SA, Palmarini GM, Weiler DA, Lyons IG, Pensa KA, Meedeniya AC, Davidson BP, Lambert NA, Condie BG (2004) Differentiation of human embryonic stem cells to dopaminergic neurons in serum-free suspension culture. Stem Cells 22(7):1218–1238

Schwartz RE, Fleming HE, Khetani SR, Bhatia SN (2014) Pluripotent stem cell-derived hepatocyte-like cells. Biotechnol Adv 32(2):504–513

Schwarz K, Simons M, Reiser J, Saleem MA, Faul C, Kriz W, Shaw AS, Holzman LB, Mundel P (2001) Podocin, a raft-associated component of the glomerular slit diaphragm, interacts with CD2AP and nephrin. J Clin Invest 108(11):1621–1629

Scott CW, Peters MF, Dragan YP (2013) Human induced pluripotent stem cells and their use in drug discovery for toxicity testing. Toxicol Lett 219(1):49–58

Shan J, Schwartz RE, Ross NT, Logan DJ, Thomas D, Duncan SA, North TE, Goessling W, Carpenter AE, Bhatia SN (2013) Identification of small molecules for human hepatocyte expansion and iPS differentiation. Nat Chem Biol 9(8):514–520

Shih NY, Li J, Cotran R, Mundel P, Miner JH, Shaw AS (2001) CD2AP localizes to the slit diaphragm and binds to nephrin via a novel C-terminal domain. Am J Pathol 159(6):2303–2308. doi:10.1016/S0002-9440(10)63080-5

Silva J, Nichols J, Theunissen TW, Guo G, van Oosten AL, Barrandon O, Wray J, Yamanaka S, Chambers I, Smith A (2009) Nanog is the gateway to the pluripotent ground state. Cell 138(4):722–737

Sirenko O, Hesley J, Rusyn I, Cromwell EF (2014) High-content assays for hepatotoxicity using induced pluripotent stem cell-derived cells. Assay Drug Dev Technol 12(1):43–54

Sivertsson L, Synnergren J, Jensen J, Björquist P, Ingelman-Sundberg M (2013) Hepatic differentiation and maturation of human embryonic stem cells cultured in a perfused three-dimensional bioreactor. Stem Cells Dev 22(4):581–594

Soleimani M (2015) The multiple roles of pendrin in the kidney. Nephrol Dial Transplant 30(8):1257–1266. doi:10.1093/ndt/gfu307

Song B, Smink AM, Jones CV, Callaghan JM, Firth SD, Bernard CA, Laslett AL, Kerr PG, Ricardo SD (2012) The directed differentiation of human iPS cells into kidney podocytes. PLoS One 7(9), e46453. doi:10.1371/journal.pone.0046453

Stacey G, Day JG (2007) Long-term ex situ conservation of biological resources and the role of biological resource centers. In: Day DG, Stacey GN (eds) Cryopreservation and freezedrying methods. Humana Press, Totowa

Stacey GN, Masters JR (2008) Cryopreservation and banking of mammalian cell lines. Nat Protoc 3:1981–1989

Stacey G, Masters JRW, Hay RJ, Drexler HG, MacLeod RAF, Freshney IR (2000) Cell contamination leads to inaccurate data: we must take action now. Nature 403:356

Stacey G, Pistollato F, Healy L, Bremer S, Young L, Strehl R, Hyllner J, Emmanuelsson K and Peschanski M on behalf of the Scr&Tox and ToxBank consortia (2012) General Quality and Regulatory Criteria for Establishment and Dissemination of human Pluripotent Stem Cell Lines (hPSCs). Poster presented at Surat-1 scientific meeting Lisbon, 2012 and published on the ToxBank data warehouse at http://www.toxbank.net/bio-wiki or http://wiki.toxbank.net/w/images/0/08/ToxBank_poster5-CellStandards-120130.pdf

Stacey G, Healy L, Kidane L (2012) Points to consider in gaining access to human tissue and cell lines. ToxBank Deliverable 4.6. http://wiki.toxbank.net/w/images/1/18/ToxBank_D4_6_final_10_04_13.pdf

Stacey G, Kidane L, Healy L, Myatt G on behalf of ToxBank consortium (2014) Deliverable D4.7. Inventory and map of European suppliers: materials, resources, facilities and standards. http://wiki.toxbank.net/w/images/c/c2/ToxBank_D4_7_Final_11.06.13.pdf

Stacey G, Kidane L, Healy L, Myatt G, Hardy B, Bremer S (2014) Coordination of data systems in SEURAT-1 and alternative testing regulatory frameworks to provide smooth and efficient translation of research developments to qualified toxicity assays. Poster presented at SEURAT-1 scientific meeting Lisbon, 2014 and published on the ToxBank data warehouse at http://www.toxbank.net/bio-wiki; http://wiki.toxbank.net/w/images/3/35/ToxBank_ToxBank_JRC_Coord_Poster_Stacey_et_al.pdf

Stanley EG, Biben C, Elefanty A, Barnett L, Koentgen F, Robb L, Harvey RP (2002) Efficient Cre-mediated deletion in cardiac progenitor cells conferred by a 3′UTR-ires-Cre allele of the homeobox gene Nkx2-5. Int J Dev Biol 46(4):431–439

Subramanian K, Owens DJ, Raju R, Firpo M, O'Brien TD, Verfaillie CM, Hu WS (2014) Spheroid culture for enhanced differentiation of human embryonic stem cells to hepatocyte-like cells. Stem Cells Dev 23(2):124–131

Sutton M, Bonfield T (2014) Stem cells: innovations in clinical applications. Stem Cells Int 2014:516278. doi:10.1155/2014/516278

Szkolnicka D, Farnworth SL, Lucendo-Villarin B, Storck C, Zhou W, Iredale JP, Flint O, Hay DC (2014) Accurate prediction of drug-induced liver injury using stem cell-derived populations. Stem Cells Transl Med 3(2):141–148

Takahashi K, Tanabe K, Ohnuki M, Narita M, Ichisaka T, Tomoda K, Yamanaka S (2007) Induction of pluripotent stem cells from adult human fibroblasts by defined factors. Cell 131:861–872

Takasato M, Er PX, Becroft M, Vanslambrouck JM, Stanley EG, Elefanty AG, Little MH (2014) Directing human embryonic stem cell differentiation towards a renal lineage generates a self-organizing kidney. Nat Cell Biol 16(1):118–126

Takebe T, Sekine K, Enomura M, Koike H, Kimura M, Ogaeri T, Zhang RR, Ueno Y, Zheng YW, Koike N, Aoyama S, Adachi Y, Taniguchi H (2013) Vascularized and functional human liver from an iPSC-derived organ bud transplant. Nature 499(7459):481–484

Takebe T, Zhang RR, Koike H, Kimura M, Yoshizawa E, Enomura M, Koike N, Sekine K, Taniguchi H (2014) Generation of a vascularized and functional human liver from an iPSC-derived organ bud transplant. Nat Protoc 9(2):396–409

Tang M, Chen W, Liu J, Weir M, Cheng L, Xu H (2014) Human induced pluripotent stem cell-derived mesenchymal stem cell seeding on calcium phosphate scaffold for bone regeneration. Tissue Eng Part A 20(7–8):1295–1305

Tarunina M, Hernandez D, Johnson CJ, Ramathas V, Jeyakumar M, Watson T et al (2014) Directed differentiation of embryonic stem cells using a bead-based combinatorial screening method. PLoS ONE 9(9), e104301. doi:10.1371/journal.pone.0104301

TheinHan W, Liu J, Tang M, Chen W, Cheng L, Xu H (2013) Induced pluripotent stem cell-derived mesenchymal stem cell seeding on biofunctionalized calcium phosphate cements. Bone Res 4:371–384

Thomson JA, Itskovitz-Eldor J, Shapiro SS, Waknitz MA, Swiergiel JJ, Marshall VS, Jones JM (1998) Embryonic stem cell lines derived from human blastocysts. Science 282(5391): 1145–1147

Ulvestad M, Nordell P, Asplund A, Rehnström M, Jacobsson S, Holmgren G, Davidson L, Brolén G, Edsbagge J, Björquist P, Küppers-Munther B, Andersson TB (2013) Drug metabolizing enzyme and transporter protein profiles of hepatocytes derived from human embryonic and induced pluripotent stem cells. Biochem Pharmacol 86(5):691–702

United States Pharmacoipoiea (2013) General Chapter. Cryopreservation of Cells, U.S. Pharmacopeia and the National Formulary (USP–NF). http://www.usp.org/council-experts-expert-committees-overview/expert-committees/general-chapters-biological-analysis

Uosaki H, Fukushima H, Takeuchi A, Matsuoka S, Nakatsuji N, Yamanaka S, Yamashita JK (2011) Efficient and scalable purification of cardiomyocytes from human embryonic and induced pluripotent stem cells by VCAM1 surface expression. PLoS One 6(8), e23657

Van Itallie CM, Rogan S, Yu A, Vidal LS, Holmes J, Anderson JM (2006) Two splice variants of claudin-10 in the kidney create paracellular pores with different ion selectivities. Am J Physiol Renal Physiol 291(6):F1288–F1299

Vinoth K, Manikandan J, Sethu S, Balakrishnan L, Heng A, Lu K, Hande M, Cao T (2014) Evaluation of human embryonic stem cells and their differentiated fibroblastic progenies as cellular models for *in vitro* genotoxicity screening. J Biotechnol 184:154–168

Wilmes A, Jennings P (2014) The use of renal cell culture for nephrotoxicity investigations. In: Predictive toxicology. Wiley-VCH Verlag GmbH & Co, KGaA, Weinheim, pp 195–216. doi:10.1002/9783527674183.ch10

Wilmes A, Aschauer L, Limonciel A, Pfaller W, Jennings P (2014) Evidence for a role of claudin 2 as a proximal tubular stress responsive paracellular water channel. Toxicol Appl Pharmacol 279(2):163–172

Xia Y, Nivet E, Sancho-Martinez I, Gallegos T, Suzuki K, Okamura D, Wu MZ, Dubova I, Esteban CR, Montserrat N, Campistol JM, Izpisua Belmonte JC (2013) Directed differentiation of human pluripotent cells to ureteric bud kidney progenitor-like cells. Nat Cell Biol 15(12):1507–1515. doi:10.1038/ncb2872

Yuan SH, Martin J, Elia J, Flippin J, Paramban RI, Hefferan MP, Vidal JG, Mu Y, Killian RL, Israel MA, Emre N, Marsala S, Marsala M, Gage FH, Goldstein LS, Carson CT (2011) Cell-surface marker signatures for the isolation of neural stem cells, glia and neurons derived from human pluripotent stem cells. PLoS One 6(3), e17540. doi:10.1371/journal.pone.0017540

Zeng X, Chen J, Deng X, Liu Y, Rao MS, Cadet JL, Freed WJ (2006) An *in vitro* model of human dopaminergic neurons derived from embryonic stem cells: MPP+ toxicity and GDNF neuroprotection. Neuropsychopharmacology 31(12):2708–2715

Zeng H, Guo M, Martins-Taylor K, Wang X, Zhang Z, Park JW, Zhan S, Kronenberg MS, Lichtler A, Liu HX, Chen FP, Yue L, Li XJ, Xu RH (2010) Specification of region-specific neurons including forebrain glutamatergic neurons from human induced pluripotent stem cells. PLoS One 5(7), e11853. doi:10.1371/journal.pone.0011853

Zhang SC, Wernig M, Duncan ID, Brüstle O, Thomson JA (2001) *In vitro* differentiation of transplantable neural precursors from human embryonic stem cells. Nat Biotechnol 19(12):1129–1133

Zhang P, Li J, Tan Z, Wang C, Liu T, Chen L, Yong J, Jiang W, Sun X, Du L, Ding M, Deng H (2008) Short-term BMP-4 treatment initiates mesoderm induction in human embryonic stem cells. Blood 111(4):1933–1941

Zhou J, Su P, Li D, Tsang S, Duan E, Wang F (2010) High-efficiency induction of neural conversion in human ESCs and human induced pluripotent stem cells with a single chemical inhibitor of transforming growth factor beta superfamily receptors. Stem Cells 28(10):1741–1750

Chapter 12
Validation of Bioreactor and Human-on-a-Chip Devices for Chemical Safety Assessment

Sofia P. Rebelo, Eva-Maria Dehne, Catarina Brito, Reyk Horland, Paula M. Alves and Uwe Marx

Abstract Equipment and device qualification and test assay validation in the field of tissue engineered human organs for substance assessment remain formidable tasks with only a few successful examples so far. The hurdles seem to increase with the growing complexity of the biological systems, emulated by the respective models. Controlled single tissue or organ culture in bioreactors improves the organ-specific functions and maintains their phenotypic stability for longer periods of time. The reproducibility attained with bioreactor operations is, *per se*, an advantage for the validation of safety assessment. Regulatory agencies have gradually altered the validation concept from exhaustive "product" to rigorous and detailed process characterization, valuing reproducibility as a standard for validation. "Human-on-a-chip" technologies applying micro-physiological systems to the *in vitro* combination of miniaturized human organ equivalents into functional human micro-organisms are nowadays thought to be the most elaborate solution created to date. They target the replacement of the current most complex models—laboratory animals. Therefore, we provide here a road map towards the validation of such "human-on-a-chip" models and qualification of their respective bioreactor and microchip equipment along a path currently used for the respective animal models.

Keywords Bioreactor • Stirred-tank bioreactors • Hollow fibre bioreactor • Microbioreactor • Human-on-a-chip (validation)

S.P. Rebelo • C. Brito • P.M. Alves (✉)
iBET, Instituto de Biologia Experimental e Tecnológica, Oeiras 2780-901, Portugal

Instituto de Tecnologia Química e Biológica, Universidade Nova de Lisboa, Oeiras 2780-157, Portugal
e-mail: marques@itqb.unl.pt

E.-M. Dehne (✉) • R. Horland • U. Marx
Department of Medical Biotechnology, Technische Universität Berlin, Institute of Biotechnology, Gustav-Meyer-Allee 25, Berlin 13355, Germany
e-mail: eva.dehne@tissuse.com

C. Eskes, M. Whelan (eds.), *Validation of Alternative Methods for Toxicity Testing*, Advances in Experimental Medicine and Biology 856,
DOI 10.1007/978-3-319-33826-2_12

1 Bioreactor Technologies for Chemical Safety Assessment

1.1 Introduction

To effectively address the effects of chemical agents in humans, it is crucial that the *in vitro* models applied are physiologically relevant and representative of the complexity observed *in vivo*. Nevertheless, this complexity must be coupled with reproducibility, control and automation for industrial and clinical applications, which can be attained with cultures in bioreactors.

Bioreactors (BRs) are devices engineered to support biological processes for multiple applications, ranging from the production of biopharmaceuticals to tissue engineering. The key feature of these systems is the high level of control over the bioprocesses, which is achieved by on-line monitorization and automated regulation of environmental culture parameters, such as temperature, pH, partial pressure of oxygen (pO_2), and nutrient and metabolite concentrations. Moreover, the dynamic conditions offered by bioreactors ensure efficient mass transfer in the culture vessel, which is a key factor to minimise oxygen and nutrient gradients and maintain a homogenous culture environment. The control, automation and efficient mass transfer simplify the transition from bench-top bioreactors to larger scales, critical to meet industrial requirements. In addition to scalability, the low contamination risk of bioreactors provides a cost-effective solution for large-scale productions in the industry.

Bioreactors used in tissue engineering are developed to deliver a cell product that restores or improves organ-specific functions. For this purpose, the bioreactor and bioprocess should be designed to recreate the *in vivo* tissue architecture and microenvironment. Tissue recapitulation often implies (1) the transition from two-dimensional (2D) to three-dimensional (3D) culture; (2) the recreation of the tissue-like cell-extracellular matrix (ECM) and cell-cell interactions—monotypic or heterotypic—and (3) the mimicking of hydrodynamic forces and the physiochemical environment (Martin et al. 2004; Griffith and Swartz 2006; Fennema et al. 2013). Multiple bioreactor designs have been developed for organ-specific applications, suitable for sustaining 3D cultures, various cell types or ECM components. Moreover, the flexibility of bioreactors enables the modulation of the culture parameters to attain precise control of a physiochemical microenvironment, e.g. for the culture of stem cells (SCs) in hypoxic conditions as revised (Serra et al. 2012). Similarly, hydrodynamic parameters can be tuned for specific cellular requirements, such as mimicking the mass transport environment of the endothelial barrier that is favourable for the maintenance of primary hepatocyte cells *in vitro* (Lee et al. 2007). The efficient mass transfer and pO_2 control provide oxygen diffusion through compact 3D structures, preventing the formation of necrotic regions within the tissue and minimising gradients of soluble factors (Griffith and Swartz 2006).

Another important aspect of bioreactors is their flexibility towards the operation mode applied. Among the available operation modes—batch, fed-batch and perfusion—the latest has been broadly applied in tissue engineering. In perfusion operation mode, fresh medium is fed to a bioreactor containing cells that are retained within the system, with gradual replacement of culture medium. In contrast to batch and fed-batch modes, perfusion allows the removal of the toxic metabolic by-products and the constant replenishment of nutrients and growth factors that contribute to elicit specific functions.

Altogether, the bioreactor characteristics mentioned above are responsible for maintaining organ-specific properties which include cell integrity, morphology, biochemical activity and biostability.

1.2 Bioreactors for Toxicological Assessment

1.2.1 Advantages of Bioreactors for Safety Assessment

The maintenance of organ-specific properties and phenotypic stability, as well as the reproducibility achieved by culture in controlled conditions, are major advantages of using bioreactors to develop *in vitro* models for safety assessment. In bioreactor cultures, stable phenotypes are extended for long periods of time, opening the door to address relevant toxicity issues, such as repeated dose toxicity at sub-acute concentrations. This type of toxicity, which has always been studied using *in vivo* models, can now take advantage of the *in vitro* model systems to understand its subjacent molecular basis that might differ significantly from the mechanisms involved in sub-chronic exposure to high doses (Slikker Jr. et al. 2004).

The scalability offered by bioreactors for drug-drug interactions provides the possibility of combining multiple compounds using the same donor material and culture conditions, comprising comprehensive amounts of data in a single bioreactor run. For toxicological assessment, the scalability is often a synonym for parallelization achieved by designing multiplexed formats. These may be coupled with high-throughput screening platforms or high-content screening tools to facilitate the phenotypic and analytical characterization of cells acting upon drug stimuli, critical to accelerate and strengthen pharmacological testing.

Another advantage of bioreactors applying dynamic conditions for drug testing is that cells are not exposed to a constant drug concentration, as occurs in static systems. This resembles more the *in vivo* processes of drug biotransformation in which the concentration of the drug constantly changes. Similarly, efficient mass transfer is important for the diffusion of chemical compounds through the tissues, which is particularly relevant for 3D cultures.

1.2.2 Existent Bioreactor Systems for Safety Assessment

Multiple bioreactor designs have been developed to support culture strategies for organ-specific systems and SCs, which are applicable for toxicological studies. Due to the central role of the liver in detoxification, most bioreactor configurations have been developed for hepatic cultures and these also represent the most mature technologies in terms of validation. Thus, the next section will cover the bioreactors developed for hepatic cultures and their validation status, taking into account that some formats can or have been adapted to other organ systems.

Perfused Monolayer Systems

The simplest formats of bioreactors are monolayer-based bioreactors applying perfusion operation modes. In the MultiChamber Modular Bioreactor (McMB), the collagen-coated polydimethylsiloxan wells support the primary cultures of human hepatocytes with constant perfusion. Expression of phase I, II and III enzymes was up-regulated when compared with static conditions and maintained for at least 2 weeks in culture (Vinci et al. 2011), reinforcing the role of perfusion in human systems. Several formats applying the same principle and presenting several adaptations are commercially available, such as Minucell (Xia et al. 2009), which includes a collagen overlay to minimise the effects of shear stress. Despite the improvements in comparison to static cultures and the simplicity of using 2D culture systems, these systems fail to recapitulate the tissue-specific architecture. In addition, there is typically no control of pO_2, pH and temperature. Thus, perfused monolayer systems present an upgrade from static culture by incorporating dynamic flow, but are not bioreactors to full extent, with automated control and monitoring of culture parameters.

Hollow Fibres Bioreactors

Hollow fibre bioreactors comprise an interwoven network of semipermeable membranes which are perfused by medium and oxygen, aiming to resemble blood capillaries *in vivo*. The cells are arranged in compact 3D structures in between the capillary systems. Applying this principle, a number of bioreactors were designed for clinical applications to support extracorporeal liver function in patients with liver failure. Two of these systems have been validated for pharmacological applications: the Modular Extracorporeal Liver System (MELS) developed by Gerlach's research group (Gerlach et al. 1994), and the AMC bioreactor developed, by Chamuleau and co-workers (Flendrig et al. 1997). In the miniaturized format of the MELS bioreactor, scaled down to 2 mL, major drug metabolizing P450 enzymes were preserved up to 23 days in primary cultures of human hepatocytes in co-culture with non-parenchymal cells (Zeilinger et al. 2011). This design has been applied more recently for the differentiation of human pluripotent SC towards hepatocyte-like cells (Miki et al. 2011). The AMC bioreactor has been validated using the hepatic cell line HepaRG, which presented phase I and II drug metabolism and production of bile salts (Nibourg et al. 2013). A major drawback of these systems is the inaccessibility to the cell compartment throughout the culture time, not allowing phenotypic monitoring and cell sampling. Furthermore, hollow fibre bioreactors fail to accurately control pH and pO_2 within the fibres.

Stirred-Tank Bioreactors

Stirred-tank bioreactors (STBs) which have long been applied in industry for the production of biopharmaceuticals, may also be used for *in vitro* cell models for pharmacological testing. In STBs, cells are inoculated as cell suspension and the hydrodynamics of the bioreactor—determined by vessel and impeller type, and agitation rate—is adjusted to elicit cell-cell contacts and promote aggregation into cell spheroids. Dynamic parameters need to be balanced to guarantee diffusion through the aggregates, preventing the formation of necrotic centres, while the shear stress is minimised. Spheroid culture of rat hepatocytes has long been reported, resulting in increased albumin production and phase I-II activity (Abu-Absi et al. 2002) and maintenance of hepatocyte polarisation (Miranda et al. 2009). More recently, primary cultures of human hepatocytes were maintained under physiological oxygen conditions and perfusion operation mode, extending culture viability and functionality for up to 3–4 weeks (Tostoes et al. 2012). Hepatocytes in this system present a functional phenotype displaying bile canalicular networks, phase I and II enzyme activities, and inducibility of CYP P450s. The use of biomaterials in STBs has also been addressed, by alginate microencapsulation of rat hepatocyte (Miranda et al. 2010; Tostoes et al. 2011) and HepaRG (Rebelo et al. 2015) cell spheroids, which represents a strategy to overcome eventual shear stress effects on stirred culture and to retain ECM and, eventually, soluble factors of the cellular microenvironment within a stirred culture. Importantly, STBs are compatible with sterile non-destructive sampling, allowing the characterization of cultured cells throughout the culture period. Although these systems are not suitable for performing high-throughput characterization during culture, STBs may be used as feeder systems to perform endpoint assays along the culture period in higher throughput platforms. As an alternative to STBs, the rotating wall vessel (RWV) bioreactors generate a dynamic laminar flow by rotating fluid, which effectively reduces diffusion limitations with low shear stress. Few studies have been performed for hepatocytes (Schwarz et al. 1992; Mitteregger et al. 1999), but RWV bioreactors have been applied for other organ systems (Navran 2008).

Microbioreactors

The possibility of miniaturizing and multiplexing bioreactor formats for toxicological assessment is a great advantage due to the minimisation of expensive culture and biological material and parallel testing of compounds of interest. A chance emerged at the beginning of this century, with the use of micro-electro mechanical systems (MEMS), for the development of dynamic micro-scale tissue culture devices, to miniaturise *in vitro* organs to the smallest possible scale. These systems are based on microchannels for the flow of media, and miniaturized cell culture compartments. Such systems support the replication of shear stress at physiological intracapillary or interstitial rates, which is mandatory in order to maintain stable protein and oxygen gradient-based microenvironments. What is the smallest possible degree of liver miniaturization on such chips? It is important to recognise that a paradigm of stringent correlation between architecture and functionality applies to all levels of biological existence on Earth. These levels of increasing biological complexity have appeared progressively within the multi-million-year process of evolution. These developments were most probably triggered by slight changes in the external environment which created the conditions for self-assembly to the next level of complexity. Molecules, intracellular organelles, cells, organoids, organs and, finally, the individual organisms themselves, were thought to represent these levels for humans. The role and function of the organoid structures in man were underestimated for a long time. Today, however, it has been proven that almost all organs and systems are build up by multiple, identical, functionally self-reliant, structural units which perform the most prominent functions of the particular organ. It is important that these organoids are of very small dimensions, from several cell layers up to a few millimetres. In the liver, these smallest functional organoids are called liver lobuli, and are built up of a variety of cell types in a defined three dimensional arrangement. The multiplication of these structures within a given organ is nature's risk management tool to prevent a total loss of functionality during partial organ damage. With regard to evolution, this concept has allowed organ size and shape to be easily adjusted to the needs of a given species—for example, the liver in mice and men—while following nearly the same arrangement to build up single functional organoids. The advent of liver microsystems began with the single cell type culture of hepatocytes on chips (Powers et al. 2002; Leclerc et al. 2004; Ho et al. 2006; Lee et al. 2007; Toh et al. 2007, 2009; Carraro et al. 2008; Park et al. 2008; Goral et al. 2010).

Some high throughput multiwell systems, applying microfluidics for somewhat complex *in vitro* models, have undergone validation for toxicological approaches. The Perfusion Array Liver System (PEARL), designed by Lee and co-workers to mimic the liver acinus, constitutes an innovative approach to design modular units with physiological relevance. With a design compatible with a 96-well plate, the system is composed of microunits of artificial liver acinus with an endothelial-like barrier, intended to simulate the mass transfer properties of the liver sinusoid. Primary cultures of human hepatocytes were maintained for 7 days in culture and were responsive to diclofenac toxicity at high concentrations (Lee et al. 2007). Khetani and Bhatia developed a multiwell system containing micropatterned structures of PDMS for the co-culture of fibroblasts and hepatocytes, which is compatible with robotic fluid handling and phenotypic screening tools. This co-culture system was validated for up to 6 weeks with maintenance of gene expression profile, phase I/II metabolism, canalicular transport, secretion of liver-specific products and susceptibility to hepatotoxins (Khetani and Bhatia 2008).

Consequently, in a next step, heterotypic co-culture systems combining multiple crucial cell types into artificial functional units were able to more realistically mimic aspects of the liver lobulus (Kane et al. 2006; Hwa et al. 2007; Khetani and Bhatia 2008). Finally, none of the single liver organ equivalents currently used *in vitro* emulate human liver lobules in a functionally and architecturally comparable manner. The roadmap toward truly human liver-on-a-chip solutions is outlined in a recent review (Materne et al. 2013).

Bioreactors for Non-hepatic Organ Systems

Although the liver is the organ on which most mature technology and validation has been performed, the advent of SCs has brought the development of protocols and culture strategies for the differentiation towards mature organ-specific systems, bringing new tools for toxicological assessment. Regarding the central nervous system, the differentiation of SCs into mature neural cells as 3D aggregates has been performed using either stirred systems (Brito et al. 2012) or cellular microarray platforms (Meli et al. 2014) with successful outcomes concerning maturity of terminally differentiated neuronal cells. For cardiac differentiation, a number of bioprocesses have been applied exploring 3D cell architecture, hydrodynamics and hypoxic conditions in bioreactors to deliver functional cardiomyocytes (Bauwens et al. 2005; Niebruegge et al. 2009; Correia et al. 2014). Nevertheless, most of the work has focused on the enhancement of the maturity of differentiated cell phenotypes, with few toxicity studies performed so far. A few bioreactors have been developed for excretory toxicity, mostly based on aggregates of kidney cell lines in stirred systems, as reviewed in (Desrochers et al. 2014). Other cell models, such as gut (Cencic and Langerholc 2010), skin and eye (Vinardell and Mitjans 2008), have been cultured, mostly in non-controlled static systems. The role of the immune system in predicted toxic and immune response for safety assessment is critical. The first dynamic bioreactor system emulating human immune response in functional artificial human lymph node cultures was developed by Giese et al. (Giese et al. 2006). The system has been qualified and is used for repeated dose substance testing at industrial scale over weeks (Giese et al. 2010). Some drugs that appear safe in animal safety tests and early-phase clinical trials exhibit adverse reactions when exposed to larger populations, frequently through activation of inflammatory signalling pathways. These signalling pathways may influence toxicity mechanisms, modulating the response to the drug in an unpredicted manner. Synergistic effects between inflammatory pathways and metabolic activation upon drug stimuli in hepatocytes have been described, leading to an increased loss of cell viability (Cosgrove et al. 2009; Kostadinova et al. 2013). Few bioreactors have incorporated the immune system cells in their configuration, with the exception of studies in transwells applying perfusion, to unravel the regulation of lymphocyte traffic through endothelial cells by hepatocytes (Edwards et al. 2005), but the existent designs might be adapted to incorporate an additional level of complexity. Static and dynamic systems modelling human immunogenicity and immunotoxicity *in vitro* were reviewed short time ago (Giese and Marx 2014).

Until recently a number of non-hepatic organ equivalents have also been established at a smallest possible chip scale. Chips working with single cell type cultures, such as endothelial cells (Young and Simmons 2010), myoblasts (Gu et al. 2004), neurons (Rhee et al. 2005), and adipose cells (Nakayama et al. 2008), have evolved towards heterotypic co-culture systems of the lung alveolus (Huh et al. 2010), the small artery (Günther et al. 2010), the intestinal villus (Sato et al. 2009; Ootani et al. 2010; Lahar et al. 2011; Sung et al. 2011; Yu et al. 2012), the central nervous system columns (Park et al. 2009), and the bone-marrow unit (Cui et al. 2007).

1.3 Requirements for Validation of Single Organ/Tissue Bioreactors

While there is a good collection of data on the detoxification metabolism for hepatic systems, it is still necessary either to adapt existent bioreactors or to perform further validation studies to collect more data on the toxicological performance for other organs. Despite the maturity of bioreactor technologies and the quantity of toxicological data available for each cellular model, the validation is still impeded by the lack of standardization concerning culture processes, biomarkers and endpoints for functional evaluation. As overviewed in this chapter, controlled single tissue or organ culture in bioreactors improves the organ-specific functions and maintains the phenotypic stability for longer periods of time, independent of bioreactor size and architecture. The reproducibility attained with bioreactor operations is, *per se*, an advantage for the validation of safety assessment. By establishing cultures in bioreactors, the entire culture process is fully characterized, with tracking of oxygen consumption, pH fluctuations, shifts in metabolite concentration, etc. Regulatory agencies have gradually altered the validation concept from exhaustive "product" characterization to rigorous and detailed process characterization, valuing reproducibility as a standard for validation. From this perspective, establishing *in vitro* cell models and performing toxicological studies in bioreactors will be key for the validation of chemical safety assessment.

Nevertheless, harmonization of culture conditions in bioreactors is necessary to perform cross-comparison between different institutions. Several computer simulations and models have been developed for parameters such as pO_2 and hydrodynamics which are associated with the mass transfer properties of the bioreactor and may affect specific pathways or define whether a chemical agent diffuses through the tissue, as discussed by Salehi-Nik and co-authors (Salehi-Nik et al. 2013). With the support of these studies and the emergence of more modelling data, it will be possible to establish directives for bioprocess operation. In dynamic systems, it is also important to set the perfusion rates, as the enzymatic biotransformation of the drug is affected by the rates to which the compound is available at a given moment. Similarly, the media components, including growth factors, cytokines, inducers of enzyme activity applied to improve tissue specific functions during culture or to differentiate towards a specific lineage, must also be standardized for toxicity assessment.

The harmonization of cellular endpoints and biomarkers of cell differentiation/functionality for *in vitro* culture also poses a complex effort, due to the range of culture strategies available (cell sources, co-cultures), as revised by Schroeder and co-authors (Schroeder et al. 2011). In addition, the definition of general guidelines for validation still require further development of on-line/non-invasive characterization tools to address parameters such as cell integrity/viability, morphology and metabolic activity (Mendhe et al. 2012), so that comparison between several culture systems, different bioreactor configurations and *in vivo* data is possible independently of sampling accessibility.

For long-term toxicity studies, it is crucial to establish time-points, doses and specific periods of exposure to chemical agents, which depend on the organ substantially, pathway and group of chemicals studied. Importantly, it is mandatory that the data collected by *in vitro* culture in bioreactors is correlated to existent *in vivo* data and that the new standards generated for validation are based on this correlation regarding levels of activity, rates of perfusion, doses, etc. The new standards generated for validation should not only encompass general features, such as basic integrity and metabolic activity, but should also be directed at specific modes of action for groups of chemicals. This implies that a combination of data from multiple tests is performed by the use of chemical categories/grouping using "Integrated Testing Strategies", as has been proposed by experts in toxicology and gradually accepted by the regulatory agencies (Lilienblum et al. 2008; Hartung et al. 2013).

The significant progress in bioreactor-based single organ modelling at ever decreasing scale "first in history" provides a basis for organ integration into systems of true organismal complexity. Being developed in this way, such "human-on-a-chip" systems might mark a translational paradigm shift in safety and efficacy assessment in the future, eventually enabling mode of action analysis and adverse outcome pathway assessment at a level currently reserved for animal testing and clinical trials in man. These novel "human-on-a-chip" strategies are overviewed in the next section of this chapter.

2 Human-on-a-Chip Devices for Chemical Safety Assessment

2.1 Introduction

Strategies to develop "human-on-a-chip" technologies are applying micro-physiological systems towards the *in vitro* combination of miniaturized human organ equivalents into functional human micro-organisms. These aim to replace systemic toxicity testing and efficacy assessment of therapeutic agents, food additives, chemicals, or environmental pollutants in laboratory animals and might generate predictive data to humans safety and efficacy evaluation prior to substance exposure to humans. Therefore, the technologies need to functionally represent normal and diseased human biology at the smallest possible scale at reproducible and viable operation under physiological or pathological conditions over long periods of time. If these are generated from the respective donor or patient tissue sources, they bear the capacity for the representation of normal and disease phenotypes and population diversity. Finally, they are amenable to high-content screening due to their small size. Such "human-on-a-chip" technologies are at an early stage of development, but a prime top-down US initiative between DARPA (Defence Advanced Research Projects Agency), NIH (National Institutes of Health) and the FDA (Food and Drug Administration), with more than USD 140 million investment into respective developments (http://www.ncats.nih.gov/research/reengineering/tissue-chip/funding/funding.html) and a number of significant development investments in Europe, has initiated an irreversible process toward success in this area.

2.2 Historical Sketch (Toward Organismal Engineering)

Over the last hundred years, scientists have been trying to emulate human tissue architecture and microenvironment *in vitro* in order to gain mechanistic knowledge and to assist with the development of new medicines. Interestingly, as early as 1912, Alexis Carrel (Rockefeller Institute for Medical Research, New York) said "On the permanent life of tissues outside of the organism" (Carrel 1912), that some *in vitro* "cultures could be maintained in active life for 50, 55 and even for 60 days". These results demonstrated that the early death of tissues cultivated *in vitro* was preventable and "therefore that their permanent life was not impossible". At that time, synthetic cell culture media, antibiotics, disposable tissue culture flasks, aseptic techniques, and bioreactors were not available. About two decades later, an avian bone more than 7 mm long and with clear signs of calcification could be produced *in vitro* from embryonic cells (Fell and Robison 1929). Interestingly, some of the early human histotypic cultures, such as Dexter and Lajatha's culture of human haematopoietic SCs on feeder layers, demonstrated the crucial importance of the interaction of different primary human cell types with each other to form human-like growth and functionality (Dexter and Lajtha 1974). It took more than half a century to recognise that static tissue cultures in flasks or petri dishes with media levels higher than 1.2 mm generate a non-physiological low level of oxygen supply for primary liver cells from humans and rodents (McLimans et al. 1968). It became obvious that true emulation of human biology *in vitro* needs to be established on primary human cells, carrying the genotypic information of their respective donors. Furthermore, embryonic SCs give rise to ectoderm, mesoderm and endoderm early in human embryonic development. Rapid pluripotent SC proliferation and cell differentiation into various tissues, which is induced by local microenvironments, continues from fertilization to beyond adolescence, during which organs mature at different rates before functional homeostasis is reached. Should a xenobiotic cause organ or tissue damage, regenerative processes attempt to restore this homeostasis by the renewal of damaged tissue. Thus, biological substrates from the early development of human individuals might provide a valid cell source for the *in vitro* modelling of organ functionality and organ regeneration. Human embryonic SC technologies (Ben-David et al. 2012), and, more recently, induced pluripotent SC technologies (Takahashi et al. 2007; Inoue et al. 2014) have provided nearly unlimited access to human tissues for the *in vitro* emulation of organs. First impressive results of human organoid self-assembly from embryonic SC sources are demonstrated in literature, for example, for the generation of miniaturized gut equivalents (Spence et al. 2011) and human mini-brains (Lancaster et al. 2013).

In a next step, it has become clear that, in addition to efficient oxygen and nutrient supply, a local microenvironment with appropriate mechanicochemical coupling achieved by regulating interstitial flow or applying external stresses is a crucial prerequisite for mimicking the *in vivo* biology of individual organs at stable homeostasis over long periods (Griffith and Swartz 2006). Finally, *in vitro*-generated individual organs should be interconnected properly to represent the functionality of a human organism. First attempts to interconnect different cell types or tissues at a miniaturized chip scale through microchannels, applying microsystem technologies were reported in literature

(Hwan et al. 2009; Zhang et al. 2009; Sung et al. 2010; Imura et al. 2010). A long-term stable homeostasis between human 3D liver spheroids and skin biopsies on a chip were demonstrated recently (Wagner et al. 2013). In the human body, organs are interconnected by a vascular network entirely lined by human endothelial cells, representing nature's blood-tissue barrier. The endothelial cell layer communicates with the tissue and signals into the blood stream to recruit, for example, leucocytes into a region of local damage in the organism. Finally, a closed endothelial cell layer prevents blood cells from bleeding into tissue and clotting. Different approaches to establish human vasculature *in vitro* at a mini-scale through the so-called BioVaSc technology (Mertsching et al. 2009; Schanz et al. 2010; Scheller et al. 2013) and at microscale on chips (Yeon et al. 2012; Kim et al. 2013; Schimek et al. 2013; Lee et al. 2014) have been published in the past. Summarizing the historical developments: After more than 100 years of *in vitro* cell cultures, all technological prerequisites to emulate a human organism at miniature scale are in place. The question remains how many organs are necessary to achieve organismal complexity and how small can we go?

2.3 Human-on-a-Chip Systems: The Concept

Nowadays, systemic single and repeated dose safety assessment, disease modelling, systemic testing, and efficacy evaluation of substances are carried out on laboratory animals and in humans due to the lack of predictive alternatives. Relevant international guidelines for chemical testing—OECD test guidelines 407, 408, 410–413, 419, and 453—demand 28-day, 90-day and 12 month test durations, and oral, dermal and inhalation exposure routes in groups of 25–50 animals per substance for safety assessment. The toxicity testing of pharmaceuticals often adheres to approximately the same number and species of animal per drug candidate, and lasts from weeks to months, whilst safety testing in humans usually requires 60–100 healthy volunteers who are exposed over days and weeks. Notably, the use of animal disease models for the efficacy evaluation of drug candidates has increased rapidly over the last decade. Once an animal model is accepted as a suitable representation of a specific human disease, substance testing is commonly carried out over weeks and months in groups of several hundred animals, similar to human patients in clinical phase 2 trials. A translational alternative to these tests and trials should ideally narrow down the phylogenetic distance between laboratory animals and human beings, and close the biosimilarity gap between the current single "organ-on-a-chip" and human beings.

The definition of spatial-temporal biological levels is of outstanding importance for "human-on-a-chip" concepts, due to the biological fidelity of a human individual during their lifespan. It is evident in substance testing and disease modelling arenas, that prenatal development, childhood and adulthood, at gender level are discrete phases of human biology in an individual's lifespan. Considering the ever-increasing human lifespan, senescence is envisaged as a new category which can hardly be modelled using laboratory animals. In addition, the period of pregnancy is also a category to be considered. Current "human-on-a-chip" developments focus on the emulation of non-pregnant adulthood, as this time span has the largest numeric relevance for safety and efficacy testing. Steadily improving concepts towards the "human-on-a-chip" have been reviewed in literature (Huh et al. 2011; Shuler 2012; Marx et al. 2012; van de Stolpe and den Toonder 2013). The aforementioned DARPA-NIH-FDA US initiative has postulated, that organs from the following ten systems should be interconnected in a biological manner to gain human organismal homeostasis *in vitro*: circulatory, endocrine, gastrointestinal, immune, integumentary, musculoskeletal, nervous, reproductive, respiratory, and urinary. With regard to the scale of chip-based organisms, the first approaches to calculate a biologically representative scale-down of human organs have been published (Moraes et al. 2013; Wikswo et al. 2013). We have recently published a possible design of such a "human-on-a-chip" with a scaling mechanism, taking into account the organoid structure of each and every organ (Giese and Marx 2014).

2.4 A Validation Roadmap for Upcoming "Human-on-a-Chip" Solutions

Once "human-on-a-chip" concepts turn into solutions capable of replacing systemic substance testing in animals, their qualification and validation strategies should adhere to the latest standards of qualification and validation and might be compared to those laboratory animal tests, which they aim to replace. There are two procedural pathways aiming at the validation of a "human-on-a-chip" based test assay within the current regulatory landscape in US and Europe.

Firstly, if the "human-on-a-chip" based model aims to replace the animal model qualified through the new US FDA validation strategy of Drug Development Tools (DDT), the validation programme should adhere to exactly the same criteria. The aforementioned FDA DDT Qualification Programme involves a "fit-for-purpose" qualification. Once an animal model is qualified for a specific context of use as a DTT, industry can use the tool for the qualified purpose during product development, and FDA reviewers can be confident in applying the DDT without the underlying supporting data. Qualification of an animal model according to this Animal Model Qualification Programme of the FDA is voluntary (i.e. not required for product approval or licensure under the Animal Rule). The qualification process is limited to animal models used for product approval under the Animal Rule. A qualified model may be used for efficacy testing in development programmes for multiple investigational drugs for the same disease or condition targeted. Such animal models are considered to be product-independent (i.e. not linked to a specific drug). The regulatory pathway above mentioned should apply equally to a "human-on-a-chip" solution, replacing the respective animal model and, therefore, should be a possible and reliable road map leading towards fast and pragmatic validation.

Secondly, for the validation of "human-on-a-chip" models aiming to replace the animal models used in the aforementioned OECD guidelines in chemical safety assessment, the adherence to existing OECD Guidance Document on Validation of test methods for hazard assessment (OECD 2005), the EMA guideline on regulatory acceptance of 3R methods and qualification of novel methodologies for drug development (EMEA/CHMP/SAWP/72894/2008 Corr1), via the EMA Scientific Advice

Working Party (SAWP) and the recommendation of the ICH Safety Topic Recommendation Working Group could be instrumental. In other words, the following validation principles should apply to validate a "human-on-a-chip" based assay, for example, to replace the current technical guideline OECD TG 410 on "Repeated Dose Dermal Toxicity: 21/28-day Study" in adult rat, rabbit or guinea pigs:

1. Bioreactor equipment operating the "human-on-a-chip" solutions should be qualified according to standard IQ, OQ and PQ procedures (installation-, operation- and performance qualification).
2. Test design should address the endpoints covered by the existing test guidelines
3. Representative groups of substances for validation should be used with prior co-ordination with the respective regulatory agency.

In addition, the validation process should consider the recommendations of the aforementioned sources.

As of today, a subsequent combination of the two validation pathways, as shown in Fig. 12.1, seems to be the most efficient road map toward validation of upcoming "human-on-a-chip" solutions.

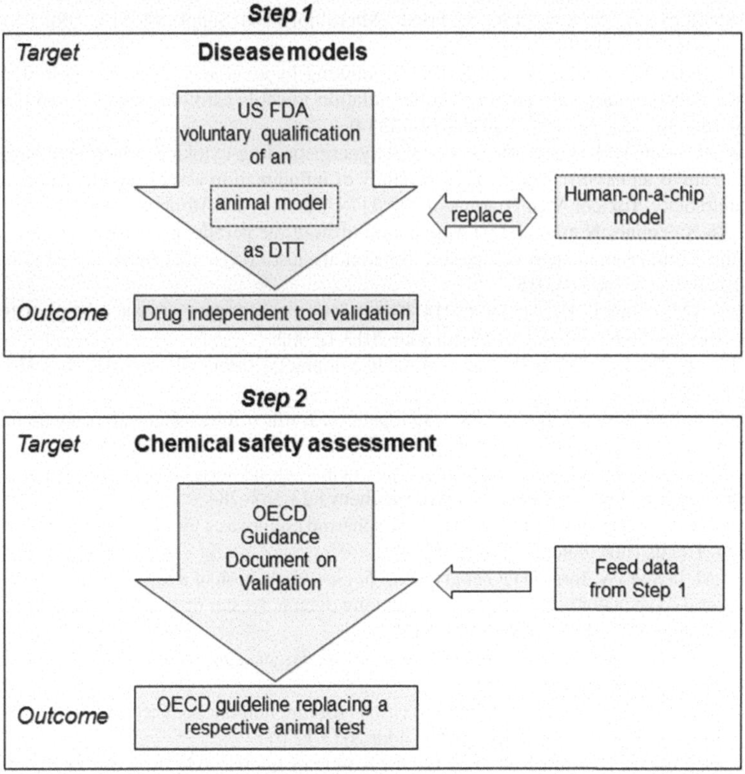

Fig. 12.1 A possible stepwise validation roadmap for upcoming human-on-a-chip solutions

Acknowledgement The authors acknowledge Fundação para a Ciência e Tecnologia, Portugal, for the financial support provided by the grants PTDC/EBB-BIO/112786/2009 and SFRH/BD/70264/2010.

References

Abu-Absi SF, Friend JR, Hansen LK, Hu W-S (2002) Structural polarity and functional bile canaliculi in rat hepatocyte spheroids. Exp Cell Res 274:56–67

Bauwens C, Yin T, Dang S et al (2005) Development of a perfusion fed bioreactor for embryonic stem cell-derived cardiomyocyte generation: oxygen-mediated enhancement of cardiomyocyte output. Biotechnol Bioeng 90:452–461

Ben-David U, Kopper O, Benvenisty N (2012) Expanding the boundaries of embryonic stem cells. Cell Stem Cell 10:666–677. doi:10.1016/j.stem.2012.05.003

Brito C, Simao D, Costa I et al (2012) 3D cultures of human neural progenitor cells: dopaminergic differentiation and genetic modification. [corrected]. Methods 56:452–460

Carraro A, Hsu W, Borenstein JT et al (2008) *In vitro* analysis of a hepatic device with intrinsic microvascular-based channels. Biomed Microdevices 10:795–805. doi:10.1007/s10544-008-9194-3

Carrel A (1912) On the permanent life of tissues outside of the organism. J Exp Med 15:516–528

Cencic A, Langerholc T (2010) Functional cell models of the gut and their applications in food microbiology—a review. Int J Food Microbiol 141(Suppl):S4–S14. doi:10.1016/j.ijfoodmicro.2010.03.026

Correia C, Serra M, Espinha N et al (2014) Combining hypoxia and bioreactor hydrodynamics boosts induced pluripotent stem cell differentiation towards cardiomyocytes. Stem Cell Rev 10(6):786–801. doi:10.1007/s12015-014-9533-0

Cosgrove BD, King BM, Hasan MA et al (2009) Synergistic drug-cytokine induction of hepatocellular death as an *in vitro* approach for the study of inflammation-associated idiosyncratic drug hepatotoxicity. Toxicol Appl Pharmacol 237:317–330. doi:10.1016/j.taap.2009.04.002

Cui ZF, Xu X, Trainor N et al (2007) Application of multiple parallel perfused microbioreactors and three-dimensional stem cell culture for toxicity testing. Toxicol *In Vitro* 21:1318–1324. doi:10.1016/j.tiv.2007.05.015

Desrochers TM, Palma E, Kaplan DL (2014) Tissue-engineered kidney disease models. Adv Drug Deliv Rev 69–70:67–80. doi:10.1016/j.addr.2013.12.002

Dexter TM, Lajtha LG (1974) Proliferation of haemopoietic stem cells *in vitro*. Br J Haematol 28:525–530

Edwards S, Lalor PF, Nash GB et al (2005) Lymphocyte traffic through sinusoidal endothelial cells is regulated by hepatocytes. Hepatology 41:451–459. doi:10.1002/hep.20585

Fell HB, Robison R (1929) The growth, development and phosphatase activity of embryonic avian femora and limb buds cultivated *in vitro*. Biochem J 23:767–784

Fennema E, Rivron N, Rouwkema J et al (2013) Spheroid culture as a tool for creating 3D complex tissues. Trends Biotechnol 31:108–115

Flendrig LM, la Soe JW, Jorning GG et al (1997) *In vitro* evaluation of a novel bioreactor based on an integral oxygenator and a spirally wound nonwoven polyester matrix for hepatocyte culture as small aggregates. J Hepatol 26:1379–1392

Gerlach JC, Encke J, Hole O et al (1994) Bioreactor for a larger scale hepatocyte *in vitro* perfusion. Transplantation 58:984–988

Giese C, Marx U (2014) Human immunity *in vitro*—solving immunogenicity and more. Adv Drug Deliv Rev 69–70:103–122. doi:10.1016/j.addr.2013.12.011

Giese C, Demmler CD, Ammer R et al (2006) A human lymph node *in vitro*—challenges and progress. Artif Organs 30:803–808. doi:10.1111/j.1525-1594.2006.00303.x

Giese C, Lubitz A, Demmler CD et al (2010) Immunological substance testing on human lymphatic micro-organoids *in vitro*. J Biotechnol 148:38–45. doi:10.1016/j.jbiotec.2010.03.001

Goral VN, Hsieh Y-C, Petzold ON et al (2010) Perfusion-based microfluidic device for three-dimensional dynamic primary human hepatocyte cell culture in the absence of biological or synthetic matrices or coagulants. Lab Chip 10:3380–3386. doi:10.1039/c0lc00135j

Griffith LG, Swartz MA (2006) Capturing complex 3D tissue physiology *in vitro*. Nat Rev Mol Cell Biol 7:211–224. doi:10.1038/nrm1858

Gu W, Zhu X, Futai N et al (2004) Computerized microfluidic cell culture using elastomeric channels and Braille displays. Proc Natl Acad Sci U S A 101:15861–15866. doi:10.1073/pnas.0404353101

Günther A, Yasotharan S, Vagaon A et al (2010) A microfluidic platform for probing small artery structure and function. Lab Chip 10:2341–2349. doi:10.1039/c004675b

Hartung T, Luechtefeld T, Maertens A, Kleensang A (2013) Integrated testing strategies for safety assessments. ALTEX 30:3–18

Ho C-T, Lin R-Z, Chang W-Y et al (2006) Rapid heterogeneous liver-cell on-chip patterning via the enhanced field-induced dielectrophoresis trap. Lab Chip 6:724–734. doi:10.1039/b602036d

Huh D, Matthews BD, Mammoto A et al (2010) Reconstituting organ-level lung functions on a chip. Science 328:1662–1668. doi:10.1126/science.1188302

Huh D, Hamilton GA, Ingber DE (2011) From 3D cell culture to organs-on-chips. Trends Cell Biol 21:745–754. doi:10.1016/j.tcb.2011.09.005

Hwa AJ, Fry RC, Sivaraman A et al (2007) Rat liver sinusoidal endothelial cells survive without exogenous VEGF in 3D perfused co-cultures with hepatocytes. FASEB J 21:2564–2579. doi:10.1096/fj.06-7473com

Hwan J, Shuler ML, Sung JH (2009) A micro cell culture analog (microCCA) with 3-D hydrogel culture of multiple cell lines to assess metabolism-dependent cytotoxicity of anti-cancer drugs. Lab Chip 9:1385–1394. doi:10.1039/b901377f

Imura Y, Sato K, Yoshimura E (2010) Micro total bioassay system for ingested substances: assessment of intestinal absorption, hepatic metabolism, and bioactivity. Anal Chem 82:9983–9988. doi:10.1021/ac100806x

Inoue H, Nagata N, Kurokawa H, Yamanaka S (2014) iPS cells: a game changer for future medicine. EMBO J 33:409–417. doi:10.1002/embj.201387098

Kane BJ, Zinner MJ, Yarmush ML, Toner M (2006) Liver-specific functional studies in a microfluidic array of primary mammalian hepatocytes. Anal Chem 78:4291–4298. doi:10.1021/ac051856v

Khetani SR, Bhatia SN (2008) Microscale culture of human liver cells for drug development. Nat Biotechnol 26:120–126. doi:10.1038/nbt1361

Kim S, Lee H, Chung M, Jeon NL (2013) Engineering of functional, perfusable 3D microvascular networks on a chip. Lab Chip 13:1489–1500. doi:10.1039/c3lc41320a

Kostadinova R, Boess F, Applegate D et al (2013) A long-term three dimensional liver co-culture system for improved prediction of clinically relevant drug-induced hepatotoxicity. Toxicol Appl Pharmacol 268:1–16. doi:10.1016/j.taap.2013.01.012

Lahar N, Lei NY, Wang J et al (2011) Intestinal subepithelial myofibroblasts support *in vitro* and *in vivo* growth of human small intestinal epithelium. PLoS One 6, e26898. doi:10.1371/journal.pone.0026898

Lancaster MA, Renner M, Martin C et al (2013) Cerebral organoids model human brain development and microcephaly. Nature 501(7467):373–379. doi:10.1038/nature12517.Cerebral

Leclerc E, Sakai Y, Fujii T (2004) Microfluidic PDMS (polydimethylsiloxane) bioreactor for large-scale culture of hepatocytes. Biotechnol Prog 20:750–755. doi:10.1021/bp0300568

Lee PJ, Hung PJ, Lee LP (2007) An artificial liver sinusoid with a microfluidic endothelial-like barrier for primary hepatocyte culture. Biotechnol Bioeng 97:1340–1346. doi:10.1002/bit

Lee H, Kim S, Chung M et al (2014) A bioengineered array of 3D microvessels for vascular permeability assay. Microvasc Res 91:90–98. doi:10.1016/j.mvr.2013.12.001

Lilienblum W, Dekant W, Foth H et al (2008) Alternative methods to safety studies in experimental animals: role in the risk assessment of chemicals under the new European Chemicals Legislation (REACH). Arch Toxicol 82:211–236. doi:10.1007/s00204-008-0279-9

Martin I, Wendt D, Heberer M (2004) The role of bioreactors in tissue engineering. Trends Biotechnol 22:80–86

Marx U, Walles H, Hoffmann S et al (2012) Human-on-a-chip developments: a translational cutting-edge alternative to systemic safety assessment and efficiency evaluation of substances in laboratory animals and man. Altern Lab Anim 40:235–257

Materne E-M, Tonevitsky AG, Marx U (2013) Chip-based liver equivalents for toxicity testing—organotypicalness versus cost-efficient high throughput. Lab Chip 13:3481–3495. doi:10.1039/c3lc50240f

McLimans WF, Crouse EJ, Tunnah KV, Moore GE (1968) Kinetics of gas diffusion in mammalian cell culture systems. I. Experimental. Biotechnol Bioeng 10:725–740. doi:10.1002/bit.260100603

Meli L, Barbosa HSC, Hickey AM et al (2014) Three dimensional cellular microarray platform for human neural stem cell differentiation and toxicology. Stem Cell Res 13:36–47

Mendhe R, Rathore AS, Krull IS (2012) Analytical tools for enabling process analytical technology applications in biotechnology. LC GC North Am 30(1):52–62

Mertsching H, Schanz J, Steger V et al (2009) Generation and transplantation of an autologous vascularized bioartificial human tissue. Transplantation 88:203–210. doi:10.1097/TP.0b013e3181ac15e1

Miki T, Ring A, Gerlach J (2011) Hepatic differentiation of human embryonic stem cells is promoted by three-dimensional dynamic perfusion culture conditions. Tissue Eng Part C Methods 17:557–568

Miranda JP, Leite SB, Muller-Vieira U et al (2009) Towards an extended functional hepatocyte *in vitro* culture. Tissue Eng Part C Methods 15:157–167. doi:10.1089/ten.tec.2008.0352

Miranda JP, Rodrigues A, Tostoes RM et al (2010) Extending hepatocyte functionality for drug-testing applications using high-viscosity alginate-encapsulated three-dimensional cultures in bioreactors. Tissue Eng Part C Methods 16:1223–1232. doi:10.1089/ten.TEC.2009.0784

Mitteregger R, Vogt G, Rossmanith E, Falkenhagen D (1999) Rotary cell culture system (RCCS): a new method for cultivating hepatocytes on microcarriers. Int J Artif Organs 22:816–822

Moraes C, Labuz JM, Leung BM et al (2013) On being the right size: scaling effects in designing a human-on-a-chip. Integr Biol (Camb) 5:1149–1161. doi:10.1039/c3ib40040a

Nakayama H, Kimura H, Komori K, et al (2008) Development of a disposable three-compartment micro-cell culture device for toxicokinetic study in humans and its preliminary evaluation, pp. 619–622

Navran S (2008) The application of low shear modeled microgravity to 3-D cell biology and tissue engineering. Biotechnol Annu Rev 14:275–296

Nibourg GAA, Hoekstra R, van der Hoeven TV et al (2013) Increased hepatic functionality of the human hepatoma cell line HepaRG cultured in the AMC bioreactor. Int J Biochem Cell Biol 45:1860–1868

Niebruegge S, Bauwens CL, Peerani R et al (2009) Generation of human embryonic stem cell-derived mesoderm and cardiac cells using size-specified aggregates in an oxygen-controlled bioreactor. Biotechnol Bioeng 102:493–507

OECD (2005) Guidance document on the validation and regulatory acceptance of new and updated test methods for hazard assessment. Series on Testing and Assessment, No.34 [ENV/JM/MONO(2005)14], OECD, Paris

Ootani A, Li X, Sangiorgi E et al (2010) Sustained *in vitro* intestinal epithelial culture within a Wnt-dependent stem cell niche. Nat Med 15:701–706. doi:10.1038/nm.1951.Sustained

Park J, Li Y, Berthiaume F et al (2008) Radial flow hepatocyte bioreactor using stacked microfabricated grooved substrates. Biotechnol Bioeng 99:455–467. doi:10.1002/bit

Park JY, Hwang CM, Lee S (2009) Ice-lithographic fabrication of concave microwells and a microfluidic network. Biomed Microdevices 11(1):129–133. doi:10.1007/s10544-008-9216-1

Powers MJ, Domansky K, Kaazempur-Mofrad MR et al (2002) A microfabricated array bioreactor for perfused 3D liver culture. Biotechnol Bioeng 78:257–269. doi:10.1002/bit.10143

Rebelo SP, Costa R, Estrada M et al (2015) HepaRG microencapsulated spheroids in DMSO-free culture: novel culturing approaches for enhanced xenobiotic and biosynthetic metabolism. Arch Toxicol 89(8):1347–1358. doi:10.1007/s00204-014-1320-9

Rhee SW, Taylor AM, Tu CH et al (2005) Patterned cell culture inside microfluidic devices. Lab Chip 5(1):102–107. doi:10.1039/b403091e

Salehi-Nik N, Amoabediny G, Pouran B et al (2013) Engineering parameters in bioreactor's design: a critical aspect in tissue engineering. Biomed Res Int 2013:762132. doi:10.1155/2013/762132

Sato T, Vries RG, Snippert HJ et al (2009) Single Lgr5 stem cells build crypt-villus structures *in vitro* without a mesenchymal niche. Nature 459:262–265. doi:10.1038/nature07935

Schanz J, Pusch J, Hansmann J, Walles H (2010) Vascularised human tissue models: a new approach for the refinement of biomedical research. J Biotechnol 148:56–63. doi:10.1016/j.jbiotec.2010.03.015

Scheller K, Dally I, Hartman N et al (2013) Upcyte microvascular endothelial cells repopulate decellularized scaffold. Tissue Eng Part C Methods 19:57–67

Schimek K, Busek M, Brincker S et al (2013) Integrating biological vasculature into a multi-organ-chip microsystem. Lab Chip 13:3588–3598. doi:10.1039/c3lc50217a

Schroeder K, Bremm KD, Alepee N et al (2011) Report from the EPAA workshop: *in vitro* ADME in safety testing used by EPAA industry sectors. Toxicol *In Vitro* 25:589–604

Schwarz RP, Goodwin TJ, Wolf DA (1992) Cell culture for three-dimensional modeling in rotating-wall vessels: an application of simulated microgravity. J Tissue Cult Methods 14:51–57

Serra M, Brito C, Correia C, Alves PM (2012) Process engineering of human pluripotent stem cells for clinical application. Trends Biotechnol 30:350–359

Shuler ML (2012) Modeling life. Ann Biomed Eng 40:1399–1407. doi:10.1007/s10439-012-0567-7

Slikker W Jr, Andersen ME, Bogdanffy MS et al (2004) Dose-dependent transitions in mechanisms of toxicity. Toxicol Appl Pharmacol 201:203–225

Spence JR, Mayhew CN, Rankin SA et al (2011) Directed differentiation of human pluripotent stem cells into intestinal tissue *in vitro*. Nature 470:105–109. doi:10.1038/nature09691.Directed

Sung JH, Kam C, Shuler ML (2010) A microfluidic device for a pharmacokinetic-pharmacodynamic (PK-PD) model on a chip. Lab Chip 10:446–455. doi:10.1039/b917763a

Sung JH, Yu J, Luo D et al (2011) Microscale 3-D hydrogel scaffold for biomimetic gastrointestinal (GI) tract model. Lab Chip 11:389–392. doi:10.1039/c0lc00273a

Takahashi K, Okita K, Nakagawa M, Yamanaka S (2007) Induction of pluripotent stem cells from fibroblast cultures. Nat Protoc 2:3081–3089. doi:10.1038/nprot.2007.418

Toh Y, Zhang C, Zhang J et al (2007) A novel 3D mammalian cell perfusion-culture system in microfluidic channels. Lab Chip 7:302–309. doi:10.1039/b614872g

Toh Y, Lim TC, Tai D et al (2009) A microfluidic 3D hepatocyte chip for drug toxicity testing. Lab Chip 9:2026–2035. doi:10.1039/b900912d

Tostoes RM, Leite SB, Miranda JP et al (2011) Perfusion of 3D encapsulated hepatocytes—a synergistic effect enhancing long-term functionality in bioreactors. Biotechnol Bioeng 108:41–49. doi:10.1002/bit.22920

Tostoes RM, Leite SB, Serra M et al (2012) Human liver cell spheroids in extended perfusion bioreactor culture for repeated-dose drug testing. Hepatology 55:1227–1236. doi:10.1002/hep.24760

Van de Stolpe A, den Toonder J (2013) Workshop meeting report Organs-on-Chips: human disease models. Lab Chip 13:3449–3470. doi:10.1039/c3lc50248a

Vinardell MP, Mitjans M (2008) Alternative methods for eye and skin irritation tests: an overview. J Pharm Sci 97:46–59. doi:10.1002/jps.21088

Vinci B, Duret C, Klieber S et al (2011) Modular bioreactor for primary human hepatocyte culture: medium flow stimulates expression and activity of detoxification genes. Biotechnol J 6:554–564

Wagner I, Materne E-M, Marx U et al (2013) A dynamic multi-organ-chip for long-term cultivation and substance testing proven by 3D human liver and skin tissue co-culture. Lab Chip 13:3538–3547. doi:10.1039/c3lc50234a

Wikswo JP, Curtis EL, Eagleton ZE et al (2013) Scaling and systems biology for integrating multiple organs-on-a-chip. Lab Chip 13:3496–3511. doi:10.1039/c3lc50243k

Xia L, Ng S, Han R et al (2009) Laminar-flow immediate-overlay hepatocyte sandwich perfusion system for drug hepatotoxicity testing. Biomaterials 30:5927–5936

Yeon JH, Ryu HR, Chung M et al (2012) *In vitro* formation and characterization of a perfusable three-dimensional tubular capillary network in microfluidic devices. Lab Chip 12:2815–2822. doi:10.1039/c2lc40131b

Young EWK, Simmons CA (2010) Macro- and microscale fluid flow systems for endothelial cell biology. Lab Chip 10(2):143–160. doi:10.1039/b913390a

Yu J, Peng S, Luo D, March JC (2012) *In vitro* 3D human small intestinal villous model for drug permeability determination. Biotechnol Bioeng 109:2173–2178. doi:10.1002/bit.24518

Zeilinger K, Schreiter T, Darnell M et al (2011) Scaling down of a clinical three-dimensional perfusion multicompartment hollow fiber liver bioreactor developed for extracorporeal liver support to an analytical scale device useful for hepatic pharmacological *in vitro* studies. Tissue Eng Part C Methods 17:549–556

Zhang C, Zhao Z, Abdul A et al (2009) Towards a human-on-chip: culturing multiple cell types on a chip with compartmentalized microenvironments. Lab Chip 9:3185–3192. doi:10.1039/b915147h

Chapter 13
Integrated Approaches to Testing and Assessment

Andrew P. Worth and Grace Patlewicz

Abstract In this chapter, we explain how Integrated Approaches to Testing and Assessment (IATA) offer a means of integrating and translating the data generated by toxicity testing methods, thereby serving as flexible and suitable tools for toxicological decision making in the twenty-first century. In addition to traditional *in vitro* and *in vivo* testing methods, IATA are increasingly incorporating newly developed *in vitro* systems and measurement technologies such as high throughput screening and high content imaging. Computational approaches are also being used in IATA development, both as a means of generating data (e.g. QSARs), interpreting data (bioinformatics and chemoinformatics), and as a means of integrating multiple sources of data (e.g. expert systems, bayesian models). Decision analytic methods derived from socioeconomic theory can also play a role in developing flexible and optimal IATA solutions. Some of the challenges involved in the development, validation and implementation of IATA are also discussed.

Keywords Integrated approach to testing and assessment (IATA) • Integrated testing strategy (ITS) • Predictive toxicology • Adverse outcome pathway • Chemical safety assessment

A.P. Worth (✉)
European Commission, Joint Research Centre (JRC), Ispra, Italy
e-mail: andrew.worth@ec.europa.eu

G. Patlewicz
Dupont Haskell Global Centers for Health and Environmental Sciences, Newark, DE, USA

National Center for Computational Toxicology (NCCT), US Environmental Protection Agency (EPA), Research Triangle Park, NC 27711, USA

C. Eskes, M. Whelan (eds.), *Validation of Alternative Methods for Toxicity Testing*,
Advances in Experimental Medicine and Biology 856,
DOI 10.1007/978-3-319-33826-2_13

1 Introduction

In order to realise the vision of Toxicology in the twenty-first Century (NRC 2007), the regulatory assessment of chemical safety needs to move away from the use of "one-size-fits-all" and largely pre-defined batteries of standard toxicity tests and exposure studies towards the use of more-focused and hypothesis-driven approaches that are tailored to the characteristics and intended use of the chemical. Furthermore, this paradigm shift in toxicology implies a transformation of the current way of conducting toxicity testing from a system based on phenotypic responses in animals towards the use of pathway-based approaches that capture our understanding of chemical distribution and fate (in the environment and biological organisms) and the physiological mechanisms underlying toxicity in exposed organisms. The move towards a more mechanistically-based risk assessment process implies a knowledge of the underlying toxicokinetic and toxicodynamic processes, the use of mechanistic data derived from high-throughput and high-content screening (HTS/HCS) assays in cell lines, cell cultures and/or tissue surrogates, combined with the application of a range of computational methods for data analysis and predictive modelling.

The integrated use of these different methodologies and data sources in a transparent and scientifically sound manner represents a considerable intellectual and practical challenge, and many solutions have been proposed. These are typically referred to as Intelligent or Integrated Testing Strategies (ITS) or (more recently) Integrated Approaches to Testing and Assessment (IATA). While various definitions have been proposed for these terms (Table 13.1), in this chapter we use the

Table 13.1 Definitions of ITS and IATA in the scientific literature

Definition/explanation of ITS or IATA	Reference
"An integrated testing strategy is any approach to the evaluation of toxicity which serves to reduce, refine or replace an existing animal procedure, and which is based on the use of two or more of the following: physicochemical data, *in vitro* data, human data (for example, epidemiological, clinical case reports), animal data (where unavoidable), computational methods (such as quantitative structure-activity relationships [QSAR]) and biokinetic models."	Blaauboer et al. (1999)
"In the context of safety assessment, an Integrated Testing Strategy is a methodology which integrates information for toxicological evaluation from more than one source, thus facilitating decision-making. This should be achieved whilst taking into consideration the principles of the Three Rs (reduction, refinement and replacement)."	Kinsner-Ovaskainen et al. (2009)
"ITS can be described as combinations of test batteries covering relevant mechanistic steps and organised in a logical, hypothesis-driven decision scheme, which is required to make efficient use of generated data and to gain a comprehensive information basis for making decisions regarding hazard or risk. We approach ITS from a system analysis perspective and understand them as decision support tools that synthesise information in a cumulative manner and that guide testing in such a way that information gain in a testing sequence is maximised.	Jaworska and Hoffmann (2010)

(continued)

Table 13.1 (continued)

Definition/explanation of ITS or IATA	Reference
This definition clearly separates ITS from tiered approaches in two ways. First, tiered approaches consider only the information generated in the last step for a decision … Secondly, in tiered testing strategies the sequence of tests is prescribed, albeit loosely, based on average biological relevance and is left to expert judgment. In contrast, our definition enables an integrated and systematic approach to guide testing such that the sequence is not necessarily prescribed ahead of time but is tailored to the chemical-specific situation. Depending on the already available information on a specific chemical the sequence might be adapted and optimised for meeting specific information targets"	
"A tiered approach to data gathering, testing, and assessment that integrates different types of data (including physicochemical and other chemical properties as well as *in vitro* and *in vivo* toxicity data). When combined with estimates of exposure in an appropriate manner, the IATA provides predictions of risk. In an IATA, unsuitable substances are screened out early in the process. This reduces the number of substances that are subjected to the complete suite of regulatory tests. Plausible and testable hypotheses are formulated based on existing information and/or information derived from lower tier testing and only targeted testing is performed in the higher tiers. Failure to satisfy the toxicity requirements at a lower tier typically precludes further testing at a higher tier."	Council of Canadian Academies (2012)
"In the context of safety assessment, an ITS is a methodology integrating information from several sources of toxicological evaluation allowing appropriate decision making."	De Wever et al. (2012)
"An integrated test strategy is an algorithm to combine (different) test result(s) and, possibly, non-test information (existing data, *in silico* extrapolations from existing data or modeling) to give a combined test result. They often will have interim decision points at which further building blocks may be considered."	Hartung et al. (2013)
An OECD working definition (as of 2015) is "a structured approach used for hazard identification (potential), hazard characterisation (potency) and/or safety assessment (potential/potency and exposure) of a chemical or group of chemicals, which strategically integrates and weighs all relevant data to inform regulatory decisions regarding potential hazard and/or risk and/or the need for further targeted testing and therefore optimising and potentially reducing the number of tests that need to be conducted"	OECD (2015a, b, c)

term IATA to refer to any such approach. Possible applications of IATA include priority setting, hazard identification/profiling, hazard classification and labelling, PBT and vPvB assessment, and risk assessment. The increasing use of IATA in the assessment of chemicals is expected to have numerous benefits, including the reduction, refinement and replacement of animal testing, increased efficiencies in testing and assessment, and the generation of more extensive and more reliable data that will inform the safe design of innovative chemical products and improve the protection of human health and the environment.

In this chapter, we present a few examples of IATA to illustrate the breadth the IATA approach, and then discuss some of the key issues relating to the development, validation and implementation of IATA in a regulatory setting.

2 Historical Perspective on IATA

Research on the development and evaluation of IATA has been published since the early 1990s. A pioneering example was the ECITTS project, which explored and illustrated the integration of biokinetic modelling with *in vitro* testing for the prediction of systemic toxicity (Blaauboer et al. 1994; DeJongh et al. 1999). Building on this work, a Task Force established by the European Centre for the Validation of Alternative Methods (ECVAM; now the European Union Reference Laboratory for Alternatives to Animal Testing—EURL ECVAM) proposed some generic and endpoint-specific strategies for assessing systemic toxicity (Blaauboer et al. 1999). Subsequent efforts by EURL ECVAM focused on the development of tiered strategies for local toxicity (skin and eye irritation/corrosion) as well as on the development of a generic approach for evaluating these strategies (Worth et al. 1998; Worth and Fentem 1999; Worth 2000, 2004).

In 2001, the European Commission's proposal for the REACH legislation stimulated widespread efforts in the EU aimed at developing and proposing IATA. A DEFRA-funded project resulted in proposed decision trees for all the major human health and environmental effects required under REACH (Grindon et al. 2008). Furthermore, several EU-funded research projects focused on the development of building blocks and testing strategies for REACH-relevant endpoints, as summarised in Table 13.2. Around the same time, the European Commission's Joint Research Centre coordinated the formulation of testing strategies for human health and environmental effects that ultimately became part of ECHA's guidance on fulfilling information requirements under REACH (ECHA 2012).

More recently, in attempts to build IATA that are mechanistically-based, there has been increasing emphasis on the use of pathway-based approaches, such as toxicity pathways (TPs) and adverse outcome pathways (AOPs). The general premise of the these approaches is that a limited set of key (and measurable) events is

Table 13.2 EU-funded research projects that have contributed to the development of IATA

Project	Endpoint(s)	References
OSIRIS http://www.ufz.de/osiris/	Skin sensitisation, repeated-dose toxicity, mutagenicity, carcinogenicity	Buist et al. (2013), Rorije et al. (2013), Vermeire et al. (2013) and Tluczkiewicz et al. (2013)
ReProTect http://www.reprotect.eu/	Reproductive and developmental toxicity	Marx-Stoelting et al. (2009) Piersma et al. (2013)
AcuteTox http://www.acutetox.eu/	Acute systemic toxicity	Clemedson et al. (2007)
SensiTiv http://www.sens-it-iv.eu/	Skin and respiratory sensitisation	Rovida and Roggen (2007)
ChemScreen http://www.chemscreen.eu/	Reproductive and developmental toxicity	Piersma et al. (2013)

sufficient for describing toxicological effects and predicting responses at multiple levels of biological organisation (cell, tissue/organ, organism, population). They are based on the assumption that a toxicant, after reaching and interacting with a biological target (in the molecular initiating event; MIE), initiates a cascade of events (intermediate effects) which may lead to an adverse outcome at the organism or population level. A TP refers to a normal cellular response pathway that, when sufficiently perturbed, is expected to result in adverse health effects (NRC 2007). An AOP refers to a sequence of events from the exposure of an individual or population to a chemical substance through a final adverse (toxic) effect at the individual level (for human health) or population level (for ecotoxicological endpoints) (OECD 2013). It is important to appreciate that an AOP is a pragmatic simplification of the biological complexity and is therefore generally depicted as a linear, unbranched pathway. However, it is recognised that, with the exception of highly specifically acting chemicals, most chemicals will perturb more than one AOP, so that AOP networks will ultimately be the functional unit of prediction for most chemically-induced adverse outcomes.

Another commonly used term is mode of action (MoA), which is the sequence of key cellular and biochemical events (measurable parameters), starting with the interaction of an agent with the target cell, through functional and anatomical changes, resulting in cancer or other adverse health effects. The AOP concept is the most broadly applicable, encompassing the concepts of TP and MoA.

For practical purposes in chemical hazard and risk assessment, the application of the AOP approach means that a detailed molecular understanding of all possible molecular interactions and effects is not necessary, and that ultimately it may be sufficient for decision making to predict the adverse outcome at organism and population level from early ("upstream") key events. Some researchers have even taken this concept a step further to propose a "region of safety" approach in which the most sensitive adverse apical effect is not known and not directly predicted, but disruption of important biological processes is measured and the dose response relationship characterised for use in a margin of exposure safety assessment (Thomas et al. 2013).

In addition to these major initiatives, various reviews and commentaries on IATA have been published (van Leeuwen et al. 2007; Ahlers et al. 2008; Schaafsma et al. 2009; Jaworska and Hoffmann 2010; Hartung et al. 2013; Patlewicz et al. 2013). For the most part, these focus on chemicals in their bulk form, but increasingly the IATA approach is being developed for the assessment of nanomaterials (Nel et al. 2013; Oomen et al. 2014; Stone et al. 2014).

3 The Building Blocks of IATA

Since there is an almost limitless choice of building blocks for IATA, they tend to be described in terms of their methodological approach (QSAR, read-across, *in chemico*, *in vitro*, exposure-based waiving) or technology (e.g. HTS). More recently, and in line with the AOP approach, the building blocks are also being described in terms of the key events they measure or compute, and the adverse outcomes they can be used to (partially) predict.

A QSAR model is mathematical relationship that (quantitatively) links chemical structure and physicochemical properties to a well-defined process, such as biological activity or reactivity. Whereas QSARs were traditionally developed to model the adverse outcome directly, a new generation of QSARs are now being developed to model key events, and in particular the MIE. Read-across is a simpler method that predicts endpoint information for one chemical by using data for the same endpoint from another similar chemical (or group of chemicals, sometimes referred to as a chemical category).

In chemico assays are cell-free experimental tests. They are usually developed to identify, and in some cases quantify, the intrinsic reactivity of substances. Most *in chemico* tests relevant to toxicity prediction have investigated the reaction of an electrophilic molecule (normally assumed to be toxicant) with a model nucleophile (representing a target on a biological macromolecule). Examples include the Direct Peptide Reactivity Assay (DPRA) (Gerberick et al. 2004, 2007), as well as direct reactivity with glutathione (Schultz et al. 2005). Such assays have found applications in the identification of skin sensitisers and the prediction of excess acute toxicity to aquatic organisms as shown by Aptula et al. (2006) and discussed in more detail elsewhere (Aptula and Roberts 2006; Roberts et al. 2008). These assays provide a means of measuring the MIE in cases where the chemicobiological interaction consists of a covalent interaction between the toxicant and molecular target. This kind of MIE is important since it is relatively non-specific and is therefore potentially implicated in multiple toxicological endpoints.

In vitro tests refer to a range of cell and tissue-based methods (primary cells, tissues, or organs; cell lines; stem cells, and reconstructed tissue models). As a means of effecting the paradigm shift in toxicology, there is an increasing interest in the use of mechanistic endpoints that are measurable *in vitro*, and also in the automation of *in vitro* test protocols by using high throughput screening (HTS) and high content imaging (HCI) technologies. Traditionally, *in vitro* tests have been used for hazard identification or classification and labelling through the use of a prediction model that translates the *in vitro* data into a prediction of hazard (Worth and Balls 2001). Increasingly, the quantitative data generated by *in vitro* tests (manual or automated) are being used not only for hazard identification/classification, but also as points of departure in risk assessment through the use of physiologically based kinetic (PBK) and dynamic (PBD) modelling (Blaauboer 2010; Thomas et al. 2013). PBK models are defined by differential mathematical equations describing the adsorption, distribution, metabolism, and excretion of a specific chemical and/or its metabolite; PBD models connect the internal dose to the dose response relationship of the adverse dynamic effect.

One way of introducing efficiencies and maximising the information gain in toxicity testing is to use exposure information in order to either conclude on the absence (unlikelihood) or presence (likelihood) of concern. These approaches are sometimes referred to as exposure-based triggering (EBT) and exposure-based waiving (EBW) of testing. The principle behind any EBW is that there are situations when human or environmental exposures are so low or infrequent that there is a very low probability that the acquisition of additional effect information may lead to an improvement in the ability to manage risk (Vermeire et al. 2010). The concept of EBW is implicit in some regulatory frameworks; for example, REACH provides the possibility to waive the toxicity testing of a substance (e.g. for inhalational toxicity) based on scenarios developed in the exposure assessment (Rowbotham and Gibson 2011).

4 Development of IATA

IATA represent a very diverse set of solutions for toxicological assessment and decision making. At one extreme, an IATA can be explicitly described and prescriptive, leaving little or no room for expert choices. At the other extreme, IATA can be loosely described and flexible, allowing multiple options for the assessor. Some IATA are endpoint-specific, whereas others address multiple types of toxic effect. Some include exposure considerations, whereas others do not. Some include the option for animal testing, whereas others are animal-free. IATA can also differ in the extent to which they apply mechanistic reasoning and use mechanistically relevant data. Furthermore, IATA can have different starting points in terms of the data needed, and can be directly or indirectly linked to regulatory consequences.

For all these reasons, and this is a key consideration in both the development and validation of IATA, the optimal choice of IATA is highly context dependent, making a one-size-fits-all solution to a given problem difficult or impossible (depending on how the problem is formulated, and how costs are weighted against benefits).

There is little guidance on the development of IATA, despite the fact that the conceptual and methodological basis for these approaches has been evolving since the early 1990s. In the design of IATA, the distinction between the battery approach and the tiered (stepwise or sequential) testing approach has been discussed (Jaworska and Hoffmann 2010; Hartung et al. 2013). A test battery refers to a series of tests usually performed at the same time or in close sequence. Each test within the battery is designed to complement the other tests and generally to measure a different component of a multi-factorial toxic effect. In a tiered approach on the other hand, results are obtained and collected in a stepwise manner, and the process stops when there is sufficient information to reach a conclusion. A possible limitation in the tiered approach is that information from a given source tends to be used only once within each step. However, the broad concept of IATA as presented here can include both the battery and tiered approaches. It can even include a combination of the two approaches—in principle, a tiered approach could be designed in which multiple information sources are used in each tier, and re-used in subsequent tiers, in such a way that increasing and more accurate types of information are used in succession.

As described above, IATA can be composed of many types of component parts, each if which may have been developed and optimised for a different purpose. It is therefore necessary to integrate the component parts on a rational basis. At one extreme, integration can be carried out "manually" for example by ordering several non-testing and testing approaches into a tiered testing and evaluation strategy. At the other extreme, mathematical and statistical algorithms can be used to optimise the use of data deriving from the component parts. Such integration tools include a wide range of machine learning approaches, including Bayesian statistics, Boolean statistics, regression modelling, principal components analysis, and neural networks. There are also intermediate solutions, for example an IATA could be designed "manually" as a tiered approach, but the prediction model applied at each tier is optimised by using machine learning methods. The mechanistic basis for such an IATA can be more or less apparent, for example, the tiers could follow the sequential order of the key events in an AOP, or they could simply make use of key events that are considered to be crucial and conserved across multiple AOPs related to the adverse outcome of interest. An interesting case concerns biologically-based models such as physiologically based biokinetic models (Gajewska et al. 2014) or systems biology models (Bhattacharya et al. 2012), which could also be regarded as a kind of IATA. In these models, the basic arrangement of the building blocks is based on the known architecture (physiology) of biological organisms, while mathematical algorithms are used to calibrate and simulate the spatiotemporal dynamics of molecular fluxes and biological perturbations at multiple physiological levels (from intracellular molecular networks to whole organisms).

In general terms, the development of an IATA can be thought of as an optimisation problem, with one or more optimisation criteria that depend on the problem formulation. From a regulatory perspective, the most usual criteria considered are the ability of the IATA to generate relevant and reliable results, and to reduce or replace animal testing. From an industry perspective, additional criteria could include costs, time and likelihood of regulatory acceptance. It is likely that there will need to be trade-offs between some of these criteria—for example, the generation of more reliable and relevant data may require the use of more expensive test systems. Thus, different solutions will be more or less suitable depending on the decision-making context, which means that an IATA and its component parts need to be "fit-for-purpose". This is a principle that has implications for the assessment of IATA performance.

The capacity of an IATA to generate relevant results can be judged on the basis of three considerations: predictivity, mechanistic basis, and protectiveness. These criteria are interlinked—in the absence of mechanistic understanding and mechanistically-based methods, confidence in the results generated by an alternative method will be based essentially on statistically derived performance measures (predictivity). In the design of validation studies, this implies the need for an extensive and heterogeneous test set of chemicals, which, in practice, is rarely available. However, with the advent of mechanistically-based *in vitro* methods, confidence can additionally be derived from the use of IATA with a sound mechanistic rationale and well-characterised applicability domain. The strengths and limitations of individual components are less important than the performance of the IATA as a whole. In addition, the degree of protectiveness is also an important consideration—for example, it is known that the Cramer classification scheme that is commonly used in the Threshold of Toxicological Concern (TTC) approach (see below) is not particularly predictive of repeat dose toxicity (since it does not discriminate well between chemicals of high, medium and low concern), but the approach is accepted by some regulatory authorities since it is known to be health-protective against the vast majority of chemicals, when judiciously applied.

The implications of costs, including direct financial costs, time to achieve regulatory acceptance, and animal welfare concerns can be factored into the design of IATA by using socioeconomic theory, and in particular approaches such as cost-benefit analysis (CBA), cost-effectiveness analysis (CEA), and value of information (VOI) analysis. In these approaches, socioeconomic "efficiency" increases if an equivalent degree of toxicological information on a chemical can be obtained with fewer monetary costs, in less time and with improved animal welfare (Gabbert and van Ierland 2010). VOI analysis has been applied to quantify the value of additional evidence from *in silico* methods in an IATA for skin sensitisation (Jaworska et al. 2010), and to determine efficient combinations of tests in an IATA for mutagenicity (Gabbert and Weikard 2013). The latter study demonstrates that the optimal order of tests in a sequential testing strategy depends on multiple mutually dependent factors such as prior information about a chemical, the diagnostic performance of the tests, monetary costs and the expected consequences of regulatory decision-making. On this basis, the authors concluded that neither the selection nor the order of tests can be pre-defined. Instead, it is concluded that the optimal IATA should be developed for each chemical individually depending on the information that is already available and the remaining knowledge gaps. In another study (Norlen et al. 2014), CEA was used to explore the cost-effectiveness of different batteries of alternative methods for predicting acute oral toxicity based on a set of four QSAR models and one *in vitro* method. The results confirmed that the *in silico* tools are more cost-effective than the *in vitro* test, but that batteries do not necessarily outperform single methods because additional information gains from the battery are easily outweighed by additional costs.

5 Validation of IATA

The need to validate IATA, and in particular the choice of validation approach, has been discussed extensively (Kinsner-Ovaskainen et al. 2009, 2012; Hartung et al. 2013). There are three main issues concerning the practical validation of IATA: (a) the extent to which a given IATA can in principle be validated, given that it can be more or less well defined in terms of one or more pre-defined prediction models; (b) the scientific feasibility of the exercise, given that most available data have generally been used already in developing the IATA and its component parts; and (c) the nature and formality of the validation process.

The first issue relates to the fact that only in rare cases does an IATA take the form of an unambiguous algorithm with well-defined inputs that are consistently translated into well-defined outputs. In such cases, the IATA can in principle be validated. In most cases, however, the "translational steps" within IATA are not completely and unambiguously defined, and multiple choices can be made by the user.

Thus, in this situation, only certain elements of the IATA (recently called "defined approaches") can be meaningfully validated, even though possible outcomes of applying of entire IATA can be simulated.

The second issue relates to the practical difficulty that even though an IATA can be associated with one or more unambiguous algorithms that are in principle "validatable", it is likely that most if not all of the available experimental data have been exhausted in discovering and optimising the underlying models. In particular, the availability of animal or human data for assessing the predictivity of adverse outcomes is likely to be limited. In addition, it is widely recognised that IATA need to be flexible, and therefore capable of evolving over time as new methods are developed, data generated, and experience gained. This begs the questions as to how much an IATA would need to change before a new validation or simulation exercise would be necessary.

The third issue concerns the formality and institutional character of the validation process. While validation is necessarily a scientific exercise, at one extreme it can be nothing more than this, resulting in a publication that should ideally have undergone some degree of independent peer review, while at the other extreme it can be an institutionalised process in which conclusions and/or official positions on the scientific validity of a method are taken by a committee and/or institution. The need to formalise and institutionalise the validation process is linked to the intended applications of the assessment method and the possible consequences of its use (in particular on public health, environmental safety, industrial competitiveness). *In vitro* toxicity tests that are intended to be used as replacements of animal studies in the regulatory assessment of chemicals typically undergo formal validation by a validation body such as EURL ECVAM, or one of its partners of the International Cooperation on Alternative Test Methods (ICATM), before being considered for regulatory acceptance (e.g. as an EU test method and/or OECD test guideline). In contrast, QSAR models are not subject to an official validation or adoption process, since they are typically developed and applied for various purposes in the safety assessment of chemicals, but not for the replacement of required animal toxicity tests (Worth 2010). The validation of *in vitro* tests and QSARs is further discussed in Chaps. 4 (Zuang et al.) and 6 (Patlewicz et al.), respectively. The validation of mechanistically based (*in vitro*) assays is further discussed in Chap. 8 (Andersen et al.).

5.1 Emerging Principles for the Evaluation of IATA

While there are different views concerning the practical validation of IATA, it is possible to identify a number of general principles that are broadly accepted as useful or essential when assessing the scientific robustness and applicability of IATA.

A project carried out under the auspices of the OECD Task Force for Hazard Assessment (TFHA) has developed guidance on the assessment of Defined Approaches to be used within IATA, including general evaluation principles and a reporting format to describe Defined Approaches in a transparent and systematic manner, according to the principles (OECD 2016). These principles refer to the need to:

(a) define the endpoint being assessed;
(b) define the purpose/application;
(c) describe the underlying rationale;
(d) describe the individual information sources used and how they are integrated to derive the final prediction;
(e) describe the predictive capacity, limitations and known uncertainties associated with the application of the approach.

The endpoint assessed refers to the property or toxicological effect that is generally used for decision making, for example in a traditional test guideline. Typically, this is the adverse outcome at the individual or population level.

The purpose refers to the intended application of the Defined Approach. In the regulatory setting, possible applications include priority setting, hazard identification/profiling, hazard classification and labelling, PBT and vPvB assessment, and risk assessment. In the industry setting, screening candidate chemicals and drugs would be an additional application.

The rationale according to which the Defined Approach is constructed should include an explanation of why particular component methods were chosen, and why they were combined in a certain way, in order to be fit for purpose. In the case of a mechanistically based approach, the choice of component methods and their method of integration should make reference to available knowledge on relevant AOPs. The component methods should also be described according to a consistent reporting format. For example, QSAR models can be summarised in terms of the QSAR Model Reporting Format (QMRF) as described in Chap. 6 (Patlewicz et al.), and efforts under the auspices of the OECD have resulted in guidance for how to characterise and document non-test guideline *in vitro* methods including high throughput screening and high content information assays (OECD 2014a).

In relation to the predictive performance of the Defined Approach, it is important to take into account the reliability of any reference test against which predictions are being compared. In particular, the variability of the animal test places an upper limit on the expected performance of any prediction model designed to predict the animal data or a (regulatory) classification based on such data (Worth and Cronin 2001; Hoffmann et al. 2010).

5.2 *Proposed Stepwise Approach to Evaluation of IATA*

Depending on the ultimate decision an IATA might be applied towards and the extent to which this is underpinned by mechanistic understanding such as that captured within an AOP, development of scientific confidence in the *in vitro* test methods and their associated prediction models is essential to assure that the use of this new kind of knowledge for decision making is both scientifically credible and relevant.

An alternative, albeit similar, framework to evaluating mechanistically based IATA has been proposed by Cox et al. known as the Scientific Confidence Framework (Cox et al. 2014; Becker et al. 2014). This is a hybridization of the two validation frameworks, the Institute of Medicine (IOM) framework (IOM 2010) and the OECD Validation principles (OECD 2007) that were both described and adapted for HT/HC assays by Patlewicz et al. (2013). The Scientific Confidence Framework contains three inter-related elements, Analytical Validation, Qualification and Utilization, which collectively enable systematic, transparent, and objective evaluation and documentation of the scientific confidence of *in vitro* test methods, their prediction models and their associated AOPs. Table 13.3 outlines how these elements would be characterised for the test methods and their prediction models.

A way of factoring AOPs into the Scientific Confidence Framework was proposed by Patlewicz et al. (2015) and is presented in Table 13.4. This provides a systematic approach to the construction and evaluation of AOPs as well as AOP-based IATA. Weight of evidence (WoE) evaluations are important in the development, analysis and application of AOPs, and these can be most readily accomplished by adapting the WoE procedures developed and employed in evaluating mode of action (Meek et al. 2014). Such evaluations will provide a measure of the maturity or completeness of the AOP which will in turn dictate the types of regulatory purposes that the AOP can be applied to and therefore the type of IATA that can be

Table 13.3 The major components of the scientific confidence framework for prediction models from *in vitro* test methods

Analytical validation	For each *in vitro* test method, there should be a defined mechanistic endpoint (e.g., the intermediate or key event in the mode of action or AOP), a defined chemical domain of applicability, documentation of assay performance characteristics (reliability, sensitivity, and specificity) and transparent data sets (to enable independent verification).
Qualification	This is an assessment of the prediction models derived from the *in vitro* screening assays. A defined algorithm for each prediction model is needed to ensure transparency. Appropriate measures of goodness-of-fit, robustness and predictivity of the prediction model need to be presented. Some prediction models may be quantitative, others may be qualitative. Known limitations of each prediction model should also be summarised. Prediction models should be characterised in sufficient detail to facilitate review, reconstruction and independent verification of results.
Utilization	This is a contextual and weight-of-evidence analysis on the (qualitative or quantitative) use of the prediction model for a given, specific purpose. This includes summarising results of the Analytical Validation and Qualification steps, defining the intended purpose of the prediction model and documenting/justifying applications, based on weight of evidence, where there is sufficient scientific confidence to support the use of the prediction model. The types of uses that the prediction models could be applied for include, but are not limited to: (1) priority setting, where the model is used to identify priority substances that will go on to more detailed evaluation; (2) screening level assessment of a biomarker, where the model is used as a surrogate data point for a biochemical endpoint or a biomarker; (3) integrated testing strategy, where the model is used to describe/predict a hazard property in lieu of conducting a traditional animal toxicity study or (4) to predict an adverse outcome

Table 13.4 Steps for incorporating the scientific confidence framework for prediction models from *in vitro* test methods into adverse outcome pathways

Step 1	Develop the AOP
Step 2	Develop new (or map existing) specific assays or *in silico* methods (such as QSARs) to key events within the AOP
Step 3	Conduct (or document) Analytical Validation of each assay
Step 4	Develop new (or map existing) models that predict a specific key event from one or more precursor key events (The input data for the prediction models comes from the assays described in Steps 2 and 3 above)
Step 5	Conduct (or document) Qualification of the prediction models
Step 6	Utilization: defining and documenting where there is sufficient scientific confidence to use one or more AOP-based prediction models for a specific purpose (e.g., priority setting, chemical category formation, integrated testing, predicting *in vivo* responses, etc.)
Step 7	For regulatory acceptance and use, processes need to be agreed upon and utilised to ensure robust and transparent review and determination of fit-for-purpose uses of AOPs. This should include dissemination of all necessary datasets, model parameters, algorithms, etc. to enable stakeholder review and comment, fully independent verification and independent scientific peer review. Whilst these processes have yet to be defined globally, in time, these should evolve to enable credible and transparent use of AOPs with sufficient scientific confidence by all stakeholders

constructed. Once an AOP has either been developed or identified, the next step is to map available test methods or *in silico* approaches such as QSARs to the AOP. This will help identify what practical tools can form the basis of an IATA. An IATA can therefore be considered as the roadmap to practically exploit the AOP by guiding the identification of relevant existing information for a given chemical or group of chemicals and targeting the generation of new information from appropriate test and non-test methods. The extent to which new information may need to be generated will depend on two factors, the quality of the information already available and more importantly its adequacy for the particular purpose of interest. Each test method, non-testing approach and associated prediction models need to be characterised and qualified to inform on the context of use. The steps form the basis of deriving an IATA toolbox from identifying what information needs to be gathered or generated for each of the elements within the IATA, and how the new information synthesised can be passed back into refining and improving the associated AOP.

6 Examples of IATA

In this section, we provide several examples of IATA to illustrate some of the similarities and differences between IATA. These IATA are summarised in Table 13.5.

IATA for skin irritation and corrosion represent tiered assessment approaches that were among the first IATA to be accepted by regulatory bodies. They are also relatively limited in terms of the endpoints predicted, and relatively well-defined in terms of how the component parts are combined and used. Many of these components are

Table 13.5 Different kinds of IATA

Application(s)	Endpoint(s)	References
Prioritisation of testing	Endocrine activity/disruption	Willett et al. (2011)
Hazard identification	Skin sensitisation	Nukada et al. (2013)
Classification and Labelling	Skin irritation and corrosion	OECD (2002), ECHA (2012) and UN (2013)
	Skin sensitisation	
Potency assessment	Skin sensitisation	Maxwell et al. (2014)
Risk assessment	Repeat dose toxicity/TTC approach	Cramer et al. (1978) and EFSA (2012)

officially accepted physicochemical and *in vitro* tests. These IATA are already used for the purposes of hazard identification and for classification and labelling.

The mechanistic basis of skin sensitisation is relatively mature and has most recently been summarised in the AOP as published by the OECD (OECD 2012). As a consequence, there have been many efforts to outline what an IATA may represent for skin sensitisation. For the early key events (haptenation, keratinocyte and dendritic cell activation), relevant *in chemico* or *in vitro* tests have been formally validated and have either undergone or are currently undergoing the process of regulatory acceptance. OECD Test Guidelines have been published for two tests which characterise the haptenation and keratinocyte activation (OECD 2015b, c). In the case of other key events (T cell activation and proliferation), *in vitro* tests are either lacking or not yet validated. A number of IATA for skin sensitisation have been proposed representing different kinds of assessment approaches, including tiered approaches (Nukada et al. 2013) and mathematical models based on Bayesian networks (Jaworska et al. 2013; Roberts and Patlewicz 2014) as well as IATA which exploit existing information from non-testing approaches (OECD 2014b; Patlewicz et al. 2014). These IATA are also being characterised as the basis for OECD guidance on the evaluation and application of IATA for skin sensitisation. In addition to their use in-house within industry, it is expected that some of these IATA will eventually be accepted by regulatory authorities for the purposes of hazard identification, classification and labelling, and in some cases potency (risk) assessment.

The Threshold of Toxicological Concern (TTC) approach is an example of a relatively fixed IATA based on the use of a decision tree that includes consideration of multiple endpoints (genotoxicity, carcinogenicity, neurotoxicity, and repeat-dose toxicities from different animal studies) and that depends on the availability of reliable exposure information (by the oral route). The TTC approach, by combining hazard (potency) information with exposure information, is potentially useful not only for the risk-based prioritisation of testing, but also for (preliminary or screening level) risk assessment. The approach is broadly accepted by the scientific community; however it has met with differing degrees of acceptance depending on the sector and intended use, i.e. whether it is applied to the assessment of chemicals in food or consumer products, industrial chemicals, residues of active ingredients in pesticide and biocide formulations, etc. (Dewhurst and Renwick 2013).

6.1 Skin Irritation and Corrosion

For both skin corrosion and irritation, a number of different testing and non-testing methods are available in addition to the traditional *in vivo* test (Draize rabbit test). Many of the available *in vitro* methods have undergone formal validation procedures and are internationally recognized (OECD Test Guidelines). IATA for skin corrosion and irritation have been published by regulatory bodies, including the OECD (2002), ECHA (2012) and the United Nations (2013). These provide guidance on how to combine information on physicochemical properties and QSARs with information from *in vitro* methods, the *in vivo* test and, where available, existing human information to support decisions on classification and labelling.

As an illustrative example, OECD Test Guideline 404 on the traditional Draize animal test (OECD 2002) contains a supplement which provides a "Testing and evaluation strategy for dermal irritation/corrosion". This supplement outlines a possible "Testing and Evaluation Strategy for Dermal Irritation/Corrosion", which was first proposed at an OECD workshop in 1996 and subsequently evaluated in relation to its ability to classify skin corrosives (Worth et al. 1998) and irritants (Hoffmann et al. 2010).

The strategy is based on the use of three main data sources: (a) available information as well structural and physicochemical data; (b) validated and accepted *in vitro*/*ex vivo* tests for skin corrosion and irritation; and (c) animal testing (i.e. traditional Draize rabbit test), if necessary. More specifically, it is a sequential approach containing eight steps that can be grouped according to the three data sources, as illustrated in Table 13.6.

Table 13.6 Stepwise testing and evaluation strategy for skin irritation and corrosion

Step	Description	Information source
1	Consideration of existing human or animal data with regard to skin corrosion/irritation	Available information (human or animal data and structural and physicochemical information)
2	Structure-Activity Relationship (SAR) data with regard to skin corrosion/irritation	
3	Use of pH measurements: extreme pH ($\leq$2 or $\geq$11.5) suggests corrosive properties	
4	Systemic toxicity data via the dermal route (can also be considered before steps 2 and 3)	
5	Validated and accepted *in vitro* or *ex vivo* test for skin corrosion	*In vitro* tests
6	Validated and accepted *in vitro* or *ex vivo* test for skin irritation	
7	TG404 test using one animal to test for corrosive effects	*In vivo* rabbit test
8	If not corrosive in step 7, testing using one or two animals to assess whether substance is corrosive or irritant or not	

6.2 Skin Sensitisation

As has been noted above, the mechanistic understanding surrounding skin sensitisation is sufficiently mature which has stimulated and resulted in the development of methods and approaches which characterise a number of the key events within the AOP. At the same time there is also a wealth of knowledge that has been derived from many *in silico* investigations. The latter have been reviewed elsewhere in more detail (Roberts and Patlewicz 2009; Roberts et al. 2008). As part of the AOP work programme and in particular the development of the OECD Toolbox (as described in Chap. 6), the AOP for skin sensitisation recently published by OECD was encoded and implemented into the OECD Toolbox to facilitate a read-across to be derived for a chemical of interest using the knowledge and data for the different key events (OECD 2014a, b). This implementation can be likened to a type of IATA that focuses solely on existing data and other non-testing approaches and marks a step change in terms of how read-across and other QSAR information can be applied. It also illustrates the sorts of QSARs that could be derived in future i.e., instead of predicting the apical endpoint or ultimate adverse outcome (AO), QSARs would be developed to model individual KEs. In the OECD implementation of the AOP for skin sensitisation, a set of so-named profilers were developed which encoded available knowledge derived for substances that had been tested in assays that characterised each of the KEs. These profilers are not necessarily direct predictors of the KEs or AOs but serve to provide a mechanistic basis for grouping chemicals to enable read-across. Profilers exist which encode known SAR information for chemicals tested in traditional *in vivo* methods such as the LLNA and GPMT as well as hypothetical SARs based on organic chemistry reaction principles (Aptula and Roberts 2006). These specific profilers are called the Protein Binding alerts by OASIS and OECD and until recently presented the only means of grouping chemicals within the Toolbox to read-across for *in vivo* sensitisation. There are also profilers for lysine and cysteine depletion based on data generated on chemicals tested in the DRPA as well as profilers for chemicals tested in the KeratinoSens™ assay (Emter et al. 2010), the human Cell Line Activation Test (h-CLAT) (Sakaguchi et al. 2007) and Myeloid U937 Skin Sensitisation Test (MUSST) (Python et al. 2007). A user starts by profiling their chemicals on the basis of protein binding alerts. If alerts are identified, then a stepwise approach of gathering data for each of the different events and investigating the feasibility of deriving a read-across prediction is evaluated. Specific thresholds to pass from one KE to the next have been encoded into the AOP implementation to help in the interpretation of the novel data. Depending on the decision being made and the availability of experimental data for the related analogues at each step of the IATA and the chemical under consideration, sufficient scientific confidence may have been reached at the MIE, alternatively other KEs may need to be considered into the evaluation. Figure 13.1 outlines the organisation of information within the Toolbox for the AOP.

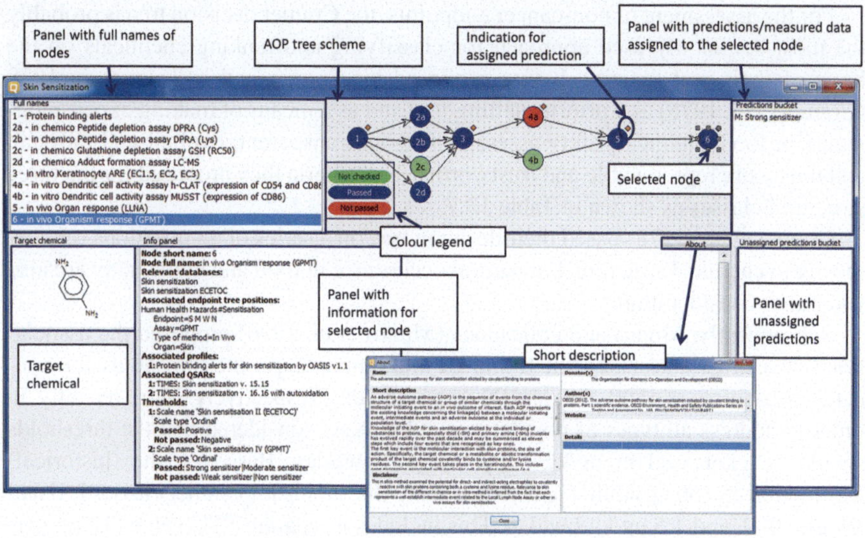

Fig. 13.1 Adverse outcome pathway for skin sensitisation in the OECD QSAR Toolbox

6.3 The Threshold of Toxicological Concern (TTC) Approach

The Threshold of Toxicological Concern (TTC) concept is based the fundamental principle that toxicity is related to the dose and duration of exposure. It is based on empirical evidence that for non-cancer effects, there are thresholds below which toxicity does not occur, whereas for cancer effects, the likelihood of tumours is zero to very small at very low exposure levels. Thus, for chemicals of unknown toxicity, human exposure thresholds can be established below which there is a low probability of adverse effects on health. Accordingly, a range of human exposure thresholds (TTC values) have been developed for both cancer and non-cancer endpoints, on the basis of data from extensive toxicological testing in animals (Hennes 2012).

The combined and stepwise use of TTC values, which may be termed the TTC approach, can be used to assess substances of unknown toxicity present at low levels in the diet or consumer products. Application of the TTC approach requires only knowledge of the chemical structure of the substance concerned and reliable information on human exposure. The extent to which the TTC is accepted depends on the regulatory application and context. In general, the approach is better accepted for the assessment of non-intentionally added substances, such as contaminants, reaction byproducts, and metabolites, for which experimental toxicity data are not available and consumer exposure is low, compared to the TTC threshold.

For the assessment of genotoxic carcinogens, a practical approach has been proposed for assessment of genotoxic carcinogens based on a TTC value of 0.15 μg/person/day, whereas non-genotoxic carcinogens may be assessed using higher TTC values.

For the assessment of non-cancer endpoints, the Cramer decision tree is probably the most commonly used approach for classifying and ranking chemicals on the basis of their oral toxicity. It was proposed by Cramer and colleagues in 1978 (Cramer et al. 1978) as a priority setting tool and as a means of making expert judgments in food chemical safety assessment more transparent, explicit and rational, and thus more reproducible and trustworthy. The criteria they proposed for the three structural classes as shown in Table 13.7.

Cramer et al. (1978) based their decision tree on a series of 33 questions relating mostly to chemical structure, but natural occurrence in food and in the body are also taken into consideration.

Subsequently, Munro and colleagues (Munro et al. 1996) proposed the association between Cramer classes I, II and III and human exposure thresholds for non-cancer endpoints of 1800, 540 and 90 μg/person/day, respectively. More recently, in order to address all types of populations, it has been considered that the thresholds should be expressed in μg/kg body weight (bw)/day. Based on the (historical) assumption of a 60 kg adult, the corresponding thresholds for Cramer classes I, II and III are 30, 9, and 1.5 μg/kg bw/day. This includes a separate TTC value (18 μg/person/day or 0.3 μg/kg bw/day) for organophosphate and carbamate neurotoxicants.

Taking into account these historical developments, along with some widely accepted exclusion categories of chemicals for which the TTC approach is not considered applicable, EFSA subsequently published a generic scheme for the application of the TTC approach (EFSA 2012). In this Chapter, the TTC approach as represented by the EFSA decision tree (Fig. 13.2) is regarded as an IATA—one that integrates the use of exposure (dietary intake) information with predictions of genotoxicity, carcinogenicity, neurotoxicity, and repeat dose toxicity.

It is worth noting that the Cramer scheme was proposed in the late 1970s, before the development of what is now understood by the TTC approach and before the advent of computer-based tools for interpreting chemical structure and applying structure-activity relationships. Since the original publication, the Cramer classification scheme has been implemented into freely available software tools such as Toxtree (Patlewicz et al. 2014; Lapenna and Worth 2011) (http://toxtree.

Table 13.7 Cramer classification scheme and associated TTC values

Cramer class	Description	TTC value (μg/kg bw/day)
Class I	Substances with simple chemical structures and for which efficient modes of metabolism exist, suggesting a low order of oral toxicity	1800 μg/person/day 30 μg/kg bw/day
Class II	Substances which possess structures that are less innocuous than class I substances, but do not contain structural features suggestive of toxicity like those substances in class III	540 μg/person/day 9 μg/kg bw/day
Class III	Substances with chemical structures that permit no strong initial presumption of safety or may even suggest significant toxicity or have reactive functional groups	90 μg/person/day 1.5 μg/kg bw/day

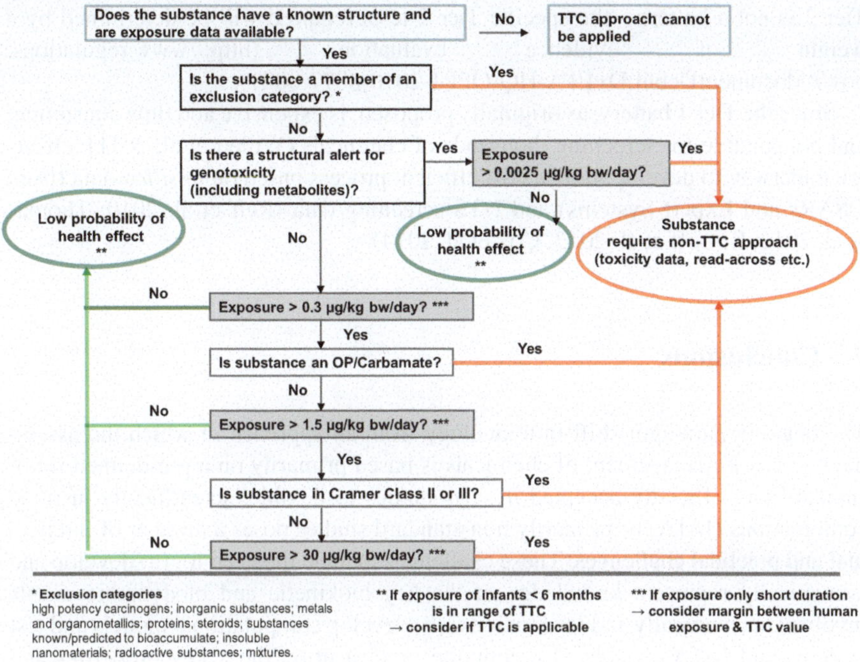

Fig. 13.2 Generic scheme for the application of the TTC approach (reproduced with permission from EFSA 2012)

sourceforge.net/index.html) and the OECD QSAR Toolbox (http://www.qsartoolbox.org/). Moreover, there have been various proposals in the scientific literature to refine the TTC approach (Tluczkiewicz et al. 2011; Kalkhof et al. 2012). This is therefore an example of an IATA which is evolving and being adapted for use in different sectors (e.g. food, cosmetics, pesticides, chemicals).

6.4 Identification of Endocrine Active Substances

With a view to identifying substances with the potential to interact with components of the endocrine system and, then, for substances with such potential, to identify dose response of adverse effects for risk assessment, the U.S. Environmental Protection Agency (EPA) launched an Endocrine Disruptor Screening Program (EDSP) in 2009. The EDSP utilizes a two-tiered approach. The Tier 1 battery consists of five *in vitro* and six *in vivo* assays that are intended to determine the potential of a chemical to interact with the estrogen (E), androgen (A), or thyroid (T) hormone pathways. Tier 2 is proposed to consist of multigenerational reproductive and developmental toxicity tests in several species and is intended to determine whether a chemical can cause adverse effects resulting from E, A, or T modulation. EDSP

Tier 2 is not a battery—the specific Tier 2 tests required will be determined by a weight of evidence evaluation (http://www.regulations.gov/#!documentDetail;D=EPA-HQ-OPPT-2010-0877-0021).

Since the Tier 1 battery, as originally proposed, is expensive and time consuming and not suitable for screening thousands of chemicals (Willett et al. 2011), efforts are underway to develop more a cost efficient process based on *in silico* data (from QSARs and Expert Systems) and HTS screening data (Reif et al. 2010; Thomas et al. 2012; Rotroff et al. 2013; Cox et al. 2014).

7 Conclusions

The ongoing paradigm shift in toxicology from an approach in which the assessment and risk management of chemicals is based primarily on a pre-defined set of standard and officially accepted *in vivo* studies to flexible, scientifically-justified combinations (IATA) of primarily non-standard studies poses a number of intellectual and practical challenges. These challenges include the need to: (a) develop and systemically represent knowledge of the key biokinetic and biodynamic events involved in chemically-induced toxicity; (b) develop computational models and test systems and IATA capable of computing or measuring these key properties and effects; (c) design IATA that integrate such computational models and test systems in a credible and practical way; and (d) generate sufficient evidence to convince regulators, product stewards, and other decision makers that a given IATA is fit for its intended purpose.

In developing integrated approaches for regulatory decision making, it is useful to distinguish between activities aimed primarily at knowledge generation and capture; the development and validation of models, *in vitro* tests and IATA; and their application in (regulatory) decision making. That said, there is inevitably an interplay between these three activity streams (Fig. 13.3). For example, AOP development should be regarded as an ongoing process, based on the evolving knowledge of key events and their interrelationships with each other and adverse outcomes of interest. Even partial knowledge of the AOP(s) underlying a given adverse outcome may be sufficient to motivate the design of mechanistically-based IATA, which should then be applied in order to gain practical experience. This experience will likely lead to refinements of the IATA, for example to incorporate new components that expand the biological and chemical applicability domains, or to recalibrate prediction models for improved accuracy of prediction. At the same time, the practical application of IATA should enable important knowledge gaps to be pinpointed, thereby setting the scene for the development of tailor-made and test systems that target key mechanisms of toxicological action. Within this iterative cycle, validation of the component parts is a key consideration, but the overriding principle is the IATA as a whole that should be fit for purpose, and from this perspective, multiple and different solutions could be equivalent. The role of AOPs in informing the development of IATA for regulatory purposes is further discussed by Tollefsen et al. (2014).

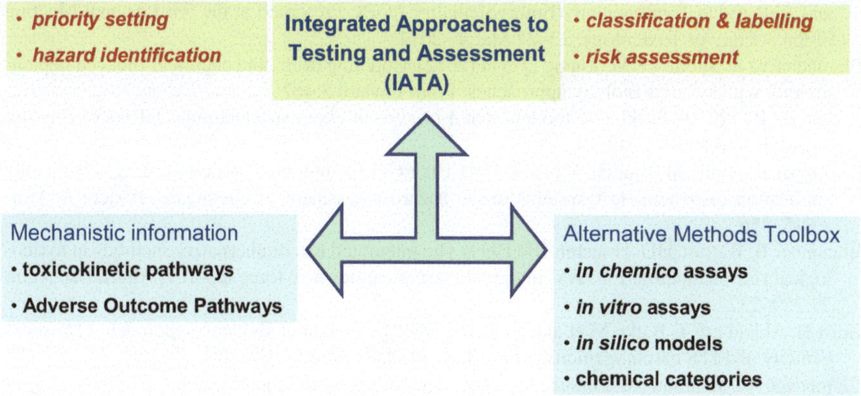

Fig. 13.3 Generation of mechanistic knowledge and its use in guiding the development of alternative methods and design of testing strategies

The constantly shifting landscape of IATA, based not only on evolving knowledge and technologies, but on different preferences for data integration, clearly poses a challenge for regulatory acceptance, which has traditionally been based on the adoption of relatively fixed solutions such as test guidelines. Documenting and communicating scientific confidence in IATA is therefore key. To address this challenge, more flexible approaches to validation and acceptance are needed. A step in this direction has already been taken by the OECD, which through its TFHA, is developing non-prescriptive guidance on the evaluation of Defined Approaches to be used within IATA. If this model proves successful, it could be expanded to establish an international forum for exchanging experience on IATA, thereby facilitating, to the extent possible, the development of harmonised approaches.

Acknowledgement The authors would like to thank Rick Becker (American Chemistry Council, Washington DC, USA) for critically reviewing this work.

References

Ahlers J, Stock F, Werschkun B (2008) Integrated testing and intelligent assessment-new challenges under REACH. Environ Sci Pollut Res Int 15:565–572

Aptula AO, Roberts DW (2006) Mechanistic applicability domains for nonanimal-based prediction of toxicological end points: general principles and application to reactive toxicity. Chem Res Toxicol 19:1097–1105

Aptula AO, Patlewicz G, Roberts DW, Schultz TW (2006) Non-enzymatic glutathione reactivity and *in vitro* toxicity: a non-animal approach to skin sensitization. Toxicol *In Vitro* 20:239–247

Becker RA, Simon T, Patlewicz G, Kennedy SW, Farhat A, Budinsky R (2014) Improving the development of adverse outcome pathways: lessons learned from the AhR Rodent Liver Tumor and AhR Avian Teratogenicity/Embryolethality AOPs. Presented at the 53rd Annual Meeting of the Society of Toxicology, 23–27 March, 2014

Bhattacharya S, Shoda LKM, Zhang Q et al (2012) Modeling drug- and chemical-induced hepatotoxicity with systems biology approaches. Front Physiol 3:462

Blaauboer BJ (2010) Biokinetic modeling and *in vitro-in vivo* extrapolations. J Toxicol Environ Health B Crit Rev 13:242–252

Blaauboer BJ, Balls M, Bianchi V et al (1994) The ECITTS integrated toxicity testing scheme: the application of *in vitro* test systems to the hazard assessment of chemicals. Toxicol *In Vitro* 8:845–856

Blaauboer B, Barratt MD, Houston JB (1999) The integrated use of alternative methods in toxicological risk evaluation. ECVAM integrated test strategies task force report 1. Altern Lab Anim 27:229–237

Buist H, Aldenberg T, Batke M et al (2013) The OSIRIS Weight of Evidence approach: ITS mutagenicity and ITS carcinogenicity. Regul Toxicol Pharmacol 67:170–181

Clemedson C, Kolman A, Forsby A (2007) The Integrated Acute Systemic Toxicity project (ACuteTox) for the optimisation and validation of alternative *in vitro* tests. Altern Lab Anim 35:33–38

Council of Canadian Academies (2012) Integrating emerging technologies into chemical safety assessment. http://www.scienceadvice.ca/en/assessments/completed/pesticides.aspx

Cox LA, Douglas D, Marty S, Rowlands JC, Patlewicz G, Goyak KO, Becker RA (2014) Developing scientific in HTS-derived prediction models for endocrine endpoints: lessons learned from an endocrine case study. Regul Toxicol Pharmacol 69:443–450

Cramer GM, Ford RA, Hall RL (1978) Estimation of toxic hazard—a decision tree approach. Food Cosmet Toxicol 16:255–276

Dejongh J, Forsby A, Houston JB et al (1999) An Integrated Approach to the Prediction of Systemic Toxicity using Computer-based Biokinetic Models and Biological *In vitro* Test Methods: Overview of a Prevalidation Study Based on the ECITTS Project. Toxicol *In Vitro* 13:549–554

De Wever B, Fuchs HW, Gaca M et al (2012) Implementation challenges for designing integrated *in vitro* testing strategies (ITS) aiming at reducing and replacing animal experimentation. Toxicol *In Vitro* 26:526–534

Dewhurst I, Renwick AG (2013) Evaluation of the Threshold of Toxicological Concern (TTC)—challenges and approaches. Regul Toxicol Pharmacol 65:168–177

ECHA (2012) Guidance on information requirements and chemical safety assessment. Chapter R.7a: Endpoint specific guidance. In: Guidance for the implementation of REACH. Version 2.0. November 2012. http://echa.europa.eu/documents/10162/13632/information_requirements_r7a_en.pdf

EFSA (2012) Scientific Opinion on exploring options for providing advice about possible human health risks based on the concept of Threshold of Toxicological Concern (TTC). EFSA J 10(7):2750, European Food Safety Authority. http://www.efsa.europa.eu/en/efsajournal/doc/2750.pdf

Emter R, Ellis G, Natsch A (2010) Performance of a novel keratinocyte-based reporter cell line to screen skin sensitisers *in vitro*. Toxicol Appl Pharmacol 245:281–290

Gabbert S, van Ierland EC (2010) Cost-effectiveness analysis of chemical testing for decision-support: how to include animal welfare? Hum Ecol Risk Assess 16(3):603–620

Gabbert S, Weikard H-P (2013) Sequential testing of chemicals when costs matter: a value of information approach. Hum Ecol Risk Assess An Int J 19:1067–1088

Gajewska M, Worth A, Urani C, Briesen H, Schramm K-W (2014) Application of physiologically-based toxicokinetic modelling in oral-to-dermal extrapolation of threshold doses of cosmetic ingredients. Toxicol Lett 227:189–202

Gerberick GF, Vassallo JD, Bailey RE et al (2004) Development of a peptide reactivity assay for screening contact allergens. Toxicol Sci 81:332–343

Gerberick GF, Vassallo JD, Foertsch LM et al (2007) Quantification of chemical peptide reactivity for screening contact allergens: a classification tree model approach. Toxicol Sci 97:417–427

Grindon C, Combes R, Cronin MTD et al (2008) Integrated testing strategies for use with respect to the requirements of the EU REACH legislation. Altern Lab Anim 36(Suppl 1):7–27

Hartung T, Luechtefeld T, Maertens A, Kleensang A (2013) Integrated testing strategies for safety assessments. ALTEX 30:3–18

Hennes EC (2012) An overview of values for the threshold of toxicological concern. Toxicol Lett 211:296–303

Hoffmann S, Kinsner-Ovaskainen A, Prieto P et al (2010) Acute oral toxicity: variability, reliability, relevance and interspecies comparison of rodent LD50 data from literature surveyed for the ACuteTox project. Regul Toxicol Pharmacol 58:395–407

IOM (2010) Evaluation of biomarkers and surrogate endpoints in chronic disease. Institute of Medicine, Washington, DC. ISBN 978-0-309-15129-0

Jaworska J, Hoffmann S (2010) Integrated Testing Strategy (ITS)—Opportunities to better use existing data and guide future testing in toxicology. ALTEX 27:231–242

Jaworska J, Gabbert S, Aldenberg T (2010) Towards optimization of chemical testing under REACH: a Bayesian network approach to Integrated Testing Strategies. Regul Toxicol Pharmacol 57:157–167

Jaworska J, Dancik Y, Kern P et al (2013) Bayesian integrated testing strategy to assess skin sensitization potency: from theory to practice. J Appl Toxicol 33:1353–1364

Kalkhof H, Herzler M, Stahlmann R, Gundert-Remy U (2012) Threshold of toxicological concern values for non-genotoxic effects in industrial chemicals: re-evaluation of the Cramer classification. Arch Toxicol 86:17–25

Kinsner-Ovaskainen A, Akkan Z, Casati S et al (2009) Overcoming barriers to validation of non-animal partial replacement methods/Integrated Testing Strategies: the report of an EPAA-ECVAM workshop. Altern Lab Anim 37:437–444

Kinsner-Ovaskainen A, Maxwell G, Kreysa J et al (2012) Report of the EPAA-ECVAM workshop on the validation of Integrated Testing Strategies (ITS). Altern Lab Anim 40:175–181

Lapenna S, Worth A (2011) Analysis of the Cramer classification scheme for oral systemic toxicity—implications for its implementation in Toxtree. JRC Scientific and Technical Report EUR 24898 EN. Publications Office of the European Union, Luxembourg. http://publications.jrc.ec.europa.eu/repository/

Marx-Stoelting P et al (2009) A review of the implementation of the embryonic stem cell test (EST). The report and recommendations of an ECVAM_ReProTect Workshop. Altern Lab Anim 37:313–328

Maxwell G, MacKay C, Cubberley R et al (2014) Applying the skin sensitisation adverse outcome pathway (AOP) to quantitative risk assessment. Toxicol *In Vitro* 28:8–12

Meek ME, Boobis A, Cote I et al (2014) New developments in the evolution and application of the WHO/IPCS framework on mode of action/species concordance analysis. J Appl Toxicol 34:1–18

Munro IC, Ford RA, Kennepohl E, Sprenger JG (1996) Correlation of structural class with no-observed-effect levels: a proposal for establishing a threshold of concern. Food Chem Toxicol 34:829–867

Norlen H, Worth AP, Gabbert S (2014) A tutorial for analysing the cost-effectiveness of alternative methods for assessing chemical toxicity: the case of acute oral toxicity prediction. Altern Lab Anim 42:115–127

Nukada Y, Miyazawa M, Kazutoshi S et al (2013) Data integration of non-animal tests for the development of a test battery to predict the skin sensitizing potential and potency of chemicals. Toxicol *In Vitro* 27:609–6188

Nel AE, Nasser E, Godwin H et al (2013) A multi-stakeholder perspective on the use of alternative test strategies for nanomaterial safety assessment. ACS Nano. 7:6422–6433

NRC (2007) Toxicity testing in the 21st century: a vision and a strategy. National Academic Press, Washington, DC. http://www.nap.edu/read/11970/chapter/1

OECD (2002) Test No. 404: Acute Dermal Irritation/Corrosion, OECD Guidelines for the Testing of Chemicals, Section 4. http://www.oecd.org/env/testguidelines
OECD (2007) Guidance document on the validation of (Quantitative) structure-activity relationships [(Q)SAR] models. ENV/JM/MONO(2007)2. http://www.oecd.org/officialdocuments/displaydocumentpdf?cote=env/jm/mono(2007)2&doclanguage=en
OECD (2012) The adverse outcome pathway for skin sensitisation initiated by covalent binding to proteins part 1: scientific evidence. Series on Testing and Assessment No. 168. ENV/JM/MONO(2012)10/PART1. http://www.oecd.org/chemicalsafety/testing/seriesontestingandassessmentpublicationsbynumber.htm
OECD (2013) Guidance document on developing and assessing adverse outcome pathways. OECD Environment, Health and Safety Publications. Series on Testing and Assessment No. 184. ENV/JM/MONO(2013)6. http://www.oecd.org/chemicalsafety/testing/seriesontestingandassessmentpublicationsbynumber.htm
OECD (2014a) Guidance document for describing non-guideline *in vitro* test methods. Series on Testing and Assessment no.211. ENV/JM/MONO(2014)35. http://www.oecd.org/chemicalsafety/testing/seriesontestingandassessmentpublicationsbynumber.htm
OECD (2014b) How to use the Toolbox AOP workflow for Skin Sensitization. http://www.oecd.org/env/ehs/risk-assessment/Tutorial_1_How%20to%20use%20AOP%20for%20Skin%20sensitization_F_28012014.pdf
OECD (2015a) Report of the workshop on a Framework for the development and use of Integrated Approaches to Testing and Assessment. ENV/JM/HA(2015)1
OECD (2015b) Test Guideline 442c: *in chemico* skin sensitisation (Direct Peptide Reactivity Assay DPRA). http://www.oecd.org/chemicalsafety/testing/oecdguidelinesforthetestingofchemicals.htm
OECD (2015c) Test Guideline 442d: *in vitro* skin sensitisation (ARE-Nrf2 luciferase test method). http://www.oecd.org/chemicalsafety/testing/oecdguidelinesforthetestingofchemicals.htm
OECD (2016) Guidance Document on the Reporting of Defined Approaches to be used within Integrated Approaches to Testing and Assessment. ENV/JM/HA(2016)10
Oomen AG, Bos PMJ, Fernandes TF et al (2014) Concern-driven integrated approaches to nanomaterial testing and assessment-report of the NanoSafety Cluster Working Group 10. Nanotoxicology 8:334–348
Patlewicz G, Simon T, Goyak K et al (2013) Use and validation of HT/HC assays to support 21st century toxicity evaluations. Regul Toxicol Pharmacol 65:259–268
Patlewicz G, Kuseva C, Kesova A, Popova I, Zhechev T, Pavlov T, Roberts DW, Mekenyan OM (2014) Towards AOP application—implementation of an integrated approach to testing and assessment (IATA) into a pipeline tool for skin sensitization. Regul Toxicol Pharmacol 69:529–545
Patlewicz G, Simon TW, Rowlands JC, Budinsky RA, Becker RA (2015) Proposing a scientific confidence framework to help support the application of adverse outcome pathways for regulatory purposes. Regul Toxicol Pharmacol 71:463–477
Piersma AH, Bosgra S, van Duursen MBM et al (2013) Evaluation of an alternative *in vitro* test battery for detecting reproductive toxicants. Reprod Toxicol 38:53–64
Python F, Goebel C, Aeby P (2007) Assessment of the U937 cell line for the detection of contact allergens. Toxicol Appl Pharmacol 220(2):113–124
Reif DM, Martin MT, Tan SW et al (2010) Endocrine profiling and prioritization of environmental chemicals using ToxCast data. Environ Health Perspect 118:1714–1720
Roberts DW, Patlewicz G (2009) Chemistry based non-animal predictive modeling for skin sensitization. In: Devillers J (ed) Ecotoxicology modeling. Springer, Heidelberg, pp 61–83
Roberts DW, Patlewicz GY (2014) Integrated testing and assessment approaches for skin sensitization: a commentary. J Appl Toxicol 34(4):436–440
Roberts DW, Aptula AO, Patlewicz G, Pease C (2008) Chemical reactivity indices and mechanism-based read-across for non-animal based assessment of skin sensitisation potential. J Appl Toxicol 28:443–454

Rorije E, Aldenberg T, Buist H et al (2013) The OSIRIS Weight of Evidence approach: ITS for skin sensitisation. Regul Toxicol Pharmacol 67:146–156

Rotroff DM, Dix DJ, Houck KA, Knudsen TB, Martin MT, McLaurin KW, Reif DM, Crofton KM, Singh AV, Xia M, Huang R, Judson RS (2013) Using *in vitro* high throughput screening assays to identify potential endocrine-disrupting chemicals. Environ Health Perspect 121:7–14

Rovida C, Roggen EL (2007) Management of an Integrated Project (Sens-it-iv) to develop *in vitro* tests to assess sensitisation. Altern Lab Anim 35:317–322

Rowbotham AL, Gibson RM (2011) Exposure-driven risk assessment: applying exposure-based waiving of toxicity tests under REACH. Food Chem Toxicol 49:1661–1673

Sakaguchi H, Ashikaga T, Kosaka N, Sono S, Nishiyama N, Itagaki H (2007) The *in vitro* skin sensitization test; human cell line activation test (h-CLAT) using THP-1 cells. Toxicol Letts 172:S93

Schaafsma G, Kroese ED, Tielemans EL et al (2009) REACH, non-testing approaches and the urgent need for a change in mind set. Regul Toxicol Pharmacol 53:70–80

Schultz TW, Yarbrough JW, Johnson EL (2005) Structure-activity relationships for reactivity of carbonyl-containing compounds with glutathione. SAR QSAR Environ Res 16:313–322

Stone V, Pozzi-Mucelli S, Tran L et al (2014) ITS-NANO—Prioritising nanosafety research to develop a stakeholder driven intelligent testing strategy. Part Fibre Toxicol 11:9

Thomas RS, Black MB, Li L, Healy E, Chu TM, Bao W, Andersen ME, Wolfinger RD (2012) A comprehensive statistical analysis of predicting *in vivo* hazard using high-throughput *in vitro* screening. Toxicol Sci 128:398–417

Thomas RS, Philbert MA, Auerbach SS et al (2013) Incorporating new technologies into toxicity testing and risk assessment: moving from 21st century vision to a data-driven framework. Toxicol Sci 136:4–18

Tluczkiewicz I, Buist HE, Martin MT et al (2011) Improvement of the Cramer classification for oral exposure using the database TTC RepDose—a strategy description. Regul Toxicol Pharmacol 61:340–350

Tluczkiewicz I, Batke M, Kroese D et al (2013) The OSIRIS Weight of Evidence approach: ITS for the endpoints repeated-dose toxicity (RepDose ITS). Regul Toxicol Pharmacol 67:157–169

Tollefsen KE, Scholz S, Cronin MT, Edwards SW, de Knecht J, Crofton K, Garcia-Reyero N, Hartung T, Worth A, Patlewicz G (2014) Applying Adverse Outcome Pathways (AOPs) to support Integrated Approaches to Testing and Assessment (IATA). Regul Toxicol Pharmacol 70:629–640

United Nations (2013) Report of the Committee of Experts on the Transport of Dangerous Goods and on the Globally Harmonized System of Classification and Labelling of Chemicals on its sixth session: amendments to the fourth revised edition of the Globally Harmonized System of Classification and Labelling of Chemicals (GHS) (ST/SG/AC.10/30/Rev.4). http://www.unece.org/fileadmin/DAM/trans/doc/2013/dgac10/ST-SG-AC10-40a3e.pdf

Van Leeuwen CJ, Patlewicz GY, Worth AP (2007) Intelligent testing strategies. In: van Leeuwen CJ, Vermeire TG (eds) Risk assessment of chemicals. An introduction, 2nd edn. Springer, Heidelberg, pp 467–509

Vermeire T, van de Bovenkamp M, de Bruin YB et al (2010) Exposure-based waiving under REACH. Regul Toxicol Pharmacol 58:408–420

Vermeire T, Aldenberg T, Buist H et al (2013) OSIRIS, a quest for proof of principle for integrated testing strategies of chemicals for four human health endpoints. Regul Toxicol Pharmacol 67:136–145

Willett CE, Bishop PL, Sullivan KM (2011) Application of an integrated testing strategy to the U.S. EPA endocrine disruptor screening program. Toxicol Sci 123:15–25

Worth AP (2000) The integrated use of physicochemical and *in vitro* data for predicting chemical toxicity. PhD thesis, Liverpool John Moores University

Worth AP (2004) The tiered approach to toxicity assessment based on the integrated use of alternative (non-animal) tests. In: Cronin MTD, Livingstone D (eds) Predicting chemical toxicity and fate. CRC Press, Boca Raton, pp 389–410

Worth AP (2010) The role of QSAR methodology in the regulatory assessment of chemicals. In: Puzyn T, Leszczynski J, Cronin MTD (eds) Recent advances in QSAR studies: methods and applications. Springer, Heidelberg, pp 367–382
Worth AP, Balls M (2001) The importance of the prediction model in the validation of alternative tests. Altern Lab Anim 29:135–144
Worth AP, Cronin MT (2001) The use of bootstrap resampling to assess the variability of Draize tissue scores. Altern Lab Anim 29:557–573
Worth AP, Fentem JH (1999) A general approach for evaluating stepwise testing strategies. Altern Lab Anim 27:161–177
Worth AP, Fentem JH, Balls M, Botham PA, Curren RD, Earl LK, Esdaile DJ, Liebsch M (1998) An evaluation of the proposed OECD testing strategy for skin corrosion. Altern Lab Anim 26:709–720

Chapter 14
International Harmonization and Cooperation in the Validation of Alternative Methods

João Barroso, Il Young Ahn, Cristiane Caldeira, Paul L. Carmichael, Warren Casey, Sandra Coecke, Rodger Curren, Bertrand Desprez, Chantra Eskes, Claudius Griesinger, Jiabin Guo, Erin Hill, Annett Janusch Roi, Hajime Kojima, Jin Li, Chae Hyung Lim, Wlamir Moura, Akiyoshi Nishikawa, HyeKyung Park, Shuangqing Peng, Octavio Presgrave, Tim Singer, Soo Jung Sohn, Carl Westmoreland, Maurice Whelan, Xingfen Yang, Ying Yang and Valérie Zuang

Abstract The development and validation of scientific alternatives to animal testing is important not only from an ethical perspective (implementation of 3Rs), but also to improve safety assessment decision making with the use of mechanistic information of higher relevance to humans. To be effective in these efforts, it is however imperative that validation centres, industry, regulatory bodies, academia and other interested parties ensure a strong international cooperation, cross-sector collaboration and intense communication in the design, execution, and peer review of validation studies. Such an approach is critical to achieve harmonized and more

J. Barroso (✉) • S. Coecke • B. Desprez • C. Griesinger • A.J. Roi • M. Whelan • V. Zuang
European Commission, Joint Research Centre (JRC), Ispra, Italy
e-mail: joao.barroso@ec.europa.eu

I.Y. Ahn • C.H. Lim • H. Park • S.J. Sohn
Toxicological Evaluation and Research Department, Korean Center for the Validation of Alternative Methods (KoCVAM), National Institute of Food and Drug Safety Evaluation, Cheongju-si, South Korea

C. Caldeira • W. Moura • O. Presgrave
Brazilian Center for Validation of Alternative Methods (BraCVAM) and National Institute of Quality Control in Health (INCQS), Rio de Janeiro, Brazil

P.L. Carmichael • J. Li • C. Westmoreland
Unilever Safety and Environmental Assurance Centre, Colworth Science Park, Sharnbrook, Bedfordshire, UK

C. Eskes, M. Whelan (eds.), *Validation of Alternative Methods for Toxicity Testing*, Advances in Experimental Medicine and Biology 856,
DOI 10.1007/978-3-319-33826-2_14

transparent approaches to method validation, peer-review and recommendation, which will ultimately expedite the international acceptance of valid alternative methods or strategies by regulatory authorities and their implementation and use by stakeholders. It also allows achieving greater efficiency and effectiveness by avoiding duplication of effort and leveraging limited resources. In view of achieving these goals, the International Cooperation on Alternative Test Methods (ICATM) was established in 2009 by validation centres from Europe, USA, Canada and Japan. ICATM was later joined by Korea in 2011 and currently also counts with Brazil and China as observers. This chapter describes the existing differences across world regions and major efforts carried out for achieving consistent international cooperation and harmonization in the validation and adoption of alternative approaches to animal testing.

Keywords International cooperation • Harmonization • ICATM • Validation • Alternative methods • ECVAM • ICCVAM • NICEATM • JaCVAM • Health Canada • KoCVAM • BraCVAM • CFDA

W. Casey
Division of the National Toxicology Program, National Institute of Environmental Health Sciences, Research Triangle Park, DC, USA

Interagency Coordinating Committee on the Validation of Alternative Methods (ICCVAM), Washington, DC, USA

R. Curren • E. Hill
Institute for *In Vitro* Sciences, Inc., Gaithersburg, MD, USA

C. Eskes
Services and Consultation on Alternative Methods (SeCAM), Magliaso, Switzerland

J. Guo • S. Peng
Evaluation and Research Centre for Toxicology, Institute of Disease Control and Prevention, Academy of Military Medical Sciences, Beijing, China

H. Kojima • A. Nishikawa
Japanasese Center for the Validation of Alternative Methods (JaCVAM), National Institute of Health Sciences, Tokyo, Japan

T. Singer
Environmental Health Science and Research Bureau, Health Canada, Ottawa, Canada

X. Yang • Y. Yang
Guangdong Province Centre for Disease Control and Prevention, Guangzhou, China

Abbreviations

AM	Alternative method
ANVISA	Brazil National Sanitary Agency
BraCVAM	Brazilian Center for Validation of Alternative Methods
BSRC	Biological Safety Research Center
CDC	Centre for Disease Control and Prevention
CFDA	China Food Drug Administration
CIQ	China Inspection and Quarantine Bureaux
CONCEA	Brazilian National Council of the Control of Animal Experimentation (Conselho Nacional de Controle da Experimentação Animal)
ESAC	EURL ECVAM Scientific Advisory Committee
ESTAF	EURL ECVAM Stakeholder Forum
EU-NETVAL	European Union Network of Laboratories for the Validation of Alternative Methods
EURL ECVAM	European Union Reference Laboratory for Alternatives to Animal Testing
GLP	Good Laboratory Practice
HC	Health Canada
ICATM	International Cooperation on Alternative Test Methods
ICCR	International Cooperation on Cosmetics Regulation
ICCVAM	U.S. Interagency Coordinating Committee on the Validation of Alternative Methods
ICH	International Conference on Harmonization of Technical Requirements for Registration of Pharmaceuticals for Human Use
JaCVAM	Japanese Center for the Validation of Alternative Methods
JRC	European Commission Joint Research Centre
KoCVAM	Korean Center for the Validation of Alternative Methods
MAD	Mutual Acceptance of Data
MFDS	Ministry of Food and Drug Safety
MHLW	Ministry of Health, Labour and Welfare
MoC	Memorandum of Cooperation
MoST	Chinese Ministry of Science and Technology
NC	National Coordinator for the OECD Test Guidelines Programme
NICEATM	U.S. National Toxicology Program Interagency Center for the Evaluation of Alternative Toxicological Methods
NIEHS	National Institute of Environmental Health Sciences
NIFDS	National Institute of Food and Drug Safety Evaluation
NIHS	National Institute of Health Sciences
NTP	National Toxicology Program
OECD	Organisation for Economic Cooperation and Development
PARERE	EURL ECVAM Network for Preliminary Assessment of Regulatory Relevance

REACH	Registration, Evaluation, Authorisation and Restriction of Chemicals
RENAMA	Brazilian National Network of Alternative Methods (Rede Nacional de Métodos Alternativos)
SACATM	Scientific Advisory Committee on Alternative Toxicological Methods
SPSF	Standard Project Submission Form
TG	Test Guideline
TGP	Test Guidelines Programme
TPF	Test Pre-submission Form
TST	Test Submission Template
VMG	Validation Management Group (means the same as VMT)
VMT	Validation Management Team (means the same as VMG)
WNT	Working Group of the National Coordinators for the Test Guidelines Programme

1 Introduction

Extensive efforts have been carried out during the last 20 years in order to develop, validate and implement reduction, refinement, and replacement alternative methods (AMs) to animal testing. Although progress has been made, international harmonization and standardization is one of the key endeavours to remove regulatory and possible trade barriers for the protection of consumers whilst ensuring the development and implementation of scientifically-based decision-making side-by-side with animal welfare considerations.

In the European Union, the Cosmetics Regulation prohibits animal testing (testing ban) on finished cosmetic products (since 2004) and cosmetic ingredients (since 2009), as well as the marketing (marketing ban) of finished cosmetic products tested on animals or containing ingredients which were tested on animals within or outside the European Union (complete marketing ban implemented in 2013) (European Commission 2009). In addition, the European chemicals regulation (REACH) calls for the use of AMs and requires for example, that *in vitro* testing is carried out for eye and skin irritation for substances marketed in volumes between 1 and 10 t/year. It also defines general rules for adaptation to the standard regimen, which comprise the use of AMs (European Commission 2006). Here again, such provisions apply not only to manufactured but also imported substance in quantities of 1 ton or more per year (European Commission 2006). As a consequence, the use of internationally accepted alternative methods to animal testing is important not only to comply with geographical regulatory requests, but also in order to favour international industrial commerce.

However, the acceptance of AMs can depend on various factors including national regulatory requirements, test method purposes, their uses and applicability.

In Europe, the Directive on the protection of laboratory animals for experimental and other scientific purposes originally adopted in 1986 stated that "An (animal) experiment shall not be performed if another scientifically satisfactory method of obtaining the result sought, not entailing the use of an animal, is reasonably and practicably available" (Directive 86/609/EEC; European Commission 1986). As such, for an AM to be regulatory accepted, it was considered critical to demonstrate that the method is scientifically satisfactory, i.e., valid, for the purpose intended. The scientific validation of AMs has since gained international acceptance, and represents nowadays a generally recognized requirement for the regulatory acceptance of a test method for safety assessment purposes. This is usually conducted through a validation process by where the scientific validity of a test method can be demonstrated.

Although having different format requests, the data requirements for establishing the scientific validity of an AM are similar across different regions of the world and follow internationally agreed principles of validation (Stokes et al. 2002; ICCVAM 2003; Hartung et al. 2004; OECD 2005). However, experience has shown that there can be room for variation in the evaluation of the scientific validity of an AM depending on e.g., the regulatory framework as well as on the background and experience of the bodies carrying out the evaluations. For example, different rates of over- and under-predictions may be considered acceptable depending upon the context and foreseeable uses of the assays. Similarly, regarding the regulatory acceptance of AMs different approaches may exist in different world regions. In the European Union for example, the existence of a legislation on animal welfare requirements (European Commission 1986, 2010) and the fact that its validation body is part of the European Commission, the central body responsible for proposing new and revised legislation, scientific valid AMs have usually been included into the European legislation. In the United States, legislation is handled independently by specific regulatory agencies, so that the acceptance of AMs may depend on the specific needs of each regulatory agency and the specificities of their regulated products (Stokes et al. 2002). Furthermore, the current mechanisms and procedures for regulatory acceptance across the world may differ also depending on the uses and purposes of the test methods (cosmetics, chemicals, pesticides, drugs), including:

- The recognition/tolerance by (control) authorities that manufacturers routinely use alternative approaches in their in-house safety assessments;
- The acceptance of scientifically valid safety alternative approaches as part of safety reviews by authoritative review bodies; and
- The formal recommendation/obligation to use certain validated AMs in the registration of chemicals.

As a consequence, international collaboration is critical to favour harmonized acceptance and validation of AMs. In addition, international harmonization can promote standardized criteria for the safety assessment and protection of consumers in different geographical locations, and can help in providing transparent criteria for the design and optimization of newly developed methods for predicting adverse effects for regulatory purposes, and accelerate validation of AMs.

For this reason, the International Cooperation on Alternative Test Methods (ICATM) has been established in 2009 by validation centres from Europe, USA, Canada and Japan, later joined by Korea. This establishment followed a recommendation made by the International Cooperation on Cosmetics Regulation, with the aim to strengthen collaboration and communication in the design, execution and peer review of validation studies. The existing differences across world regions and major efforts carried out for the cooperation and international harmonization of the validation of AMs will be described in this chapter.

2 Creation and Purpose of the International Cooperation on Alternative Methods (ICATM)

The International Cooperation on Cosmetics Regulation (ICCR), an international group of cosmetics regulatory authorities consisting of the European Commission's Directorate General for Internal Market, Industry, Entrepreneurship and SMEs (DG GROW), the U.S. Food and Drug Administration (FDA), the Ministry of Health, Labour and Welfare of Japan, Health Canada and, since 2014, the Brazil National Sanitary Agency (ANVISA), met for the first time on 26–28 September 2007. During this first meeting, the ICCR recognized the importance of replacing, reducing, and refining (less pain and distress) animal testing (3Rs) and recommended the enhancement of international collaboration and communication in the design, execution, and peer review of validation studies on AMs. To this end, the ICCR invited four validation organizations coordinating validation studies and test method evaluations in the ICCR member countries to propose options to increase international cooperation, namely (1) the European Centre for the Validation of Alternative Methods (ECVAM), (2) the U.S. National Toxicology Program Interagency Center for the Evaluation of Alternative Toxicological Methods (NICEATM) and the U.S. Interagency Coordinating Committee on the Validation of Alternative Methods (ICCVAM), (3) the Japanese Center for the Validation of Alternative Methods (JaCVAM), and (iv) the Environmental Health Science and Research Bureau within Health Canada. These validation organizations agreed that a framework was needed and, in response to ICCR, developed the International Cooperation on Alternative Methods (ICATM) to facilitate increased and consistent international cooperation, coordination, and communication on AMs. The ICATM framework was formally established on the 27th of April 2009 through the signature of a Memorandum of Cooperation (MoC) by representatives of these four international organizations. On the 8th of March 2011, a fifth partner organization, the Korean Center for the Validation of Alternative Methods, joined the ICATM and an updated MoC was signed by all partners.

Through the MoC, ICATM partners agreed to promote consistent and enhanced voluntary international cooperation, coordination, and communication on 3R approaches in order to:

1. Ensure the optimal design and conduct of validation studies that will support national and international regulatory decisions on the usefulness and limitations of AMs proposed for regulatory testing;
2. Ensure high quality and consistent independent scientific peer reviews of AMs that incorporate transparency and the opportunity for stakeholder involvement;
3. Enhance the likelihood of development of harmonized recommendations by national validation organizations on the usefulness and limitations of AMs proposed for regulatory testing;
4. Achieve greater efficiency and effectiveness by avoiding duplication of effort and leveraging limited resources;
5. Support the timely international adoption of AMs;
6. Ensure that new alternative test methods/strategies adopted for regulatory use will provide equivalent or improved protection for people, animals, and the environment, while replacing, reducing, or refining (causing less pain and distress) animal use where scientifically feasible.

Thus, the ICATM collaboration addresses four critical areas of cooperation: test method validation studies, independent peer review of the validation studies, development of formal and harmonized recommendations, and communication to stakeholders in order to ensure worldwide regulatory acceptance of alternative methods and strategies.

Collaborations among ICATM partners have existed and have steadily increased for more than a decade. Prior to ICATM, however, coordination of interactions occurred on an *ad hoc* informal basis only. Since the establishment of ICATM, all partners meet once a year in order to reinforce their cooperation, address further the terms and practicalities of this collaboration and present updates on their activities.

3 Mapping of ICATM Partner Organizations' Functioning: Mandate, Legislative Context, Drivers, Structure and Workflow

3.1 The European Union Reference Laboratory for Alternatives to Animal Testing (EURL ECVAM)

The European Commission's first initiative focusing on the validation of alternative approaches to animal testing started in 1991, with the creation of ECVAM. ECVAM was originally established following a Communication from the European Commission to the Council and the European Parliament in October 1991 (European Commission 1991), pointing to Article 23 of former Directive 86/609/EEC. This

Article required the Commission and the Member States to actively support the development, validation and acceptance of AMs which could reduce, refine or replace the use of laboratory animals and would provide the same level of information as that obtained in experiments using animals. This was followed by the formal establishment in 2011 of the European Union Reference Laboratory for Alternatives to Animal Testing (EURL ECVAM) due to the increasing need for new methods to be developed, validated and regulatory accepted in the European Union. EURL ECVAM's legal basis is laid out in Directive 2010/63/EU on the protection of animals used for scientific purposes. Under this Directive, EURL ECVAM's mandate has been broadened to cover the entire life cycle of AMs, i.e. from development over validation to regulatory acceptance, international recognition and proper scientific use. The main duties and tasks of EURL ECVAM are:

- To co-ordinate and promote the development and use of alternatives including in the areas of basic and applied research and regulatory testing;
- To co-ordinate the validation of alternative approaches at EU level;
- To act as a focal point for the exchange of information on the development of alternative approaches;
- To set up, maintain and manage public databases and information systems on alternative approaches and their state of development; and
- To promote dialogue between legislators, regulators, and all relevant stakeholders, in particular, industry, biomedical scientists, consumer organizations and animal-welfare groups, with a view to the development, validation, regulatory acceptance, international recognition, and application of alternative approaches.

EURL ECVAM is an integral part of the Joint Research Centre of the European Commission. EURL ECVAM is embedded in the Chemicals Safety and Alternative Methods Unit of the JRC. The JRC has extensive activities supporting the development and implementation of European Union chemicals policy. EURL ECVAM's main focus is on advancing safety assessment science using AMs to support regulatory decision making. EURL ECVAM comprises about 65 staff divided in three competence groups focusing on Assay Validation, Predictive toxicology, and Toxicity Pathways. EURL ECVAM also operates an *in vitro* Good Laboratory Practice (GLP) test facility and a High Throughput Screening (HTS) facility. The current work of EURL ECVAM is associated with five main themes:

- Combined exposures and chemical mixtures
- Integrated Approaches to Testing and Assessment (IATA)
- Endocrine disruptors
- Information systems supporting safety assessment and advancement of AMs
- Standardization and international harmonization of AMs

EURL ECVAM has a long tradition in the validation of methods which reduce, refine or replace the use of animals for safety assessment and efficacy/potency testing of chemicals, biologicals and vaccines. Validation is a key step towards the regulatory acceptance and use of a test method. The EURL ECVAM validation process allows for consistency in the execution of all EURL ECVAM validation studies and

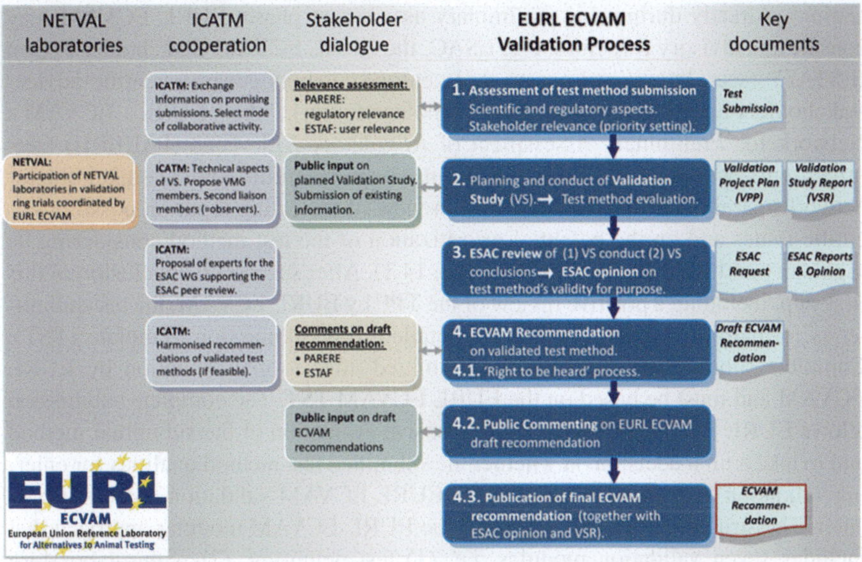

Fig. 14.1 The EURL ECVAM validation workflow. The process (*blue*) comprises four key steps. Interactions with stakeholders (PARERE and ESTAF), ICATM and EU-NETVAL laboratories are indicated to the *left* of the process. Key output documents on the *right*

comprises four key steps: (1) the assessment of a test method submission, (2) the planning and conduct of a validation study, (3) the independent scientific peer-review of the validation study and its conclusions and (4) the development of an ECVAM Recommendation on the validity status of the test method (Fig. 14.1). During this process, EU regulatory experts (PARERE), stakeholders (ESTAF), ICATM partners, the test method submitters and even the public are consulted at key stages (see below for more details), so that their views but also their technical and scientific input are taken into account to the extent possible. More detailed information on the validation process can be found in Chap. 4 of this book.

Research laboratories can submit AMs that they have developed to EURL ECVAM for scientific validation. Finalized validation studies performed by industry or other stakeholders can also be submitted to EURL ECVAM for evaluation and eventual scientific peer review by EURL ECVAM's Scientific Advisory Committee (ESAC) (see: https://eurl-ecvam.jrc.ec.europa.eu/about-ecvam/scientific-advice-stakeholders-networks/ecvam-scientific-advisory-committee-esac). The spontaneous submission process itself follows a two-step procedure, starting with a first concise "pre-submission" that may be followed by a more extensive "complete submission" upon a positive assessment of the pre-submission by EURL ECVAM. In the first step of the submission process, method developers are required to complete a Test Pre-submission Form (TPF), which allows EURL ECVAM to perform a preliminary assessment of the status of development, optimization and/or validation of the test method, its relevance for a human health(-related) or environmental effect, and its potential impact on the 3Rs (replacement, reduction, refinement of animal testing). Already during this preliminary assessment phase, EURL ECVAM may consult its Advisory Structure, e.g. ESAC, the EURL ECVAM Stakeholder Forum (ESTAF) (see: https://eurl-ecvam.jrc.ec.europa.eu/about-ecvam/scientific-advice-stakeholders-networks/estaf-ecvam-stakeholder-forum), the EURL ECVAM's Network for *P*reliminary *A*ssessment of *RE*gulatory *RE*levance (PARERE) (see: https://eurl-ecvam.jrc.ec.europa.eu/about-ecvam/scientific-advice-stakeholders-networks/parere) and ICATM partners, in view of collecting input on possible scientific issues and on the potential prioritization of the test method considering its regulatory and stakeholder relevance (Fig. 14.1). After successful conclusion of this first step, including a positive review of the TPF by EURL ECVAM, the test submitter is formally invited to complete a detailed Test Submission Template (TST). Complete submissions can only be submitted after formal invitation by EURL ECVAM and must be based on the EURL ECVAM TST. The complete submission allows EURL ECVAM to do a comprehensive evaluation of the submitted method and to take a final decision on whether the submitted test method qualifies for entering validation and at which stage of the EURL ECVAM validation process it may enter. The structure of the TST follows the EURL ECVAM modular approach that includes seven validation modules, i.e. (1) test definition, (2) within-laboratory reproducibility, (3) transferability, (4) between laboratory reproducibility, (5) predictive capacity, (6) applicability domain, and (7) performance standards. The TST thus allows for a comprehensive description of the test method's status with respect to its development, optimization and/or validation. The information submitted in the TST also allows making decisions on the prioritization of the test method, which is generally done in close dialogue with the EURL ECVAM Advisory Structure and ICATM (Fig. 14.1). Ideally, test submissions are assessed in the context of EURL ECVAM strategy papers that were, or are being defined in various toxicological areas such as e.g. skin sensitization, genotoxicity, acute systemic toxicity, fish toxicity and toxicokinetics (see: https://eurl-ecvam.jrc.ec.europa.eu/eurl-ecvam-strategy-papers). In defining these strategy documents, EURL ECVAM aims to be transparent and explicit about the formulation of strategic aims and associated objectives, and to be as inclusive as possible in taking into account the views and suggestions of various stakeholders and ICATM partners.

Beside the normal test submission process mentioned above, EURL ECVAM may also launch public calls for *in vitro* method nominations in areas identified as of regulatory priority and where *in vitro* methods may have been sufficiently developed and optimized to enter the EURL ECVAM validation workflow.

Prioritized test methods may enter a validation study coordinated by EURL ECVAM in order to assess their reliability and relevance for a particular purpose. Validation studies (i.e. multi-laboratory trials) coordinated by EURL ECVAM are performed, where possible, with laboratories of the European Union Network of Laboratories for the Validation of Alternative Methods (EU-NETVAL). EU-NETVAL was created by EURL ECVAM to address provisions in Directive 2010/63/EU whereby Member States of the European Union are requested to contribute to the validation of AMs. Currently there are a total of 37 members of EU-NETVAL, selected against pre-defined eligibility criteria (including 36 test facilities from EU Member States plus the European Commission's own *in vitro* GLP test facility operated by EURL ECVAM, which coordinates the network). In agreement with EURL ECVAM validation principles and those outlined in Guidance Document No. 34 of the Organisation for Economic Co-operation and Development (OECD 2005), an international and independent Validation Management Group (VMG) is set up by EURL ECVAM for each validation study to oversee all aspects of the study. ICATM partners may propose VMG members and may participate themselves in the VMG as observers to provide scientific, technical and regulatory input. More detailed information on the design and conduct of a multi-laboratory validation trial and EU-NETVAL can be found in Chap. 5 of this book.

EURL ECVAM also organizes and coordinates the scientific peer review of validation studies conducted or evaluated by EURL ECVAM. Completed validation studies undergo independent scientific peer review by the EURL ECVAM Scientific Advisory Committee (ESAC). Specialized Working Groups (WGs) of the ESAC are set up on an *ad hoc* basis to prepare ESAC peer reviews and document their findings in "ESAC WG Reports". WGs are typically composed of ESAC members and external scientists nominated by ESAC, EURL ECVAM and ICATM partners. The output of ESAC consists in "ESAC Opinions" which summarize the scientific advice given to EURL ECVAM.

At the end of the validation process, EURL ECVAM issues a Recommendation on the test method (see: https://eurl-ecvam.jrc.ec.europa.eu/eurl-ecvam-recommendations). The aim of a EURL ECVAM Recommendation is to:

1. Provide EURL ECVAM's views on the validity status of the test method in question, its mechanistic relevance, performance, limitations and applicability taking into account the ESAC Opinion, EURL ECVAM's own evaluation and other relevant information;
2. Advise on possible regulatory applicability and proper scientific use of the test method;
3. Suggest possible follow-up activities in view of addressing knowledge gaps.

During the development of its Recommendations, EURL ECVAM consults PARERE, ESTAF, other Commission services and its international validation partner organizations of ICATM (Fig. 14.1). EURL ECVAM also invites comments from the general public and, if applicable, from the test method submitters before finalising its Recommendations and publishing them on its website.

Following adequate validation demonstrating the usefulness and limitations of a test method/approach, it may be considered for adoption by regulatory authorities. Many of the AMs evaluated and/or validated by EURL ECVAM have been taken up into EU law and in international programmes, such as the Test Guidelines Programme (TGP) of the Organisation for Economic Co-operation and Development (OECD). Often, EURL ECVAM takes an active role in this process of regulatory acceptance by taking leadership in the drafting of new or updated Test Guidelines and/or Guidance Documents.

Of key importance is also the dissemination of information across stakeholders. EURL ECVAM takes an active role in promoting the dissemination of information on the development and validation of alternative methods and approaches, their application in industry and their acceptance by regulators. A tracking system of AMs towards regulatory acceptance provides an overview of all test methods that were submitted to EURL ECVAM for validation and/or peer review from 2008 up to now (see: http://ihcp.jrc.ec.europa.eu/our_labs/eurl-ecvam/test-submission). This tracking system is currently under revision to add information on the validation and regulatory acceptance process of submitted test methods.

3.2 The U.S. Interagency Coordinating Committee on the Validation of Alternative Methods (ICCVAM) and the U.S. National Toxicology Program Interagency Center for the Evaluation of Alternative Toxicological Methods (NICEATM)

ICCVAM was first established as an *ad hoc* committee in 1994 in response to a mandate to the Director of the National Institutes of Health (NIH) under the NIH Revitalization Act of 1993 (Public Law 103-43, Sec. 205, pp. 25–27; accessible at: http://history.nih.gov/research/downloads/PL103-43.pdf). Later in 1997, ICCVAM was designated as a standing committee by the National Institute of Environmental Health Sciences (NIEHS). In 2000, the ICCVAM Authorization Act (Public Law 106-545, 42 USC 2851-5; accessible at: http://history.nih.gov/research/downloads/PL106-545.pdf) established ICCVAM as a permanent interagency committee of the NIEHS under NICEATM. NICEATM was established in 1998 and is funded by NIEHS through the National Toxicology Program (NTP). The NIEHS is one of the 27 institutes composing the NIH. NICEATM provides scientific and operational support for ICCVAM.

ICCVAM is composed of representatives from 7 U.S. Federal regulatory agencies (Consumer Product Safety Commission, Department of Agriculture, Department of the Interior, Department of Transportation, Environmental Protection Agency, Food and Drug Administration, and Occupational Safety and Health Administration) and 8 U.S. Federal research agencies (Agency for Toxic Substances and Disease Registry, National Institute for Occupational Safety and Health, National Cancer Institute, National Institute of Environmental Health Sciences, National Library of Medicine, National Institutes of Health, Department of Defence, and Department of Energy) that require, use, generate or disseminate toxicological and safety testing information. ICCVAM's mission is to "*facilitate development, validation, and regulatory acceptance of new and revised regulatory test methods that reduce, refine, and replace the use of animals in testing while maintaining and promoting scientific quality and the protection of human health, animal health, and the environment.*" The ICCVAM Authorization Act of 2000 formally established the mandate of ICCVAM, defining its purposes as follows:

- To increase the efficiency and effectiveness of U.S. Federal agency test method review;

- To eliminate unnecessary duplicative efforts and share experiences between U.S. Federal agencies;
- To optimize utilization of scientific expertise outside the U.S. Federal Government;
- To ensure that new and revised test methods are validated to meet the needs of U.S. Federal agencies; and
- To reduce, refine, or replace the use of animals in testing, where feasible.

The ICCVAM Authorization Act of 2000 also directs ICCVAM to carry out the following duties:

1. Review and evaluate new or revised or alternative test methods… that may be acceptable for specific regulatory uses, including [those]… of interagency interest.
2. Facilitate appropriate interagency and international harmonization of acute or chronic toxicological test protocols that encourage the reduction, refinement, or replacement of animal test methods.
3. Facilitate and provide guidance on the development of validation criteria, validation studies and processes for new or revised or alternative test methods and help facilitate the acceptance [and awareness] of such scientifically valid test methods… by Federal agencies and other stakeholders.
4. Submit ICCVAM test recommendations for the test method reviewed by ICCVAM…to each appropriate Federal agency…
5. Consider for review and evaluation, petitions received from the public that (A) identify a specific regulation, recommendation, or guideline regarding a regulatory mandate; and (B) recommend new or revised or alternative test methods and provide valid scientific evidence of the potential of the test method.
6. Make available to the public final ICCVAM test recommendations to appropriate Federal agencies and the responses from the agencies regarding such recommendations.
7. Prepare [biennial] reports to be made available to the public on [ICCVAM] progress under this Act.

Finally, the ICCVAM Authorization Act established a standing NTP Scientific Advisory Committee, now designated the Scientific Advisory Committee on Alternative Toxicological Methods (SACATM), to advise ICCVAM and NICEATM regarding ICCVAM activities. SACATM members are appointed by the NIEHS Director and include stakeholders from regulated industries, animal protection organizations, academia, U.S. state or international regulatory bodies, and companies or organizations that develop, market, or use test methods. ICCVAM members who represent the 15 ICCVAM agencies serve as non-voting SACATM members.

ICCVAM relies on stakeholders to carry out AM research, development, and validation, providing guidance to test method developers where needed/requested. As such, ICCVAM does not carry out validation studies on behalf of test method developers but evaluates alternative toxicological test methods nominated/submitted by stakeholders and reviews validation data, organizes expert peer reviews of promising methods, and makes recommendations on the use of reviewed test methods to appropriate U.S. Federal agencies. ICCVAM therefore welcomes nominations/submissions of innovative test methods or approaches that may be acceptable for specific regulatory use and that align with ICCVAM member agency needs and priorities. Submissions/nominations accepted for review by ICCVAM will generally need to be supported by at least one federal agency, which takes the role of 'sponsor' for the proposed project, thereby ensuring that work done by ICCVAM is aligned with the needs of the agencies. ICCVAM is responsible for determining the relevance and priority of test method nominations/submissions, whereas NICEATM facilitates their scientific review by making use of its competencies in validation study design, computational toxicology, chemoinformatics, data management and data analysis.

To optimize utilization of resources and avoid duplication of effort, ICCVAM coordinates test method evaluation activities with ICATM partner organizations. International collaboration is also facilitated through ICCVAM agency participation in the OECD TGP. The U.S. National Coordinator for the OECD TGP (NC) has become an *ad hoc* member of ICCVAM in 2013 and uses the Committee as an interface both to provide frequent updates to federal agencies on topics of international interest, thus increasing agency awareness of international 3R efforts, and to gather feedback on those activities for the OECD. Broader engagement with the scientific community and stakeholders is achieved through focused workshops, webinars and face-to-face forums.

Both ICCVAM and NICEATM were reorganized in 2013. The vision for the 2013 reorganization of ICCVAM is described on the NTP website (see: http://ntp.niehs.nih.gov/pubhealth/evalatm/iccvam/mission-and-vision/index.html), and in the draft document issued by ICCVAM, "A New Vision and Direction for ICCVAM" (see: http://ntp.niehs.nih.gov/ntp/about_ntp/sacatm/2013/september/iccvamnewvision_aug2013_508.pdf). This document presents ICCVAM's (1) areas of priority and scientific focus for immediate resource investment, (2) plans to improve communications with stakeholders and the public, and (3) interest in exploring new paradigms for the validation and utilization of alternative toxicological methods. NICEATM's mandate was also updated in 2013, and now includes other NTP activities, e.g., supporting NTP's participation in the interagency Tox21 initiative, in addition to its original purpose as the scientific and operational support for ICCVAM.

3.3 Health Canada

There are significant differences between Health Canada (HC) and the other ICATM partners. HC is not at 'arms-length' from regulatory bodies like the other ICATM partners because HC is a regulatory body itself. It has broad responsibilities as a regulator of foods, biologics, consumer products, medical devices, natural health products, pesticides, pharmaceuticals, and chemical substances. HC is made up of 12 Branches, Offices and Bureaux as well as four Agencies. Branches involved in regulation of potentially harmful exposures include the Healthy Environments and Consumer Safety Branch, the Health Products and Food Branch, and the Pest Management Regulatory Agency. HC's Environmental Health Science and Research Bureau coordinates Canada's participation in ICATM.

Unlike the other ICATM partners, Canada does not have a national validation centre or a specific legislative mandate to work on 3Rs. As such, HC does not initiate any validation work or receive nominations of AMs for evaluation, but may 'sponsor' such work in collaboration with an ICATM partner. Priorities are drawn from national regulatory needs, international commitments and future scientific directions, overlaid with domestic expertise and existing partnerships. HC also has limited institutional expertise with peer reviews and therefore, does not lead international peer reviews of AMs/validation studies like some of the other ICATM partners do. Even though there is no formal programme on AMs or a validation body in Canada, HC does contribute to research and method development, to validation lab-work including international collaborative studies, to validation management committees, to international peer reviews (by providing experts as members of a peer review panel), and to the OECD TGP in the area of alternatives. The primary interest is to ensure that new alternative methods or strategies offer equivalent or better protection of human health while respecting the 3Rs. HC is also a permanent member of the Canadian Council on Animal Care, which oversees the ethical use of animals in science in Canada.

In terms of test method recommendations, HC works mostly via the OECD TGP to disseminate information on acceptable AMs. The Canadian OECD TGP National Coordinator is from HC.

HC will continue contributing expertise to the validation of AMs. Where there are opportunities to participate and adequate resources, HC will contribute to the design, management and conduct of validation studies, to peer reviews of validated AMs and to the development of recommendations on their suitability and limitations.

3.4 The Japanese Center for the Validation of Alternative Methods (JaCVAM)

JaCVAM was established by the Ministry of Health, Labour and Welfare (MHLW) in November 2005 as a part of the National Institute of Health Sciences (NIHS)'s Division of Pharmacology at the Biological Safety Research Center (BSRC). JaCVAM is charged with the following roles and responsibilities: (1) to promote the use of AMs to animal testing in regulatory studies in Japan, thereby replacing, reducing, or refining (the 3Rs) the use of animals wherever possible while meeting the responsibility of the BSRC to ensure the protection of the general public by assessing the safety of chemicals and other materials, as stipulated in the regulations of the NIHS; (2) to ensure new methods originated in Japan are validated, peer

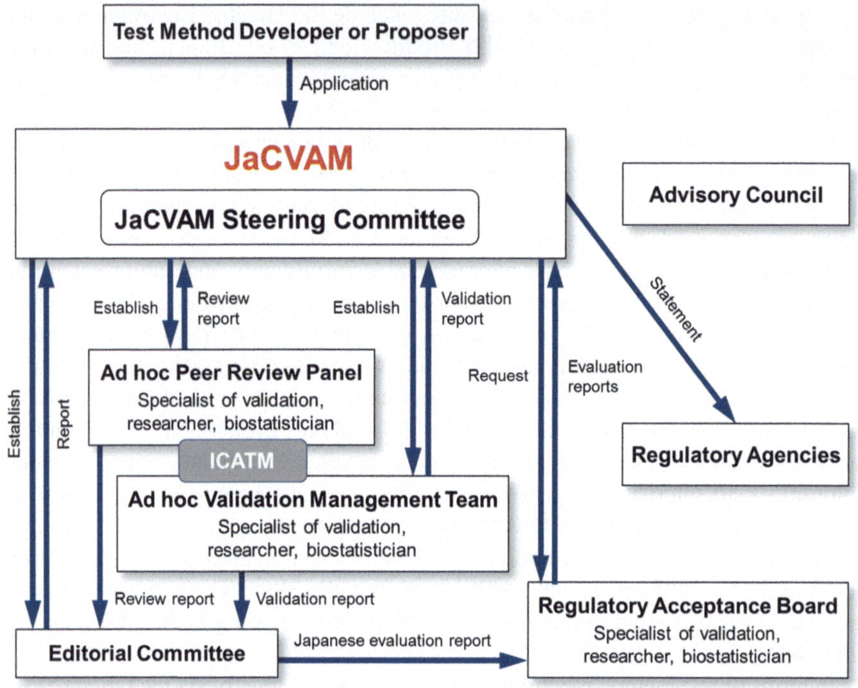

Fig. 14.2 Framework for validation, peer review and regulatory acceptance of AMs in Japan

reviewed, accepted by regulatory agencies, and made internationally compatible; and (3) to establish guidelines for new alternative experimental methods through international collaboration. To accomplish this, JaCVAM assesses the utility, limitations, and suitability for use in regulatory studies of AMs for determining the safety of chemicals and other materials and also performs validation studies when necessary. In addition, JaCVAM cooperates and collaborates with similar organizations in related fields, both in Japan and internationally. The Regulation on the Foundation of JaCVAM of May 1, 2007 (accessible at: http://www.jacvam.jp/files/regulations120813.pdf) establishes JaCVAM's Rules of Operation.

The framework for the validation, peer review and regulatory acceptance of AMs in Japan is summarized in Fig. 14.2. JaCVAM is organized around a JaCVAM Steering Committee composed of the NIHS Director General, the BSRC Director (chairperson) and other BSRC division representatives, a representative from the MHLW, a representative from the Pharmaceuticals and Medical Devices Agency, and the head of the Section for the Evaluation of Novel Methods in the Division of Pharmacology of the BSR. The JaCVAM Steering Committee deliberates on the selection/acceptance of novel and modified methods for study by JaCVAM, based on documentation submitted by a test method developer/proposer. If required, the Steering Committee proposes the organization of a validation study and commissions a new Validation Management Team (VMT), appointing its chairperson.

Other members of the VMT are selected in consultation with ICATM partners. If necessary, the Steering Committee also finalizes budgetary and manpower allocations necessary to implement evaluation and determine the scientific validity of the methods selected for study. Validation Management Teams are responsible for planning and implementing validation processes. In addition, Validation Management Teams deliberate on results obtained from validation processes and prepare validation reports that include recommended protocols for submission to the Steering Committee. A Peer Review Panel is also commissioned by the Steering Committee each time a test method is submitted for consideration, and consists of experts on the safety of chemical substances and statistical analysis, who were not involved in the development or validation of the test method under consideration. After the Steering Committee appoints a chairperson, additional panel members are named by the chairperson in consultation with the secretariat and the ICATM partners. Peer Review Panels evaluate test methods under consideration from a scientific and independent point of view, and prepare draft evaluation reports. When necessary, a Peer Review Panel can propose implementation of additional validation work and recommends topics of interest for the validation. After deliberating on the results of a validation study of an AM, the Peer Review Panel issues a report to the Steering Committee.

An Editorial Committee is also commissioned by the JaCVAM Steering Committee each time a test method is submitted for consideration, and consists of domestic experts on the safety of chemical substances and statistical analysis, who were not involved in the development or validation of the test method under consideration. After the Steering Committee appoints a chairperson, additional panel members are named by the chairperson in consultation with the JaCVAM Secretariat and the ICATM partners. The Editorial Committee is responsible for preparing a Japanese evaluation report based on the validation reports, peer review report and other background information, which is submitted to a Regulatory Acceptance Board for examination. The Regulatory Acceptance Board examines the reports as mentioned above in order to deliberate on the scientific validity, regulatory utility, and potential for acceptance by society in general of the test method under consideration, after which it issues a final report that undergoes public commenting. The Regulatory Acceptance Board consists of representatives of NIHS, experts of AMs, toxicologists, representatives recommended by industry groups, biostatisticians, and regulatory officers. Once finalized, the report from the Regulatory Acceptance Board is submitted to the JaCVAM Steering Committee for final deliberation. The Steering Committee is responsible for establishing the official policy of JaCVAM regarding test methods judged to be suitable for regulatory studies, and issues documentation of the results of these activities, which are then submitted to relevant agencies at the MHLW and/or other ministries as well as made available to the public.

All of JaCVAM activities are overseen by an Advisory Council composed of the NIHS Director General (chairperson), the BSRC Director, administrators from relevant governmental agencies, experts on animal welfare, representatives from related academic societies, and industry representatives, as well as other persons

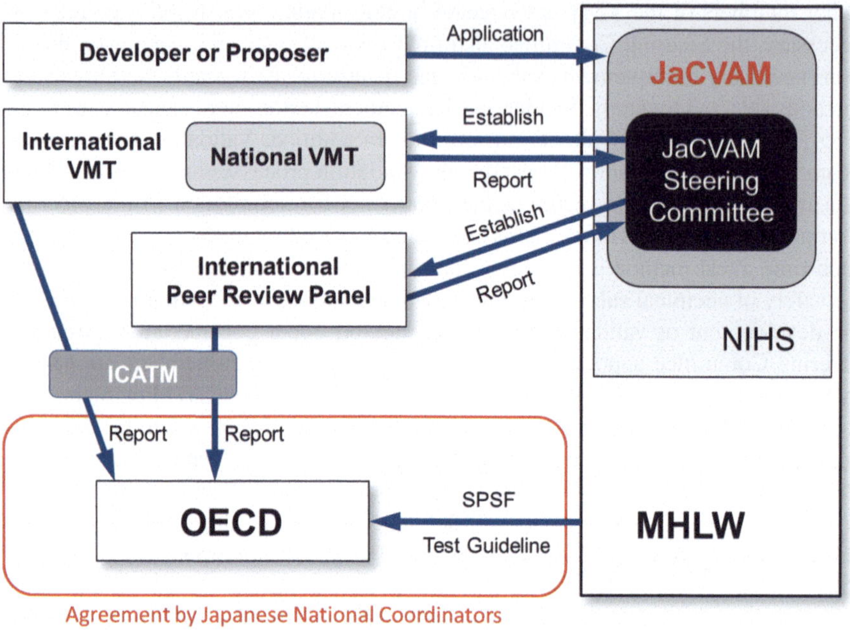

Fig. 14.3 Framework for the validation and international regulatory acceptance of AMs

judged necessary by the chairperson. The Advisory Council receives reports from the Steering Committee once or more each year, on which it deliberates and for which it provides advice.

Japan is a member of the OECD, and therefore "accepts" the use of OECD Test Guidelines for regulatory testing purposes. Japan has several NCs represented in the OECD Working Group of the National Coordinators for the Test Guidelines Programme (WNT), coming different ministries, namely the MHLW, the Ministry of Economy, Trade and Industry (METI), the Ministry of Agriculture, Forestry and Fisheries (MAFF), the Ministry of the Environment (MOE) and the Ministry of Foreign Affairs (MOFA). They aim at developing harmonized consensus views in Japan on specific test methods and promoting their adoption at OECD level. New project proposals (SPSFs—Standard Project Submission Forms) and new draft TGs on AMs proposed by JaCVAM/MHLW need to be agreed by all NCs before submission to the OECD (Fig. 14.3). ICATM partners are also consulted on every new Japanese proposal to the OECD for adoption of new AMs.

To promote the 3Rs in Japan, the director of JaCVAM collects information about the 3Rs from Japan and abroad and broadcasts this information through scientific societies, web sites, publications, and symposia. Additionally, JaCVAM distributes publications that summarize JaCVAM's activities to the relating organizations. To facilitate the conduct of its activities, JaCVAM also cooperates with several scien-

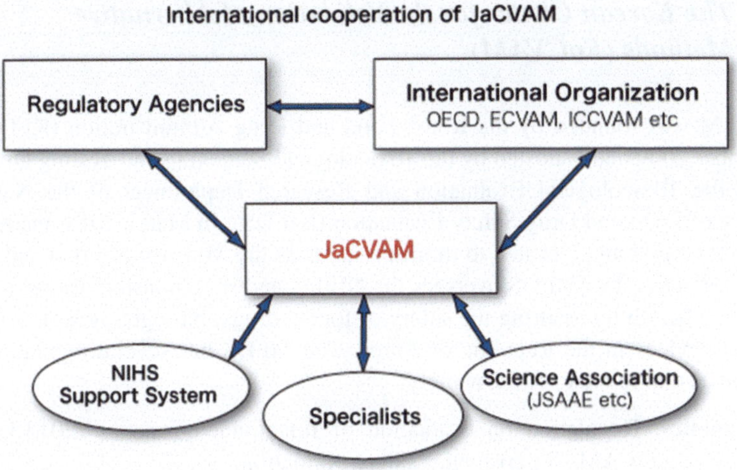

Fig. 14.4 International cooperation of JaCVAM

tific organizations such as the Japanese Society for Alternatives to Animal Experiments (JSAAE) (Fig. 14.4). The JSAAE, founded in 1989, is an academic research organization with the purpose "to promote research, development, education, and studies related to animal welfare and alternatives to animal experiments, disseminating the results of these activities." JaCVAM and the JSAAE have also provided important leadership in the establishment of AMs in other Asian countries. They collaborated with Korea in launching of the Korean Center for the Validation of Alternative Methods (KoCVAM).

The English version of the JaCVAM website, launched in August 2009, provides stakeholders outside of Japan with easier access to their activities. The site provides details on the organization of JaCVAM, and describes the committees and groups that participate in advising JaCVAM and in the review of AMs in Japan. The "Update on JaCVAM" section of the official JaCVAM website (see: http://www.jacvam.jp/en_effort/index.html) provides a table that describes the validation/peer review/regulatory acceptance status of different AMs that are undergoing, or have undergone, evaluation by JaCVAM. Ongoing validation studies coordinated by JaCVAM are: the SIRC-CVS assay, the Vitrigel-EIT and a "me too" reconstructed human cornea-like epithelium (LabCyte CORNEA model) for eye irritation testing; the IL-8 Luc assay for predicting skin sensitization; and a stably transfected transcriptional activation (STTA) antagonist assay for androgen disruptor screening (AR-EcoScreen). The latter is being supported by the OECD "Validation Management Group—Non Animal" and is foreseen to be included into a Performance-Based Test Guideline (PBTG) together with the AR-CALUX test method currently being validated by EURL ECVAM.

3.5 The Korean Center for the Validation of Alternative Methods (KoCVAM)

KoCVAM was founded by the Korea Food and Drug Administration (KFDA) in November 2009 and is hosted by the Toxicological Screening and Testing Division within the Toxicological Evaluation and Research Department of the National Institute of Food and Drug Safety Evaluation (NIFDS). In March 2013, the KFDA was restructured and elevated to ministry status as the Ministry of Food and Drug Safety (MFDS). The MFDS oversees the NIFDS and is responsible for protecting the public health by assuring the safety of foods, drugs, biologics, cosmetics, and medical devices in the Republic of Korea. The MFDS has several regulations in these areas that relate to AMs, namely:

- **Cosmetics**: Regulation on evaluation of functional cosmetics (2013.12.31), which accepts AMs for cosmetics safety evaluation;
- **Drugs**: Guidance on Nonclinical Safety Studies for the Conduct of Human Clinical Trials and Marketing Authorization for Pharmaceuticals (2012.06.29) and Guideline on Photosafety evaluation of Pharmaceuticals (2014.10.24), which both reflect the 3R principles (Replacement, Reduction, Refinement);
- **Medical devices**: Standards for biological safety evaluation of medical devices (2014.4.24), which accepts the Local Lymph Node Assay and *in vitro* skin irritation tests using reconstructed human epidermis for safety evaluation of medical devices.

KoCVAM was established based on the Korean Laboratory Animal Act of March 2008, which states in Article 5 (6) that the "establishment of policies and their execution on the development and approval of alternative test methods" is the duty of the minister of MFDS. KoCVAM was therefore established with the following mission:

- To support policies on development and approval of alternative methods to animal testing;
- To manage the overall process of validating AMs, conducting peer reviews, providing recommendations to regulatory authorities, and proposing test guidelines on validated AMs;
- To promote cooperation and joint research on AMs with domestic and international organizations;
- To disseminate AMs in Korea by providing educational programs and holding workshops.

KoCVAM seeks to institutionalize AMs through various activities, to build cooperative relationships with both domestic and foreign organizations and to review and validate proposed alternatives. KoCVAM also intends to keenly respond to these global trends by globally promoting AMs developed by Korean organizations. On this basis, KoCVAM's main activities focus on the coordination of validation studies on AMs, the implementation in Korea of internationally validated AMs, the development and recommendation of international test guidelines, and the dissemination of AMs in and out of Korea.

The KoCVAM's validation workflow can be summarized into the following consecutive steps:

1. **Test method developer/proposer**: Submission to KoCVAM of a candidate AM and validation study proposal;
2. **KoCVAM**: Review of the proposed AM, decision whether or not to initiate a validation study on the proposed method, and establishment of an *ad-hoc* Validation Management Team (VMT), including experts in chemistry and biostatistics;
3. **VMT**: Drafting of a project plan for the validation study, establishment of test chemicals and biostatistics teams, selection of test chemicals (proposed by test chemicals team), approval of participating laboratories and approval of the test method's protocol/SOP;
4. **Lead laboratory**: Transfer of the proposed test method to the participating laboratories and finalization of the test method's protocol/SOP (if required);
5. **KoCVAM**: Distribution of coded test chemicals to the participating laboratories;
6. **Participating laboratories**: Proficiency testing of coded test chemicals followed by intra- and inter-laboratory validation (assessment of reproducibility and predictive capacity) conducted by the lead laboratory and other participating laboratories;
7. **KoCVAM**: Collection of data and QA audit on data and study documentation;
8. **Bio-statistics team**: analyses and evaluation of the data and drafting of a statistics report;
9. **VMT**: Preparation of a validation study report (the VMT can propose an additional study if found necessary);
10. **KoCVAM**: Establishment of a Peer Review Panel;
11. **Peer Review Panel**: Peer review of the study on the basis of the validation study report and preparation of a peer review report (the Peer Review Panel can propose an additional study if found necessary);
12. **KoCVAM**: Evaluation of the peer review report, decision whether or not to propose the method for regulatory use and organization of an *ad-hoc* Regulatory Recommendation Team;
13. **Regulatory Recommendation Team**: Provide recommendations based on the peer review report;
14. **KoCVAM**: Recommendation of the test method to regulatory authorities.

Korea also participates in several international validation studies and peer-reviews coordinated by ICATM partners (Table 14.1). The process usually starts with an ICATM organization making a formal request to KoCVAM for Korean experts and/or laboratories to join an international validation study or peer-review panel. KoCVAM then notifies relevant national academic institutions, research institutes, GLP facilities, CRO or other laboratories and requests for applications.

Table 14.1 Participation of KoCVAM/Korean experts/Korean laboratories in international validation studies and peer-reviews of AMs

Leading organization	Endpoint	Test method	KoCVAM activities
NICEATM/ICCVAM	Endocrine disrupter screening	CertiChem MCF-7 cell proliferation assay	Validation study
		BG1Luc Estrogen Receptor Transactivation assay	Peer Review Panel
		–	Reference chemicals international working group
	Ocular toxicity	Short Time Exposure assay	Peer Review Panel
EURL ECVAM	Skin sensitization	KeratinoSens, DPRA and h-CLAT	Peer Review Panel
JaCVAM	Genotoxicity	*In vitro* Comet assay	Validation study, VMT
	Ocular toxicity	SIRC cytotoxicity test	Validation study, Peer Review Panel
		Vitrigel-EIT method	VMT
	Phototoxicity	ROS assay	VMT, Peer Review Panel
	Skin sensitization	IL-8 reporter gene assay	VMT, Peer Review Panel
		Vitrigel-SST method	VMT
	Teratogenicity	Hand1-Luc EST assay	VMT
NC3Rs	Inhalation toxicity	Fixed concentration procedure for acute inhalation study	Collaborative study
EDQM	Biological test	*In vitro* acellular pertussis vaccine assay	Collaborative study

All applications are reviewed by KoCVAM to assess the qualifications of the expert(s) or lab(s) and the selected ones are then nominated/recommended to the ICATM partner.

Korea joined the OECD in 1996. Since its establishment, KoCVAM has constantly adopted OECD TGs in Korea so that AMs can be widely used by academia and industry in the safety evaluation of cosmetics. Indeed, KoCVAM directly supports the implementation and application of accepted OECD TGs in the area of cosmetics. This is achieved by first optimizing the test guideline and its implementation in a laboratory by conducting the test at KoCVAM. KoCVAM then prepares a draft national test guideline, solicits public comments, and finally makes a recommendation to regulatory authorities. KoCVAM is thus playing a leading role in promoting regulatory use of AMs in Korea. The evaluation and acceptance of the new test guideline is done by the Cosmetics Evaluation Division of MFDS. In Korea, 11 AMs are officially accepted by the MFDS from OECD alternative TGs (Table 14.2).

Table 14.2 MFDS acceptance for cosmetics

Test method	International acceptance	National acceptance
In vitro 3T3 NRU phototoxicity test	OECD TG 432 (2004)	2007
Skin sensitization: Local Lymph Node Assay	OECD TG 429 (2007)	2007
Acute oral toxicity—Fixed Dose Procedure	OECD TG 420 (2002)	2008
Acute oral toxicity—Acute Toxic Class Method	OECD TG 423 (2002)	2008
Skin absorption: *in vitro* method	OECD TG 428 (2004)	2009
Bovine Corneal Opacity and Permeability test method	OECD TG 437 (2009)	2011 2014 (revision)
Skin sensitization: Local Lymph Node Assay: DA	OECD TG 442A (2010)	2013
Skin sensitization: Local Lymph Node Assay: BrdU-ELISA	OECD TG 442B (2010)	2013
In vitro skin irritation: Reconstructed human Epidermis test method	OECD TG 439 (2010)	2014
Isolated Chicken Eye test method	OECD TG 438 (2009)	2015
Acute oral toxicity—Up-and-Down Procedure (UDP)	OECD TG 425 (2008)	2015

Korea has a very active (science) policy towards AMs for chemical testing and is spending considerable funding on test method development and validation. The budget over the last 7 years amounts to 8.5 million US dollars, with funding going significantly up since 2013 (1.6 million in 2013, 2.2 million in 2014 and 2.2 million in 2015). Currently, KoCVAM is validating five test methods: two "me too" skin irritation test methods (the MCTT KeraSkin and the Tego Neoderm-ED); a non-radioactive LLNA based on flow cytometry (BrdU) for skin sensitization testing (using the Performance Standards of OECD TG 429); a method using inflammatory mediators (IL1a and IL6) for predicting skin sensitization; and a reconstructed human cornea-like epithelium (MCTT HCE model) for eye irritation testing. Several facilities within Korea are conducting AMs and participating in validation studies forming a national network of validation laboratories, including MFDS/NIFDS, the Korean Testing and Research Institute (KTR), the Seoul National University Hospital Biomedical Research Institute, the AMOREPACIFIC R&D center, the Biotoxtech Co. and the Catholic University of Daegu GLP center.

3.6 *Points in Common and Differences Between the Different ICATM Organizations*

Table 14.3 summarises the points in common and the differences between the different international organizations that are currently members of ICATM and that were described in the previous paragraphs.

Table 14.3 Mapping of ICATM partners activities, workflow and functioning

Activities	EURL ECVAM	ICCVAM	NICEATM	JaCVAM	KoCVAM	Health Canada
Promoting 3Rs	Yes	Yes	Yes	Yes	Yes	3Rs considered but HC not focused on it
Chemical Safety	Yes	Yes	Yes	Yes	Yes	Yes
Legislative Mandate to Validate Methods	Yes, EURL ECVAM is legally entitled to conduct validation of alternative methods	No, ICCVAM works on request by federal agencies	NICEATM is part of the NTP overseen by the NIEHS (NIEHS is 1 of the 27 NIH institutes)	JaCVAM works for the NIHS which is overseen by the MHLW	KoCVAM is established under 'Laboratory Animal Act of 2008'	Health Canada has a broad mandate to protect health and acts primary as a regulator. HC is a federal department and is a regulatory body
Fitting into International Requirements/Topics/ Activities	Yes	Yes	Yes, through ICCVAM	Yes	Yes	Yes
Interaction and Cooperation with International Organizations, e.g. OECD	Yes, e.g. participation in OECD expert groups and WNT; development and coordination of OECD AOP Knowledge Base; development of OECD TGs, GDs and IATA.	Yes, the U.S. NC for the OECD is part of ICCVAM	Yes, through ICCVAM	Yes, focus on OECD TGP. Japan has several National Coordinators. The NCs are organized by ministries (Ministry of Health, Ministry of Environment…). – A SPSF has to be agreed by all NCs – NCs hold a coordination meeting before the WNT meeting	Yes, NIFDS has an OECD NC	Yes, HC has decided to base its recommendations on the OECD TGP
Regulatory Acceptance	Normally through OECD TGP first. Adopted OECD TGs are then translated into EU Test Methods and adopted in the EU Test Methods Regulation (TMR)	Regulatory acceptance has to fit with national/federal agencies needs as they trigger the process of TM evaluation. Validation at the OECD level is not mandatory but additional		Through OECD TGP	Through OECD TGP	Through OECD TGP; HC and Environment Canada share the responsibility of regulatory acceptance, however the Canadian NC at the OECD is from HC

Integrated Approaches, Shift of Paradigm in Toxicity Testing	Yes	Yes	Yes, through ICCVAM	Yes	Yes	Yes
Dialogue with						
– Regulators	Yes via PARERE	Not directly, agencies do	Yes, through ICCVAM	Yes via Reg. Acc. Board	Yes	HC is a regulatory body itself
– Stakeholders	Yes via ESTAF	Yes (biggest driver)		Yes also via Reg. Acc. Board	Yes	
– Agencies	Yes	Yes		Yes	Yes	
Validation	Required by law with respect to the 3Rs, and protection of human health and environment	Contribution/ Advice via Peer Reviews	Validation Process under consideration	Yes	Yes	No
Conducting Peer Reviews	Yes, via ESAC. PR Panels are composed of independent experts. EURL ECVAM organizes pure PRs w/out stakeholders (≠ other VAMs). Stakeholders can however place inputs/objections after PR opinion (ESTAF level). ESAC Working Group Report and ESAC Opinion issued together with EURL ECVAM Recommendation	Yes, on request of federal agencies. PR Panels are composed of experts and stakeholders. Peer Review Report issued together with ICCVAM Recommen-dation	Yes	Yes, Peer Review Panel established for each method by the Steering Committee; PR Panel delivers PR Report to the Steering Committee; PR Panels are composed of independent experts	Yes, KoCVAM organizes Peer Review Panels that produce PR reports	No

(continued)

Table 14.3 (continued)

Activities	EURL ECVAM	ICCVAM	NICEATM	JaCVAM	KoCVAM	Health Canada
Validation Workflow	1. Assessment of Test Method Submission *[Submission Assessment Report]* 2. Validation Study*[Validation Report]* 3. ESAC Peer Review*[ESAC Report and Opinion]* 4. Draft ECVAM Recommendation and commenting 5. Final Recommendation	In relation with NICEATM validation procedure under finalization (see next column)	*Validation process under consideration:* 1. Test Method Nominations and Submissions 2. Validation Study *[Test Method Evaluation Report]* 3. Peer Review*[Peer Review Report]* 4. Draft ICCVAM Recommendation and commenting 5. Final ICCVAM Recommendation, will trigger Agency Response	– JaCVAM activities overseen by the JaCVAM Steering Committee (SC), Reg. Acc. Board, and an Advisory Council – SC deliberates on the selection/acceptance of methods for validation based on submissions – SC establishes a Validation Management Team (VMT) – SC establishes a Peer Review Panel; PR conducted after completion of validation. study similar to EURL ECVAM and ICCVAM/ NICETAM – SC places requests to the Reg. Acc. Board to deliberate on the scientific validity and regulatory utility of a method – JaCVAM Statements provided to Regulatory Agencies	1. Submission of candidate method and assessment by KoCVAM 2. Establishment of VMT in charge of coordinating and conducting the validation study (delivers validation study report) 3. Establishment of Peer Review Panel by KoCVAM (delivers a peer-review report) 4. Establishment of Regulatory Recommendation Team by KoCVAM 5. Peer-review report is reviewed by Regulatory Recommendation Team (delivers a Draft Recommendation) 6. The Final Recommendation is communicated to the regulatory authorities via KoCVAM	Not applicable

Prioritization of Submitted Test Methods	Yes	Yes	Through ICCVAM	Yes	Yes	Not applicable
Network/Partnership with Laboratories	Yes, via EU-NETVAL	Yes, using agencies' laboratories	Part of the NTP, and through ICCVAM	Yes	Yes	No
Dissemination and Communication Systems/Databases	1. TSAR: Tracking System of Alternative methods towards Regulatory acceptance 2. DB-ALM: DataBase service on ALternative Methods to animal experimentation 3. (Q)SAR Model database	1. Database on *in vivo* data on endocrine disruption 2. NICEATM LLNA database 3. Database on *in vivo* acute oral and dermal toxicity data		JaCVAM Secretariat is in charge of tracking the validation/peer review/ regulatory acceptance status of alternative methods on the JaCVAM website	Activities on education and dissemination of alternative methods	No system related to alternative methods present
Funding	Through European Commission	No budget. However each ICCVAM agency is capable of funding validation studies	Through NTP, NIEHS and NIH. Each ICCVAM agency is capable of funding validation studies	Through MHLW	Through MFDS	–

4 Recent Developments in Countries with Observer Status at ICATM

Two countries, namely Brazil and China, have currently an observer status at ICATM with the intention of becoming full members in the near future. This section will briefly describe recent developments in AMs in Brazil and China focusing on training, validation and implementation in legislation over the last few years.

4.1 Brazil

The idea to create a Brazilian Center for Validation of Alternative Methods (BraCVAM) arose during a round table discussions at the I EMALT (Brazilian Meeting on Alternative Methods to Animal Use for Regulatory Purposes) (Presgrave and Bhogal 2005). Later, BraCVAM's creation was embraced by researchers from academia, industry and regulatory bodies and an embryonic format of structure was proposed (Presgrave 2008; Eskes et al. 2009; Presgrave et al. 2010).

After a long period without a specific law about animal experimentation, in 2008, Brazil published the Law 11,794 that regulates the use of laboratory animals in experimentation and education. Two important actions were implemented: (1) the creation of the National Council of the Control of Animal Experimentation (CONCEA—Conselho Nacional de Controle da Experimentação Animal); and (2) that all institutions that use animals for experimentation or education are obliged to have an Ethics Committee on Animal Use (CEUA—Comissão de Ética no Uso de Animais) (Cardoso and Presgrave 2010; Brasil 2008). Laws 11,794/2008 and 9605/98 (Law against environmental crimes) state that when an AM is available, it must to be used and the original methods are not allowed to be performed.

Besides BraCVAM and CONCEA, Brazil counts on the National Network of Alternative Methods (RENAMA—Rede Nacional de Métodos Alternativos) to conduct the process of validation. In general, BraCVAM identifies the need of validating a method, organizes the peer-review process and recommends the scientific validity of an assay to CONCEA; laboratories from RENAMA execute the assays; and CONCEA becomes the method official in Brazil.

The validation process in Brazil follows the OECD Guidance Document No. 34 (OECD 2005). BraCVAM aims to interact and collaborate with Brazilian and international partners, such as participating at the ICATM.

In 2014, as one of the first important acts, BraCVAM indicated 17 already scientifically validated and internationally accepted test methods to CONCEA for their official acceptance in Brazil. These methods were those that cover skin corrosion/irritation (OECD TGs 430, 431, 435 and 439), serious eye damage/eye irritation (TGs 437, 438, 460), phototoxicity (TG 432), skin permeation/absorption (TG 428), skin sensitization (TGs 429, 442A and 442B), acute toxicity (TGs 129, 420, 423 and 425) and genotoxicity (TG 487).

In 2015, the first Brazilian validation study was initiated, funded by the Brazilian Ministry of Science, Technology and Innovation (MCTI—Ministério da Ciência, Tecnologia e Inovação) and aiming at the validation of HET-CAM (Hen's Egg Test—Chorion-Allantoic Membrane). It counts on international experts as part of the Management Group and mainly aims, besides validating HET-CAM, to get hands-on experience of the validation process.

4.2 China

Chinese regulations for the toxicity testing of chemicals still rely primarily on traditional animal testing methods. Many of these regulations were established prior to 2000 and to date, *in vitro* OECD Test Guideline (TG) methods have not been adopted into many Chinese chemical related regulations as alternatives to using animal tests. Some alternatives, such as Quantitative Structure Activity Relationships (QSAR) and read-across may be used when chemicals of interest are not amenable to animal testing.

In 1997, four Chinese ministries (i.e., the Chinese Ministry of Science and Technology (MoST), Agriculture, Health and Food Drug) proposed the first development plan for AMs in China linked to the 3Rs principles of laboratory animal sciences (MoST of P. R. China 1997). Further effort was evident in an animal welfare science policy issued by MoST in 2001 as well as the inclusion of China as an observer within a number of international programmes related to AMs (e.g. the OECD TGP, ICATM). A particular focus within these discussions was raising awareness of AMs and their application within Chinese authority laboratories (e.g. China Food Drug Administration (CFDA), China Inspection and Quarantine Bureaux (CIQ), Centre for Disease Control and Prevention (CDC)).

The last few years have seen significant progress in both the science and evaluation of AMs (i.e. in the application of *in vitro* OECD TG methods) in China. Knowledge and capability development in the use of AMs across many of the authority laboratories have been greatly improved following many years' domestic efforts led by a few Chinese pioneer scientists in the field as well as rising international support from foreign governments, companies and NGOs across the globe (Curren and Jones 2012). Moreover, scientific initiatives within China to develop new approaches to integrate existing AMs (Yang et al. 2010) and to initiate research activities aligned with the 'Toxicity Testing in the Twenty-First Century' vision (National Research Council 2007) have been growing steadily across many academic, research institute and regulatory laboratories, following an increase in government funding in these areas from both the National Natural Science foundation of China and the Key Science Programmes of MoST.

As well as increased uptake and awareness of AMs, regulatory changes have also taken place in China which will reduce the overall numbers of animals used for safety testing. For example, from July 2014, domestically manufactured "non-special" cosmetics, such as shampoo, soap, and some skin care products, can be marketed without mandatory product testing in animals, where safety information based on risk assessment is acceptable (CFDA 2011, 2014).

4.2.1 Status of Training and Acceptance of OECD Test Guidelines on Alternative Methods

China has been recognized as a key partner by the OECD since May 2007, though as yet China is not an OECD member country. The current regulatory climate in China essentially requires that validated test methods and competent domestic laboratories be in place before acceptance of non-animal tests can be accomplished. OECD TG methods (e.g. those for skin irritation, phototoxicity and eye irritation) provide an optimal starting point for non-OECD member countries. Indeed, over the last 8 years, Chinese scientists at some leading regulatory labs (e.g. Guangdong (GD)-CDC, GD-CIQ) have devoted significant effort to the evaluation of these TG methods in their local labs (see those evaluated AMs in Table 14.4).

Chinese laboratory capability for development in this area has been reinforced with interaction from many external stakeholders, including foreign governments, NGOs, trade associations and multinational consumer goods companies through numerous dialogues and scientific meetings in the past few years (e.g. see Table 14.5 for major events). Given the wide awareness of TG methods and acceptance of their scientific robustness amongst a large proportion of the scientific community and regulatory bodies in China, the remaining challenge for regulatory adoption of a TG method is to develop domestic laboratories trained in these methods across national and provincial levels with direct responsibility for the notification, registration and post market surveillance of chemicals/cosmetics products. This would require intensive hands-on training, development of documentation tools (Standard Operating Procedures (SOPs), protocols, etc.), availability of reagents, equipment and test substrates as well as guidance on interpretation of the resulting non-animal data. A number of recent training events (see Table 14.4) illustrate the considerable effort from multiple stakeholders in this area. Since 2014, Chinese regulatory bodies have been accelerating their engagement with international stakeholders. For example, in May of 2014, the National Institute for Food and Drug Control (NIFDC, Beijing, a division of CFDA), signed a memorandum of understanding with US-based Institute for *In Vitro* Sciences (IIVS) to focus on hands-on training of national and provincial regulators to help accelerate the adoption of *in vitro* methods in China.

China has also established several national standards and industrial standards (see Table 14.6) relating to TG methods issued by the Standardization Administration of China (SAC) in public under the General Administration of Quality Supervision, Inspection and Quarantine (AQSIQ)—a governing authority of all local CIQs across provinces in charge of entry-exit commodity inspection, certification and accreditation, standardization.

Table 14.4 Selected major events of training and evaluation of OECD TG methods or selected other methods in China between 2008 and 2015

Date	Methods	Organizers	Evaluation/Training
Mar. 2008	3T3 Neutral Red Uptake (NRU) Phototoxicity Test (OECD 432) (Yang et al. 2009b)	GD-CDC	Evaluation (5 labs)
Jan. 2010	Eye Irritation Tests (i.e., Hen's Egg-Chorioallantoic Membrane (HET-CAM), Fluorescein Leakage Assay (FLT) (OECD 460), Chorioallantoic Membrane-Trypan Blue Staining (CAM-TBS)); Acute Oral Toxicity Tests (i.e., Acute Toxic Class Method (OECD 423), Fixed Dose Procedure (OECD 420), Up-And-Down Procedure (OECD 425))[a]; Transcutaneous Electrical Resistance Test (TER) (OECD 430); Skin Sensitization: Local Lymph Node Assay (LLNA) (OECD 429)[a] (Yang et al. 2009a, 2010)	GD-CDC and China Ministry of Health	Evaluation and Pre-evaluation (5 labs)
13th–14th Apr. 2011	HET-CAM	GD-CDC	Training (15 labs)
Dec. 2011	3T3 NRU Phototoxicity Assay for cosmetic ingredients	CFDA	Evaluation and Training (5 labs)
Apr. 2012	Chorioallantoic Membrane Vascular Assay (CAMVA), Bovine Corneal Opacity and Permeability (BCOP, OECD 437) and EpiSkin for Skin Irritation (OECD 439)	GD-CDC	Training (18 labs)
June 2012	CAMVA and BCOP	GD-CIQ and MB Research Labs (U.S.)	Training (11 labs)
23rd–25th Oct. 2012	BCOP, 3T3 NRU Phototoxicity Test	Beijing Technology and Business University (BTBU) and IIVS	Training
18th–22nd Mar. 2013	BCOP, CAMVA, EpiSkin for Skin Irritation (OECD TG439), 3T3 NRU Phototoxicity Test	IIVS, GD-CIQ, L'Oréal	Training

(continued)

Table 14.4 (continued)

Date	Methods	Organizers	Evaluation/Training
May 2012–June 2013	EpiSkin	Beijing CIQ, Shanghai-CIQ, GD-CIQ, L'Oréal China	Training and evaluation (5 labs)
23rd–27th Sept. 2013	BCOP, CAMVA and Tecskin for Skin Irritation	NIFDC and IIVS	Training
Nov. 2014	BCOP, CAMVA and Tecskin for Skin Irritation	NIFDC and IIVS	Training
7th–11th Apr. 2015	BCOP, CAMVA, EpiSkin, *In vitro* Reconstructed Skin Micronucleus Test	GD-CIQ, SUN YAT-SEN University	Training

[a]Not non-animal methods, but tests which give 3R benefits of refinement and reduction, compared to standard animal tests

Table 14.5 Selected major meetings in China on AMs between 2011 and 2015

Time	Meeting	Organizers
14th–15th Mar. 2011	Toxicity Testing in the Twenty-First Century Symposium, Shanghai	Unilever
10th–12th Apr. 2011	The First International Symposium on Cosmetics—Alternatives to Animal Experimentation for Cosmetics, Beijing	Beijing Technology and Business University; CFDA
14th–15th Apr. 2011	International Symposium on Technology and Application of Alternatives to Animal Testing, Guangzhou	GD-CDC
Nov. 2011	International Workshop on Validation and Application of 3T3 NRU Test in China, Guangzhou	GD-CDC
24th–25th June 2013	Workshop on non-animal approaches to cosmetics safety, Shanghai	Unilever and Shanghai-FDA
13th–15th Nov. 2013	"The Future of Toxicology: Twenty-First Century Safety Sciences"—continuing education programme at the 6th C-SOT congress, Guangzhou	C-SOT, GD-CDC and Unilever
13th–14th Oct. 2014	Workshop on Mitochondrial Toxicity and Pathway-Based Chemical Safety Assessment—An inaugural symposium of the Society of Toxicological Alternatives and Translational Toxicology, CSOT, Beijing	C-SOT and Unilever
20th Mar. 2015	Joint Workshop on Safety Assessment of Cosmetics, Beijing	European Commission, UK Government and CFDA
8th–10th Apr. 2015	Workshop of Cosmetic Risk Assessment and Alternatives to Animal Testing, Guangzhou	GD-CDC, C-SOT and Unilever

Table 14.6 Major national and industrial standards of AMs issued by SAC (see Chinese governmental website: http://cx.spsp.gov.cn/)

ID of standards	Name
GB (Guo Biao—national standards) for chemicals	
GB/T 21827-2008	Skin allergy test—local lymph node testing[a]
GB/T 21769-2008	*In vitro* 3T3 NRU phototoxicity test
GB/T 21757-2008	Acute oral toxicity—acute toxic class method[a]
GB/T 21804-2008	Acute oral toxicity—fixed dose procedure[a]
GB/T 21826-2008	Acute oral toxicity: up-and-down procedure[a]
GB/T 27829-2011	*In vitro* membrane barrier test method for skin corrosion
GB/T 27830-2011	*In vitro* human skin model for skin corrosion
Professional standards for entry exit inspection and quarantine under AQSIQ	
SN/T 2285-2009	Cosmetics—GLP for *in vitro* alternative tests
SN/T 2328-2009	Cosmetics—acute toxicity of keratinocyte cytotoxicity test
SN/T 2329-2009	Cosmetics—ocular corrosive and irritant HET-CAM test
SN/T 2330-2009	Cosmetics—embryotoxicity and developmental toxicity test: EST (mice embryonic stem cells) test
SN/T 3715-2013	Cosmetics—developmental toxicity test: WEC (whole embryo culture) test
SN/T 3824-2014	Cosmetics of phototoxicity: combined RBC (red blood cell) test
SN/T 3899-2014	Cosmetics—GLP for *in vitro* alternative tests: cell culture and sample preparation
SN/T 3898-2014	Cosmetics—validation procedures for *in vitro* alternative tests of cosmetics
SN/T 3882-2014	Chemicals—skin sensitization of chemicals: LLNA (local lymph node assay) for Brdu-ELISA[a]

[a]Not non-animal methods, but tests which give 3R benefits of refinement and reduction compared to standard animal tests

4.2.2 New Investment in the Twenty-First Century Safety Sciences

Whilst non-animal alternatives and OECD guidelines exist for several toxicity endpoints, there are still important human safety endpoints for which no validated alternative approaches exist (Adler et al. 2011). In 2007, the U.S. National Research Council's publication 'Toxicity Testing in the Twenty-First Century: A Vision and a Strategy' (National Research Council 2007) outlined an approach to safety assessment that "could transform toxicity testing from a system based on whole-animal testing to one founded primarily on *in vitro* methods that evaluate changes in biologic processes using cells, cell lines, or cellular components, preferably of human origin". Chinese scientists are increasingly part of the global research effort to turn this vision into a reality (e.g. website: http://tt21c.org/site/symposium.html). Increasing research investment from a number of government funding bodies has been focussed in this area. It was estimated that over 50-million RMB was invested from 2010 to 2014 years on the R&D of AMs and an additional ten million RMB on investigating toxicity mechanisms of chemicals and predictive toxicology in 2014 alone. There are currently a number of ongoing toxicity pathway-based research programmes across institutes, universities and industries. For example, a research collaboration on mitochondrial toxicity pathways between the Chinese Academy of Military Medical Sciences (AMMS), Unilever and the Hamner Institutes for Health Sciences (U.S.) has now started to generate promising findings assessing adverse effects without the use of animals (Guo et al. 2013; Yuan et al. 2016). Additional initiatives that are focused on mechanistic-based safety decisions without recourse to animal tests include work at the Beijing Proteome Research Centre exploring kinase sensors for stress pathways, and cutting-edge research using *in silico* approaches to predict ligand-receptor interactions at the Research Centre for Eco-Environmental Sciences, part of the Chinese Academy of Sciences and a State Key Laboratory. These research initiatives exemplify the willingness of top Chinese researchers, from different and diverse disciplines, to engage on the big challenges faced in non-animal approaches for systemic toxicology; applying world-class technological resources to these challenges within China. Increasingly the same Chinese researchers, regulators and policy-makers are being asked to contribute to and help shape international programmes addressing non-animal approaches, such as playing a part in the EU Horizon 2020 proposals.

To facilitate the exchange of research experience in this area, two new scientific societies were established in China in 2014: the Chinese Society of Toxicological Alternatives and Translational Toxicology (TATT) under the China Society of Toxicology (C-SOT) and the Society of Toxicity Testing and Alternatives (STTA) under the Chinese Environmental Mutagen Society (CEMS). The establishment of both societies marked another milestone in China's commitment to the development of non-animal approach for assuring safety. Both societies will provide useful scientific platforms with a focus on drawing together both domestic and international efforts to promote the new safety science as well as to facilitate acceptance of this new science in potential regulations of the future. Furthermore, key educational initiatives promoting non-animal approaches have begun in China, including a 'Talent Development' programme at the School of Public Health, Peking University and a number of summer training programmes in academia (e.g. Fudan University and AMMS).

4.2.3 Challenges Ahead and Future Opportunities

Despite the progress mentioned above, there are still many challenges remaining for China (and indeed for the rest of the world) in fully embedding non-animal approaches for assuring safety. Increased laboratory capability development for conducting AMs is still required for many regions of China who have less experience with these techniques. Likewise, many domestic companies within China also lack expertise in applying AMs to chemical safety in the context of risk assessments for consumer safety. Animal testing remains the principal approach demanded for toxicity testing in chemical regulations. In current cosmetics regulations, for example, animal testing on "special" cosmetics and imported cosmetics is still a mandatory requirement. Likewise, for a new chemical to be used in China, animal data must be provided to the Ministry of Environmental Protection according to the Chinese new chemical substance notification.

Recent progress has shown China's willingness to embark on the same journey regarding the use of non-animal approaches for the assurance of safety to that which many scientists and regulators are currently engaged with worldwide. We believe that China will continue on this journey driven by several different factors: (1) a drive to harness advances in science and technology for Chinese innovation; (2) world-wide regulatory changes including animal testing bans in cosmetics regulations across the globe (e.g. EU, Israel, New Zealand and India); (3) China's desire to embrace a global economy and a rising awareness of animal welfare and finally; (4) ongoing economic reform within China with an increased emphasis on markets playing a 'decisive' role in the economy. Given that since 2013, China has become the world's second-largest economy and the second only to the United States in research and development (R&D) spending (Ni 2015), we have many reasons to believe that China may become an important player in the global community developing new non-animal approaches and applying AMs in regulations.

5 Strengthening International Harmonization and Cooperation on the Validation and Acceptance of Alternative Methods

The ICATM partners and observers are continuously discussing ways to improve harmonization and strengthen their collaboration on the validation and acceptance of AMs. Several steps have been taken to avoid duplication, increase efficiency in the validation of AMs and to have a greater impact in their acceptance. Important efforts have been and continue to be made to approximate several aspects of test method validation in the different regions in order to facilitate cross-border acceptance of validation study outputs and outcomes and increase global acceptance of validated approaches. The following sections describe several aspects of test method validation and acceptance for which increased harmonization/collaboration across ICATM organizations has been/is being discussed and implemented, including test method selection/prioritization, conducting validation studies, setting-up and using validation laboratory networks, conducting peer-reviews, issuing test method recommendations, disseminating information on AMs, and supporting international acceptance.

5.1 Test Method Selection/Prioritization

The ICATM organizations have agreed to harmonize the criteria used for selection and prioritization of submitted test methods for validation/peer-review, such as:

- *Regulatory impact/relevance* (top-down evaluation considering regulatory needs/gaps);
- *Scientific value* (bottom-up evaluation considering the scientific/mechanistic merits of the method);
- *Budget/Cost*;
- *Availability*;
- *Impact on 3Rs*, although this criterion is of higher importance in Europe than in other regions due to legislative requirements.

The ICATM partners have also committed to better informing each other and stakeholders/regulators on method prioritization, e.g., by using, to the extent possible, a common communication and dissemination platform (see Sect. 5.6 below).

Moreover, in addition to continue allowing spontaneous submissions of test methods by developers, the ICATM partners agreed that it would be useful to promote a system to communicate needs/gaps to method developers and to organize common calls on specific toxicological areas to cover those needs/gaps. For example, in 2014 EURL ECVAM announced a call on *in vitro* methods for estimating human hepatic metabolic clearance/stability in view of identifying methods that can contribute to the development of harmonized standards and associated international toxicokinetics test guidelines. This type of call facilitates the identification of suitable methods addressing existing needs as well as their selection/prioritization for eventual evaluation/validation, where required. The implementation of an evaluation/validation process at ICATM level of prioritized methods could then follow two different routes, i.e., leadership by one ICATM partner of a specific area or full collaboration in a specific area by all interested partners.

5.2 Conducting Validation Studies

The ICATM organizations usually collaborate actively in the conduct of international validation studies coordinated by one of the partners. Indeed, all include in their validation processes the requirement to officially request from the other partners the nomination of experts to serve in *ad-hoc* VMGs/VMTs. Depending on the required expertise, ICATM partners may actually participate in VMGs as active members. However, the possibility of having liaisons from ICATM serving as observers in VMGs is also offered to facilitate collaboration and flow of information without active participation. Due to resource limitations, participation as liaisons for consultation only on critical aspects is often preferable to full VMG membership since liaisons are usually involved in all discussions at the discretion of the VMG. Agreed aspects to cover in VMGs include: prior experience in the coordination of validation studies, understanding of regulatory requirements (regulators/toxicologists), expertise in the area at stake, technical knowledge in the methodology at stake, knowledge in biostatistics, and (computational) chemistry expertise. The early establishment of VMGs and its involvement on all aspects of the study is advisable and therefore early collaboration between ICATM organizations in setting-up VMGs is of key importance.

Several key elements of a validation study can hugely benefit from an active collaboration between different validation bodies during a validation study. Indeed, wider consultation on key aspects during a validation study could facilitate study progress and conclusion. The most critical points are: (1) tackling technical aspects, e.g., evaluation of SOPs; (2) test chemicals selection; (3) selecting participating laboratories, and (4) validation study design and data analysis. Test method protocols/SOPs may be shared between ICATM partners and observers for scientific commenting before embarking on a multi-laboratory trial. EURL ECVAM goes further and actually transfers the method to its own GLP laboratory before initiating the validation study in order to assess if the SOP is well described and sufficiently developed. These steps decrease the probability of failure of a validation study and help to avoid entering a validation study with an underdeveloped test method. Ultimately, they increase the efficiency of the validation of AMs with regard to both time and resources spent.

The collaboration on test chemicals selection is usually done in the context of validation studies at the level of VMGs/Chemicals Selection Groups. The members of ICATM have nevertheless agreed that it would be extremely useful to collaborate out of the context of a validation study in the development of reference databases for different key areas. As such, several databases have been or are currently being developed, namely for: skin sensitization, genotoxicity, serious eye damage/eye irritation, endocrine disruption, and acute oral and dermal toxicity. An active collaboration in the selection of chemicals for validation studies brings several important benefits, such as the selection of the best possible chemicals for a given study (based on reference data and chemical properties), coverage to the extent possible of various classification systems of importance to the different ICATM regions (although ultimately it may be advantageous and critical to work under the same classification system, e.g. UN GHS), and the selection of the same chemical sets in multiple validation studies to facilitate downstream data integration activities. The involvement and collaboration of multiple ICATM organizations in validation studies also facilitates the selection of participating laboratories from different countries/continents, thus rendering the study more international and its conclusions more easily acceptable across regions. Finally, harmonization in the design of validation studies and in the statistical analysis of the generated data is also desirable to simplify and expedite the peer-review and regulatory acceptance of validated AMs.

Nowadays, an increasing number of validation studies are being conducted externally to validation bodies and being submitted to ICATM organizations for evaluation and eventual peer-review. This may boost the number of AMs undergoing validation, which is certainly positive; however, validation studies are highly complex endeavours with regard to both scientific and logistical aspects. A proper conduct and management of the study is crucial to its success. It is therefore important that stakeholders engaging on validation studies of AMs with the intention of having them proposed for regulatory use have a thorough understanding of good validation practice. Having this in mind, the ICATM partners agreed that it would be important to create guidance on the validation of AMs to support the practice of external validation studies and facilitate proper study planning and conduct. EURL ECVAM has recently initiated the drafting of such guidance in collaboration with the other ICATM organizations and its advisory networks.

5.3 Validation Laboratory Networks

The establishment and use of validation laboratory networks can increase the validation capacity of a country or region due to increased available resources. Such networks also facilitate higher standardization across laboratories, increasing the quality and potential for success of multi-laboratory validation trials. EURL ECVAM established the European Union Network of Laboratories for the Validation of Alternative Methods (EU-NETVAL) in response to the provisions of the European Union Directive 2010/63/EU, which requests that EU Member States assist the European Commission in the validation of AMs. While setting up EU-NETVAL, EURL ECVAM developed a set of eligibility criteria that need to be met by any laboratory that wishes to become a member. These criteria guarantee the quality and suitability of all laboratories participating in validation studies coordinated by EURL ECVAM. Terms of Reference for EU-NETVAL have also been developed. They detail the legislative anchor, the establishment of the network and the maintenance of its membership, the tasks of the members of the network and of EURL ECVAM in support of validation studies, the allocation of tasks to members, and the financing of network activities. The EU-NETVAL eligibility criteria and ToR have been shared with the other ICATM partners and observers and may therefore be used to support the selection of appropriate laboratories for participation in validation studies outside of EU. KoCVAM and BraCVAM have also established their own networks of validation laboratories (called RENAMA in Brazil). Mooving forward, ICATM may wish to maximize the utility of the available resources and create an ICATM network of validation laboratories on the basis of the existing regional networks that could be used by all members for validation purposes and/or for testing adequacy of SOPs before initiation of a validation study.

5.4 Conducting Peer Reviews

Independent peer reviews of AMs and their validation are necessary before the methods are considered for regulatory acceptance (e.g., as an OECD TG). They are conducted to verify that the validation study was properly conducted, to perform a scientific review of the results and conclusions of the study, and to develop a scientific independent view to support a validation body in taking a position and formulating recommendations on the AM to regulators and stakeholders. Four ICATM partners currently organize and conduct independent peer reviews (counting on external expertise), namely EURL ECVAM, ICCVAM, JaCVAM and KoCVAM. Health Canada does not conduct its own peer reviews but accepts the outcomes of those organized by others if they are conducted rigorously and independently. In the past, the peer-review processes of the different partners were significantly different, leading to the need to conduct shadow peer-reviews of methods already peer reviewed by another partner. The ICATM partners have however made significant efforts to harmonize their peer-review processes. Some small differences still exist, but for the majority, the processes are very similar. Existing differences are mostly on the requirements for the composition of peer review panels. While all ICATM organizations have similar requirements in terms of expertise coverage, all allow for nomination of experts by other ICATM partners and none objects the use of experts from academia, only ICCVAM includes interest groups (industry, NGOs and regulators) in all their peer-review panels and has to achieve gender balance and wide geographic distribution of the panellists to the extent possible. EURL ECVAM, for example, organises a pure scientific review, generally without stakeholder involvement, which is rather done at the level of developing a recommendation. Nevertheless, stakeholders may sometimes serve in peer-review panels due to the need for specific expertise. Thus, today, there are no longer fundamental issues to impede the ICATM organizations using each other's peer review outputs to inform the development of recommendations.

5.5 Test Method Recommendations

All of the validation bodies that are members of ICATM issue official public recommendations/statements on AMs at the end of their validation process in order to communicate the outcome of the validation study/peer-review and facilitate the regulatory acceptance of the AM. Health Canada is slightly different in this respect since, being a regulatory authority, the acceptance of a given method follows an internal process without the need for an own recommendation. On this basis, the ICATM partners and observers have a clear desire to work towards the development of "Harmonized Recommendations", which should include the necessary information to suit the needs of all regional and international regulatory contexts. Currently, the ICATM organizations already consult each other during the drafting of their recommendations to include all relevant regional information, but this may be further strengthened in the future by agreeing on a common structure that would make test method recommendations more global and useful to all parties.

5.6 Dissemination and Communication

In a recent ICATM meeting, all partners expressed a mutual interest to better disseminate to stakeholders and the public updated information on alternative methods being evaluated and considered for regulatory use. To date, each partner has been using its own approach to communicate to the public the progress made in the area of alternatives to animal testing. EURL ECVAM maintains three related systems, namely the Tracking System of Alternative methods towards Regulatory acceptance (TSAR), the DataBase service on ALternative Methods to animal experimentation (DB-ALM) and the (Q)SAR Model database. TSAR is a tool that provides a transparent view on the status of AMs as they progress from purely scientific protocols submitted for validation to being actively used in a regulatory context. This tracking system intends to cover all steps, from the initial submission for validation until final adoption by inclusion in the EU legislation and/or international guidelines (e.g. OECD, ICH). The DB-ALM is a public, factual database service that provides evaluated information on development and applications of advanced and alternative methods to animal experimentation in biomedical sciences and toxicology, both in research and for regulatory purposes. EURL ECVAM also disseminates information through its advisory, regulatory and stakeholder networks (ESAC, PARERE and ESTAF, respectively), through dedicated workshops and through the publication of early status reports (see: https://eurl-ecvam.jrc.ec.europa.eu/eurl-ecvam-status-reports). ICCAVM also disseminates information through workshops, through its dedicated website, through the issuing of biennial progress reports and by maintaining curated databases that are made available to the public (e.g., database on *in vivo* data on endocrine disruption, NICEATM LLNA database, *in vivo* acute oral and dermal toxicity data). JaCVAM tracks the validation/peer review/regulatory acceptance status of different AMs that are undergoing, or have undergone, evaluation by JaCVAM through its website. KoCVAM organizes activities on education and dissemination of alternative methods. Health Canada does not have any specific information systems promoting regulatory acceptance of alternative methods.

On this basis, the ICATM partners and observers agreed that there is potential for better harmonization and for improvement in the way information is disseminated to the public. A joint communication strategy and the use of a single platform by all partners could facilitate achieving these goals and would also allow for an objective presence of ICATM on the web. The ICATM organizations are thus currently exploring the possibility of using a common tool for communication and dissemination of information on AMs, which will serve two main objectives:

- *Facilitate communication within ICATM* on new developments, to increase collaboration and effectiveness, and avoid duplication;
- *Improve communication to stakeholders and regulators* on new developments and collaborations at ICATM level.

It was agreed to pursue a major overhaul of TSAR, led by EURL ECVAM but with input from ICATM partners, with the intention of increasing and adapting its functionality to make it the common ICATM dissemination/communication platform. The new TSAR is expected to be publically released towards the end of 2016.

5.7 International Programmes and Regulatory Acceptance

The ICATM partners and observers have extensively addressed opportunities for sustained cooperation with regard to their involvement in international regulatory programmes such as the OECD, the ICCR, and the International Conference on Harmonization of Technical Requirements for Registration of Pharmaceuticals for Human Use (ICH). This includes getting reciprocally informed whenever possible on submissions of new project proposals (e.g., for the development of a new OECD Test Guideline or Guidance Document) in order to foster mutual support, enhance synergies and avoid unnecessary duplication. It should however be noted that the ICATM organizations may not necessarily be aware of all incoming SPSFs that are submitted to the OECD by the NCs from their country/region, because the NCs are in several cases outside of ICATM. The exceptions are Health Canada, whose Director is also the Canadian NC, and EURL ECVAM since the European Commission NC is also located at the JRC. As such, EURL ECVAM knows of all SPSFs submitted by the European Commission NC, but this is many times not the case with SPSFs submitted to the OECD by NCs from the EU Member States. EURL ECVAM therefore uses the PARERE network to inform and get informed of SPSFs being prepared by ICATM and EU Member States on health and environmental safety assessment (including both *in vitro* and *in vivo* methods). ICCVAM has also taken steps to improve its awareness of all U.S. OECD activities with the U.S. NC becoming an *ad hoc* member of ICCVAM in 2013. Where SPSFs are known to an ICATM organization, there is the commitment to exchange information with the other partners and observers ahead of the official submission to the OECD. Further consultation usually occurs before the SPSFs are discussed/approved by the WNT in April each year.

ICATM has also communicated its interest to engage with the "Joint Committee for Medicinal Products for Veterinary Use (CVMP)/Committee for Medicinal Products for Human Use (CHMP) *Ad-hoc* Expert Group on the Application of the 3Rs in Regulatory Testing of Medicinal Products" (JEG 3Rs) and to support the development by the JEG 3Rs of an European Medicines Agency (EMA) Guideline on Regulatory Acceptance of 3R (Replacement, Reduction, Refinement) Testing Approaches (see: http://www.ema.europa.eu/docs/en_GB/document_library/Scientific_guideline/2014/10/WC500174977.pdf). In fact, EURL ECVAM maintains an observer status at the JEG 3Rs with the purpose of providing expertise and support to the group on matters related to the 3Rs, avoiding duplication across sectors and facilitating consideration of methods evaluated by ICATM partners by ICH and vice versa. At the same time, the JEG 3Rs coordinates EMA responses to requests from EURL ECVAM to PARERE on potential regulatory relevance of AMs.

6 Concluding Remarks

Alternative approaches to animal testing are gaining momentum as more and more test methods achieve worldwide acceptance thanks in large part to the efforts of international collaborations such as ICATM and OECD and the commitment of validation bodies around the world (EURL ECVAM, ICCVAM, JaCVAM, KoCVAM and BraCVAM). The regulatory adoption of the 3Rs principles across the world and the adoption of harmonized principles of validation are critical to ensure that different countries and sectors approach the questions related to human safety assessment in a harmonized and standardized manner.

These approaches are relevant to a wide range of stakeholders involved in the development, validation, and evaluation of data generated by alternative methods including scientists from academia, industry, validation bodies and regulatory agencies. Implementation of internationally harmonized standards permits as such to strive, in a joint effort, towards limiting animal testing while establishing scientifically-driven decision-making approaches for the safety assessment of chemicals and protection of consumers.

Such efforts are of great value to ensure a global understanding of the importance of progressing the protection of human beings and environmental species, side-by-side with the development and implementation of scientifically-based decision-making as well as the replacement, reduction and refinement of animal testing, whenever possible. The several international cooperation bodies that exist today such as the ICATM and the OECD allow for collaborative efforts in validating and adopting alternative methods, and it is hoped that new countries and other areas of safety assessment can join these initiatives so that international harmonization increases steadily and in parallel with the global demands.

References

Adler S et al (2011) Alternative (non-animal) methods for cosmetics testing: current status and future prospects. Arch Toxicol 85:367–485

Brasil (2008) Lei No. 11.794, de 08 de outubro de 2008. Diário Oficial [da República Federativa do Brasil], de 09 de outubro de 2008, Brasília, Brasil: República Federativa do Brasil (In Portuguese)

Cardoso C, Presgrave O (2010) Princípios éticos na experimentação animal. In: Biologia, Manejo e Medicina de Primatas Não Humanos na Pesquisa Biomédica (Orgs Andrade A, Andrade MCR, Marinho AM, Filho JF). Editora FIOCRUZ, Rio de Janeiro, pp 435–449 (In Portuguese)

CFDA (2011) Note on the management and the notification process of domestic non-special use cosmetics. Record: [2011] No. 181, issued on 21st April 2011 (In Chinese)

CFDA (2014) A further clarification letter on the implementation issues of cosmetic registration and notification. Record: [2014] No. 70, issued on 11th April 2014 (In Chinese)

Curren R, Jones B (2012) China is taking steps toward alternatives to animal testing. Altern Lab Anim 40:1–2

Eskes C, Sá-Rocha V, Nunes J, Presgrave O, de Carvalho D, Masson P, Rivera E, Coecke S, Kreysa J, Hartung T (2009) Proposal for a Brazilian centre on alternative test methods. ALTEX 26:303–306

European Commission (1986) Council Directive 86/609/EEC of 24 November 1986 on the approximation of laws, regulations and administrative provisions of the Member States regarding the protection of animals used for experimental and other scientific purposes. Off J Eur Union L358:1

European Commission (1991) Establishment of a European Centre for the Validation of Alternative Methods (CEVMA). Communication from the Commission to the Council and the European Parliament, 6pp. Brussels, Belgium: Commission of the European Communities

European Commission (2006) Regulation (EC) No 1907/2006 of the European Parliament and of the Council of 18 December 2006 concerning the Registration, Evaluation, Authorisation and Restriction of Chemicals (REACH), establishing a European Chemicals Agency, amending Directive 1999/45/EC and repealing Council Regulation (EEC) No 793/93 and Commission Regulation (EC) No 1488/94 as well as Council Directive 76/769/EEC and Commission Directives 91/155/EEC, 93/67/EEC, 93/105/EC and 2000/21/EC. Official Journal of the European Union, L396, 1

European Commission (2009) Regulation (EC) No 1223/2009 of the European Parliament and of the Council of 30 November 2009 on cosmetic products. Off J Eur Union L342:59–209

European Commission (2010) Directive 2010/63/EU of 22 September 2010 on the protection of animals used for scientific purposes. Off J Eur Union L276:33–79

Guo J et al (2013) Role of PGC-1α and mitochondrial biogenesis in cardiovascular diseases. Chinese J Pharmacol Bullet 29:1–5 (In Chinese)

Hartung T, Bremer S, Casati S, Coecke S, Corvi R, Fortaner S, Gribaldo L, Halder M, Hoffmann S, Roi AJ, Prieto P, Sabbioni E, Scott L, Worth A, Zuang V (2004) A modular approach to the ECVAM principles on test validity. Altern Lab Anim 32:467–472

ICCVAM (2003) ICCVAM guidelines for the nomination and submission of new and revised alternative test methods. NIH publication No. 03-4508. National Institute of Environmental Health Sciences, Research Triangle Park, North Carolina, 50 pp

MoST of P. R. China (1997) "Opinions on laboratory animal science development", the Ninth Five-Year Plan on National Economy and Social Development. Policy No (1997—No. 432) (In Chinese)

National Research Council (2007) Toxicity testing in the twenty-first century: a vision and a strategy. The National Academies Press, Washington, DC

Ni X (2015) China's research and development spend. Nature 520:S8–S9. doi:10.1038/520S8a

OECD (2005) Guidance document on the validation and international acceptance of new or updated test methods for hazard assessment. OECD Series on Testing and Assessment No. 34. ENV/JM/MONO(2005)14, 96pp. Organisation for Economic Co-operation and Development, Paris

Presgrave O (2008) The need for the establishment of a Brazilian Center for the Validation of Alternative Methods (BraCVAM). Altern Lab Anim 36:705–708

Presgrave O, Bhogal N (2005) EMALT: a Brazilian meeting on alternative methods to animal use for regulatory purposes. Altern Lab Anim 33:670–672

Presgrave O, Eskes C, Presgrave R, Alves E, Freitas J, Caldeira C, Gimenes I, Silva R, Nogueira S, Nunes J, Rivera E, Sá-Rocha V, Coecke S, Hartung T (2010) A proposal to establish a Brazilian Center for Validation of Alternative Methods (BraCVAM). ALTEX 27:47–51

Stokes WS, Schechtman LM, Hill RN (2002) The Interagency Coordinating Committee on the Validation of Alternative Methods (ICCVAM): a review of the ICCVAM test method evaluation process and current international collaborations with the European Centre for the Validation of Alternative Methods (ECVAM). Altern Lab Anim 30(S2):23–32

Yang X et al (2009a) Preliminary study on neutral red uptake assay as an alternative method for eye irritation test. AATEX 14(Special Issue):509–514

Yang Y et al (2009b) Establishment and use of 3T3 NRU assay for assessment of phototoxic hazard of cosmetic products. AATEX 14(Special Issue):515–518

Yang Y et al (2010) Combined *in vitro* tests as an alternative to *in vivo* eye irritation tests. Altern Lab Anim 38:303–314

Yuan H et al (2016) A PGC-1α-mediated transcriptional network maintains mitochondrial redox and bioenergetic homeostasis against doxorubicin-induced toxicity in human cardiomyocytes: implementation of TT21C. Toxicol Sci 150(2):400–417

Chapter 15
Evolving the Principles and Practice of Validation for New Alternative Approaches to Toxicity Testing

Maurice Whelan and Chantra Eskes

Abstract Validation is essential for the translation of newly developed alternative approaches to animal testing into tools and solutions suitable for regulatory applications. Formal approaches to validation have emerged over the past 20 years or so and although they have helped greatly to progress the field, it is essential that the principles and practice underpinning validation continue to evolve to keep pace with scientific progress. The modular approach to validation should be exploited to encourage more innovation and flexibility in study design and to increase efficiency in filling data gaps. With the focus now on integrated approaches to testing and assessment that are based on toxicological knowledge captured as adverse outcome pathways, and which incorporate the latest *in vitro* and computational methods, validation needs to adapt to ensure it adds value rather than hinders progress. Validation needs to be pursued both at the method level, to characterise the performance of *in vitro* methods in relation their ability to detect any association of a chemical with a particular pathway or key toxicological event, and at the methodological level, to assess how integrated approaches can predict toxicological endpoints relevant for regulatory decision making. To facilitate this, more emphasis needs to be given to the development of performance standards that can be applied to classes of methods and integrated approaches that provide similar information. Moreover, the challenge of selecting the right reference chemicals to support validation needs to be addressed more systematically, consistently and in a manner that better reflects the state of the science. Above all however, validation requires true partnership between the development and user communities of alternative methods and the appropriate investment of resources.

M. Whelan (✉)
European Commission, Joint Research Centre (JRC), Ispra, Italy
e-mail: maurice.whelan@ec.europa.eu

C. Eskes
SeCAM Services & Consultation on Alternative Methods, Magliaso, Switzerland

C. Eskes, M. Whelan (eds.), *Validation of Alternative Methods for Toxicity Testing*,
Advances in Experimental Medicine and Biology 856,
DOI 10.1007/978-3-319-33826-2_15

Validation is intrinsic to scientific endeavour where there is desire to apply knowledge. It is an important pursuit in many different fields of science and engineering and thus has numerous definitions that are conceptually consistent but vary somewhat in terminology and emphasis. Such semantic differences are often important since they reflect the context of validation within a particular community. Within the domain of toxicity testing for safety assessment of chemical substances, the Organisation for Economic Co-operation and Development (OECD) defines validation to be "the process by which the reliability and relevance of a particular approach, method, process or assessment is established for a defined purpose" (OECD 2005). Although originally proposed over 20 years ago (Balls et al. 1990), this definition is still appropriate and useful with regard to the validation of new ways to evaluate and predict toxicity based on the rational combination of *in vitro*, computational and other alternative methods. In essence though, validation of new methods is about demonstrating their credibility in order to gain acceptance by the relevant parties who ultimately depend on the toxicological information derived from them.

We can view validation as an essential step in the translation of knowledge and associated methodology from a development community to a user community. A development community consists of scientific and technical experts who use their knowledge and technology to produce new methods and processes, whereas a user community consists of experts knowledgeable in problem definition who are looking for solutions to apply in their domain. The developer and user communities differ in the way they think and work, but both have an equally important stake in the validation process. The developers need validation to demonstrate the utility of what they have produced, and the users need validation to understand if they can trust what the developers are offering. Therefore the correct framing and design of a validation study relies on input from both communities, a fact that unfortunately is not always the case. In this respect there has been a tendency for the development community to undertake validation studies without sufficient prior engagement of the user community, only to find that the outcome fails to impress. Adding to this however, the regulatory user community has often been reluctant or has found it difficult to define and articulate their expectations of a validation study *a priori*, tending to only realise what they consider appropriate or not after the event. Going forward, more proactive and effective cooperation is needed to properly define validation frameworks and individual studies to ensure that the time and resources invested in validation yield the maximum benefit to all concerned.

Assembling the evidence during a validation study to demonstrate the reliability and relevance of a method or approach can be tackled in different ways. As proposed by the then European Centre for the Validation of Alternative Methods (ECVAM), now the EU Reference Laboratory for Alternatives to Animal Testing (EURL ECVAM), validation can be conceptually portrayed as being modular in nature (Hartung et al. 2004), where the necessary information on test definition, transferability, within and between laboratory reproducibility, predictive capacity, applicability domain and proposed performance standards can be gathered somewhat independently, and can combine retrospective analysis of existing information with the prospective generation of new data, as appropriate. This modular approach promotes consistency in the type and organisation of information that should be gathered through validation but offers flexibility in the design of studies to suit specific circumstances. Although the modular concept has been around for many years, it has yet to be fully exploited since most validation studies conducted have followed a more traditional sequential process i.e. test definition—transferability—reproducibility—predictivity—performance standards. However this is changing. For example, exploiting robotic High Throughput Screening (HTS) technology for *in vitro* method validation gives the possibility to generate data on large sets of reference chemicals in order to understand predictive capacity, before embarking on a multi-laboratory ring trial to assess transferability and reproducibility. In addition, HTS can also be employed as a follow-up to a validation study to further assess method performance (Bouhifd et al. 2012). In this respect, the ToxCast (Judson et al. 2014) and Tox21 (Huang et al. 2016) programmes in the USA have proven invaluable in identifying relevant assays with predictive value, and those lacking it. Moreover, these programmes have shown that validation with respect to modules on transferability and between laboratory reproducibility may not be required if the intention is to produce data with assays and systems that for one reason or another are not intended for widespread use by third parties in their own laboratories, or envisioned for development into standard test guidelines recognised across different regulatory jurisdictions (Judson et al. 2013). Restricting validation to the modules of test definition, within laboratory reproducibility and predictive capacity does not necessarily limit the regulatory applicability of the data produced, as clearly illustrated by the commitment of the US Environmental Protection Agency within its Endocrine Disrupter Screening Program to use the predictions derived from a battery of *in vitro* assays and associated computational models to satisfy information requirements in lieu of data derived from conventional animal tests, with respect to the screening of chemicals for their potential to disrupt an endocrine pathway (Browne et al. 2015).

Another example of how the modular approach to validation can be better exploited in the future concerns the evolution of the concept of performance standards for *in vitro* methods. Performance standards are usually proposed for a validated method in cases where other technically similar methods are expected to appear which produce equivalent information on the same endpoint (sometimes referred to as "me too" methods). In the context of the OECD test guidelines programme for the testing of chemicals, performance standards comprise (i) essential test method components that define technical similarity (ii) reference chemicals for which the relevant toxicological property is known and (iii) performance targets with respect to reproducibility and accuracy of prediction. The proposal of performance standards is usually motivated by the desire to encourage innovation in the development and use of alternative methods for regulatory purposes while at the same time avoiding potential commercial monopolies. Thus performance standards facilitate the development of a test guideline based on a method that incorporates proprietary elements, including those protected by formal intellectual property rights. If a developer can demonstrate through a "catch-up" validation study that their method complies with the minimum performance standards derived from the validated reference method(s), then their method can be deemed to be equally valid and thus suitable for potential inclusion in the associated (performance based) test guideline.

As we embrace the progressive change from observational to predictive toxicology, *in vitro* methods and their associated performance standards should also be seen in a new light. Integrated Approaches to Testing and Assessment (IATA) are emerging that rationally exploit toxicological knowledge to weight and combine various types of information, including *in vitro* data, in order to draw a conclusion on the potential hazard or risk of a chemical (OECD 2014a, 2016). Since a particular piece of mechanistic or hazard information can often be obtained from different *in vitro* methods, the emphasis within IATA is shifting towards prescribing what type of information is required for an assessment, rather than what actual methods should be used to generate the information. Practically speaking therefore, in relation to their potential utilisation within IATA, *in vitro* methods can be considered as belonging to the same class if they produce equivalent mechanistic or hazard information. In this context then, performance standards may be associated in some cases with a class of methods rather a single method.

A good example of a well-defined class of *in vitro* methods comprises Estrogen Receptor Transactivation Assays (ERTA), covered by OECD Test Guideline 455 (OECD 2012a). The mechanistic information produced by this ERTA class relates to the potential of a chemical to interfere (agonism/antagonism) with the nuclear receptor mediated estrogen signalling pathway, while the primary technical attribute defining the class is a test system based on a genetically engineered gene-reporter cell line capable of generating a luminescent signal in response to the activation of the pathway. The associated performance standards (OECD 2012b) specify how a method should respond (positive/negative) to selected reference chemicals (i.e. accuracy of response) and what level of reproducibility should be attainable. How broad or narrow the mechanistic and technical domains of a certain class should be defined is difficult to say without particular context, and thus can often be arrived at through quite arbitrary and pragmatic reasoning. However both domains are obviously related in that as the mechanistic information derived from a method class becomes more specific, the technical attributes that define the class are likely to be more specific too. In the case of ERTA for example, initially a test system based on mammalian cells was proposed as a defining technical attribute of the class (i.e. an "essential test method component" in OECD terminology), but this was later relaxed to include cells from other species when it was acknowledge that similar mechanistic information could be derived from methods utilising bacteria (Arnold et al. 1996).

As a consequence of the increasing importance of standards to provide a validation framework for *in vitro* methods, the primary output of formal validation studies should change from being a single validated method to validated standards applicable to multiple methods within a class. These standards can then be used by various parties in various contexts, with the appropriate degree of formality, to characterise their method in order to compare it to similar methods in its class and to demonstrate its fitness for a particular purpose, assuming the performance requirements are known. Of course the development and validation of standards cannot be purely a theoretical exercise and requires the generation of data using *in vitro* methods representative of the class of interest. An important by-product of the validation process therefore will be a set of methods validated against the standards.

This approach is in fact being actively pursued by EURL ECVAM. The first validation study to be conceived and designed along these lines is being conducted within the context of an OECD project to produce a set of validated methods and associated performance standards for a class of Androgen Receptor Transactivation Assays (ARTA) suitable for testing chemicals for their potential to interfere with the AR signalling pathway, and which are intended to eventually form the basis of a new Performance Based Test Guideline (PBTG) for ARTA that can be applied throughout OECD member countries. A second study being undertaken by EURL ECVAM aims to develop standards for a class of methods that provides information on *in vitro* human metabolic clearance, essential in the understanding and predication of the toxicokinetic profile of a chemical *in vivo*. A comprehensive review of methods described in the literature together with a public survey to find unpublished methods helped to identify 11 similar methods falling into the class. A meta-analysis of these methods revealed the salient technical attributes or components that could be used to define the class and has highlighted a range of functional parameters that should be characterised through validation. This has then led to the proposal of reference chemicals and associated experimental procedures that could be used for prospective method validation. The study is now entering the laboratory phase to practically evaluate the proposed standards and protocols using clearance methods representative of the class. Ultimately the intention is to propose a comprehensive set of validation standards for this class of *in vitro* clearance methods that will guide (i) reporting (i.e. indicate how a method should be properly described), (ii) which reference chemicals to use (i.e. to experimentally characterise a method and to evaluate its performance), and (iii) how to actually use the reference chemicals (i.e. provide operating procedures that instruct on how to actually conduct the experimental assessment of a method).

The concepts, principles and definitions underpinning IATA are still being discussed and will likely evolve further in the years ahead. However, one important premise recently put forward by the OECD is the distinction between IATA and what is termed, "Defined Approaches" (DA) to testing and assessment (OECD 2016). In essence, IATA deliver an assessment conclusion typically through weight-of-evidence reasoning and thus inevitably include an element of subjective (non-formalised) expert judgement, whereas DA are rule-based (formalised) approaches that generate predictions intended to be used within IATA. The draft OECD guidance document on the reporting of DA to be used within IATA (OECD 2016) includes a detailed reporting template for DA, composed of sections for describing elements such as: the *in vitro* or computational methods used to generate the primary data; the algorithm or data interpretation procedure devised to predict the toxicological effect of concern; the reference chemicals used to develop and evaluate the global performance of the approach; estimates of its predictive capacity based on a set of reference chemicals; the strengths and limitations of the approach; and descriptions of the sources of uncertainty associated with the toxicity prediction.

A number of conclusions and recommendations regarding the validation of integrated approaches have been put forward by ECVAM and the EPAA, the European Partnership for Alternative Approaches to Animal Testing (Kinsner-Ovaskainen et al. 2012). Although the term, "Integrated Testing Strategies (ITS)" was used then to refer to all types of integrated approaches, it was recognised that one needs to essentially differentiate between rule-based ITS and judgement-based ITS when considering validation. Like judgement-based ITS, an IATA does not lend itself to validation in the traditional sense. An IATA developed and applied to satisfy a particular regulatory information requirement for a chemical can only be really deemed valid or acceptable by the regulatory body to which the IATA is submitted. This is exemplified by an IATA based primarily on a read-across prediction used to address an information requirement under the European Union's REACH legislation (REACH 2006), where the reasoning used is very much case-specific and thus cannot be validated *a priori* in a generic manner. Instead, the European Chemicals Agency (ECHA) may decide to apply its Read Across Assessment Framework (RAAF 2015) to systematically evaluate such an IATA to ensure the expectations of the Agency are met in terms of quality, thoroughness and credibility.

On the other hand, when considering a DA that predicts a toxicological property of a chemical that could be used within an IATA, conducting some degree of validation of the DA *a priori* using a set of reference chemicals is usually feasible and indeed desirable in order to understand its general predictive performance, possible limitations and likely applicability domain. Although such validation cannot reflect all potential use-cases of the DA and does not guarantee its acceptance by a particular regulatory body, it does help considerably in building confidence and encouraging uptake by end-users and regulatory authorities. Recently the OECD Task Force for Hazard Assessment identified a number of different DA for predicting the potential skin sensitisation hazard of a chemical and contributors have now incorporated them into the newly proposed reporting template. This makes it more straightforward to contrast and compare different approaches in light of their potential regulatory application. The expectation is that no particular DA will likely dominate as the best option for all situations, but instead each one will have different attributes that a prospective user will consider in order to decide if it is suitable or not for their particular needs. The outcome of this OECD activity also provides a good basis to explore the possibility of defining performance standards for this class of DA, since one can imagine that systematic comparison of all the case studies will indicate, for example, what set of reference chemicals would be optimal for assessing various aspects of predictive performance of any new DA that might be proposed in the future. Another case that would support the idea of developing performance standards for DA is the HTS screening approach for endocrine disrupters recently proposed by the US EPA (Judson et al. 2015). It predicts interference of a chemical with the estrogen signalling pathway and is based on a battery of *in vitro* assays combined with a rule-based data interpretation procedure (computational prediction model). Thus it can be appropriately described as a DA. Performance standards already exist for ERTA (OECD 2012b) but unfortunately they are too restrictive in their current form to be applied to this DA. This is because although ERTA and the DA predict more or less the same effect (i.e. interference with the estrogen signalling pathway), the DA has a much broader technical basis (i.e. a battery of HTS assays which use a variety of measurement technologies) than ERTA (i.e. a single *in vitro* method) that is obviously not foreseen in the essential test method components of the current ERTA performance standards.

Validation therefore can and should be pursued for both methods and DA. At the method level, the emphasis should be on the definition of performance standards that can be used to validate *in vitro* methods belonging to a particular class. The standards should allow the thorough assessment of the experimental reliability of an *in vitro* method and its relevance in terms of being able to determine any association of a chemical with a particular toxicological pathway, mode of action or hazard effect. The outcome of the validation should also identify the operational boundaries and technical limitations of a method (class) including the types of chemicals that can be tested. A description of the method and the results obtained from the validation study should be reported appropriately, for example by following the OECD guidance on describing non-guideline *in vitro* methods (OECD 2014b), and made public, for example via DB-ALM, the EURL ECVAM database on alternative methods (http://ecvam-dbalm.jrc.ec.europa.eu). Validation of a DA on the other hand should focus on assessing its overall capacity to provide information on a toxicological endpoint of regulatory concern and on characterising the uncertainties associated with the underlying assumptions and predictions. The emergence of performance standards targeted at the validation of DA will no doubt be useful in this respect. As mentioned above, a comprehensive and harmonised description of a DA and its validation should be provided using the new OECD guidance and reporting template (OECD 2016). Ideally, completed templates should be made publically available via a suitable on-line repository, such as the DB-ALM when its planned extension to accommodate DA is complete.

A central consideration in the validation of either methods or DA is the selection of suitable reference chemicals. This is typically very challenging and has significant consequences for the execution and potential outcome of any validation study. It has been approached practically and scientifically in different ways for different studies (Brown 2002; Eskes et al. 2007; Casati et al. 2009; Pazos et al. 2010; Jennings et al. 2014) since each study has its own particular context and scope and to-date no general framework or guidance on chemical selection has been put forward. Notwithstanding this, chemicals are usually selected based on a variety of attributes such as: toxicological properties; chemical class or structural features; physicochemical properties; product or sectorial use; availability; and cost. If we consider each of these attributes as representing a single dimension in a multidimensional "chemical space", then we can view chemical selection as a process to optimally sample this space in a way that the subset of reference chemicals chosen adequately represents the greater population of chemicals occupying the space. Naturally, as the number of attributes to be considered increases and their individual range expands, then more chemicals have to be included in the reference set to be able to cover the whole space. Operationally, sampling of the defined chemical space is influenced heavily by practical issues including the fact that in reality chemicals do not uniformly and continuously cover chemical space and the number of chemicals that can be actually tested in a study depends very much on the time and resources available.

Retrospective analysis of how chemical selection has been made for different validation studies shows that the type and range of attributes selected differ quite considerably, as does the priority given to each. In certain cases, a clear and well described rationale underpinning the choice and prioritisation of attributes and the design of the selection process has been lacking prior to the commencement of a study, with the consequence that the selection of reference chemicals required explanation and often some defence after the study was completed. In other cases, expectations from the user community have been unrealistic due perhaps to a lack of awareness of the practical barriers encountered in selecting chemicals, so that also then some debate about the chemical selection was necessary after the completion of the study. Thus there is a clear need for the elaboration of a conceptual framework for chemical selection supported by enhanced exchanges between the developer and user communities. Ideally the framework should also be complemented by practical guidance to increase efficiency, consistency, and awareness of the selection process across studies of various types. Although the overall approach to chemical selection should be the same for validation studies addressing both methods and DA, the basis for chemical selection and thus the attributes chosen will likely differ. In the case of validation of methods, emphasis is more on selecting chemicals which probe the technical and biological characteristics of a method, such as exploring the potential for experimental artefacts or interference to produce a false reading, or assessing how sensitive and specific the response of a cellular test system is to the toxicological mechanisms it is intended to detect. On the other hand, validation of DA or individual methods that aim at predicting a regulatory hazard endpoint requires the selection of chemicals that have a known toxicological hazard profile defined in regulatory terms (e.g. hazard classification following the UN's Globally Harmonised System) in order to demonstrate its likely predictive performance in a particular regulatory context. It is imperative that the chemical selection process is designed to be as inclusive as possible to ensure sufficient consultation with appropriate experts. In general, the selection of chemicals for the validation of individual methods should involve assay specialists and toxicologists knowledgeable in the scientific and technical aspects of the class of *in vitro* method being validated and the toxicological pathways concerned. On the other hand, chemical selection for the validation of DA or individual methods that aim at predicting a specific hazard endpoint should engage regulatory toxicologists and risk assessors who are familiar with regulatory information requirements and current approaches to satisfying them. Tackling chemical selection in this more systematic, consultative and transparent manner will increase the relevance and impact of validation studies and will lead to the establishment of recognised chemical validation standards that can be reutilised for methods and DA, as illustrated by some recent initiatives by EURL ECVAM (Kirkland et al. 2016) and NICETAM (Kleinstreuer et al. 2016).

The identification and characterisation of sources of uncertainty associated with methods that test for the toxicological hazard of chemicals is recognised as being fundamentally important to ensure a robust and reliable risk assessment that is accepted by risk managers and stakeholders. Extensive guidance has been developed by international bodies (WHO 2014) and agencies (EFSA 2015) which describe uncertainty analysis in great detail and which provide practical tools and examples to support the process. Not surprisingly, the focus to date regarding hazard has been on sources of uncertainty associated with animal tests, such as: extrapolating from early to late effects; determining points of departure in a dose–response experiment;

deducing effects at low doses from observations made at high doses; accounting for differences in physiology between species; and estimating inter-species variability (WHO 2014). Expression of sources of uncertainty and their potential impact on a hazard assessment can be qualitative or quantitative, with the latter type of information being more desirable in order to formulate the output of a risk assessment in probabilistic terms. Uncertainty analysis can also be approached by first considering sources of uncertainty related to the inputs of an assessment (i.e. the sources of primary data and the methods used to generate them) followed by examination of the procedure or algorithm employed to combine the inputs to produce a conclusion or prediction (EFSA 2015). Such a framework could be easily adapted for the systematic analysis of the uncertainty associated with new predictive toxicology approaches that integrate *in vitro* and computational methods. Moreover, the design of validation studies should include provision for generating and reporting the necessary information and data needed to support a thorough uncertainty analysis to facilitate the eventual uptake and use of predictive approaches in the regulatory domain.

In any discussion about the validation of new approaches in the context of human safety assessment there is usually an elephant hanging around at the back of the room wanting to raise the issue of the relevance of using animal data as a reference or benchmark. It is a difficult subject to broach and often raises quite different views and opinions depending on who is in the room at the time. It is a fact that in the EU at least, regulatory frameworks for managing the risk that chemicals may pose to human health and the environment rely heavily on generic risk considerations based on toxicological hazard classification, as prescribed for example by the Classification, Labelling and Packaging Regulation (CLP 2008). Chemicals can be classified for a wide variety of toxicological hazards such as eye or skin corrosion and irritation, skin sensitisation, acute oral toxicity, chronic specific target organ toxicity, reproductive toxicity and carcinogenicity. Classification in some of the more hazardous classes can result in a chemical being automatically subject to various risk management provisions in downstream sectorial legislation ranging from requirements for specific labelling to inform consumers about the hazard, to restricted or even prohibited use of the chemical in certain products and for certain uses. Regarding human health, the majority of hazard classes are defined with respect to effects measured in conventional animal studies, these being rodent studies for the most part. As a consequence, most of the currently available standardised and reliable reference data come from animal studies usually carried out to satisfy regulatory information requirements. Thus validation studies have typically made use of this data to assess how good an alternative method is in predicting hazard classification, taking the established animal-derived hazard classifications of the reference chemicals used in the study as the benchmark. This paradigm is not unreasonable but its relevance depends considerably on a number of factors. The first is the actual reliability of the animal data. Many investigations have shown that animal data for even for the same chemical and (guideline) test can be highly variable. This variability can often lead to uncertainty in classification which is poorly characterised and rarely taken into account in generating performance statistics for the validated method. Another factor on which the traditional paradigm hinges is the actual relevance of the alternative method to the animal test used to classify the reference chemical. In the case where the toxicological mechanisms and effects underpinning the endpoint measured in the animal test are captured by the alternative test method then it is reasonable to expect that good correlation is at least possible between the classifications derived from both tests. However, with the development of novel human-based biological test systems (e.g. derived from induced pluripotent stem cells), novel alternative methods may prove to be a better surrogate for the human situation than an established animal test. In this case, discordance of classification is to be expected between the human-like alternative method and the reference animal test, especially when the relevant toxicological mode-of-action is not actually captured by the animal model. This issue is not typically accounted for in the traditional validation paradigm, and is often compounded by a lack of mechanistic information and suitable human hazard data on the reference chemicals used.

This issue of choosing the right reference or benchmark datasets to be used for validation is a tractable problem if approached with the same scientific thinking that is at the heart of new approaches to toxicity testing. For the definition of standards to be used in the validation of *in vitro* methods, the reference data that really matter with respect to characterising the predictive utility of a method are those that mechanistically relate or associate a reference chemical to the toxicity pathways and key events that the method is expected to model. This association can be qualitative, in terms of expected positive or negative outcomes, or quantitative in terms of potency of effect or concentration-response. Thus the actual animal-derived hazard classification data for the reference chemicals are for the most part irrelevant in such a validation context, as are the data describing their sectorial or product use. However it could well be that although animal data are not used directly as the validation reference, they could provide information to determine the modes of action of the reference chemicals, assuming they were the same in humans. In addition to mechanistic toxicological data on reference chemicals, data associating reference chemicals with technical aspects of performance are also useful and necessary to characterise a method in terms of potential limitations in testing certain types or classes of chemicals. For example, such limitations could be due to the possibility of chemicals with certain physicochemical or optical properties to interfere with the detection assay or technique employed by the method. When considering appropriate reference datasets for the validation of DA or methods that aim at predicting a hazard endpoint, the focus shifts to using data that associate a reference chemical with apical health effects that are related to the endpoint of concern. In this case of course the hazard classification of reference chemicals based on animal tests and the relevant regulatory frameworks (GHS 2015) should be considered. However, this needs to be complemented with other important data that help portray a more comprehensive hazard profile of the reference chemicals in order to ensure that the validation exercise leads to a proper and comprehensive characterisation of performance. Obviously any available data on reported toxicological effects in humans that might be relevant to the endpoint should be included, as should biokinetics data and information on the toxicity of structurally similar analogues. Ultimately, expressing the validity of the approach for a particular purpose will not be simplistic in terms, but instead will be a multifaceted judgement arrived at through the weighting of multiple streams of evidence and accounting for all the relevant sources of variability and uncertainty associated both with the approach being validated and the reference data that it is being compared to.

As outlined here, it is imperative that the principles and practice of validation are continuously examined to ensure that they evolve appropriately to keep pace with the development of new alternative approaches to toxicity testing. In addition however, the actual process of validation must be carefully considered as well, to make certain that it is flexible, efficient and makes the best use of available resources. How validation might be approached needs be considered more frequently and systematically by test developers in the early phases of research and development (R&D). Likewise, user communities need to be clearer about their anticipated requirements and desired performance. In terms of investment and planning, validation of a method or DA should be seen by all stakeholders as being as important as the R&D that produced it. To facilitate this, parties undertaking validation need access to dedicated financing to run their studies. Other practical support is also required such as guidance on aspects of validation, lists of recommended reference chemicals and associated databases, and input and advice from validation experts where needed. The level of formality adopted in a validation study should be adequate to ensure objectivity, rigour and credibility but has to be appropriate to the aims of the study and should not unduly burden the process. Developers and users should seek to cooperate with each other in setting up validation studies, sharing knowhow and establishing working standards. In this context, academic societies, trade associations and other networks with a stake in promoting alternative approaches to animal testing have an important contribution to make by providing their members with the support and facilitation they require. Of course validation bodies such as EURL ECVAM and its partners in the International Cooperation on Alternative Test Methods (ICATM) must continue to play a central role as hubs for regional and international coordination and knowledge sharing, providers of practical support and guidance, and as champions in progressing new approaches towards regulatory use. Only working together can we build a dynamic international validation community that is committed to accelerate the translation of a new generation of scientifically advanced alternative methods into modern toxicology practice.

References

Arnold SF, Robinson MK, Notides AC, Guillette LJ, McLachlan JÁ (1996) A yeast estrogen screen for examining the relative exposure of cells to natural and xenoestrogens. Environ Health Perspect 104(5):544–548

Balls M, Blaauboer B, Brusick D, Frazier J, Lamb D, Pemberton M, Reinhardt C, Roberfroid M, Rosenkranz H, Schmid B, Spielmann H, Stammati A-L, Walum E (1990) Report and recommendations of the CAAT/ERGATT workshop on the validation of toxicity test procedures. Altern Lab Anim 18:313–337

Bouhifd M, Bories G, Casado J, Coecke S, Norlén H, Parissis N, Rodrigues R, Whelan M (2012) Automation of an *in vitro* cytotoxicity assay used to estimate starting doses in acute oral systemic toxicity tests. Food Chem Toxicol 50:2084–2096

Brown NA (2002) Selection of test chemicals for the ECVAM international validation study on *in vitro* embryotoxicity tests. European Centre for the Validation of Alternative Methods. Altern Lab Anim 30(2):177–198

Browne P, Judson R, Casey W, Kleinstreuer N, Thomas R (2015) Screening chemicals for estrogen receptor bioactivity using a computational model. Environ Sci Technol 49(14):8804–8814

Casati S, Aeby P, Kimber I, Maxwell G, Ovigne JM, Roggen E, Rovida C, Tosti L, Basketter D (2009) Selection of chemicals for the development and evaluation of *in vitro* methods for skin sensitisation testing. Altern Lab Anim 37(3):305–312

CLP (2008) Regulation (EC) No 1272/2008 of the European Parliament and of the Council of 16 December 2008 on classification, labelling and packaging of substances and mixtures, amending and repealing Directives 67/548/EEC and 1999/45/EC, and amending Regulation (EC) No 1907/2006

EFSA (2015) EFSA Scientific Committee Opinion: draft guidance on uncertainty in EFSA scientific assessment, European Food Safety Authority (EFSA), Parma, Italy

Eskes C, Cole T, Hoffmann S, Worth A, Cockshott A, Gerner I, Zuang V (2007) The ECVAM international validation study on *in vitro* tests for acute skin irritation: selection of test chemicals. Altern Lab Anim 35:1–17

GHS (2015) Globally harmonized system of classification and labelling of chemicals (GHS)—Sixth revised edition United Nations Economic Commission for Europe (UNECE). http://www.unece.org/trans/danger/publi/ghs/ghs_rev06/06files_e.html#c38156

Hartung T, Bremer S, Casati S, Coecke S, Corvi R, Fortaner S, Gribaldo L, Halder M, Hoffmann S, Roi AJ, Prieto P, Sabbioni E, Scott L, Worth A, Zuang V (2004) A modular approach to the ECVAM principles on test validity. Altern Lab Anim 32:467–472

Huang R, Xia M, Sakamuru S, Zhao J, Shahane SA, Attene-Ramos M, Zhao T, Austin CP, Simeonov A (2016) Modelling the Tox21 10 K chemical profiles for *in vivo* toxicity prediction and mechanism characterization. Nat Commun 7:10425

Jennings P, Schwarz M, Landesmann B, Maggioni S, Goumenou M, Bower D, Leonard MO, Wiseman JS (2014) SEURAT-1 liver gold reference compounds: a mechanism-based review. Arch Toxicol 88(12):2099–2133

Judson R, Kavlock R, Martin M, Reif D, Houck K, Knudsen T, Richard A, Tice RR, Whelan M, Xia M, Huang R, Austin C, Daston G, Hartung T, Fowle JR 3rd, Wooge W, Tong W, Dix D (2013) Perspectives on validation of high-throughput assays supporting 21st century toxicity testing. ALTEX 30(1):51–56

Judson R, Houck K, Martin M, Knudsen T, Thomas R, Sipes N, Shah I, Wambaugh J, Crofton K (2014) *In vitro* and modelling approaches to risk assessment from the U.S. Environmental Protection Agency ToxCast programme. Basic Clin Pharmacol Toxicol 115(1):69–76

Judson R, Magpantay F, Chickarmane V, Haskell C, Tania N, Taylor J, Xia M, Huang R, Rotroff D, Filer D, Houck K, Martin M, Sipes N, Richard A, Mansouri K, Setzer W, Knudsen T, Crofton K, Thomas R (2015) Integrated model of chemical perturbations of a biological pathwayusing 18 *in vitro* high throughput screening assays for the estrogen receptor. Toxicological Sciences, Society of Toxicology, pp. 1–42

Kinsner-Ovaskainen A, Maxwell G, Kreysa J, Barroso J, Adriaens E, Alépée N, Berg N, Bremer S, Coecke S, Comenges JZ, Corvi R, Casati S, Dal Negro G, Marrec-Fairley M, Griesinger C, Halder M, Heisler E, Hirmann D, Kleensang A, Kopp-Schneider A, Lapenna S, Munn S, Prieto P, Schechtman L, Schultz T, Vidal JM, Worth A, Zuang V (2012) Report of the EPAA-ECVAM workshop on the validation of Integrated Testing Strategies (ITS). Altern Lab Anim 40(3):175–181

Kirkland D, Kasper P, Martus HJ, Müller L, van Benthem J, Madia F, Corvi R (2016) Updated recommended lists of genotoxic and non-genotoxic chemicals for assessment of the performance of new or improved genotoxicity tests. Mutat Res Genet Toxicol Environ Mutagen 795(2016):7–30

Kleinstreuer NC, Ceger PC, Allen DG, Strickland J, Chang X, Hamm JT, Casey WM (2016) A curated database of rodent uterotrophic bioactivity. Environ Health Perspect 124(5):556–562. doi:10.1289/ehp.1510183

OECD (2005) Guidance Document No.34: The validation and international acceptance of new or updated test methods for hazard assessment, OECD Series on Testing and Assessment. Organization for Economic Cooperation and Development, Paris

OECD (2012a) Test No. 455: Performance-based test guideline for stably transfected transactivation *in vitro* assays to detect estrogen receptor agonists, OECD guidelines for the testing of chemicals, section 4, OECD Publishing, Paris. doi:10.1787/9789264185388-en

OECD (2012b), Performance Standards No. 173: Performance standards for stably transfected transactivation *in vitro* assays to detect estrogen agonists for TG 455. OECD Series on Testing and Assessment, OECD Publishing, Paris

OECD (2014a) Guidance document on integrated approaches to testing and assessment of skin irritation/corrosion. Series on Testing and Assessment, No. 203, OECD, Paris

OECD (2014b) Guidance Document No. 211: Guidance document for describing non-guideline *in vitro* test methods. Series on Testing and Assessment, OECD Publishing, Paris

OECD (2016) Guidance Document on the Reporting of Defined Approaches to be used within Integrated Approaches to Testing and Assessment, Task Force on Hazard Assessment. Draft January 2016

Pazos P, Pellizzer C, Stummann T, Hareng L, Bremer S (2010) The test chemical selection procedure of the European Centre for the Validation of Alternative Methods for the EU Project ReProTect. Reprod Toxicol 30(1):161–199

RAAF (2015) Read across assessment framework. European Chemicals Agency, Helsinki, http://echa.europa.eu/documents/10162/13628/raaf_en.pdf

REACH (2006) Regulation (EC) No 1907/2006 of the European Parliament and of the Council of 18 December 2006 concerning the Registration, Evaluation, Authorisation and Restriction of Chemicals (REACH), establishing a European Chemicals Agency, amending Directive 1999/45/EC and repealing Council Regulation (EEC) No 793/93 and Commission Regulation (EC) No 1488/94 as well as Council Directive 76/769/EEC and Commission Directives 91/155/EEC, 93/67/EEC, 93/105/EC and 2000/21/EC

WHO (2014) Guidance document on evaluating and expressing uncertainty in hazard characterization, WHO International Programme on Chemical Safety: Harmonization Project Document No.11. http://www.who.int/ipcs/methods/harmonization/uncertainty_in_hazard_characterization.pdf

Index

C. Eskes, M. Whelan (eds.), *Validation of Alternative Methods for Toxicity Testing*, Advances in Experimental Medicine and Biology 856,
DOI 10.1007/978-3-319-33826-2

W

The manufacturer's authorised representative in the EU is Springer Nature Customer Service Centre GmbH, Europaplatz 3, 69115 Heidelberg, Germany. If you have any concerns regarding our products, please contact ProductSafety@springernature.com

Printed and bound by CPI Group (UK) Ltd, Croydon, CR0 4YY
30/09/2026
02225651-0013